Lecture Notes in Computer Scien

Commenced Publication in 1973
Founding and Former Series Editors:
Gerhard Goos, Juris Hartmanis, and Jan van Leeuwen

José Manuel Ferrández Vicente
José Ramón Álvarez Sánchez
Félix de la Paz López
Fco. Javier Toledo Moreo (Eds.)

Natural and Artificial Computation in Engineering and Medical Applications

5th International Work-Conference on the Interplay
Between Natural and Artificial Computation, IWINAC 2013
Mallorca, Spain, June 10-14, 2013
Proceedings, Part II

 Springer

Volume Editors

José Manuel Ferrández Vicente
Fco. Javier Toledo Moreo
Universidad Politécnica de Cartagena
Departamento de Electrónica, Tecnología de Computadoras y Proyectos
Pl. Hospital, 1
30201 Cartagena, Spain
E-mail: {jm.ferrandez; javier.toledo}@upct.es

José Ramón Álvarez Sánchez
Félix de la Paz López
Universidad Nacional de Educación a Distancia
E.T.S. de Ingenería Informática
Departamento de Inteligencia Artificial
Juan del Rosal, 16
28040 Madrid, Spain
E-mail: {jras; delapaz}@dia.uned.es

ISSN 0302-9743   e-ISSN 1611-3349
ISBN 978-3-642-38621-3   e-ISBN 978-3-642-38622-0
DOI 10.1007/978-3-642-38622-0
Springer Heidelberg Dordrecht London New York

Library of Congress Control Number: 2013938949

CR Subject Classification (1998): F.1, F.2, I.2, G.2, I.4-5, J.3-4, J.1

LNCS Sublibrary: SL 1 – Theoretical Computer Science and General Issues

Typesetting: Camera-ready by author, data conversion by Scientific Publishing Services, Chennai, India

Printed on acid-free paper

Springer is part of Springer Science+Business Media (www.springer.com)

Preface

Living Nature Computing

A trend in computing is the disappearing computer, that is, computers become part of the environment, part of our daily lives, sharing our personal spaces, our amusements, even our own bodies - e.g., embedded systems, accompanying robotics, electronic neuroprosthetics. This implies that computing will also become part of living nature and that computing will be performed by living nature. We believe it is essential to articulate the relationship between different areas of science in order to collaborate and foster a discussion on the interplay of computing with living natural systems.

Natural computing has two sides: computing inspired by nature and computing in nature. These two sides are usually seen as complementary, but our meeting is situated at the intersection/interplay of these two sides. We call this intersection "Living Nature Computing." The term living nature should be taken in a broad sense: from cells via animals to humans. Living nature also puts a focus on natural computing: evolutionary computing and neural networks are examples of natural computing inspired by living nature. DNA computing is included as far as the computation takes place inside living nature. A further focus is the type of computation: we consider engineered computation as inspired by living nature.

Six interesting areas are selected, which share aspects of living nature computing, that we would like to promote in our meetings.

- Cognitive Robotics: Cognitive robotic systems, apart from being practical engineering challenges, are also excellent platforms for experimenting with situated and embodied cognitive theories and models, posing interesting and hard challenges to theoretical and fundamental cognition issues. As Prof. Mira said: "Robotics is the most complete paradigm in Artificial Intelligence: it includes, perception, reasoning, and action."
- Natural Computing: Natural computing refers to computational processes observed in nature, and to human-designed computing inspired by nature. When complex natural phenomena are analyzed in terms of computational processes, our understanding of both nature and the essence of computation is enhanced. Characteristic for human-designed computing inspired by nature is the metaphorical use of concepts, principles, and mechanisms underlying natural systems. Natural computing includes evolutionary algorithms, neural networks, molecular computing, quantum computing, neural modelling, plasticity studies, etc.
- Wetware computation: Wetware computation refers to an organic computer built from living neurons. Silicon-based microchips have physical limits and also power dissipation problems. Wetware computing uses biochemistry

instead of silicon for finding better solutions to be used in future electronics and in information and communications technologies. Molecules or living organisms may carry electrical charge and may perform computing functions with less integration or power dissipation. A wetware computer may be built from leech neurons and be capable of performing simple arithmetic operations and simple pattern-recognition tasks. Another wetware computation is based on cellular cultures. Cells could be neurons from dissociated hippocampus or cortical tissue, neuroblastoma cells, or even PC12 cells, a cell line derived from the rat adrenal medulla.

– Quality of Life Technologies: During the last few years, there is an increasing interest in technologies oriented to improve the quality of life of people. Quality of life technologies (QoLTs) cover a broad area of research including engineering, computer science, medicine, psychology, or social sciences. Typical applications of QoLTs include assistance technologies for people with some kind of disability as, for example, assistive robots, elderly care technologies, or smart homes. However, QoLTs include powerful tools to improve the well-being of individuals and societies in general.

– Biomedical and Industrial Perception Applications: Image understanding is a research area involving both feature extraction and object identification within images from a scene, and a posterior treatment of this information in order to establish relationships between these objects with a specific goal. In biomedical and industrial scenarios, the main purpose of this discipline is, given a visual problem, to manage all aspects of prior knowledge, from study start-up and initiation through data collection, quality control, expert independent interpretation, to the design and development of systems involving image processing capable of tackle with these tasks.

– Web Intelligence and Neuroscience: The study of a user's brainwaves applied to reveal the impact of personalized Web content opens the door for powerful collaborations between Web intelligence and neuroscience. The purpose of this area is to discuss the potential applications of neuroscience methods and concepts in order to evaluate, design, and develop user satisfaction models for intelligent websites. Some of the potential research areas to be analyzed are personalization of Web content and Web application design. On the basis of empirical recordings, it is possible to reveal which features are relevant for each user, therefore learning how to personalize both content and business logic in order to improve the overall user satisfaction. Brainwave analysis can provide the mechanisms to record and measure relevant user behavior patterns regarding memory, attention, and other cognitive aspects. The data obtained by brainwave analysis could guide the development of user satisfaction models, i.e., conceptual and mathematical models that could explain not only why users are satisfied with the provided content, but also predict future satisfaction.

We want to study living nature computing using a systems approach. Computing is not an add-on to the living system, but it is an integral part that is in symbiosis with the system. In fact, it is often not possible to distinguish the computational parts from the non computational parts. This wider view of the computational paradigm gives us more elbow room to accommodate the results of the interplay between nature and computation. The IWINAC forum thus becomes a methodological approximation (set of intentions, questions, experiments, models, algorithms, mechanisms, explanation procedures, and engineering and computational methods) to the natural and artificial perspectives of the mind embodiment problem, both in humans and in artifacts. This is the philosophy of IWINAC meetings, the "interplay" movement between the natural and the artificial, facing this same problem every two years. This synergistic approach will permit us not only to build new computational systems based on the natural measurable phenomena, but also to understand many of the observable behaviors inherent to natural systems.

The difficulty of building bridges between natural and artificial computation is one of the main motivations for the organization of IWINAC 2013. The IWINAC 2013 proceedings contain the works selected by the Scientific Committee from more than 100 submissions, after the refereeing process. The first volume, entitled *Natural and Artificial Models in Computation and Biology*, includes all the contributions mainly related to the methodological, conceptual, formal, and experimental developments in the fields of neurophysiology and cognitive science. The second volume, entitled *Natural and Artificial Computation in Engineering and Medical Applications*, contains the papers related to bioinspired programming strategies and all the contributions related to the computational solutions to engineering problems in different application domains, especially health applications, including the CYTED "Artificial and Natural Computation for Health" (CANS) research network papers.

An event of the nature of IWINAC 2013 cannot be organized without the collaboration of a group of institutions and people who we would like to thank now, starting with *UNED* and *Universidad Politécnica de Cartagena*. The collaboration of the *UNED associated center* was crucial, as was the efficient work of the Local Organizing Committee, Miguel Angel Vázquez Segura, and Francisco J. Perales López with the close collaboration of the *Universitat de les Illes Balears*. In addition to our universities, we received financial support from the Spanish *CYTED, Red Nacional en Computación Naturally Artificial* and *APLIQUEM s.l.*.

We want to express our gratefulness to our invited speakers, Rodolfo Llinás, from the Department of Physiology and Neuroscience at New York University, Dario Floreano, from the Laboratory of Intelligent Systems at EPFL (Switzerland), and Pedro Gómez-Vilda, from the Oral Communication Lab "Robert Wayne Neucomb" in UPM (Spain), for accepting our invitation and for their magnificent plenary talks.

We would also like to thank the authors for their interest in our call and the effort in preparing the papers, condition sine qua non for these proceedings. We thank the Scientific and Organizing Committees, in particular the members of these committees who acted as effective and efficient referees and as promoters and managers of pre organized sessions on autonomous and relevant topics under the IWINAC global scope.

Our sincere gratitude goes also to Springer and to Alfred Hofmann and his collaborators, Anna Kramer and Elke Werner, for the continuous receptivity, help efforts, and collaboration in all our joint editorial ventures on the interplay between neuroscience and computation.

Finally, we want to express our special thanks to ESOC S.L., our technical secretariat, and to Victoria Ramos and Nuria Pastor, for making this meeting possible, and for arranging all the details that comprise the organization of this kind of event.

All the authors of papers in this volume, as well as the IWINAC Program and Organizing Committees, would like to commemorate the memory of Professor Mira, who passed away five years ago, both as a great scientist, with an incredible dissemination profile, and as best friend. We still carry his memory deep inside our hearts.

June 2013

José Manuel Ferrández Vicente
José Rámon Álvarez Sánchez
Félix de la Paz López
Fco. Javier Toledo More

Organization

General Chair

José Manuel Ferrández Vicente, UPCT, Spain

Organizing Committee

José Ramón Álvarez Sánchez, UNED, Spain
Félix de la Paz López, UNED, Spain
Fco. Javier Toledo Moreo, UPCT, Spain

Local Organizing Committee

Miguel Ángel Vázquez Segura, UNED, Spain
Francisco J. Perales López, UIB, Spain

Invited Speakers

Rodolfo Llinás, USA
Dario Floreano, Switzerland
Pedro Gómez-Vilda, Spain

Field Editors

Manuel Arias Calleja, Spain
Manuel Luque Gallego, Spain
Juan José Pantrigo, Spain
Antonio Sanz, Spain
Daniel Ruiz, Spain
Javier De Lope Asiaín, Spain
Darío Maravall Gómez-Allende, Spain
Antonio Fernández Caballero, Spain
Rafael Martínez Tomás, Spain
Oscar Martínez Mozos, Spain
María Consuelo Bastida Jumilla, Spain
Rosa María Menchón Lara, Spain
Luis Martínez Otero, Spain
Eduardo Sánchez Vila, Spain

International Scientific Committee

Andy Adamatzky, UK
Michael Affenzeller, Austria
Abraham Ajith, Norway
Amparo Alonso Betanzos, Spain
Jose Ramon Alvarez-Sanchez, Spain
Diego Andina, Spain
Davide Anguita, Italy
Manuel Arias Calleja, Spain
José M. Azorín, Spain
Margarita Bachiller Mayoral, Spain
Antonio Bahamonde, Spain
Dana Ballard, USA
Emilia I. Barakova, The Netherlands
Alvaro Barreiro, Spain
Senen Barro Ameneiro, Spain
M-Consuelo Bastida-Jumilla, Spain
Francisco Bellas, Spain
Guido Bologna, Switzerland
Paula Bonomini, Argentina
Juan Botia, Spain
François Bremond, France
Giorgio Cannata, Italy
Enrique J. Carmona Suarez, Spain
German Castellanos-Dominguez, Colombia
Joaquin Cerda Boluda, Spain
Alexander Cerquera, Colombia
Enric Cervera Mateu, Spain
Antonio Chella, Italy
Santi Chillemi, Italy
Eris Chinellato, Spain
Carlos Colodro-Conde, Spain
Ricardo Contreras, Chile
Erzsebet Csuhaj-Varju, Hungary
Jose Manuel Cuadra Troncoso, Spain
Felix de la Paz Lopez, Spain
Javier de Lope, Spain
Erik De Schutter, Belgium
Angel P. del Pobil, Spain
Ana E. Delgado García, Spain
Jose Dorronsoro, Spain
Gerard Dreyfus, France
Richard Duro, Spain
Reinhard Eckhorn, Germany
Patrizia Fattori, Italy

Juan Pedro Febles Rodriguez, Cuba
Paulo Félix Lamas, Spain
Eduardo Fernandez, Spain
Manuel Fernández Delgado, Spain
Antonio J. Fernández Leiva, Spain
Antonio Fernández-Caballero, Spain
Abel Fernandez-Laborda, Spain
José Manuel Ferrandez, Spain
Kunihiko Fukushima, Japan
Cipriano Galindo, Spain
Cristina Gamallo Solórzano, Spain
Jose A. Gamez, Spain
Jesus Garcia Herrero, Spain
Juan Antonio Garcia Madruga, Spain
Francisco J. Garrigos Guerrero, Spain
Tom D. Gedeon, Australia
Charlotte Gerritsen, The Netherlands
Marian Gheorghe, UK
Pedro Gomez Vilda, Spain
Juan M Gorriz, Spain
Manuel Graña, Spain
Francisco Guil-Reyes, Spain
John Hallam, Denmark
Juan Carlos Herrero, Spain
Cesar Hervas Martinez, Spain
Tom Heskes, The Netherlands
Eduardo Iáñez, Spain
Roberto Iglesias, Spain
Aleksander Igor, UK
Fernando Jimenez Barrionuevo, Spain
Jose M. Juarez, Spain
Joost N. Kok, The Netherlands
Kostadin Koroutchev, Spain
Elka Korutcheva, Spain
Yasuo Kuniyoshi, Japan
Ryo Kurazume, Japan
Petr Lansky, Czech Republic
Jorge Larrey-Ruiz, Spain
Maria Longobardi, Italy
Maria Teresa Lopez Bonal, Spain
Ramon Lopez de Mantaras, Spain
Pablo Lopez Mozas, Spain
Tino Lourens, The Netherlands
Max Lungarella, Japan
Manuel Luque, Spain

Table of Contents – Part II

Table of Contents – Part I

Ant Colony Algorithms for the Dynamic Vehicle Routing Problem with Time Windows

Barry van Veen, Michael Emmerich, Zhiwei Yang,
Thomas Bäck, and Joost Kok

LIACS, Leiden University, Niels Bohrweg 1, 2333-CA Leiden, The Netherlands

Abstract. The Vehicle Routing Problem with Time Windows relates to frequently occuring real world problems in logistics. Much work has been done on solving static routing problems but solving the dynamic variants has not been given an equal amount of attention, while these are even more relevant to most companies in logistics and transportation. In this work an Ant Colony Optimization algorithm for solving the Dynamic Vehicle Routing Problem with Time Windows is proposed. Customers and time windows are inserted during the working day and need to be integrated in partially committed solutions. Results are presented on a benchmark that generalizes Solomon's classical benchmark with varying degrees of dynamicity and different variants, including pheromone preservation and the min-max ant system.

1 Introduction

With recent developments in mobile communication and positioning systems, it is now possible for companies in transportation to view and change their planning during the day. This leads to a new group of dynamic routing problems for which algorithms have to be designed. In this paper we will extend the ant algorithm described in [7] for the standard vehicle routing problem with time windows (VRPTW) to dynamic problems (DVRPTW). As described by Psaraftis [12], a problem is dynamic when some part of the input is revealed to the solver during optimization. This means that we can not build a fixed solution and we have to adjust the solution while the problem changes. In the DVRPTW, the task is to schedule a fleet of vehicles in a working day and new orders (clients that need to be visited) are introduced during the day. This specific problem has not been solved with ant colony algorithms, yet, although it is very relevant in practice. Some features of the bio-mimetic ant-colony optimization algorithm [3] seem to well support dynamic adaptations of delivery routes, as results for the related TSP problem indicate [5]. To test our new approach a set of benchmark problems is introduced as a generalization of a common VRPTW benchmark. To make our results reproducible, the benchmark as well as the C code of the developed algorithms will be made public for other users.

In Section 2 of this paper we describe the static and dynamic VRPTW problem and existing solvers for both problems from literature. Section 3 provides a detailed description of the novel Ant-based DVRPTW solver. Section 4 deals with

J.M. Ferrández Vicente et al. (Eds.): IWINAC 2013, Part II, LNCS 7931, pp. 1–10, 2013.

the benchmark and performance studies for different variations of the algorithms. Finally, Section 5 concludes the work with a summarizing discussion.

2 Problem Description and Related Work

An important part of the ant algorithm is the way in which a tour is represented. Each tour should visit the depot more than once, which represents the start of a new vehicle. To accomplish this the depot node is copied a number of times and all copies can be visited individually. Hence we get n customer nodes with a demand and time window and up to n depot nodes(duplicates). This is used in the problem definition:

Problem 1. Capacitated Vehicle Routing Problem with Time Windows(VRPTW). Let $V = \{v_1, \ldots, v_{2n}\}$ define a set of n customer nodes and up to n depot nodes (duplicates). Let Q denote the maximal capacity of each vehicle, and q_i denote the demand and $[e_i, l_i]$ the time window, and s_i the service time at node v_i, if any. A solution is a tuple $(\pi_1, \ldots, \pi_x)$ with a maximal length of $2n$ and of which n nodes refer to customers and up to n nodes are identical with the depot. Each cycle that returns to the depot is a single tour (vehicle). In addition, $d_{i,j}$ is the traveling distance from node i to node j. The goal is to minimize the traveling distance

$$\sum_{i=1}^{x-1}(d_{\pi_i,\pi_{i+1}}) + d_{\pi_x,\pi_1} \qquad (1)$$

and minimize, with priority, the number of vehicles needed.

To be able to solve a **dynamic problem** we first have to simulate a form of dynamicity. Kilby, Prosser and Shaw [10] have described a method to do this, which is also used by Montemanni et al. [11]. The notion of a working day of T_{wd} seconds is introduced, which will be simulated by the algorithm. Not all nodes are available to the algorithm at the beginning. A subset of all nodes are given an *available time* at which they will become available. This percentage determines the degree of dynamicity of the problem. At the beginning of the day a tentative tour is created with a-priori available nodes. The working day is divided into n_{ts} time slices of length T_{wd}/n_{ts}, also notated with t_{ts}. At each time slice the solution is updated. This allows us to split up the dynamic problem into n_{ts} static problems, which can be solved consecutively. A different approach would be to restart the algorithm every time a node becomes available. This could have a very disruptive effect on the algorithm because it could be stopped before a good solution is found. Note, that the goal is similar than stated in Problem 1, except that some customers and their time windows are unknown a priori and parts of solutions might already have been committed.

In general VRP and VRPTW are considered to be intractable problems, because they generalize the NP complete traveling salesperson problem. Heuristic algorithms for static VRP problems include deterministic [2] and bio-inspired ant-based methods [1,3].

Ant-based methods were first proposed with the Ant System method [3] and simulate a population of ants which use pheromones to communicate with each other and collectively are able to solve complex path-finding problems – a phenomenon called *stigmergy*. For the VRPTW, an ant-based method was proposed by [7]. The paradigm of ant algorithms fits well to dynamic problems [9] including TSP [5] and special types of VRP, where vehicles do not have to return to the depot [11]. However, they have not been applied yet to DVRPTW. Existing work is restricted to tabu search [8], where, as opposed to MACS-VRPTW soft time windows are used.

3   Dynamic Ant Algorithm

The plan is to extend the state-of-the-art ant algorithm for VRPTW to the dynamical case. To our best knowledge, [7] is the only ant algorithm for the VRPTW with a description that allows to reproduce results, and it shows a good performance on standard benchmark problems by Solomon http://web.cba. neu.edu/ msolomon/heuristi.htm Due to space limitations, we will directly describe the new dynamic version of this algorithm and indicate changes.

The **controller** is the central part that reads the benchmark data, initializes data structures, builds an initial solution and starts the colonies. The **nearest neighbor heuristic** [6] is used to find initial solutions for the entire algorithm and the ACS-VEI colony, but it was adjusted in two ways. First the constraints on time windows and capacity are checked to make sure no infeasible tours are created. Besides that a limit on the number of vehicles is passed to the function. Because of these limitations it is not always possible to return a tour that incorporates all nodes. In that case a tour with less nodes is returned.Only nodes that are available at $t = 0$ are considered. After initialization, a timer is started that keeps track of t, the used CPU time in seconds. At the start of each time slice the controller checks if any nodes became available during the last time slice. If so, these nodes are inserted using the **InsertMissingNodes** method to make T^* feasible again. Then all necessary nodes are committed. If v_i is the last committed node of a vehicle in the tentative solution, v_j is the next node and t_{ij} is travel time from node v_i to node v_j, the v_j is committed if $e_j - t_{ij} < t + t_{ts}$. When the necessary commitments have been made two ant colony systems (ACS) are started. If a new time slice starts, the colonies are stopped and the controller repeats its loop. A pseudo-code of the controller can be seen in Algorithm 1.

ACS contains two colonies, each one of the which tries to improve on a different objective of the problem. The ACS-VEI colony searches for a solution that uses less vehicles than T^*. The ACS-TIME colony searches for a solution with a smaller traveling distance than T^* while using at most as many vehicles. The two objectives have a fixed priority: a solution with less vehicles is always preferred over a solution with a smaller distance.

There are a few differences between the two colonies. ACS-VEI keeps track of the best solution found by the colony (T^{VEI}), which does not necessarily

Algorithm 1. Controller

1: Set time $t = 0$; Set available nodes n
2: $T^* \leftarrow$ NearestNeighbor(n); $\tau_0 \leftarrow 1/(n \cdot$ length of T^*);
3: Start measuring CPU time t
4: Start ACS-TIME(vehicles in T^*) in new thread
5: Start ACS-VEI(vehicles in $T^* - 1$) in new thread
6: **repeat**
7:     **while** colonies are active and time step is not over **do**
8:         Wait until a solution T is found
9:         **if** vehicles in $T <$ vehicles in T^* **then**
10:             Stop threads
11:             $T^* \leftarrow T$
12:     **if** time-step is over **then**
13:         **if** new nodes are available or new part of T^* will be defined **then**
14:             Stop threads
15:             Update available nodes n
16:             Insert new nodes into T^*
17:             Commit to nodes in T^*
18:     **if** colonies have been stopped **then**
19:         Start ACS-TIME(vehicles in T^*) in new thread
20:         Start ACS-VEI(vehicles in $T^* - 1$) in new thread
21: **until** $t \geq T_{wd}$
22: **return** T^*

---

incorporate all nodes. As T^{VEI} also contributes to the pheromone trails it helps ACS-VEI to find a solution that covers all nodes with less vehicles. ACS-TIME does not work with infeasible solutions and does not have a colony-best solution. Unlike ACS-VEI, it performs a local search method called **Cross Exchange** [15] shown in Figure 1. The maximum number of vehicles that may be used is given as an argument to each colony. During the construction of a tour this number may not be exceeded. This may lead to infeasible solutions that do not incorporate all nodes. If a solution is not feasible it can never be send to the controller. Both colonies work on separate pheromone matrices and send their best solutions to the controller. Pseudo-codes for ACS-VEI and ACS-TIME can be found in Algorithm 2 and 3 respectively.

Algorithm 4 describes the **construction of a tour** by means of artificial ants. A tour starts at a randomly chosen depot copy and is then iteratively extended with available nodes. The set $\mathcal{N}_i^k$ contains all available nodes which have not been committed for ant k situated at node i. Committed parts of T^* have to be incorporated in every tour. Inaccessible nodes due to capacity or time window constraints are excluded from $\mathcal{N}_i^k$. In order to decide which node to chose, the probabilistic transition rules by Dorigo and Gambardella [4] are applied. For ant k positioned at node v_i, the probability $p_j^k(v_i)$ of choosing v_j as its next node is given by the following transition rule:

Algorithm 2. ACS-VEI(v)

1: **Input:** v is the maximum number of vehicles to be used
2: **Given:** τ_0 is the initial pheromone level
3:
4: Initialize pheromones to τ_0
5: Initialize IN to 0
6: $T^{\text{VEI}} \leftarrow$ NearestNeighbor(v)
7:
8: **repeat**
9:     **for each** ant k **do**
10:        $T^k \leftarrow$ ConstructTour(k, IN)
11:        **for each** nodes $i \notin T^k$ **do**
12:           $\text{IN}_i = \text{IN}_i + 1$
13:        Local pheromone update on edges of T^k using Equation 3
14:        $T^k \leftarrow$ InsertMissingNodes(k)
15:
16:     Find ant l with most visited nodes
17:     **if** nodes in $T^l >$ nodes in T^{VEI} **then**
18:        $T^{\text{VEI}} \leftarrow T^l$
19:     Reset IN to 0
20:     **if** T^{VEI} is feasible **then**
21:        **return** T^{VEI} to controller
22:
23:     Global pheromone update with T^* and Equation 4
24:     Global pheromone update with T^{VEI} and Equation 4
25: **until** controller sends stop signal

Algorithm 3. ACS-TIME(v)

1: **Input:** v is the maximum number of vehicles to be used
2: **Given:** τ_0 is the initial pheromone level
3:
4: Initialize pheromones to τ_0
5:
6: **repeat**
7:     **for each** ant k **do**
8:        $T^k \leftarrow$ ConstructTour(k, 0)
9:        Local pheromone update on edges of T^k using Equation 3
10:        $T^k \leftarrow$ InsertMissingNodes(k)
11:        **if** T^k is a feasible tour **then**
12:           $T^k \leftarrow$ LocalSearch(k)
13:
14:     Find feasible ant l with smallest tour length
15:     **if** length of $T^l <$ length of T^* **then**
16:        **return** T^l to controller
17:
18:     Global pheromone update with T^* and Equation 4
19: **until** controller sends stop signal

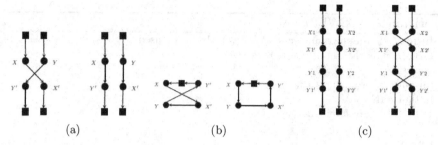

Fig. 1. Examples of 2-opt edge replacements. Squares represent depots, circles represent nodes. (a) demonstrates a move with edges from different tours. (b) is an example of a move within a single tour.(c) shows the process of cross exchange.

$$p_j^k(v_i) = \begin{cases} \arg\max_{j \in N_i}\{[\tau_{ij}]^\alpha \cdot [\eta_{ij}]^\beta\} & \text{if } q \leq q_0 \text{ and } j \in N_i^k \\[2mm] \dfrac{[\tau_{ij}]^\alpha \cdot [\eta_{ij}]^\beta}{\sum_{m \in N_i^k}[\tau_{im}]^\alpha \cdot [\eta_{im}]^\beta} & \text{if } q > q_0 \text{ and } j \in N_i^k \\[4mm] 0 & \text{if } j \notin N_i^k \end{cases} \tag{2}$$

with τ_{ij} being the pheromone level on edge (i,j), η_{ij} the heuristic desirability of edge (i,j), α the influence of τ on the probabilistic value, β the influence of η on the probabilistic value, N_i^k the set of nodes that can be visited by ant k positioned at node v_i, and $\tau_{ij}, \eta_{ij}, \alpha, \beta \geq 0$. Moreover q denotes a random number between 0 and 1 and $q_0 \in [0,1]$ a threshold.

During the ConstructTour process of ACS-VEI, the IN array is used to give greater priority to nodes that are not included in previously generated tours. The array counts the successive number of times that node v_j was not incorporated in constructed solutions. This count is then used to increase the attractiveness η_{ij}. The IN array is only available to ACS-VEI and is reset when the colony is restarted or when it finds a solution that improves T^{VEI}. ACS-TIME does not use the IN array, which is equal to setting all values in the array to zero.

The local pheromone update rule from [4] is used to decrease pheromone levels on edges that are traversed by ants. Each time an ant has traversed an edge (i,j), it applies Equation 3. By decreasing pheromones on edges that are already traveled on, there is a bigger chance that other ants will use different edges. This increases exploration and should avoid too early stagnation of the search.

$$\tau_{ij} = (1 - \rho) \cdot \tau_{ij} + \rho \cdot \tau_0 \tag{3}$$

The global pheromone update rule is given in Equations 4. To increase exploitation, pheromones are only evaporated and deposited on edges that belong to the best solution found so far and $\Delta\tau_{ij}$ is multiplied by the pheromone decay parameter ρ.

$$\tau_{ij} = (1 - \rho) \cdot \tau_{ij} + \rho \cdot \sum_{k=1}^{m} \Delta\tau_{ij}^k, \quad \forall(i,j) \in T^* \text{ and } \Delta\tau_{ij}^k = 1/L^* \tag{4}$$

where T^* is the best tour found so far and L^* is the length of T^*.

Algorithm 4. ConstructTour(k, IN)

1: **Input:** k is the ant we construct a tour for
2: **Input:** IN is an array containing the number of times nodes have not been incorporated in tours
3: **Given:** $\mathcal{N}_i^k$ is a set of nodes and depot duplicates that are reachable by ant k in node i
4:
5: Current vehicle $x \leftarrow 0$
6: Select a random depot duplicate i
7: $T^k \leftarrow \langle i \rangle$ ▷ Add vehicle i to tour k
8: current time$_k \leftarrow 0$
9: load$_k \leftarrow 0$
10: **for each** committed node v_i of the xth vehicle of T^* **do**
11:    $T^k \leftarrow \langle i \rangle$
12:    current time$_k \leftarrow$ delivery time$_i$ + service time$_i$
13:    load$_k \leftarrow$ load$_k + q_i$
14:
15: **repeat**
16:    **for each** $j \in \mathcal{N}_i^k$ **do**   ▷ The part below is taken from [4]
17:      delivery time$_j \leftarrow$ max(current time$_k + t_{ij}, e_j$)
18:      delta time$_{ij} \leftarrow$ delivery time$_j -$ current time$_k$
19:      distance$_{ij} \leftarrow$ delta time$_{ij} \times (l_j -$ current time$_k$)
20:      distance$_{ij} \leftarrow$ max(1.0, (distance$_{ij} -$ IN$_j$))
21:      $\eta_{ij} \leftarrow 1.0/$ distance$_{ij}$
22:
23:    Pick node j using Equation 2
24:    $T^k \leftarrow T^k + \langle j \rangle$
25:    current time$_k \leftarrow$ delivery time$_j +$ service time$_j$
26:    load$_k \leftarrow$ load$_k + q_j$
27:    **if** j is a depot copy **then**
28:      current time$_k \leftarrow 0$
29:      load$_k \leftarrow 0$
30:      $x \leftarrow x + 1$
31:      **for each** committed node v_i of the xth vehicle of T^* **do**
32:        $T^k \leftarrow \langle i \rangle$
33:        current time$_k \leftarrow$ delivery time$_i$ + service time$_i$
34:        load$_k \leftarrow$ load$_k + q_i$
35:    $i \leftarrow j$
36: **until** $\mathcal{N}_i^k = \{\}$
37:
38: **return** T^k

---

4   Benchmark and Results

MACS-VRPTW and its extension MACS-DVRPTW was tested on the 56 bench-mark problems by Solomon [13]. These problems are divided into six categories: C1, C2, R1, R2, RC1 and RC2. The C stands for problems with clustered nodes, the R problems have randomly placed nodes and RC problems have both. Problems of type 1 have a short scheduling horizon, only a few nodes can be serviced by a single vehicle. Problems of type 2 have a long scheduling horizon.

To simulate dynamics we modified the VRPTW problems by Solomon. A certain percentage of nodes is only revealed during the working day. A dynamicity of $X\%$ means that each node has a probability of $X\%$ to get a non-zero available time. Time intervals are assigned randomly within the available times using a method by Gendreau et al. [8]. Available time are generated on the interval $[0, \overline{e_i}]$, where $\overline{e_i} = \min(e_i, t_{i-1})$. Here, t_{i-1} is the departure time from v_i's predecessor in the best known solution. These best solutions are taken from the results of our MACS-VRPTW implementation (see table 1) – for the solutions we refer to the support material available on http://natcomp.liacs.nl/index.php?page=code. By generating available times on this interval, optimal solution can still be attained, enabling comparisons with MACS-VRPTW. For DVRPTW the Solomon prob-

Table 1. Comparison of results reported for the original MACS-VRPTW [7] and our implementation for the Solomon benchmark

	C1		C2		R1		R2		RC1		RC2	
	Dist	Vei	Dist	Vei	Dist	Vei	Dist	Vei	Dist	Vei	Dist	Vei
Original	828.40	10.00	593.19	3.00	1214.80	12.55	971.97	3.05	1395.47	12.46	1191.87	3.38
Avg	828.67	10.00	591.00	3.00	1226.05	12.52	992.49	3.00	1381.20	12.25	1165.51	3.35
Best	828.37	10.00	589.85	3.00	1216.70	12.33	949.69	3.00	1362.58	12.00	1146.89	3.25

Table 2. Average results and Standard Deviations (Stdev) for 10 runs and 56 Problems of different MACS-DVRPTW variants and dynamicity levels (Dyn).

	Normal			IIS			WPP			MMAS		
Dyn	Vei	Dist	Stdev	Vei	Dist	Stdev	Vei	Dist	Stdev	Vei	Dist	Stdev
0%	7.39	1046.06	21.72	7.35	1035.86	20.14	7.35	1043.13	20.22	7.40	1050.06	22.29
10%	7.91	1095.10	28.95	7.93	1087.06	28.39	7.93	1087.98	26.11	7.95	1093.66	31.66
20%	8.37	1131.47	29.59	8.38	1131.41	31.13	8.39	1127.67	26.52	8.43	1133.99	36.00
30%	8.79	1180.36	34.84	8.78	1177.96	34.37	8.79	1175.14	35.32	8.88	1183.02	34.59
40%	9.03	1216.72	36.73	9.02	1212.11	37.12	9.04	1210.38	37.80	9.08	1212.48	39.64
50%	9.32	1241.32	38.09	9.36	1236.36	39.64	9.34	1235.90	38.52	9.34	1235.90	39.06

lems in Table 1 were computed 10 runs of MACS-DVRPTW. Our implementation was executed on a Intel Core i5, 3.2GHz CPU with 4GB of RAM memory. The controller stops after 100 seconds of CPU time. We set the following default parameters according to literature: $m = 10$, $\alpha = 1$, $\beta = 1$, $q_0 = 0.9$, $\rho = 0.1$ (cf. [7]), $T_{wd} = 100$, and $n_{ts} = 50$ (cf. [11]). Four variants of the algorithms were tested: (1) default settings as described above, (2) spending 20 CPU seconds before the starting of the working day to construct an improved initial solution (IIS), (3) with pheromone preservation (WPP)[11] ($\tau_{ij} = \tau_{ij}^{old}(1 - \rho) + \rho\tau_0$), $\rho = 0.3$, and (4) min-max pheromone update [14]. For MMAS, we set $\rho = 0.8$. The values used are: $\tau_{max} = 1/(\rho T^*)$, $\tau_{min} = \tau_{max}/(2 \cdot \#\text{AvailableNodes})$, $\tau_0 = \tau_{max}$. These are updated every time a new improvement of T^* is found.

Average results for IIS and MMAS are almost identical to the original results. The reason for this seems to be that although the initial solution is greatly improved, it is more difficult to insert new nodes into the current best solution. Figure 2 shows results for different types of problems in more detail. WPP improves distance results for 10% dynamicity and MMAS for 50% dynamicity, both for the price of slightly more vehicles. Another finding is that for 10% dynamicity solution quality declines by up to 20% and for 50% by up to 50%.

10%	C1		C2		R1		R2		RC1		RC2	
	Dist	Vei	Dist	Vei	Dist	Vei	Dist	Vei	Dist	Vei	Dist	Vei
Static VRP	828.67	10	591	3	1226	12.52	992.49	3	1381.2	12.25	1165.51	3.35
DVRP,no wpp, no IIS	944.1	10.85	632.8	3.67	1283	13.1	1038.1	3.52	1450.76	12.75	1222.05	3.61
DVRP,0.3 wpp no IIS	947.04	10.87	629.2	3.67	1270	13.17	1023.4	3.55	1438.17	12.8	1219.73	3.56
DVRP,no wpp, IIS	943.1	10.88	628.28	3.68	1268	13.19	1022.65	3.54	1446.8	12.8	1213.7	3.51
DVRP, MMAS	954.55	10.87	632.31	3.68	1283	13.25	1013.8	3.54	1458.08	12.82	1219.99	3.57
Decline [%]	13.8089	8.5	6.308	22.3	3.409	4.633	2.14712	17.3	4.12467	4.082	4.13467	4.78
50%	C1		C2		R1		R2		RC1		RC2	
	Dist	Vei	Dist	Vei	Dist	Vei	Dist	Vei	Dist	Vei	Dist	Vei
Static VRP	828.67	10	591	3	1226	12.52	992.49	3	1381.2	12.25	1165.51	3.35
DVRP,no wpp, no IIS	1175.86	12.31	756.48	4.92	1367	14.33	1146.55	4.53	1581.72	14.26	1420.15	5.6
DVRP,0.3 wpp no IIS	1166.81	12.46	761.6	4.96	1361	14.25	1138.83	4.5	1571.06	14.21	1415.77	5.7
DVRP,no wpp, IIS	1167.09	12.48	751.26	4.91	1365	14.35	1145.02	4.46	1580.63	14.23	1409.61	5.73
DVRP,MMAS	1179.03	12.4	740.36	4.87	1378	14.42	1111.33	4.62	1586.22	14.37	1386.35	5.78
Decline [%]	40.8051	23.1	25.272	62.3	11.04	13.82	11.9739	48.7	13.746	16	18.9479	67.2

Fig. 2. Averaged results of 6 Solomon categories using different variants in 10% and 50% dynamicity. The yellow mark is for the best for each problem. Also the decline of solution quality for the dynamic problem as compared to the static problem is reported based on the best DVRP result.

5 Conclusion and Outlook

The MACS-DVRPTW is proposed as a first ant-based solver for the DVRPTW problem. Our results show that results with smaller than 20% of the original solution quality can be achieved for small dynamicity and 50% decline for 50% dynamic insertions. Different variants were tested and pheromone preservation and min-max improved distances, for the price of using on average slightly more vehicles. Future work will have to deepen the study of the conflict between these two objectives, and extend the algorithm with deletion of nodes and dynamically changing travelling times.

Support material (C-implementation of MACS-DVRPTW, benchmark, and best solutions) is available at http://natcomp.liacs.nl/index.php?page=code

Acknowledgement. The authors gratefully acknowledge financial support by Agentschap NL, The Netherlands within the project ' Deliver'. Zhiwei Yang acknowledges financial support from CSC.

References

1. Bell, J.E., McMullen, P.R.: Ant colony optimization techniques for the vehicle routing problem. Advanced Engineering Informatics 18(1), 41–48 (2004)
2. Clarke, G., Wright, J.W.: Scheduling of vehicles from a central depot to a number of delivery points. Operations Research 12(4), 568–581 (1964)
3. Colorni, A., Dorigo, M., Maniezzo, V.: Distributed optimization by ant colonies. In: Varela, F., Bourgine, P. (eds.) Proceedings of the First European Conference on Artificial Life, Paris, France, pp. 134–142. Elsevier (1991)
4. Dorigo, M., Gambardella, L.M.: Ant colony system: a cooperative learning approach to the traveling salesman problem. IEEE Transactions on Evolutionary Computation 1(1), 53–66 (1997)
5. Eyckelhof, C.J., Snoek, M.: Ant systems for a dynamic TSP. In: Dorigo, M., Di Caro, G.A., Sampels, M. (eds.) ANTS 2002. LNCS, vol. 2463, pp. 88–99. Springer, Heidelberg (2002)
6. Flood, M.M.: The traveling-salesman problem. Operations Research 4(1), 61 (1956)
7. Gambardella, L.M., Taillard, É., Agazzi, G.: MACS-VRPTW: A multiple ant colony system for vehicle routing problems with time windows. In: New Ideas in Optimization, ch. 5, pp. 63–76. McGraw-Hill (1999)
8. Gendreau, M., Guertin, F., Potvin, J.-Y., Taillard, É.: Parallel tabu search for real-time vehicle routing and dispatching. Transportation Science 33(4), 381 (1999)
9. Guntsch, M., Middendorf, M.: Applying population based ACO to dynamic optimization problems. In: Dorigo, M., Di Caro, G.A., Sampels, M. (eds.) ANTS 2002. LNCS, vol. 2463, pp. 111–122. Springer, Heidelberg (2002)
10. Kilby, P., Prosser, P., Shaw, P.: Dynamic VRPs: a study of scenarios. Technical Report APES-06-1998, University of Strathclyde (September 1998)
11. Montemanni, R., Gambardella, L.M., Rizzoli, A.E., Donati, A.V.: Ant colony system for a dynamic vehicle routing problem. Journal of Combinatorial Optimization 10, 327–343 (2005)
12. Psaraftis, H.N.: Dynamic vehicle routing: Status and prospects. Annals of Operations Research 61, 143–164 (1995)
13. Solomon, M.M.: Algorithms for the vehicle routing and scheduling problems with time window constraints. Operations Research 35(2), 254–265 (1987)
14. Stützle, T., Hoos, H.: Max-min ant system and local search for the traveling salesman problem. In: IEEE International Conference on Evolutionary Computation, pp. 309–314 (April 1997)
15. Taillard, É., Badeau, P., Gendreau, M., Guertin, F., Potvin, J.-Y.: A tabu search heuristic for the vehicle routing problem with soft time windows. Transportation Science 31(2), 170 (1997)

Pattern Detection in Images Using LBP-Based Relational Operators

José María Molina-Casado and Enrique J. Carmona

Dpto. de Inteligencia Artificial, ETSI Informática,
Universidad Nacional de Educación a Distancia (UNED),
Juan del Rosal 16, 28040, Madrid, Spain
jmolina_79@hotmail.com,
ecarmona@dia.uned.es

Abstract. This paper describes two new pattern detection image operators, $\Re_1^{riu2}$ and $\Re_2$, called, in a generic way, LBP-based relational operators (LBP-RO). The former is rotational invariant and allows searching for a particular pattern disposes in any direction, the later is a binary operator designed to find image patterns that can be modeled by a pattern function. Both of them are invariants against any monotonic transformation of the image gray scale. We have applied these operators in a case study dedicated to segment the ONH in eye fundus color photographic images. The new segmentation method, called GA+LBP-RO, was compared to a competitive ONH segmentation method in the literature and the results obtained by our method proved to be equal to or better.

Keywords: Local Binary Pattern (LBP), Relational Operator, Genetic Algorithm, ONH Segmentation.

1 Introduction

Local Binary Patterns (LBP) are a type of features used for classification in computer vision [6]. The calculation of that feature consists in comparing the intensity of a pixel, g_c, with its neighboring pixels, g_p, and considering the result of each comparison as a bit in a binary string. In that comparison, only the sign, $s(x)$, is considered:

$$s(x) = \begin{cases} 1, & x \geq 0 \\ 0, & x < 0 \end{cases} \tag{1}$$

By assigning a binomial factor 2^p for each comparison, $s(g_p - g_c)$, we obtain a unique $LBP_{P,R}$ number that characterizes the spatial structure of the local image texture:

$$LBP_{P,R} = \sum_{p=0}^{P-1} s\left(g_p - g_c\right) 2^p \tag{2}$$

J.M. Ferrández Vicente et al. (Eds.): IWINAC 2013, Part II, LNCS 7931, pp. 11–20, 2013.

where P controls the number of neighboring pixels, and R determines the radial distance of the these pixels to the central pixel. As a advantage, $LBP_{P,R}$ operator is by definition invariant against any monotonic transformation of the image gray scale.

A extension to the LBP original operator is $LBP_{P,R}^{riu2}$, described in [7]:

$$LBP_{P,R}^{riu2} = \begin{cases} \sum_{p=0}^{P-1} s\,(g_p - g_c) & if\ U\,(LBP_{P,R}) \leq 2 \\ P+1 & otherwise \end{cases} \tag{3}$$

where the superscript $riu2$ reflects the use of rotation invariant uniform patterns that have U value of at most 2, and $U(.)$ is a uniformity measure defined by:

$$U(LBP_{P,R}) = |s\,(g_0 - g_c) - s\,(g_{P-1} - g_c)| + \sum_{p=1}^{P-1} |s\,(g_p - g_c) - s\,(g_{p-1} - g_c)| \tag{4}$$

Thus, if $U(LBP_{P,R}) \leq 2$, then $LBP_{P,R}^{riu2}$ is calculated by counting the number of ones in the binary string $LBP_{P,R}$; otherwise all the other patterns are labeled as "miscellaneous" and collapsed into one value $P+1$. This operator is an excellent measure of the spatial structure of local image texture and is invariant to rotations and against any monotonic transformation of the gray scale.

The idea of the LBP operator can be also extended by using relational functions [8]:

$$\Re\,(x, y, r_1, r_2, \phi, n) \to [0, 1]^n \tag{5}$$

This function is calculated on a point (x, y) of the image, comparing the intensity of n equidistant pairs of points located, respectively, to a distance r_1 and r_2 from the point (x, y). The radial vectors, r_1 and r_2, always form a constant angle defined by ϕ. Based on different combinations of r_1, r_2 and ϕ, local information at different scales and orientations can be captured. However, that function does not provide rotational invariance.

Taking the above defined operators as inspiration, we describe in this paper two new operators, denoted by $\Re_1^{riu2}$ and $\Re_2$. The former is a rotational invariant extension of $\Re$ and the later is a LBP-based binary operator designed to find image patterns that can be modeled by a function.

The article is organized as follows. Section 2 describes the two new operators mentioned. In section 3, we test the performance of such operators to detect patterns in a case study: segmentation of the optic nerve head in eye fundus images. Finally, section 4 presents the conclusions and future work.

2   LBP-Based Relational Operators

In this section, we describe two new image operators, $\Re_1^{riu2}$ and $\Re_2$, based on the concepts of LBP [6,7] and relational function [8]. They will be called, in a

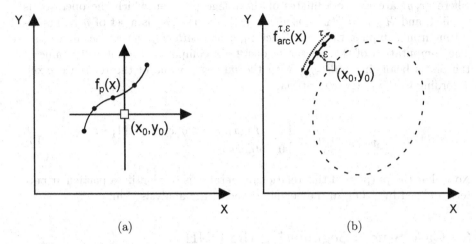

Fig. 1. Examples of pattern functions, $f_p(x)$, defined in a image point (x_0, y_0): (a) generic function (b) ellipse arc function, $f_{arc}^{\tau,\varepsilon}(x)$

generic way, LBP-based relational operators (LBP-RO). On the one hand, $\Re_1^{riu2}$ corresponds to an extension of (5) and is defined by:

$$\Re_1^{riu2}(x, y, r_1, r_2, \phi, n) \rightarrow \{0, 1, ..., (n+1)\} \tag{6}$$

Here, the idea is to add rotational invariance, using the concept of uniformity measure defined in (4). Thus, the computation of this operator is given by:

$$\Re_1^{riu2}(x, y, r_1, r_2, \phi, n) = \begin{cases} \sum_{k=0}^{n-1} R_{1k} & if \ U(R_{1k}) \leq 2 \\ n+1 & otherwise \end{cases} \tag{7}$$

where

$$R_{1k} = s(g(x_2, y_2) - g(x_1, y_1)), \quad k = 0, 1, \ldots, (n-1) \tag{8}$$

$$U(R_{1k}) = \mid R_{10} - R_{1(n-1)} \mid + \sum_{k=1}^{n-1} \mid R_{1k} - R_{1(k-1)} \mid \tag{9}$$

and the coordinates of each pair of points to compare is given by

$$(x_1, y_1) = (x + r_1 \cdot cos(k \cdot 2\pi/n), y + r_1 \cdot sin(k \cdot 2\pi/n) \tag{10}$$

$$(x_2, y_2) = (x + r_2 \cdot cos(k \cdot 2\pi/n + \phi), y + r_2 \cdot sin(k \cdot 2\pi/n + \phi) \tag{11}$$

On the other hand, the second relational operator, $\Re_2$, is defined by

$$\Re_2(x, y, \mathcal{H}_{x_i, y_i}, f_p(x), n) \rightarrow \{0, 1\} \tag{12}$$

where (x, y) are the coordinates of the image pixel on which the operator is applied, and $\mathcal{H}_{x_i, y_i} = \{(x_i, y_i) \,|\, y_i = f_p(x_i), \, i = 1, ..., n\}$ is a set of n points resulting from sampling the $f_p(x)$ function, called *pattern function* (see figure 1a). The computation of this operator consists in comparing the intensity value of the pixels belonging to $\mathcal{H}_{x_i, y_i}$ with the intensity value of the reference pixel, according to the following criteria:

$$\Re_2(x, y, \mathcal{H}_{x_i, y_i}, f_p(x), n) = \begin{cases} 1 & if \; g(x_i, y_i) < g(x, y), \forall i \epsilon \{1, ..., n\} \\ 0 & otherwise \end{cases} \tag{13}$$

Note that the purpose of this relational operator is to identify a particular pattern, defined by $f_p(x)$, in the neighborhood of the analysis point.

3   Case Study: Segmenting the ONH

The segmentation of the optic nerve head (ONH) is of critical importance in retinal image analysis because ONH disturbances can be an initial symptom of serious eye diseases. The ONH, also called optic disk or papilla, is oval-shaped and is located in the area where all the retina nerve fibres come together to form the start of the optic nerve that leaves the back of the eyeball. There is an area without any nerve fibres called excavation (the centre of the papilla) and around it another area can be found, the neuroretinal ring, whose external perimeter delimits the papillary contour.

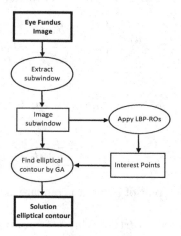

Fig. 2. Block diagram summarizing the proposed method to segment the ONH

The segmentation method here used is a variation of the method described in [3] and with which will be compared (see Section 3.2). The main difference between both methods lies in how to get the set of interest points (IPs).

Here, the IPs are obtained using LBP-ROs and, in the original method, they are obtained using a domain-knowledge-based *ad-hoc* method. First of all, in order to reduce the processing computational cost and the number of distractor patterns, the process begins by automatically extracting a sub-window from the original image that is approximately centered at a point of the papillary area [5]. Next, the two LBP-ROs above defined are applied to the image sub-window[1] and the result is a set of IPs points that have a high probability of belonging to the papillary contour. Then, we use a genetic algorithm (GA) to find an ellipse whose contour is formed by the maximum number of interest points. Finally, we select the best genetic ellipse (papillary contour) from the final population as the solution to our problem. The figure 2 summarizes the process.

3.1   LBP-RO Instantiation

Taking advantage of the characteristics of the operator $\Re_2$, we proceed as follows. Given an generic ellipse traced in the retinal image, $\Re_2$ is applied only and exclusively to those pixels belonging to the contour of this ellipse. To do that, a function denoted by $f_{arc}^{\tau,\varepsilon}(x)$ is defined for each pixel of the contour. That function corresponds to an ellipse arc which is parallel to the considered ellipse, has a length, τ, and is separated a distance, ε, (towards its outer side) from the analysis pixel (x, y) (see figure 1b). This operator, thus instantiated, will be hereinafter denoted by $\Re_{2\_0}$:

$$\Re_{2\_0} = \Re_2(x, y, \mathcal{H}_{x_i, y_i}, f_{arc}^{\tau,\varepsilon}(x), n) \tag{14}$$

The better the ellipse approximates the contour papillary, the greater the number of pixels belonging to the ellipse contour that satisfy $\Re_{2\_0} = 1$. This follows from the definition of this operator and the property of the papillary contour: frontier that separates the papilla (bright area) from the retina (darker area). However, since infinite ellipses can be traced in the image, we will use a genetic algorithm in order to search the optimal ellipse. To code this type of solutions, the phenotypic space is transformed into a genotypic space consisting of real vectors of five variables $[x, y, a, b, \theta]$, where (x, y) represents the centre of the genetic ellipse, (a, b) the magnitudes of its major and minor semi-axis respectively and θ the angle that its major axis forms with the x-axis. Finally, by applying recombination, mutation and selection operators, a population of chromosomes (ellipses) evolves until a finalization criterion is achieved. The degree of approximation to the solution of each ellipse can be calculated using the following fitness function:

$$f_{fitness_1} = \sum_{(x_k, y_k) \in GEC} \Re_{2\_0}(x_k, y_k, \mathcal{H}_{x_i, y_i}, f_{arc}^{\tau,\varepsilon}(x), n) \tag{15}$$

where *GEC* is the genetic ellipse contour. That is, this operator computes the number of pixels belonging to *GEC* that satisfy $\Re_{2\_0} = 1$.

---

[1] Specifically, the operators are applied to the V channel as result of transforming the original RGB color space into HSV space.

However, it is possible to refine the fitness function defined in (15). One must take into account that the vessels are also delimited by boundaries separating the retina (brighter areas) from the vessels (darker areas). So if we only applied the fitness function defined in (15), the papillary contour could be formed by a genetic ellipse whose contour pixels belong to vessel-retina boundaries or the papillary contour. In order to reduce this possibility, we will use the $\Re_1^{riu2}$ operator, instantiated by $\Re_{1\_0}^{riu2}$:

$$\Re_{1\_0}^{riu2} = \Re_1^{riu2}(x, y, r, r, \pi, 8) \tag{16}$$

where r is a value bigger than the width of any vessel in the image. Here we use the geometric property by which the width of a vessel is always smaller than the width of the papilla. Then it is not difficult to check that, if a pixel belongs to the papillary contour, we will obtain, with high probability, a value belonging to the set $\{4, 5\}$, as result of applying $\Re_{1\_0}^{riu2}$ to that pixel. Note that these two values correspond, respectively, to patterns 00001111 (or their equivalents rotated) and 00011111 (or their equivalents rotated). On the other hand, if the same operator is applied to a pixel belonging to the contour of a vessel, assuming uniform retinal intensities and linear vessel shapes in the zone in which the operator acts, we will obtain, with high probability, a value belonging to the set $\{7, 8\}$. Note that these two values correspond respectively to patterns 01111111 (or their equivalents rotated) and 11111111. The interesting thing about these two pattern sets is that they are disjoint. However, in real situations, there is noise in the image, that is, the retina is not always uniform, there are papillary contour zones that are traversed by vessels and these vessels can be formed by curved paths or branches. Therefore, the final idea is to use the two operators, $\Re_{1\_0}^{riu2}$ and $\Re_{2\_0}$, together to promote the synergy of detecting only papillary contour points and avoid the occurrence of false positives. From the point of view of the GA, it only involves changing the fitness function defined in (15) by this other:

$$f_{fitness_2} = \sum_{i=1}^{n} and\left(\Re_{1\_\{4,5\}}^{riu2}, \Re_{2\_0}\right) \tag{17}$$

where

$$\Re_{1\_\{4,5\}}^{riu2} = \begin{cases} 1 & if\ (\Re_{1\_0}^{riu2}) \in \{4, 5\} \\ 0 & otherwise \end{cases} \tag{18}$$

Finally, figure 3 shows two examples of how work the two LBP-OP in two different genetic ellipses (upper row and lower row). The figures 3a and 3b are the result of applying the operators $\Re_{2\_0}$ and $\Re_{1\_\{4,5\}}^{riu2}$, respectively, to the same genetic ellipse, and figure 3c shows the result of applying the *and* operator. Similar comments apply to lower row figures. It is easy to check that the total number of points that verify both criteria ($\Re_{2\_0} = 1$ and $\Re_{1\_\{4,5\}}^{riu2} = 1$) is greater for the upper row ellipse (figure 3c) than the lower row one (figure 3f). Therefore, the former will be a better approach to the papillary contour than the later one.

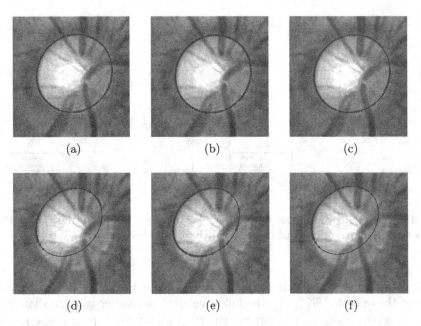

Fig. 3. Example of applying LBP-ROs to two different genetic ellipses (upper row and lower row): (a)&(d) $\Re_{2\_0}$; (b)&(e) $\Re_{1\_\{4,5\}}^{riu2}$; (c)&(f) $and\ \left(\Re_{2\_0}, \Re_{1\_\{4,5\}}^{riu2}\right)$

3.2 Results and Evaluation

To measure the performance of our algorithm, denoted by GA+LBP-RO, we used DRIONS [1,3] and ONHSD [2,4] databases. In order to do the segmentation results quantitatively reproducible, we measured the average discrepancy between the points of the contour obtained with the segmentation method and a gold standard defined from a contour that was the result of averaging several contours, each of them traced by an expert (two experts in DRIONS and four in ONHSD). Here we use the concept of discrepancy, δ, defined in [4]:

$$\delta^j = \frac{\sum_{i=1}^{N}((\mid m_i^j - \mu_i^j \mid)/(\sigma_i^j + \varepsilon))}{N} \qquad (19)$$

where δ^j is the discrepancy measurement for the image j, $i = 1, ..., N$, with N the number of angularly equidistant radial segments used for each measurement, m_i^j is the length of the radius defining the i-th point of the ellipse proposed for the image j, μ_i^j and σ_i^j are the mean and typical deviation, respectively, of the lengths of the radii defining the i-th point of the contours traced by the experts and belonging to the image j, and $\varepsilon = 0.5$ is a small factor to prevent division by zero when the experts are in exact agreement. To facilitate the visualization of the results, a discrepancy curve is plotted, namely, the percentage of images with discrepancy less than δ (y-axis) versus δ (x-axis).

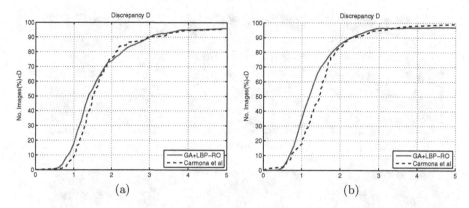

Fig. 4. Accumulated discrepancy results for our method, GA+LBP-RO, versus the segmentation method described in [3]: (a) DRIONS and (b) ONHSD databases

The parameter configuration used for the GA was the same as that used in [3]. For the operator $\Re^{riu2}_{1\_\{4,5\}}$, the following final parameter configuration was chosen: $\Re^{riu2}_{1\_\{4,5\}}(x, y, 10, 10, \pi, 8)$ for DRIONS and $\Re^{riu2}_{1\_\{4,5\}}(x, y, 8, 8, \pi, 8)$ for ONHSD. For the $\Re_{2\_0}$ operator and for both databases, the final parameter configuration was $\Re_{2\_0}(x, y, \mathcal{H}_{x'_i, y'_i}, f^{8,2}_{arc}(x), 8)$.

The results obtained were compared with the segmentation method described in [3]. Thus, the figure 4 shows the discrepancy curves obtained from applying the two methods to each image database. Owing to the stochastic nature associated with the GA, each curve of discrepancy showed (both methods use a GA-based approach) correspond to the result of averaging five curves obtained as a result of executing the GA five times.

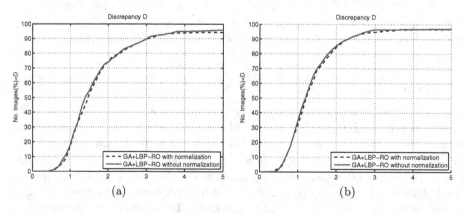

Fig. 5. Discrepancy curves, using GA+LBP-RO, with and without normalizing the image databases: (a) DRIONS (b) ONHSD

Looking at the figure 4, we can say that practically our method improves or equals the performances of the other method in both databases. Beside, it must be said that the procedure used by GA+LBP-RO to obtain the interest points is more simple than that one used in [3]. Finally, we must emphasize that, unlike the method with which is compared, GA+LBP-RO does not use any kind of normalization in the images database[2]. That is possible because the two LBP-RO used are invariant against any monotonic transformation of the images gray scale. We have made experiments applying and not applying this normalization phase (see figure 5). It is easy to see that, in both databases, the normalization phase does not contribute to improve the discrepancy curves.

4   Conclusions and Future Work

We have described two new image operators, $\Re_1^{riu2}$ and $\Re_2$, called LBP-based relational operators. The former is rotational invariant and allows searching for a particular pattern disposes in any direction and computed from a symmetric neighborhood. The later is not rotational invariant but allows searching for patterns modeled by an asymmetric neighborhood characterized by a pattern function, $f_p(x)$. Both of them are invariant against any monotonic transformation of the gray scale. We have applied these operator successfully in a segmentation method, called GA+LBP-RO, used to segment the ONH in eye fundus color photographic images. The performances of GA+LBP-RO were compared to a competitive method in the literature and the results obtained by our method proved to be equal to or better. As future work, we propose to investigate other types of LBP-RO and apply them to other image databases and other domains.

Acknowledgment. This work was supported in part by funds of the Advanced Artificial Intelligence Master Program of the Universidad Nacional de Educación a Distancia (UNED), Madrid, Spain.

References

1. DRIONS-DB: Digital retinal images for optic nerve segmentation database (January 2013), http://www.ia.uned.es/personal/ejcarmona/DRIONS-DB.html
2. ONHSD: Optic nerve head segmentation dataset (January 2013), http://reviewdb.lincoln.ac.uk/Image%20Datasets/ONHSD.aspx
3. Carmona, E.J., Rincón, M., García-Feijoo, J., Martínez-de-la Casa, J.M.: Identification of the optic nerve head with genetic algorithms. Artificial Intelligence in Medicine 43, 243–259 (2008)
4. Lowell, J., Hunter, A., Steel, D., Basu, A., Ryder, R., Fletcher, E.: Optic nerve head segmentation. IEEE Transaction on Medical Imaging 23(2), 256–264 (2004)

---

[2] In figure 4, the discrepancy curves of GA+LBP-RO were obtained without normalizing the databases.

5. Molina, J.M., Carmona, E.J.: Localization and segmentation of the optic nerve head in eye fundus images using pyramid representation and genetic algorithms. In: Ferrández, J.M., Álvarez Sánchez, J.R., de la Paz, F., Toledo, F.J. (eds.) IWINAC 2011, Part I. LNCS, vol. 6686, pp. 431–440. Springer, Heidelberg (2011)

6. Ojala, T., Pietikäinen, M., Harwood, D.: A comparative study of texture measures with classification based on featured distributions. Pattern Recognition 29, 51–59 (1996)

7. Ojala, T., Pietikäinen, M., Mäenpää, T.: Multiresolution gray-scale and rotation invariant texture classification with local binary patterns. IEEE Transactions on Pattern Analysis and Machine Intelligence 24, 971–987 (2002)

8. Setia, L., Teynor, A., Halawani, A., Burkhardt, H.: Grayscale medical image annotation using local relational features. Pattern Recognition Letters 29, 2039–2045 (2008)

Cancer Stem Cell Modeling
Using a Cellular Automaton

Ángel Monteagudo and José Santos Reyes

Computer Science Department, University of A Coruña, Spain
jose.santos@udc.es

Abstract. We used a cellular automaton model for cancer growth simulation at cellular level, based on the presence of different cancer hallmarks acquired by the cells. The rules of the cellular automaton determine cell mitotic and apoptotic behaviors, which are based on the acquisition of the hallmarks in the cells by means of mutations. The simulation tool allows the study of the emergent behavior of tumor growth. This work focuses on the simulation of the behavior of cancer stem cells to inspect their capability of regeneration of tumor growth in different scenarios.

1 Introduction

Although there are more than 200 different types of cancer that can affect every organ in the body, they share certain features. Hanahan and Weinberg described the phenotypic differences between healthy and cancer cells in a landmark article entitled "The Hallmarks of Cancer" [6] and their recent update [7]. The six essential alterations in cell physiology that collectively dictate malignant growth are: self-sufficiency in growth signals, insensitivity to growth-inhibitory (antigrowth) signals, evasion of programmed cell death (apoptosis), limitless replicative potential, sustained angiogenesis, and tissue invasion and metastasis.

One of the traditional approaches to model cancer growth was the use of differential equations to describe avascular, and indeed vascular, tumor growth [11]. Nevertheless, the approaches relying on Cellular Automata (CA) make easy the modeling at cellular level, where the state of each cell is described by its local environment. CAs have been used in Artificial Life to study and characterize the emergent behavior in complex dynamic systems, as it can be considered tumor growth in cellular systems. Thus, different works have appeared which used the CA capabilities for different purposes in tumor growth modeling [12].

The cancer stem cell (CSC) theory suggests that tumor cells include a minority population of cells (CSCs) responsible for the initiation of tumor development, growth, and tumor's ability to metastasize and reoccur [3][5]. Cancer Stem Cells can divide either symmetrically to yield two CSCs, or asymmetrically to produce a CSC and a non-stem cancer cell with limited proliferation capacity (Differentiated Cancer Cell, DCC). The probability of symmetric division is very low, which reflects the correspondingly low fraction of cancer stem cells reported throughout the literature. For example, Enderling and Hahnfeldt [3]

J.M. Ferrández Vicente et al. (Eds.): IWINAC 2013, Part II, LNCS 7931, pp. 21–31, 2013.

considered a probability $p_s = 0.01$ to determine symmetric CSC division. Figure 1 summarizes the CSC model. Such CSCs were proposed to persist in tumors as a distinct population and cause relapse and metastasis by giving rise to new tumors. So, different works tried to simulate their behavior, taking into account their main characteristics, such as their capacity to divide indefinitely [17]. If current treatments of cancer do not properly destroy enough CSCs, the tumor will reappear, so it is important to understand their behavior and effects.

Along this line, Enderling and Hahnfeldt [3] used a hybrid mathematical-cellular automaton model that simulated growth of a heterogeneous solid tumor comprised of cancer stem cells and non-stem cancer cells and focused on cell proliferation, cell migration and cell death. In [10], the authors showed that although CSCs are necessary for progression, their expansion and consequently tumor growth kinetics are surprisingly modulated by the dynamics of the non-stem cancer cells. Their simulations revealed that slight variations in non-stem cancer cell proliferative capacity can result in tumors with distinctly different growth kinetics.

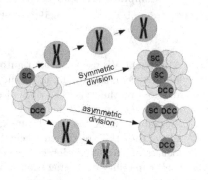

Fig. 1. In the Cancer Stem Cell (CSC) model these cells can divide symmetrically or asymmetrically to produce Differentiated Cancer Cells (DCCs) with limited proliferative capability

Vainstein et al. [16] used a 2D CA model which emphasized the distinction between CSCs and DCCs and assumed that CSC fate is governed by quorum sensing, i.e., the ability of a cancer stem cell to decide whether to differentiate, on the basis of the number of cancer stem cells in its neighborhood. Finally, Sottoriva et al. [14] showed that tumors modeled in a CSC context more faithfully resemble human malignancies and show invasive behavior and with higher tumor heterogeneity.

In previous works we applied a CA to model cancer growth behavior using a 3D grid-like environment [9][13]. In our proposal the CA rules are designed to model the mitotic and apoptotic behaviors in each cell from the information of the cell state and from its surrounding environment, together with the hallmarks acquired by a cell as consequence of mutations. In this paper we simulate the tumor initiation and development in relation to the cancer stem cell theory and, on the contrary to the previous works commented, we focused on the importance of the cancer hallmarks to generate cancer growth under the CSC model.

2   Event Model for Cancer Simulation

In the simulation each cell resides in a site in a cubic lattice and has a "genome" associated with different cancer hallmarks. The essential alterations in cell physiology that collectively dictate malignant growth are [4][6][7]:

SG. Self-Growth: Growth even in the absence of normal "go" signals. Most normal cells wait external messages (growth signals from other cells) before dividing. Cancer cells often counterfeit their own pro-growth messages [4].

IGI. Ignore Growth Inhibit: As the tumor expands, it squeezes adjacent tissue, which sends out chemical messages that would normally bring cell division to a halt. Malignant cells ignore the commands, proliferating despite anti-growth signals issued by neighboring cells.

EA. Evasion of apoptosis: In healthy cells, genetic damage above a critical level usually activates a suicide program (programmed cell death or apoptosis). Cancer cells bypass this mechanism.

AG. Ability to stimulate blood vessel construction: Tumors need oxygen and nutrients to survive. They obtain them by co-opting nearby blood vessels to form new branches that run throughout the growing mass (angiogenesis).

EI. Effective immortality: Healthy cells can divide no more than several times (< 100). The limited replicative potential arises because, with the duplication, there is a loss of base pairs in the telomeres (chromosome ends which protect the bases), so when the DNA is unprotected, the cell dies. Malignant cells avoid the telomere shortening, so such cells overcome the reproductive limit.

MT. Power to invade other tissues and spread to other organs: Cancers usually become life-threatening only after they somehow disable the cellular circuitry that confines them to a specific part of the organ in which they arose. New growths appear and eventually interfere with vital systems.

GI. Genetic instability: It accounts for the high incidence of mutations in cancer cells, allowing rapid accumulation of genetic damage. It is an enabling characteristic of cancer [7] since, while not necessary in the progression from neoplasm to cancer, makes such progression much more likely [2]. In the simulation, the cells with this factor will increase their mutation rate.

Each cell genome indicates if any hallmark is activated as consequence of mutations. Metastasis and angiogenesis are not considered, as we are interested in the first avascular phases of tumorigenesis. So, every cell has its genome which consists of five hallmarks plus some parameters particular to each cell. All the parameters are commented in Table 1. The parameters *telomere length* and *base mutation rate* can change their values in a particular cell over time, as explained in the table. The cell's genome is inherited by the daughter cells when a mitotic division occurs. The default values indicated in Table 1 are the same as those used in [1], except the initial telomere length and the value of g. In the case of the telomere length, we used the default value indicated in [15] as it corresponds to the Hayflick limit [6]. As other parameters are more difficult to set in order to have a direct analogy in nature, we used a standard value of $g = 30$, which implies a lower possibility of escaping the inhibition by contact mechanism with respect to the default value used in [1] ($g = 10$) and [15] ($g = 5$). Nevertheless, we will reason about the consequences of using different values in the parameters.

Regarding the mutation rate, we must take into account two considerations: The rate affects the abstract model of hallmarks and not the individual bases (which was experimentally determined as less than 10^{-8} per base pair). Secondly, as indicated in [1], with small cell population sizes in the simulations, large mutation rates are necessary to obtain the expected incidence of cancer. So, we used the same default value considered in [1] for the probability of acquisition of hallmarks in the mitotic divisions. Also, Basanta et al. [2] worked with parameters, such as base mutation rate (10^{-5}) and mutation rate increase for cells with acquired genetic instability ($i = 100$), with the same default values.

We used an event model, similar to that used by Abbott et al. [1], which uses an event queue for storing possible future mitotic events. A mitosis is scheduled several times in the future, between 5 and 10 time steps, simulating the variable duration of the cell life cycle (between 15 and 24 hours). Taking into account these time intervals, each iteration represents an average time of 2.6 hours, so, for example, 5000 iterations in the simulation imply an average time of 77.4 weeks. The main aspects of the model can be summarized in the following steps:

Table 1. Definition of the parameters associated with the hallmarks

Parameter name	Default value	Description
Initial telomere length (tl)	50	Every time a cell divides, the length is shortened by one unit. When it reaches 0, the cell dies, unless the hallmark EI is ON.
Evade apoptosis (e)	10	A cell with n hallmarks mutated has an extra n/e likelihood of dying each cell cycle, unless the hallmark EA is ON.
Base mutation rate (m)	10^5	Each gene (hallmark) is mutated (when the cell divides) with a $1/m$ chance of mutation.
Genetic instability (i)	10^2	There is an increase of the base mutation rate by a factor of i for cells with this mutation (GI).
Ignore growth inhibit (g)	30	As in [1], cells with the hallmark IGI activated have a probability $1/g$ of killing off a neighbor to make room for mitosis.
Random cell death (a)	10^3	In each cell cycle every cell has a $1/a$ chance of death from several causes.

Start: A mitosis is scheduled for the initial cells of the grid (push mitotic events in the event queue between 5 and 10 time instants in the future).

After the new daughter cells are created (in unoccupied adjacent sites), mitosis is scheduled for each of them, and so on. Random errors occur in this copying process, so some hallmarks can be activated when copying the genetic information, taking into account that once a hallmark is acquired in a cell, it will be never repaired by another mutation [1].

Pop event: The events are ordered on event time. Pop event from the event queue with the highest priority (the nearest in time).

Random cell death test: Cells undergo random cell death with low probability ($1/a$ chance of death, where a is a tunable parameter).

Genetic damage test: The larger the number of hallmark mutations, the greater the cell death rate. If "Evade apoptosis" (EA) is ON, death is not applied.

Mitosis tests:

1. **Replicative potential checking**: If telomere length is 0, the cell dies, unless the hallmark "Effective immortality" (EI) is mutated (ON).

2. **Growth factor checking**: As in [1] and [2], cells can perform divisions only if they are within a predefined spatial boundary, which represents a threshold in the concentration of growth factor; beyond this area (95% of the inner space in each dimension in our simulations, which represents 85.7% of the 3D grid inner space) growth signals are too faint to prompt mitosis (unless hallmark SG is ON).

3. **Ignore growth inhibit checking**: If there are not empty cells in the neighborhood, the cell cannot perform a mitotic division. As in [1][15], if the "Ignore growth inhibit" hallmark (IGI) is ON, then the cell competes for survival with a neighbor cell and with a likelihood of success $(1/g)$.

If the three tests indicate possibility of mitosis:

- Increase the base mutation rate if Genetic Instability (GI) is ON.
- Add mutations to the new cells according to base mutation rate $(1/m)$.
- Decrease telomere length in both cells.
- Push events. Schedule mitotic events (push in event queue) for both cells: Mother and daughter, with the random times in the future.

If mitosis cannot be applied:

- Schedule a mitotic event (in queue) for mother cell.

Frequently, cells are unable to replicate because of some limitation, such as contact inhibition or insufficient growth signal. Cells overcome these limitations through mutations in the different hallmarks. Additionally, as indicated in [9], the final results are independent of the starting initial condition, being the emergent behavior the same beginning the simulation with only one healthy cell at the center of the grid (as in most previous works), or if the grid is initially full of healthy cells, although this second and more realistic strategy needs more time iterations to reach a stable number of healthy or cancer cells.

3   Results

To understand the behavior of Cancer Stem Cells (CSCs) in different scenarios and also the main implications of the hallmarks, first we can check the multicellular system evolution introducing a small percentage of cells with a given hallmark activated. We selected different representative situations to show the capability of different hallmarks in cancer growth, and trying to isolate their influence on the behavior of the cellular system evolution.

The simulations begin with the grid full of healthy cells. After a given number of temporal steps we introduce a number of cells with a given hallmark and in random positions of the inner area of the grid with growth factor. If the position is occupied by a cell, such cell is replaced by the new incorporated one. We selected *Effective immortality, evade apoptosis* and *ignore growth inhibit* for this first analysis, since *genetic instability* is not a hallmark that can act separately (only increases the base mutation rate in the cells) and *self-growth* has only advantage when the cells have reached the limits with growth factor.

3.1   Hallmark Effects

The capability of the hallmark *Effective immortality* (EI) to initiate and promote tumor growth is clear as such cells do not have a limit for dividing. In Figure 2.a we can see the evolution of cancer cells and healthy cells when we introduce, at iteration 100, cells with the hallmark EI activated (1% of the grid size) in the inner area of the multicellular system. We used the same grid size (125000) as in [1]. The evolution was run without mutations in the cell's genome, so no more hallmarks can be acquired during the simulation run. In all graphs, a cell is considered as cancerous if any of the hallmarks is present. Also, given the stochastic nature of the problem, the graphs are an average of 5 different runs.

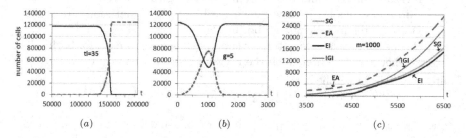

(a)   (b)   (c)

Fig. 2. (a,b) Evolution through time iterations of the number of healthy cells (continuous lines) and cancer cells (dashed lines) in different scenarios when incorporating, at $t = 100$, a number of cells with a hallmark acquired (1% of the grid size): a) incorporated cells with EI acquired, initial telomere lengths in all cells=35, no more mutations; b) incorporated cells with IGI acquired, $g = 5$, no more mutations; c) Number of cells with a given hallmark, incorporating, at $t = 100$, cells with EA acquired, high mutation rate ($m = 1000$)

The telomere length was set to 35 in all cells (original healthy cells of the grid as well as the incorporated cells), while the rest of parameters were set to their standard values. The healthy cells begin to die when their telomere length reaches the value 0 after the maximum number of 35 divisions, beginning their decrease and leaving room for the (immortal) cancer cells. The Figure shows only the time iterations in which the transition occurs. With shorter values in the telomere length the transition would be in less iterations, and the contrary with larger values of the initial telomeres. This can be considered as obvious. However, note that, even we are using only this hallmark in the incorporated cells, the other limits to cancer cell proliferation are active, in particular the apoptotic process that can eliminate such cells (with a given probability). But the apoptotic probability (default value) is not sufficient to stop the rapid proliferation of the cells with EI acquired. So, given the interrelation between the hallmarks, the simulation is the perfect tool to inspect the final behavior.

In the second case (Figure 2.b), the same number of cells (1% of the grid size) was introduced at $t = 100$, but with the hallmark *ignore growth inhibit* (IGI) activated. Again, to isolate at maximum its effect, the evolution was run without mutations in the hallmarks. We selected a representative situation with $g = 5$, which determines the probability of escaping the inhibition by contact mechanism in the cells with the hallmark IGI activated, whereas the rest of parameters were set to their standard values (grid size = 125000).

Once incorporated, the cancer cells begin to grow because these cells can replace a neighbor to make room for mitosis, and with a probability given by $1/g$ [1][15]. That is, the incorporated cells begin to divide, as they can proliferate in the full space, with the decrease of the healthy cells. Note again that the apoptotic process is not sufficient to stop the proliferation. Nevertheless, the number of cancer cells reaches a limit in which they begin to decrease rapidly. This is because, at this stage (after iteration 1000), most of the cancer cells reached the maximum number of divisions, given by the initial telomere length (standard value $tl = 50$), and because of the rapid proliferation of such cells. On the contrary, the healthy cells begin now to increase as such cells were not able to divide until such stage, as practically all the grid sites were occupied. That is, the healthy cells increase again thanks to the free sites which suddenly appear, free positions provided by the cancer cells when these reach the maximum number of mitoses. With larger values of g, cancer proliferation will be slower or could be stopped by the apoptotic process if the proliferation is not sufficiently fast.

The third case is when cells with the hallmark *evade apoptosis* (EA) are introduced at such iteration $t = 100$ (Figure 2.c). In order to see the advantage and capability of this hallmark, we can test the inclusion of those cells in an environmental situation where the base or hallmark mutation rate $(1/m)$ is high. In this case, to understand the advantage of this hallmark, the other hallmarks are considered in the simulation, so different cancer hallmarks can be acquired in all the cells. Figure 2.c shows the evolution of some hallmarks using $m = 1000$, while the other parameters were set to their default values and the grid size was 125000.

We only showed the evolution of four hallmarks at the beginning of cancer growth. As the Figure denotes, the first hallmark that begins to increase is *evade apoptosis*, avoiding the apoptotic death and allowing the proliferation of cancer cells. When a cell has EA acquired, then the same cell can acquire other hallmarks thanks to the mutations in the different mitotic divisions. Note that in this case, even for the cells with EA acquired, there are other limits that interact in the proliferation, like the inhibition by contact, the necessity of being in the area with growth factors or the maximum number of divisions. However, as the mutation rate is high, it is easy to acquire the hallmarks that allow escaping such limits, so there is a fast increase of the cancer cells (not shown in the Figure).

3.2   Cancer Stem Cells Simulation

Once the capabilities of the analyzed hallmarks were shown with some of their interrelations, we can test the capability of CSCs to renew or promote tumor growth. CSCs have two defining hallmarks acquired: *Effective immortality*, so they will have no limit in their proliferation capacity, in addition to their resistance to apoptosis (EA) [5]. The CSCs will divide symmetrically, with the same probability used in [3] ($p_s = 0.01$) or asymmetrically to produce a CSC and a non-stem cancer cell (DCC) and with probability $1 - p_s$. In the asymmetric division, the differentiated non-stem cancer cell acquires (randomly) one of the 5 hallmarks considered in the simulation as well as the initial telomere length which defines its finite replicative potential (unless it acquires the hallmark EI). When these DCCs divide later, the resultant non-stem cancer cells can acquire more mutations or hallmarks.

For the simulation, as in previous cases, we introduce a number of CSCs in the multicellular system evolution. Figure 3 includes a representative example

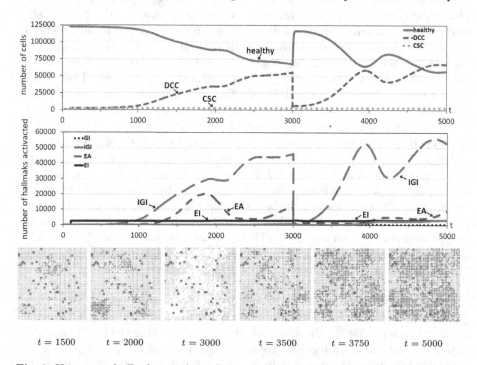

$t = 1500$ $t = 2000$ $t = 3000$ $t = 3500$ $t = 3750$ $t = 5000$

Fig. 3. Upper graph: Evolution through time iterations of the number of healthy cells (continuous line), non-stem cancer cells (dashed line) and Cancer Stem Cells (CSCs) (pointed line) with the grid initially full of healthy cells. CSCs are introduced at time iteration 100 (2% of the grid size). Center graph: Evolution through time iterations of the different hallmarks in the cells. The bottom part shows snapshots of 2D central sections of the multicellular system evolution corresponding to different time iterations (Colors: Gray - healthy cells, Blue - DCCs, Red-enlarged size - CSCs).

with a run of the system evolution and using default values (except $g = 5$). At time iteration $t = 100$ a number of CSCs corresponding to a 2% of the grid size (125000) was introduced at random points in the inner area with growth factor. These incorporated CSCs are not subject to apoptosis [5], so their presence is maintained according to the cancer stem cell theory. Moreover, as the probability of symmetric division is very low ($p_s = 0.01$), the number of CSCs remains stable during the 5000 iterations shown in Figure 3.

Even with the low mutation rate, if non-stem cancer cells with the hallmark *ignore growth inhibit* (IGI) acquired appear, then they can proliferate. As more mutations are appearing in this proliferation of DCCs, then hallmark *evade apoptosis* (EA) appears in many DCCs. Between iterations 1100 and 3000 there are several factors that act simultaneously to stop a rapid proliferation of both cell types: there is practically no free space to replicate, the cancer cells only proliferate thanks to the acquisition of the hallmark IGI (we set the value of g to obtain this proliferation in such number of iterations) and, finally, several non-stem cancer cells are ending their maximum number of divisions.

CSCs have been suggested to be more resistant to therapeutic interventions such as chemotherapy or irradiation compared with their differentiated counterparts [14] or, as indicated by Vainstein et al. [16], because the low efficacy of cytotoxic therapy on this cell population (CSCs). Hence, at time iteration 3000, the 100% of non-stem cancer cells are killed, simulating a (perfect) cancer therapy, causing the drastic drop of the DCCs. The healthy cells that have not performed the maximum number of divisions fill rapidly the space, but the DCCs recover quickly because the CSCs not killed produce again DCCs, repeating a similar evolution pattern like the one at the beginning of the simulation. Note that this second proliferation on non-stem cancer cells (DCCs) is faster as they have more opportunities to divide just after killing the non-stem cancer cells, while the space is not full, in contrast to the initial situation. Nevertheless, the same evolution pattern of proliferation is obtained again. The bottom part of Figure 3 shows cross-sections of this multicellular system evolution at different time iterations, showing the progressive expansion of the tumor cells (DCCs), with the colonization of many areas previously initialized with some DCCs and thanks to the continuous presence of CSCs. This presence of island-like formations outside the main tumor expansion is present in many malignancies, as indicated in the examples shown in [14].

4 Conclusions

In this paper we simulated tumor development according to the cancer stem cell theory using a simulation tool based on a cellular automaton. The emergent behavior of growing multiclonal tumors is almost impossible to infer intuitively and, as commented by Kansal et al. [8], the growing tumor must be investigated and treated as a self-organizing dynamic system. This cannot be done with currently available in vitro/in vivo models or common mathematical approaches [8], needing novel computational models to simulate the mechanistic complexity of solid tumor growth and invasion.

The different hallmarks interfere or influence on the global behavior of the system, so we tried to isolate and study the capability for initiating tumor growth in different conditions, and which can be incorporated in a simulation of CSCs. The CSCs can regenerate the proliferation of cancer cells when the DCCs, with the appropriate hallmarks acquired, find advantageous scenarios. The simulations show how the proliferation of non-stem cancer cells is faster after a therapy that specifically targets such differentiated cancer cells, given more opportunities to CSCs to differentiate. This is in agreement with the clinical observations describing increased growth speed and enhanced invasion in the relapsing malignancy [14] or with the observations by Vainstein et al. [16], when they point out that accelerated death of differentiated cancer cells decreased the number of such cells, but increased the number of CSCs.

As the elimination of CSCs, the root of cancer origin and recurrence according the CSC theory, has been thought as a promising approach to improve cancer survival, our study can help to determine the potential of the main cancer hallmarks to generate or re-generate tumor growth in different circumstances.

Acknowledgements. This work was funded by the Ministry of Science and Innovation of Spain (project TIN2011-27294).

References

1. Abbott, R.G., Forrest, S., Pienta, K.J.: Simulating the hallmarks of cancer. Artificial Life 12(4), 617–634 (2006)
2. Basanta, D., Ribba, B., Watkin, E., You, B., Deutsch, A.: Computational analysis of the influence of the microenvironment on carcinogenesis. Mathematical Biosciences 229, 22–29 (2011)
3. Enderling, H., Hahnfeldt, P.: Cancer stem cells in solid tumors: Is 'evading apoptosis' a hallmark of cancer? Progress in Biophysics and Molecular Biology 106, 391–399 (2011)
4. Gibbs, W.W.: Untangling the roots of cancer. Scientific American 289, 56–65 (2003)
5. Gil, J., Stembalska, A., Pesz, K.A., Sasiadek, M.M.: Cancer stem cells: the theory and perspectives in cancer therapy. J. App. Genet. 49(2), 193–199 (2008)
6. Hanahan, D., Weinberg, R.A.: The hallmarks of cancer. Cell 100, 57–70 (2000)
7. Hanahan, D., Weinberg, R.A.: Hallmarks of cancer: The next generation. Cell 144(5), 646–674 (2011)
8. Kansal, A.R., et al.: Simulated brain tumor growth dynamics using a three-dimensional cellular automaton. Journal of Theoretical Biology 203, 367–382 (2000)
9. Monteagudo, Á., Santos, J.: A cellular automaton model for tumor growth simulation. In: Rocha, M.P., Luscombe, N., Fdez-Riverola, F., Rodríguez, J.M.C. (eds.) 6th International Conference on PACBB. AISC, vol. 154, pp. 147–155. Springer, Heidelberg (2012)
10. Morton, C.I., et al.: Non-stem cancer cell kinetics modulate solid tumor progression. Theoretical Biology and Medical Modelling 8, 48 (2011)
11. Patel, M., Nagl, S.: The role of model integration in complex systems. An example from cancer biology. Springer (2010)

12. Rejniak, K.A., Anderson, A.R.A.: Hybrid models of tumor growth. WIREs Syst.
    Biol. Med. 3, 115–125 (2010)
13. Santos, J., Monteagudo, Á.: Study of cancer hallmarks relevance using a cellular
    automaton tumor growth model. In: Coello, C.A.C., Cutello, V., Deb, K., Forrest,
    S., Nicosia, G., Pavone, M. (eds.) PPSN 2012, Part I. LNCS, vol. 7491, pp. 489–499.
    Springer, Heidelberg (2012)
14. Sottoriva, A., et al.: Cancer stem cell tumor model reveals invasive morphology
    and increased phenotypical heterogeneity. Cancer Research 70(1), 46–56 (2010)
15. Spencer, S.L., Gerety, R.A., Pienta, K.J., Forrest, S.: Modeling somatic evolution
    in tumorigenesis. PLoS Computational Biology 2(8), 939–947 (2006)
16. Vainstein, V., et al.: Strategies for cancer stem cell elimination: Insights from math-
    ematical modeling. Journal of Theoretical Biology 298, 32–41 (2012)
17. Wodarz, D., Komarova, N.: Can loss of apoptosis protect against cancer? Trends
    in Genetics 23(5), 232–237 (2007)

Improved Polar Scan-Matching Using an Advanced Line Segmentation Algorithm

Israel Navarro Santosjuanes, José Manuel Cuadra-Troncoso,
Félix de la Paz López, and Raúl Arnau Prieto

Dpto. de Inteligencia Artificial, UNED, Madrid, Spain
israeln@alum.mit.edu, {jmcuadra,delapaz}@dia.uned.es,
rarnau@ctcomponentes.com

Abstract. This work presents an enhanced polar scan-matching procedure (E-PSM) that obtains its inputs from the application of an
advanced line segmentation algorithm to the laser range returns. Additionally, a set of alternative methods based on local and global optimization algorithms is introduced. Results from robot simulation tests
are provided for different ranges of laser range return noise and odometry sensor error levels. The results show that the proposed E-PSM algorithm and one of the methods based on global optimization yield good
robot pose estimation precision while keeping computational costs at a
reasonable level.

1 Introduction

Laser scan matching is an approach to solve the problem of mobile robot Simultaneous Localization and Mapping (SLAM). SLAM provides a robot the
capability to perform its tasks autonomously in an unknown environment by
creating a map of its surrounding environment and at the same time locating
itself [1,2]. Scan matching can be performed in 2D as well as in 3D and can be
applied to static and dynamic environments. Also, scan matching approaches
can be local [3], or global [4]. In local scan matching the robot pose is estimated by comparing a pair of laser scans [5]. Starting from a known position,
the matching process is repeated at regular intervals. For each segment traveled,
the scan taken at the initial position (reference scan) is stored. Additionally, the
scan taken at the end of the segment (called current scan) is matched against
the reference scan; from this comparison, the robot pose is calculated relatively
to its initial pose. This process provides position and orientation corrections to
offset errors in the robot pose calculation obtained from odometry data alone.
In global scan matching [6,7], the current scan is matched against a global map
of features (determined *a priori* or built while exploring the environment) or a
database of scans.

The core of the scan matching process is the methodology employed to as-
sociate and then compare the data obtained in the current scan to the original
reference scan data. Some of the methods are: feature to feature, points to line,
and point to point. In feature to feature approaches, elements like line segments

J.M. Ferrández Vicente et al. (Eds.): IWINAC 2013, Part II, LNCS 7931, pp. 32–44, 2013.

[8], corners or range extrema [9] are extracted from laser scans and then matched. These features do not have to be concrete physical items; for example, another scan matching approach is based on the use of a cross correlation function [5]. In this case, both scans are replaced by stochastic representations (histograms) and the matching is solved by finding the maximum of a cross correlation function. In points to line approaches [7], the points of the current scan are matched against an *a priori* polygonal world model consisting exclusively of line segments. Some of the most prominent point to point matching approaches are: iterative closest point [10], iterative matching range point, and the iterative dual correspondence [3]. These three approaches are defined in a Cartesian coordinate frame and do not take advantage of the native polar coordinate system of a laser scan. Diosi [11] provides a novel method called Polar Scan Matching (PSM). PSM's main advantage is its ability to search for point associations by simply matching points with the same bearing and thus eliminating the search for corresponding points.

From this discussion, it should be apparent that any factor degrading the precision of the laser scan readings can have a detrimental effect on the performance of any scan matching algorithm. Therefore. it is imperative to understand and manage accordingly all sources of noise. Several sources of laser scan range data noise can be distinguished: (i) intrinsic errors in the measurement devices themselves, (ii) calibration errors, and (iii) environmental irregularities (dependency on surface reflective properties, uneven surfaces, etc). The effect of the first two can be minimized or at least evaluated using the device calibration procedures and technical specifications. The last case can be the main source of error and hardest to control; it is, after all, the result of the physical nature of the environment. Cuadra et al. [12] presented an algorithm for line extraction that makes use of a generic noise model; this feature will allow applying the algorithm to worlds with surfaces of different characteristics.

2   Procedure Overview

This work presents an enhanced PSM procedure (referred in this article as E-PSM) based on the original work of Diosi [11]. Thus, the scheme consists of a relative, point-to-point scan-matching algorithm in polar coordinates. Several improvements over the original PSM proposal are introduced. First of all, an advanced line segmentation algorithm is applied to the laser range returns. Second, an additional criteria to select scan points suitable for matching is added. Third, several robot pose estimation accuracy improvements are implemented both in the orientation and in the translation search tasks. As an alternative for this third modification, the possibility of using local or global optimization techniques is included.

The scan-matching algorithm is applied after the robot detects a significant robot pose change (as measured by its odometry sensors) or after a certain time interval has been elapsed. The ensuing procedure can be decomposed in four main steps:

- Filtering and segmentation of the reference and current scans
- Projection and interpolation of the current scan in the reference frame of coordinates
- Pruning of the scan data to discard range returns unfit for the matching process
- Application of a pose search method; either an alternating orientation and position search or a simultaneous orientation and position search is used.

2.1   Filtering and Line Segmentation Algorithm

In [12], Cuadra et al. present a novel algorithm for line extraction in polar coordinates. In order to allow the management of varying scan noise levels, this algorithm uses a noise model in which the errors have a standard deviation proportional to a known function of the expected measure.

Figure 1 (the robot travels from the left to the right) presents an example of how this algorithm is applied to a pair of reference and current scans, being the standard deviation of laser measurements proportional to the expected measure, the proportionally constant was 20mm/m. In the rest of the paper values for standard deviation of laser measurements have to be understood as standard deviation when expected distance is 1m. Notice how most of the noisy signals are replaced by accurately placed straight segments.

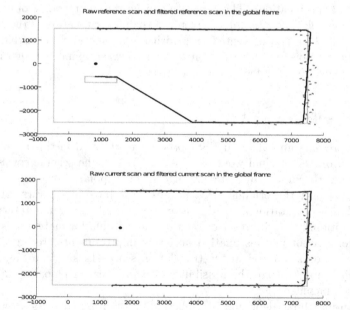

Fig. 1. Scan filtering and segmentation example. Raw measures, points, filtered measures, black line, and true map, gray line. Units are millimeters.

3   Improved Scan-Matching Procedure

3.1   Scan Projection

Once we have de-noised the reference and current scans, we must get into the position to be able to compare them. For this purpose, we must find a common viewpoint; in this work, the point selected is the reference scan viewpoint. This means that we have to determine the values of the range readings obtained in the final robot position (i.e. the current scan) as if they had been obtained instead from the reference viewpoint. Following Diosi [11], for each of the range readings $r_{cur,i}$ obtained at the final robot position, the range measurements and bearings from the reference robot position (refer to Figure 2) are given by equations 1 and 2 (note subindex "∗" is used here instead of subindex "cur" for more compact notation).

$$r^c_{*,i} = \sqrt{(r_{*,i} \cdot \cos(\theta_{rob} + \phi_{*,i}) + x_{rob})^2 + (r_{*,i} \cdot \sin(\theta_{rob} + \phi_{*,i}) + y_{rob})^2} \quad (1)$$

$$\phi^c_{*,i} = \arctan 2(r_{*,i} \cdot \sin(\theta_{rob} + \phi_{*,i}) + y_{rob}, \; r_{*,i} \cdot \cos(\theta_{rob} + \phi_{*,i}) + x_{rob}) \quad (2)$$

where: $r^c_{*,i}$ is the current range reading computed from the reference viewpoint corresponding to the measurement taken at the bearing angle $\phi_{*,i}$; $\phi^c_{*,i}$ is the computed bearing corresponding to $r^c_{*,i}$; $r_{*,i}$ is the range reading obtained from the final robot position at the bearing angle $\phi_{*,i}$; $\phi_{*,i}$ is the bearing angle at which $r_{*,i}$ was obtained; and x_{rob}, y_{rob}, and θ_{rob} are the robot Cartesian position coordinates and orientation in the final position.

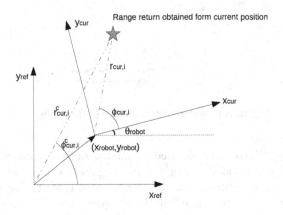

Fig. 2. Reference and actual frames

3.2 Interpolation and Pruning

The following step consists of interpolating the newly computed range data at the bearings $\phi^c_{cur,i}$ $(= \phi^c_{*,i}$ in eq. 2) to the bearings at which the original reference range data were obtained, $\phi_{ref,i}$. Additionally, we must identify which points, either in the current scan or in the reference scan are not fit for being part of the matching process that will follow next. The result of this process must be a set of pairs $\left\{ r_{ref,i}, r^f_{cur,i} \right\}$ for each reference bearing $\phi_{ref,i}$ for which there is valid reference and current laser range returns.

For the points contained in a line identified by the segmentation algorithm[1], a basic linear interpolation scheme is applied. Figure 3 presents a sample scenario for a robot with only seven laser sensors equally spaced every 30 degrees in its forward direction. For this simple example, the interpolation task would consist of calculating at $\phi_{ref,30}$ and $\phi_{ref,60}$ the corresponding $r^f_{cur,30}$ and $r^f_{cur,60}$ using the calculated ranges $r^c_{cur,0}$, $r^c_{cur,30}$, $r^c_{cur,60}$, and $r^c_{cur,90}$ (corresponding to angles $\phi^c_{cur,0}$, $\phi^c_{cur,30}$, $\phi^c_{cur,60}$, and $\phi^c_{cur,90}$).

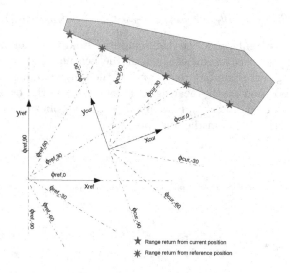

Fig. 3. Range data interpolation and pruning

Several strategies have been devised to handle occlusion cases. In the first place, when several interpolated range value $r^f_{cur,i}$ are computed for the same reference bearing $\phi_{ref,i}$, only the smaller range value is kept. The present work introduces a new criteria to prune out those reference scan range returns that

---

[1] In this work, we do not apply the kind of segmentation heuristics used by Diosi and other authors. The segmentation algorithm ensures bad data points (like those over the maximum range) or outliers are not assigned to any segment and are therefore not taken into account in the scan matching process.

cannot be observed from the final robot position (i.e. they are out of the area spanned by the current scan). For the particular case in which the robot has sensors only in the area 180 degrees ahead of itself, this criterion can be simplified to calculating the x-coordinate of the reference range return in the frame defined by the current robot position, as given by Equation 3 (note subindex "#" is used here instead of subindex "ref" for more compact notation).

$$x^{cur}_{\#,i} = (r_{\#,i} \cdot \cos(\phi_{\#,i}) - x_{rob}) \cdot \cos(\theta_{rob}) + (r_{\#,i} \cdot \sin(\phi_{\#,i}) - y_{rob}) \cdot \sin(\theta_{rob}) \quad (3)$$

where $x^{cur}_{\#,i}$ is x-coordinate value of the reference scan range return in the current frame.

With this definition, all reference range returns with an $x^{cur}_{ref,i}$ ($= x^{cur}_{\#,i}$ in eq. 3) below zero can be assumed to be invisible from the current position and discarded from the scan-matching process. An example of how the overall process works can be seen in Figure 4. Notice that the upper graph provides (in polar coordinates) the values of the $r^c_{cur,i}$; in the same graph, the dashed line corresponds to the reference scan against which the $r^c_{cur,i}$ are matched. Note that these $r^c_{cur,i}$ data points marked with a cross if they are to be used in the scan-matching process. Those marked with a circle were discarded for not having a correspondence in the current scan or by the segmentation algorithm (outliers); those marked with an inverted triangle were discarded using the new criterion just defined. The lower graph simply provides the same information but presented in the x-y plane together with the initial and final robot positions (hollow and filled black dots respectively).

3.3 Robot Position and Orientation Update

The updated robot pose is to be determined so it minimizes the error in the match between the reference and projected current scans found previously. For this purpose, we must first establish an error measurement. Recall that after the scan projection and pruning process, we will have (for those reference scan bearings for which valid reference and current scan data is available) a one-to-one correspondence between $r_{ref,i}$ and $r^f_{cur,i}$. From, this we can define the following objective function f to be minimized:

$$f = \sum_{i=1}^{n} (r_{ref,i} - r^f_{cur,i})^2 \quad (4)$$

Two procedures are proposed to solve this problem.

Alternating Orientation and Position Search. Following the general approach in [11], we are going to first attempt to update the robot orientation θ_{rob} and then the robot position (x_{rob}, y_{rob}). This process is repeated until convergence is reached or a maximum number of iterations are consumed.

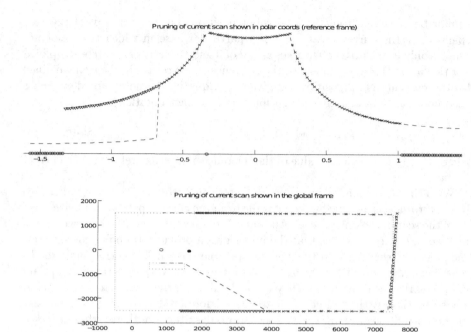

Fig. 4. Scan pruning example

The orientation search process is based on a brute-force approach. Starting from the initial orientation guess provided by the odometry sensors, a systematic re-evaluation of the objective function f is performed every 0.5 degrees on the interval $[\theta_{rob} - 10, \theta_{rob} + 10]$. If no relative minimum is found, the systematic evaluation process is expanded in the direction of decreasing error until successful. Note that every evaluation for each proposed θ_{rob} implies performing the projection, interpolation and pruning process again.[2] Once a relative minimum is found, an additional fine search is performed in the 1 degree interval around this value at every 0.01 degrees[3].

For the position update, the difference between the reference scan and the projected current scan is modeled as a first order linear approximation [11]:

$$\overrightarrow{r_{cur}^{f}} - \overrightarrow{r_{ref}} = H \begin{bmatrix} \delta x_{rob} \\ \delta y_{rob} \end{bmatrix} + \overrightarrow{v} \tag{5}$$

---

[2] Unlike in [11], the current scan interpolation and pruning is not assumed to remain invariant with the shifting process associated to the θ_{rob} search process. The presented approach allows to re-calculate the objective function at arbitrary points along the θ_{rob} line.

[3] In [11], a parabolic approximation is used to implement this fine search. This approach has not been used in this work as it was found to be a source of errors.

where the $(\delta x_{rob}, \delta y_{rob})$ are the increments over the current robot position estimate, H is the Hessian matrix and $\overrightarrow{v}$ is the noise vector. The linear least squares
problem is well known, see [13], Equation 6 provides its solution[4].

$$\begin{bmatrix} \delta x_{rob} \\ \delta y_{rob} \end{bmatrix} = (H^T H)^{-1} H^T (\overrightarrow{r_{cur}^f} - \overrightarrow{r_{ref}}) \tag{6}$$

**Simultaneous Position and Orientation Optimization Using Global
or Local Algorithms.** In this alternative approach the objective function f,
see Equation 4, will be minimized using a suite of different global and local
(both gradient and non-gradient based) optimization methods. In all cases, the
nlopt library is used. Refer to [13] and [14] for detailed information about these
algorithms.

4   Results

Several test cases have been devised to be run in a robot simulator (the map
used corresponds to the one shown in Figure 1) that cover the factors impacting
the overall algorithms performance. The algorithms are first tested without any
laser range return noise or initial error from the odometry sensors. In this case,
the initial robot pose guess corresponds to the exact pose solution; additionally,
the robot is provided with ideal laser range returns. This test provides a baseline
to check the basic algorithm stability. Secondly, the amount of noise introduced
by the odometry sensors is introduced at different levels (1%, 5% and 10%)
while keeping perfect laser range returns. The noise from the odometry sensors
introduces errors in the initial robot pose estimation. As this initial error grows
large, this set-up allows us to test the algorithms capability to search the solution
space without getting stuck in local minima. Thirdly, noise in the laser scan
is introduced (with standard deviations of 5mm and 20mm); in this occasion,
the odometry error is set to zero. In this test, the initial guess is already the
solution but we now need to verify how the map distortions influence the robot
pose search. This allows us to test the capabilities provided by the filtering
and segmentation algorithm. Lastly, a test combining both types of errors is
introduced. This is meant to represent the more realistic case against which the
scan-matching algorithms are designed. Note that the results to be provided
hereafter represent the results of letting the simulator run for one hundred scan-
matching cycles.

---

[4] In this work the Hessian matrix is found numerically using the central differences
method. [11] provides a closed form solution for the Hessian matrix that assumes
implicitly that there is change in $\overrightarrow{r_{cur}^f}$ as result of the interpolation and pruning
process. This assumption has not been used here.

4.1   Alternating Orientation and Position Search

Three scan-matching algorithms using the alternating orientation and position search approach are tested: (i) The original PSM algorithm as presented in [11], (ii) the E-PSM algorithm proposed in section 3 and, (iii) a brute-force approach method. For this last case, the orientation search is identical to the one used in E-PSM. However, the position search is performed by making a complete sweep of the area around the robot (800mm by 800mm area in 10mm intervals). This method is provided only to assess the other two algorithms performance and not as a practical solution.

For each of the test cases discussed previously, Table 1 provides the results (average and maximum) obtained for five different variables: δx_{rob} (error in the position update measured along x_{ref} in mm), δy_{rob} (error in the position update measured along y_{ref} in mm), $\delta\theta_{rob}$ (error in the orientation update measured in degrees), f (objective function as defined in section 4.3) and the computational time used by the complete scan-matching function for each segment traveled (measured in seconds in a computer equipped with a Intel Core2 Duo processor).

Taking as a reference the $\delta\theta_{rob}$ error results in Table 1, it can be clearly seen that the application of the original PSM algorithm results in large orientation estimation errors (average error between 0.2 and 0.7 degrees and maximum errors between 2 and 10 degrees). Similar trends can be seen for errors in δx_{rob}, δy_{rob} and f. The E-PSM and brute-force algorithms obtain much better precision (average error between 0.03 and 0.27 degrees and maximum errors between 0.6 and 1.4 degrees). The computational time usage recorded shows a significant difference among the three algorithms. The PSM algorithm on average takes about 0.003 seconds to provide a pose estimation, in contrast, the E-PSM algorithm takes about 0.07 seconds and the brute force algorithm about ten seconds. This result show that the E-PSM algorithm represents an excellent alternative as it obtains only slightly worse precision than the brute-force algorithm with a two-orders of magnitude lower computational costs.

4.2   Simultaneous Orientation and Position Search

The development of this alternative solution method was motivated by the presence of errors in the brute-force algorithm discussed previously beyond what was expected given the intervals at which the objective function f was evaluated (the evaluation was made every 0.01 degrees and 10 mm while resulting error presented values an order of magnitude greater than these). It was considered that the approach of alternating position and orientation searches could be the source of errors since the search was likely to get stuck in local minima. This suspicion was also the origin of implementing in the E-PSM orientation search algorithm the rule by which the evaluation does not proceed to the fine orientation search phase until finding a relative minimum; this feature improved the E-PSM precision noticeably in some test cases.

To analyze systematically this proposal to search simultaneously the robot orientation and position using the nlopt library, several groups of algorithms were selected:

Table 1. Alternating orientation and position search results

			No noise	Odo 1%	Odo 5%	Odo 10%	σ_s 5mm	σ_s 20mm	σ_s 5mm / Odo 5%
Error	PSM	avg	3.09	3.32	7.38	8.64	10.51	33.16	10.34
δx_{rob}		max	33.95	35.25	132.21	231.84	269.39	337.98	169.69
(mm)	E-PSM	avg	0.51	0.20	0.71	0.84	2.95	15.68	3.02
		max	15.98	2.40	30.70	9.06	14.66	122.87	16.21
	B.force	avg	0.89	0.82	1.57	2.14	2.42	14.04	2.83
		max	10.21	9.99	10.76	33.75	10.97	120.08	21.65
Error	PSM	avg	4.51	4.70	5.20	5.13	7.38	20.34	6.72
δy_{rob}		max	40.15	40.60	79.51	89.49	134.69	324.52	53.65
(mm)	E-PSM	avg	0.44	0.27	0.86	1.71	2.04	8.46	2.51
		max	24.36	7.04	8.99	38.63	16.27	96.37	27.75
	B.force	avg	0.80	0.61	1.26	1.43	0.92	5.49	1.81
		max	9.89	10.26	19.56	9.89	10.22	69.92	29.88
Error	PSM	avg	0.18	0.22	0.28	0.38	0.41	0.67	0.39
$\delta\theta_{rob}$		max	1.33	2.27	3.65	6.35	7.70	10.11	5.00
(deg)	E-PSM	avg	0.034	0.027	0.041	0.091	0.069	0.265	0.094
		max	0.58	0.59	0.45	1.08	0.68	1.12	1.36
	B.force	avg	0.026	0.025	0.033	0.062	0.068	0.219	0.083
		max	0.32	0.40	0.32	0.88	0.59	1.17	0.57
f	PSM	avg	8.42	9.93	14.36	19.16	20.82	45.00	19.37
		max	106.37	126.28	241.15	593.00	391.50	476.51	321.89
	E-PSM	avg	2.08	1.34	1.39	3.77	6.75	23.36	6.08
		max	34.09	30.71	15.46	128.27	153.22	144.98	56.20
	B.force	avg	1.65	1.31	1.72	2.47	6.13	20.98	4.67
		max	17.35	16.10	15.03	20.75	153.56	144.18	22.20
Time	PSM	avg	0.0020	0.0021	0.0025	0.0025	0.0026	0.0034	0.0028
(s)		max	0.0065	0.0061	0.0082	0.0085	0.0094	0.0099	0.0085
	E-PSM	avg	0.069	0.068	0.074	0.087	0.067	0.081	0.075
		max	0.29	0.27	0.09	0.34	0.28	0.16	0.19
	B.force	avg	10.72	10.48	10.22	11.11	10.97	12.59	10.64
		max	15.55	15.12	13.53	16.25	16.31	22.96	16.24

- Global optimization algorithms: DIRECT-L, CRS2, ISRES and MLSL-LDS
- Local non gradient-based algorithms: COBYLA, BOBYQA, and SBPLX
- Local gradient-based algorithms: SLSQP, BFGS, NEWTON, and VAR2

Refer to [14] for detailed information in each of these optimization algorithms.

Table 2 details, for each of the test cases used before, the average value obtained for two different variables: f (objective function as defined in section 4.3) and the computational time used by the complete scan-matching function for each segment traveled. The data obtained for the objective function show that the original PSM algorithm (and in second place the three non gradient-based algorithms COBYLA, BOBYQA, and SBPLX) provides the worst precision results. The local gradient-based algortithms SLSQP, BFGS, NEWTON, and VAR2 fail provide adequate precision consistently. In some cases, they reach

Table 2. Simultaneous orientation and position search results

		No noise	Odo 1%	Odo 5%	Odo 10%	σ_s 5mm	σ_s 20mm	σ_s 5mm / Odo 5%
f	PSM	8.42	9.93	14.36	19.16	20.82	45.00	19.37
	E-PSM	2.08	1.34	1.39	3.77	6.75	23.36	6.08
	B.force	1.65	1.31	1.72	2.47	6.13	20.98	4.67
	DIRECT-L	1.19	1.64	1.89	2.06	3.73	19.50	3.85
	CRS2	1.37	1.46	1.29	1.49	3.61	19.73	3.53
	ISRES	1.15	1.21	1.31	1.59	3.52	19.47	3.57
	MLSL_LDS	1.16	0.98	0.99	1.36	3.47	19.39	3.50
	COBYLA	2.31	6.29	11.75	18.21	9.58	23.96	15.85
	BOBYQA	2.49	7.00	13.07	14.79	9.58	26.72	14.00
	SBPLX	1.43	1.86	3.65	5.90	3.77	19.92	6.81
	SLSQP	1.82	1.63	7.21	12.06	4.75	22.24	12.99
	BFGS	1.88	1.68	3.84	4.11	5.94	24.38	12.46
	NEWTON	1.88	1.68	2.03	3.85	6.09	24.39	12.27
	VAR2	1.88	1.67	3.85	3.12	5.99	24.35	12.50
Time (s)	PSM	0.002	0.0021	0.0025	0.0025	0.0026	0.0034	0.0028
	E-PSM	0.0689	0.0676	0.0739	0.087	0.0665	0.0808	0.0752
	B.force	10.72	10.48	10.22	11.11	10.97	12.59	10.64
	DIRECT-L	2.11	3.90	4.66	4.79	11.60	15.44	11.60
	CRS2	0.0576	0.0946	0.0918	0.0838	0.1127	0.1194	0.1252
	ISRES	0.21	0.77	0.90	0.87	1.74	1.91	1.51
	MLSL_LDS	9.29	6.16	6.31	9.34	53.00	60.66	45.43
	COBYLA	0.0048	0.0067	0.0078	0.008	0.0083	0.0078	0.0083
	BOBYQA	0.0052	0.0068	0.0087	0.0089	0.0077	0.0079	0.0086
	SBPLX	0.0076	0.0177	0.0266	0.0319	0.0265	0.0354	0.0378
	SLSQP	0.0056	0.0075	0.0102	0.0121	0.0127	0.0172	0.0173
	BFGS	0.0088	0.0115	0.0143	0.018	0.0176	0.0222	0.0197
	NEWTON	0.0091	0.0131	0.0199	0.0245	0.0195	0.0217	0.0242
	VAR2	0.0086	0.0119	0.0153	0.0162	0.0181	0.0199	0.0199

results on a par with the global schemes but the errors in the results become unacceptable for the combination test case (σ_s 5mm / Odo 5%). Overall, it seems apparent that neither the PSM algorithm nor the local optimization approaches provide consistently precise robot pose estimations. As expected, the four global optimization algorithms obtain the best precision in all test cases (among them, MLSL_LDS obtains the edge in most test cases). Regarding the E-PSM algorithm proposed in this work, the results show it provides similar results than its global counterparts (except for the Odo 10% and σ_s 5mm test cases in which the E-PSM does show worse results). Regarding the computational time usage, the following groups can be readily identified in order of increasing resource requirements: PSM (around 0.003s per pose estimation process), local optimization algorithms (between 0.004s and 0.05s), E-PSM and CRS2 (between 0.07s and 0.1s), and finally the rest of global algorithms as well as the brute-force scheme (in the order of several seconds or more). These results show that the application of a simultaneous orientation and position search can be advantageous in terms of the precision obtained but only when global methods are applied. Of these, the CRS2 presents itself as the most promising one since it provides in general better precision than the E-PSM procedure with only marginally higher computational time requirements.

5 Conclusions

The simulation test results provided for different ranges of laser range return noise and odometry sensor error levels show that the proposed E-PSM algorithm and the CRS2-based global optimization algorithm yield good robot pose estimation performance while keeping computational costs at a reasonable level. On the other hand, the PSM and the local optimization algorithms are not able to reach high-precision pose estimations consistently. The global optimization algorithms do obtain the best precision results; unfortunately, except for the CRS2 case, the computational costs associated to their use do not seem reasonable for most applications. Therefore, it can be said that the E-PSM and CRS2 algorithms afford the best-balanced behavior of all the algorithms tested thus far. Nevertheless, additional testing in other simulated maps as well as in real robots must be done to confirm these results. Considering the work performed and the limitations observed in the scan-matching techniques developed, several lines of future work have been identified:

- Optimization of computational resources used by the algorithm E-PSM
- Review of the scan interpolation procedures
- Combination of local and global optimization techniques, or combination of the E-PSM algorithm with local or global optimization techniques
- Extension of the E-PSM algorithm to dynamic environments and non-polygonal worlds

Acknowledgements. This work was partially supported by the Spanish Government through Project TIN2010-20845-C03-02.

References

1. Leonard, J.J., Durrant-Whyte, H.F.: Simultaneous map building and localization for an autonomous mobile robot. In: Proc. IEEE Int. Workshop on Intelligent Robots and Systems, Osaka, Japan, pp. 1442–1447 (1991)
2. Smith, R., Self, M., Cheeseman, P.: Estimating uncertain spatial relationships in robotics. In: UAI, pp. 435–461 (1986)
3. Lu, F., Milios, E.: Robot pose estimation in unknown environments by matching 2d range scans. In: Proceedings of the 1994 IEEE Computer Society Conference on Computer Vision and Pattern Recognition, CVPR 1994, pp. 935–938 (June 1994)
4. Tomono, M.: A scan matching method using euclidean invariant signature for global localization and map building. In: Proceedings of the 2004 IEEE International Conference on Robotics and Automation, ICRA 2004, vol. 1, pp. 866–871 (2004)

5. Weiss, G., Puttkamer, E.: A map based on laserscans without geometric interpretation. In: Proceedings of Intelligent Autonomous Systems 4 (IAS-4), pp. 403–407. IOS Press (1995)
6. Borthwick, S., Durrant-Whyte, H.: Simultaneous localisation and map building for autonomous guided vehicles. In: Proceedings of the IEEE/RSJ/GI International Conference on ntelligent Robots and Systems 1994. Advanced Robotic Systems and the Real World, IROS 1994, vol. 2, pp. 761–768 (September 1994)
7. Cox, I.: Blanche-an experiment in guidance and navigation of an autonomous robot vehicle. IEEE Transactions on Robotics and Automation 7(2), 193–204 (1991)
8. Gutmann, J.S.: Robuste Navigation autonomer mobiler Systeme. PhD thesis, Albert-Ludwigs-Universitat Freiburg, Institut fur Informatik (2000)
9. Lingemann, K., Surmann, H., Nuchter, A., Hertzberg, J.: Indoor and outdoor localization for fast mobile robots. In: Proceedings of the 2004 IEEE/RSJ International Conference on Intelligent Robots and Systems, IROS 2004, vol. 3, pp. 2185–2190 (2004)
10. Besl, P., McKay, H.: A method for registration of 3-d shapes. IEEE Transactions on Pattern Analysis and Machine Intelligence 14(2), 239–256 (1992)
11. Diosi, A., Kleeman, L.: Fast laser scan matching using polar coordinates. Int. J. Rob. Res. 26, 1125–1153 (2007)
12. Cuadra Troncoso, J.M., Álvarez-Sánchez, J.R., de la Paz López, F., Fernández-Caballero, A.: Improving area center robot navigation using a novel range scan segmentation method. In: Ferrández, J.M., Álvarez Sánchez, J.R., de la Paz, F., Toledo, F.J. (eds.) IWINAC 2011, Part I. LNCS, vol. 6686, pp. 233–245. Springer, Heidelberg (2011)
13. Nocedal, J., Wright, S.J.: Numerical Optimization. Springer (1999)
14. Johnson, S.G.: The nlopt nonlinear-optimization package, http://ab-initio.mit.edu/nlopt

Reactive Navigation and Online SLAM
in Autonomous Frontier-Based Exploration

Raúl Arnau Prieto, José Manuel Cuadra-Troncoso,
José Ramón Álvarez-Sánchez, and Israel Navarro Santosjuanes

Dpto. de Inteligencia Artificial, UNED, Madrid, Spain
rarnau@ctcomponentes.com, {jmcuadra,jras}@dia.uned.es, israeln@alum.mit.edu

Abstract. This paper describes an autonomous exploration algorithm
for mobile robots. The method implements a frontier-based exploration
strategy that relies on a reactive navigation system and a SLAM algorithm. Despite its strong biological inspiration, the navigation method
is specially well suited for the exploration task since it is able to accept
and follow higher level position targets while guaranteeing the integrity
of the robot. The SLAM module is intended for online execution, but
it is able to solve the entire path of the robot in real-time. The hierarchical nature of the SLAM algorithm allows for drift modeling and
reduction, which achieves very good resolution maps directly from laser
measurements, without extracting landmarks or correction steps. The exploration strategy attempts to exploit the benefits of both the mapping
and the navigation algorithms, providing a basic framework for more
sophisticated autonomous behaviors.

1 Introduction

This article describes the design of an algorithm for autonomous mobile robots
whose objective is the efficient exploration of an unknown environment. Exploration is one of the most challenging problems in autonomous robotics. Solving
optimally the exploration strategy of an unknown environment is still considered nowadays an intractable problem in the literature [1]. The approach here
presented tackles exploration as a greedy search over the available world information. The algorithm takes as an input the actual knowledge of the world and
decides where to move next in order to get as much new information as possible, with the minimum displacement cost. The proposed exploration method
relies on two algorithms: a simultaneous localization and mapping algorithm, or
SLAM, and a navigation method.

The autonomous exploration of the environment is based on the concept of
unexplored frontier. As described in [2], frontiers are regions dividing the explored
and unexplored areas in the robot environment. By moving to a frontier the
robot can look into the unexplored region and gain new knowledge, pushing
the frontier back as its internal representation of the world extends. Frontier
based exploration consists on visiting iteratively the subsequent frontiers until
the robot eventually registers all the world into its map. The core of this strategy

J.M. Ferrández Vicente et al. (Eds.): IWINAC 2013, Part II, LNCS 7931, pp. 45–55, 2013.
© Springer-Verlag Berlin Heidelberg 2013

is selecting the best frontier region according to a criteria of efficiency. In our approach this criteria is based on a tradeoff between the expected information gain (*reward*) and the penalization of moving to the frontier region (*cost*). In order to test the efficiency of our algorithm, several experiments are conducted in a simulated environment and compared to other methods. The comparison uses the map coverage evolution with time as the main metric.

This paper is organized as follows. Section 2 describes the reactive navigation module. Section 3 includes a brief overview of the SLAM problem and describes the selected algorithm and its adaptation. In section 4 the proposed exploration strategy is presented. Experimental results are summarized in section 5. Section 6 contains the main conclusions and future work.

2   Reactive Navigation

Reactive systems can solve gracefully the secure navigation problem even in changing environments, although the ability of moving the robot towards a determined goal may usually require some form of planning. The Area Center (AC) navigation algorithm [3,4,5] used in this approach implements both capabilities in a purely reactive fashion.

A very simple wandering algorithm has been implemented in order to evaluate the navigation behavior provided by the AC method. This naive method basically drives the robot ahead in straight line, with random angle turns when an obstruction is detected. Both algorithms are tested using a simulation environment. During each experiment the robot is allowed to wander randomly for one hour while its path is being recorded. This procedure is repeated 15 times with each algorithm and the result is shown in figure 1 in the form of position histogram. The robot is equipped with a laser rangefinder.

These results show that the AC algorithm outperforms the naive wandering method in terms of smoothness and robot integrity. However, it can be seen that both methods tend to visit some regions more frequently than others. The simple wandering algorithm provides a more random distribution of the robot position over the map. This could be confused as a desired exploration behavior since,with enough time, this kind of random-walk strategy will eventually cover the full open space [6]. The AC method, on the other hand, provides a more natural (human-like) way of navigation which is indeed better suited for an exploration task: using the AC method the robot is able to visit all the rooms, but does not waste time in the vicinity of obstacles. In section 4 it will be shown that the robot is also able to register all the obstacles in the map, but this is done from a security distance.

3   SLAM

In order to build a map of its environment a robot must be able to localize itself. While building iteratively a map from known robot positions is quite straightforward, and localizing the robot given a correct map can also be solved easily,

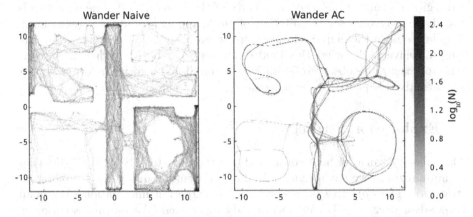

Fig. 1. 2D histograms of the robot position for each method: simple wandering and AC. Each figure contains the error-free robot trajectories for fifteen independent simulations of one hour. Units in axes are in meters.

solving both concurrently is one of the most challenging problems in autonomous robotics [7]. SLAM is precisely the problem of determining simultaneously both the map and the robot position using the robot sensors and odometry.

Being able to solve the SLAM problem in real-time is very important in the proposed exploration procedure, since occupancy maps are taken as the input for the target selection and path planner modules. The selected algorithm, DP-SLAM [8], is a metric approach which uses directly laser range measurements, without any landmark detection step. It is capable of solving the complete SLAM problem in real time while obtaining accurate and high-detailed maps, without any offline correcting step. Like FastSLAM algorithm [9], real time execution is accomplished exploiting the conditional independence of the objects in the world given the robot position [10,11,12]. Besides, DPSLAM uses a mapping technique (named *Distributed Particle Mapping*) which is able to maintain efficiently a high number of hypothesis with a complexity that does not scale with the number of iterations of the particle filter.

DPSLAM handles filters drift by means of another particle filter [8]. Experimental results with the hierarchical implementation of DPSLAM show an improvement in map accuracy, but at the expense of great increase in computational complexity. This hinders real-time execution because during the higher level filter computation the robot might remain static. In order to cope with this difficulty, the hierarchical algorithm has been modified in this work to run both filters in parallel instead of sequentially. This modification allows the algorithm to run online but imposes some restrictions in the configuration of the algorithms (and a bit of careful tuning) since the higher level layer must finish execution faster in order to avoid waiting.

In the exploration algorithm here presented, navigation and map building are treated as independent modules, so no explicit interaction between them is considered. Other approaches in the literature use active exploration and SLAM strategies in order to improve the quality of the maps [13,14]. The objective is usually to overcome some limitation of the mapping module, i.e. revisiting a singular point in the map or closing loops to avoid the robot getting lost during long maneuvers. However, this kind of strategy has been avoided in the current approach since observed limitations in the SLAM algorithm do not impact significantly the quality of the maps, and it would penalize modularity.

4   Exploration Strategy

The robotic algorithm here considered has the only task to explore and map an initially unknown environment. To do so efficiently requires good exploration strategies. The objective is to maximize the expected information gain (map knowledge) over time [1] and, as already mentioned, the optimal solution is computationally intractable.

According to a frontier-based exploration strategy [2], the best objective in terms of new information gain would be to visit the limit between the already explored obstacle-free region and unexplored territory. When a robot moves to a frontier it can see into unexplored territory and obtain new information that expand its knowledge of the world. The repetition of this behavior is called frontier-based exploration.

A frontier exploration strategy implies both, finding the frontiers for the current map and selecting the best one according to some criterion of efficiency. In this work, a tradeoff between displacement cost and information gain is considered, similar to the concept of bids used by Simmons in [1]. Cost is estimated in the proposed method using the A* optimal path, which provides the shortest path assuming deterministic and holonomic motion. Estimating gains is more difficult, actual information gain is impossible to predict, since depends on the physical structure of the environment.

In this work it is assumed that information gain is constant over frontier cells, so the value of each frontier region depends only on the number of cells contained in it. More sophisticated approaches are found in the literature: i.e. Simmons and [15]. However, the proposed algorithm using the simpler gain estimation works well enough for a single-robot exploration task.

Frontier regions are obtained using computer vision techniques over the global gray scale map computed by the SLAM module. The process is quite straightforward: first, an edge detection operation is performed to obtain all the limits between regions. Then, regions near occupied cells are extracted, figure 2b). Remaining pixels correspond to explored region cells that are in contact with unexplored territory. Frontier regions are formed by grouping adjacent cells into clusters, figure 2c).

4.1   AC + Best Blob

Figure 2d) shows a tentative exploration strategy that combines frontier detection and AC navigation. This approach, AC + Best blob, exploits the ability

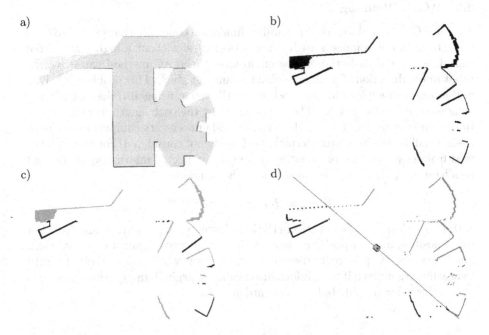

Fig. 2. Frontier extraction process: a) original gray scale map, b) detected frontiers, c) frontier regions and d) simple exploration strategy, AC + Best blob, combining frontier detection and AC navigation

of the AC algorithm to accept an external preference during the split point selection process. The direction preference is computed using the current pose of the robot to draw a line that divides the frontier cells into two groups: left and right. The total number of pixels on each side of the robot is considered as a rough estimation of the information gain associated with the turn preference. This method ignores completely the actual topology of the world, such as walls or obstacles, and groups together frontier regions which could be physically disjoint. This simplification is not intended to provide an accurate estimation of the information gain but just a hint to help the robot in the turn selection. With simple enough scenarios this method could be a good alternative to wandering, being able to improve exploration without penalizing execution time. This exploration technique can be extended to use the expected exploration gain. Instead of merging all the frontier cells or selecting the closest region, the modified method can select the frontier region with best expected information gain and use its centroid as a position target for the navigation algorithm. In this work the exploration value of a frontier region is directly obtained from the number of cells, so the higher the score the bigger the region. This reduces the method to seeking the biggest cell cluster.

4.2   AC + Planning

Testing AC + Best blob technique under simulation conditions has proved to have limitations in environments with concave obstacles: without a higher algorithm considering the distribution of the environment, the AC method can drive the robot in the direction of a concave obstacle and get stuck. This problem is solved in the proposed exploration method, AC + Planning, using also the cost of traveling to each frontier region. The cost is given by the path length from the robot to the center of mass of the cell cluster. To avoid unnecessary computation of path costs, small blobs (in relative terms) are directly not considered. For each considered frontier region, a score relating its reward (expected information gain) and penalization (path cost) is obtained using the equation:

$$Value_i = Reward_i \cdot \exp^{-\lambda \cdot Cost_i}$$

with $Reward_i$ the number of cells in the i-th cluster, $Cost_i$ its path length from the robot position and λ a positive constant that weights cost against expected gain. Small λ values give priority to the gain of information, while high ones give priority to nearby targets even if their information gain is marginal. In experiments we use $\lambda = 1$, this value roughly balances reward and cost.

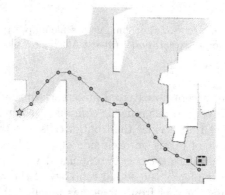

Fig. 3. AC + Planning: robot trajectory is shown in green, the sequencer waypoints are plotted as green dots and current target as blue rectangles

The pathplanner module computes a minimal cost path from the robot to the center of each candidate frontier region. The exploration utility for each region is computed according to this expression and the region with the higher score is selected. To drive the robot to the target frontier, the already computed minimal path is converted into a sequence of waypoints, which are used as targets positions for the navigation system. Iteration through the list of waypoints is triggered by the distance to the current target, letting the robot wander randomly when no more points are available. The trajectory sequencer ensures that the AC method, which only handles local knowledge, is able to drive the robot to the target proximity while avoiding potentially risky obstacles, as illustrated in figure 3.

5   Experimental Results

Several experiments have been conducted in order to check the feasibility of the different exploration strategies. These tests are performed under a simulation environment (Player-Stage, [16]) using a real world laboratory layout. Robot odometry and laser measurements are logged with a 200 ms period. The internal representation of the world provided by the SLAM module is also stored with every update of the higher level algorithm. A reference map is used for evaluating the total area covered with each method. This map is considered as the perfectly explored map (or ground truth) for the comparison. The comparison method ignores the spatial information and computes the normalized intersection of the histogram as the main metric.

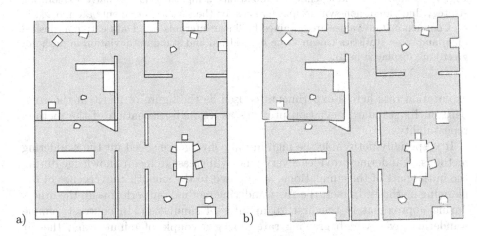

a)   b)

Fig. 4. Possible reference maps for the exploration rate evaluation. a) Laboratory layout used as the simulation input, map length and width are 14m. b) One of the maps obtained by the SLAM algorithm (selected subjectively).

Figure 4 shows the laboratory layout and the reference map used in the comparison. Notice that the reference map contains small inaccuracies originated in the mapping process. At first glance it should seem that the laboratory layout could be a better reference for comparison, but it contains regions that the robot is not able to sense. In practice both images give similar results, with a slightly lower maximum value for the floor plan.

Since all the experiments have the same duration (one hour), the final coverage is enough to compare the different methods. However, it is interesting to analyze the evolution with time of the exploration rate. A common time reference is obtained across independent experiments using linear interpolation between update instants. This allows to compute some statistics over the set of experiments for each particular method, as illustrated in figure 5. Repeating this procedure for each method yields the results showed in figure 6. The mean value of the

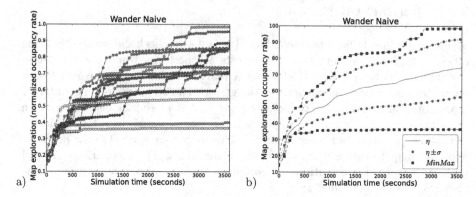

Fig. 5. Results of 15 independent simulations using the wander naive exploration strategy. Individual non-synchronized values for the set of experiments are plotted in a). Common instant values are obtained by linear interpolation. Data extrema (dashed blue) and some statistics (mean value in solid red and standard deviation in dash-dot green) are displayed in b).

exploration rate across experiments is used as the figure of merit in the comparison. Its standard deviation provides as a rough estimation of the method repeatability.

It is specially noticeable the high initial convergence speed int the wandering methods. Wandering provides a very quickly increase in mean knowledge during the initial part of the simulations, as opposed to the smooth convergence of the planning methods. In fact, the AC wandering method outperforms all the others during approximately the first 20 minutes of simulation. In contrast, simple wandering gives a high growing rate during a couple of minutes and then it tends to stabilize to a low value. This method is clearly the less favorable for an exploration task: its expected exploration rate is the worst and it has the biggest deviation.

The exploration method that selects the target frontier attending only to the expected information reward and ignoring the penalization, performs in average worse than just wandering with the AC method. This is a consequence of the obstacle distribution for the current map, extremely unfavorable for a method that does not consider the presence of concave obstacles. Instead, this method relies on the ability of the navigation method to find a path to the goal, which could do the job in less complex environments.

After the first 20 minutes of simulation the planning method becomes the best choice. It gives a mean exploration rate near 98%, while wandering with the AC method remains below a 90%. Selecting the best frontier in terms of information gain gives a mean exploration of 83% and the naive wandering control reaches a 74%. According to these results, the complete exploration solution does not seem to perform in average much better than the other methods. The real difference is that it not only gives the highest mean exploration, but also the best repeatability. This fact is better illustrated in figure 7. It can be seen that most methods

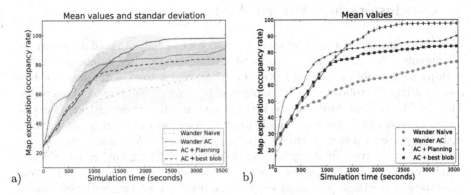

Fig. 6. Evolution of the exploration rate mean values and standard deviation for each method

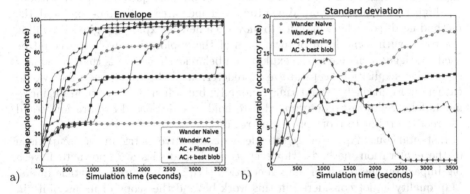

Fig. 7. a) Exploration rate envelope (extreme values) for all the simulations in each method and b) instantaneous standard deviation for each method

present a variance that keeps growing during the whole simulation time. The two exceptions are the AC wandering method and frontier-based exploration. However, wandering has a more random behavior and its deviation keeps fluctuating slightly around a constant value after the first 10 minutes of simulation. Frontier-based exploration reaches high values of deviation for the first half an hour, but then decreases to a value near zero and keeps almost constant. This is explained by the fact that most experiments with this method were able to explore the world to a high extent towards the second half of the simulation.

In order to give a measure of method performance, we compute the mean of the robot velocity and the robot stopped time over 15 one hour simulations. The mean velocity was 15.4 cm/s and the robot stopped time was 17.4% of the total time. Using an optimized method implementation we expect to considerably decrease robot stopped time.

6   Conclusions and Future Work

This article describes a method for the autonomous exploration of unknown environments. The algorithm uses frontier-based exploration in combination with a path-planner to obtain an estimation of the best region to explore. Reactive navigation drives the robot through the planned path while avoiding local obstacles. This combination has proven to be very efficient in comparison with other exploration approaches. It provides the highest exploration rate during simulations, with a expected value very close to 100%. It is also the less uncertain: 60% of the simulations give a final exploration rate in the 97.75%±0.5% range. Using a wander exploration strategy the exploration rate decreases to a 90.35%±7.35%.

In spite of the these promising exploration results, evolution with time shows that wandering provides a better performance during the first minutes of simulation. This suggest that the AC wandering method is specially well suited for the exploration of an open region, and that it can be done quite efficiently without any higher level intervention. When the given open area tends to be explored, the method needs some help in order to move to a new unexplored region. Providing the robot with some form of awareness of the exploration evolution could be used to trigger a change in the exploration behavior. The idea is that, when new knowledge acquisition keeps above a predetermined threshold, the robot would gather information by wandering randomly, but when new information increments falls bellow a predetermined threshold, frontier-based exploration would redirect the robot to more promising regions.

Instantaneous map coverage provides a quantitative metric in the comparison of the exploration methods. However, control strategies are known to have a big impact on the result of map generation [17], specially using particle filters. Map quality is not considered in this work beyond the scope of its usability in the exploration task, but it became obvious during testing that some navigation behaviors yield higher quality maps. This could be considered in order to guide exploration while at the same time trying to favor the map generation process.

Acknowledgments. This work was partially supported by the Spanish Government through Project TIN2010-20845-C03-02.

References

1. Simmons, R., Apfelbaum, D., Burgard, W., Fox, D., Moors, M., et al.: Coordination for multi-robot exploration and mapping. In: Proceedings of the AAAI National Conference on Artificial Intelligence. AAAI (2000)
2. Yamauchi, B.: A frontier-based approach for autonomous exploration. In: Proceedings of the IEEE International Symposium on Computational Intelligence, Robotics and Automation, pp. 146–151 (1997)
3. Sánchez, J.R.Á., de la Paz López, F., Troncoso, J.M.C., de Santos Sierra, D.: Reactive navigation in real environments using partial center of area method. Robotics and Autonomous Systems 58, 1231–1237 (2010)

4. Álvarez-Sánchez, J.R., de la Paz Lépez, F., Troncoso, J.M.C., Sánchez, J.I.R.: Partial center of area method used for reactive autonomous robot navigation. In: Mira, J., Ferrández, J.M., Álvarez, J.R., de la Paz, F., Toledo, F.J. (eds.) IWINAC 2009, Part II. LNCS, vol. 5602, pp. 408–418. Springer, Heidelberg (2009)
5. Álvarez, J.R., de la Paz, F., Mira, J.: On Virtual Sensory Coding: An Analytical Model of the Endogenous Representation. In: Mira, J. (ed.) IWANN 1999. LNCS, vol. 1607, pp. 526–539. Springer, Heidelberg (1999)
6. Thrun, S.B.: Efficient exploration in reinforcement learning. Technical report, Carnegie-Mellon University (1992)
7. Thrun, S., Burgard, W., Fox, D., Hexmoor, H., Mataric, M.: A probabilistic approach to concurrent mapping and localization for mobile robots. Machine Learning, 29–53 (1998)
8. Eliazar, A.I., Parr, R.: Hierarchical linear/constant time slam using particle filters for dense maps. In: Advances in Neural Information Processing Systems 18, pp. 339–346 (2006)
9. Montemerlo, M., Thrun, S., Koller, D., Wegbreit, B.: Fastslam: A factored solution to the simultaneous localization and mapping problem. In: Proceedings of the AAAI National Conference on Artificial Intelligence, pp. 593–598. AAAI (2002)
10. Murphy, K.P.: Bayesian map learning in dynamic environments. In: NIPS, pp. 1015–1021 (1999)
11. Thrun, S.: Particle filters in robotics. In: Darwiche, A., Friedman, N. (eds.) Proceedings of the 18th Conference in Uncertainty in Artificial Intelligence, UAI 2002, University of Alberta, Edmonton, Alberta, Canada, August 1-4, pp. 511–518. Morgan Kaufmann (2002)
12. Montemerlo, M., Thrun, S., Koller, D., Wegbreit, B.: FastSLAM 2.0: An improved particle filtering algorithm for simultaneous localization and mapping that provably converges. In: Proceedings of the Sixteenth International Joint Conference on Artificial Intelligence, IJCAI, Acapulco, Mexico. IJCAI (2003)
13. Thrun, S.B., Möller, K.: Active exploration in dynamic environments (1992)
14. Fox, D., Burgard, W., Thrun, S.: Active markov localization for mobile robots. Robotics and Autonomous Systems 25, 195–207 (1998)
15. Visser, A., Xingrui-Ji, van Ittersum, M., González Jaime, L.A., Stancu, L.A.: Beyond frontier exploration. In: Visser, U., Ribeiro, F., Ohashi, T., Dellaert, F. (eds.) RoboCup 2007. LNCS (LNAI), vol. 5001, pp. 113–123. Springer, Heidelberg (2008)
16. Gerkey, B.P., Vaughan, R.T., Howard, A.: The player/stage project: Tools for multi-robot and distributed sensor systems. In: Proceedings of the 11th International Conference on Advanced Robotics, pp. 317–323 (2003)
17. Eliazar, A., Parr, R.: Dp-slam: Fast, robust simultaneous localization and mapping without predetermined landmarks. In: Proc. 18th Int. Joint Conf. on Artificial Intelligence, IJCAI 2003, pp. 1135–1142. Morgan Kaufmann (2003)

Multiscale Dynamic Learning
in Cognitive Robotics

Pilar Caamaño, Andrés Faíña, Francisco Bellas, and Richard J. Duro

Integrated Group for Engineering Research,
Universidade da Coruña, 15403, Ferrol, Spain
{pcsobrino,afaina,fran,richard}@udc.es
http://www.gii.udc.es

Abstract. This paper is concerned with the dynamics of Cognitive Developmental Robotic architectures and how to produce structures that allow these types of architectures to deal with the different time scales a robot must cope with. The most important types of dynamics that occur in different time scales are defined and different mechanisms within a particular cognitive architecture, the Multilevel Darwinist Brain, are suggested to model each one of them. The paper also proposes a novel neuroevolutionary technique, called τ-NEAT, in order to capture processes based on precise temporal cues. This technique is analyzed when addressing dynamic environments in a real robotic test.

Keywords: Cognitive Robotics, Dynamic Learning, Neuroevolution, NEAT, Delay-Based ANN.

1 Introduction

Cognitive architectures in robotics have been the focus of a large amount of research in the last few years. These architectures are the computational implementation of cognitive models [1], and as such, constitute the substrate of functionalities like perception, attention, action selection, learning, reasoning, etc. The most remarkable cognitive architectures can be found in the Cognitive Developmental Robotics (CDR) subfield [2]. The bases of CDR were articulated by Weng in [3] who indicated that "a developmental architecture requires not only a specification of processors and their interconnections, but also their online, incremental, automatic generation from real-time experience". In other words, the control structure of the robot in this paradigm cannot be produced through some explicit design in which the designer is basically incorporating his/her take on the operation and perception of the robot. This control structure has to be obtained autonomously by the robot through interaction with its environment. Quite a few architectures that try to address these characteristics have been proposed in the literature. These go from SASE [4], which was the first cognitive architecture to include autonomous development for its main modules [3], to more recent examples such as the Epigenetic Robotics Architecture (ERA) [5], the iCUB cognitive architecture [6],the Intelligent Adaptive Curiosity [7] or the

J.M. Ferrández Vicente et al. (Eds.): IWINAC 2013, Part II, LNCS 7931, pp. 56–65, 2013.
© Springer-Verlag Berlin Heidelberg 2013

Multilevel Darwinist Brain (MDB) [8], which is the base architecture we will use throughout this paper.

One of the fundamental topics within this fast moving field is that of dynamic interaction with the world. Developmental cognitive architectures base their operation on implicit or explicit models of the world and of the individual itself that are learnt through interaction with the environment. In the end it is these models that provide the individual agent with the capability of projecting into the near future and deciding on the appropriate actions it must take in order to fulfill its motivations. Consequently, it is important to identify what types of dynamics the models must be able to address and provide structures that are designed to capture these dynamics so that the models obtained using CDR based architectures really reflect the operation of the world.

In this paper we define the most important types of dynamics in different time scales that models within a cognitive architecture must capture, and then suggest specific mechanisms that would be appropriate for this purpose in a particular cognitive architecture, the Multilevel Darwinist Brain (MDB). One of these mechanisms is a novel neuroevolutionary technique, the τ-NEAT, which is presented here and analyzed when addressing dynamic environments in a real robotic test.

2   Dynamic Learning Scales in a Cognitive Architecture

Capturing the dynamics of a world by a cognitive architecture and efficiently using them for its purposes, that is, learning them, is a multilevel process. There are different types of dynamic processes and, consequently different types of learning skills a cognitive architecture must show in order to deal with them. As a first approximation, a division can be made into four main types of learning processes any cognitive architecture must support to address dynamics over different time scales. These processes are:

1. Modeling environments that only depend on the state of the world in the previous instant of time.
2. Modeling environments that depend on the history of events that occurred. Two subcases may be contemplated here:
   (a) Modeling dynamic processes whose evolution depends on sequences of events or values where what is important is the order or sequence of the events and not really their precise timing.
   (b) Modeling dynamic processes whose evolution depends on sequences of events or values that occur with particular and precise temporal patterns.
3. Adapting models to changing environments and dynamics. It is important to note here that an important decision is when to change a model in order to follow the changes of the environment and when to simply expand the model so that it considers a more general or larger piece of the evolution of the environment.
4. Reusing models that correspond to environments or dynamics seen in the past, thus avoiding the need to relearn them over and over.

Consequently, for a robot cognitive architecture to be able to efficiently operate in a dynamic world it must possess the tools to support the four learning scales mentioned above and do it in an integrated manner.

2.1   The Multilevel Darwinist Brain

The Multilevel Darwinist Brain (MDB) is a cognitive architecture that has been applied to different real robot learning problems in static and non-static tasks [8]. The cognitive model on which it is based uses three types of functions called world, internal and satisfaction models to select the action it will execute through an internal reasoning process that must lead to the maximization of the satisfaction.These models are not pre-defined and they must be learnt during the robot's lifetime while it is interacting with its environment. The distinctive feature of the MDB is that the on-line adaptive learning of the models is performed by means of evolutionary algorithms (EA) which construct the artificial neural networks (ANN) that represent the models.

Fig. 1 displays a functional diagram of the current version of the MDB. As we can see, the MDB has been organized into two different time scales, one devoted to the execution of the actions in the environment in real time (execution scale) and the other to the learning of the models (learning scale). The operation of the MDB can be described in terms of these two scales.

Starting from the learning scale, let us assume that the robot has executed an action in the environment and a new iteration starts on time t. We have a new *episode* (made up of the perceptual values and applied action) that is introduced in the *Short-Term Memory* (STM) and that is a candidate for storage in the *Episodic Buffer* (EB) of each type of model if it passes an *attention mechanism* that acts as a filter and replacement mechanism with the aim of storing the most relevant episodes for learning in each case. Once the episode is stored or discarded, the *model evolution* starts for each type of model using as population the models stored in the *working memory*. This evolution is carried out during a small number of generations to avoid a premature convergence of the models towards a specific content of the episodic buffer in a given iteration. The evolved models are ANNs that are randomly initialized and stored in their corresponding working memory in the first iteration and reused in subsequent iterations. After the partial evolutionary process finishes, the best models are selected as *current world, internal and satisfaction models* and they make up the internal representation of the robot's world in iteration t. These models are passed on to the action-selection module in the execution time scale that performs an internal selection of the action, choosing the one that provides the higher predicted satisfaction in iteration t according to the predictions of the current models. These models are updated each iteration of the MDB, but as the time scales are different, the same set of current models can be used in several iterations of the MDB until newer and better ones are obtained.

The action-selection module must provide actions as they are demanded by the robot using the last version of models that the learning scale provides, which, in the first iterations of the architecture, will be random leading to an initial random

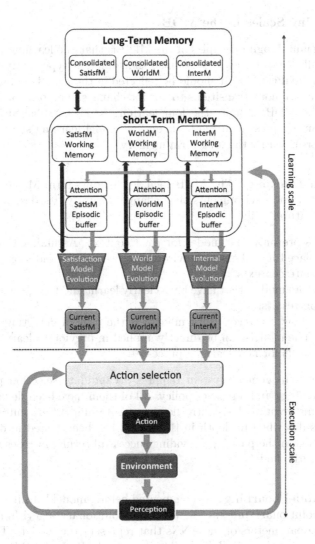

Fig. 1. MDB functional diagram

action selection. Anyway, every time the robot executes an action, successful or not, a new episode from the real environment is obtained. As more iterations take place, the MDB acquires more information and thus the model learning processes should produce better models and, consequently, the actions selected using them will be more appropriate to maximize the satisfaction.

To finish the review of the MDB working principles, we must comment the *long-term memory* element, which stores the models that perform successfully in a given context, consolidated models in Fig. 1, to be reused avoiding the repetition of a previous learning process.

2.2 Learning Scales in the MDB

One of the main design principles of the MDB is that the learning process must be intrinsically adaptive as it is assumed that different types of dynamics could appear in the real world that should be reflected in the models in real time operation: from continuous transitions to abrupt changes. The previously described version of the MDB includes the four learning scales established in section 2, three of them are present in the basic on-line model learning structure and the fourth one provided by the long-term memory.

Long-Term Learning. The MDB includes a Long-Term Memory (LTM) to deal with dynamic environments that change abruptly. Its design is based on three basic features [14]:

1. It must store only one model for each context. Assuming that the robot will be placed in a dynamic environment, we must avoid storing models of intermediate contexts.
2. It must store models that have been clearly learnt and that provide successful prediction results.
3. It must allow the recovery of models in the case of detecting a change of context. The models can be directly applied in previously learnt contexts or serve as seeds for new learning processes.

The memory management system requires a candidate selection policy, a replacement strategy and a recover policy. All of them have been developed based on the learning error analysis with the aim of obtaining a fully-autonomous system. LTM is described in depth in [14] and it has been tested in different real robot problems in the past [8], providing successful results in cases with abrupt changes in static models.

On-Line Model Learning. As commented before, model learning in the MDB is a neuroevolutionary process [8], with an evolutionary algorithm devoted to adjusting the parameters of the ANNs that represent the models. This learning is carried out using the episodes stored in the episodic buffer (EB) for each type of model, which are obtained progressively as the robot operates in its environment. Consequently, the model learning process is characterized by two interrelated elements: the episode management strategy and the neuroevolutionary process.

Even in the case of dealing with a static environment or task, due to the limited nature of the STM, it must be assumed that the episodes stored in the EB in a given iteration represent only a partial view of the reality that must be learnt and, consequently, a gradual learning of the different EBs is required. In the MDB this is achieved by preserving the population of the evolutionary algorithms between iterations and evolving just a few generations per iteration. It seems clear that the episodes that are stored in the EB for each model are critical for the learning process. The quality of the learnt models depends on what is stored in this memory and the way it changes. The influence of the selected features on the resulting models has been analyzed in previous work [9] considering static environments and models that did not depend on time. In dynamic models the most relevant feature is the time at which the episode was obtained and its relation to the previous episodes.

Regarding the neuroevolutionary algorithm used to learn the models, in previous versions of the MDB, the Promoter Based Genetic Algorithm (PBGA) was applied providing a successful response in adaptive learning from episodes [8], covering the requirements 1 and 3 proposed in section 2. In order to deal with time-dependent models (requirement 2 established in section 2), a type of ANN that can manage temporal information is required, which is not the case of the PBGA. To this end, we have selected the NEAT neuroevolutionary algorithm [10], a widely tested approach mainly in dynamic learning problems in several cases [11,12]. The NEAT algorithm is able to manage time dependent phenomena through recurrent connections between neurons, so the dynamic processes that depend on sequences of events are intrinsically supported (requirement 2.a in section 2). However, dynamic processes characterized by precise temporal patterns are very hard to model using recurrent connections alone and therefore are not adequately covered by NEAT. Again, this is a requirement that was clearly established in section 2 for learning in any cognitive architecture (point 2.b) and, as such, mechanisms must be provided to deal with it.

The τ-NEAT algorithm: the embedding theorem [13] states that when we have a dynamic system characterized by a measured signal, it is only necessary to embed this signal in a higher dimensional space of dimension D by taking D samples of the signal spaced by periods τ in order to make it predictable or its unambiguous modeling feasible. The challenge here is to obtain these values autonomously for a signal or process that is being modeled by our cognitive architecture. Basically, how to obtain the number of points it must consider and the temporal spacing between them when they are regularly spaced. Going one step further, one could hypothesize that in many cases lower dimensional embedding spaces could be used if the samples were not evenly spaced in time and, consequently, if one considered an uneven distribution of delays a lower number of points would be necessary to disambiguate many dynamic processes. In fact several authors have already hinted towards this conclusion [13].

This is the approach we have followed here to provide the cognitive architecture with a structure that can capture the dynamics of any process it is presented with. It has been implemented by introducing variable and trainable *delays* in the synaptic connections of the networks that are evolved as a sort of representation of the different lengths these connections could present, which would have a bearing on the time signals would take to traverse them. To this end, the original NEAT algorithm was extended to be able to work with trainable synaptic delays thus creating the τ-NEAT algorithm. τ-NEAT is basically a neuroevolutionary algorithm for growing neural networks that may include recurrent connections and synaptic delays.

The structure of a τ-NEAT ANN changes slightly from the original [10]. Now it includes a synaptic delay τ_{ij} in addition to the classical synaptic weight W_{ij}

Fig. 2. Snapshots of the Aibo robot "safe crossing" experiment

corresponding to neurons i and j, and a buffer in each synapse containing the set of the last n input values to that synapse. The synaptic delay is included in the chromosome and its value is applied over the buffer establishing a sort of length of the synaptic connection. Thus, with the application of this τ-NEAT algorithm in the learning processes of the MDB, requirement 2.b established in section 2 should be fulfilled. The response of the MDB with τ-NEAT will be shown in the next section with a real robot experiment.

3   A Simple Learning Experiment

The robotic experiment that has been implemented to analyze the relevance of the multiscale dynamic learning capabilities provided by the application of the τ-NEAT algorithm in the MDB is simple but illustrative. Fig 2 displays six snapshots of the "safe crossing" experiment. We have an Aibo ERS-7 robot and an e-puck robot with a pink ball on its top that crosses in front of the Aibo. The MDB objective is to learn the models required by the Aibo to advance without running over the e-puck. The desirable situation is that shown is the bottom left image while an undesirable one is that displayed in the bottom right image.

A schematic overview of this setup can be observed in the left image of Fig. 3. As we can see, the robot is placed at a fixed distance from the e-puck, which has a continuous and linear movement in front of the robot that follows a temporal pattern. The Aibo must select the appropriate instant to move and cross without

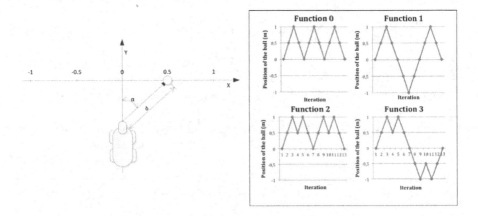

Fig. 3. Left: "Safe crossing" experiment schematic overview. Right: E-puck position at iterations in the four movement patterns considered.

running over the e-puck. Depending on the temporal pattern followed by the e-puck, this selection could be very complex requiring a precise temporal modeling to anticipate the e-puck position.

In this configuration, the robot has a permanent vision of the ball from its starting position. The Aibo employs its camera, placed in its head, to obtain an estimation of the distance d and the angle α to the ball. Specifically, in each iteration, the robot moves its neck from 90° left to 90° right having a complete view of the environment in front of it. If a ball is detected during this neck displacement, the robot centers this ball in the camera image. After that, the distance is calculated as a function of the size of the detected ball and the angle to the ball is the angle of the neck actuator. The robot can perform two actions: move forward or not move. If the robot decides to move forward when the ball is crossing the "y" axis, it catches the ball and, as a consequence, the distance and the angle are zero. In any other case, when the robot moves forward, it reaches the origin of coordinates but cannot see the ball, so the distance the sensor returns after processing the image is 3 meters (out of range). Otherwise, the specific values of distance and angle are in a continuous range from 0 to 1 (distance) and from -1 to 1 (angle).

We have implemented four different patterns of the ball's movement to illustrate the MDB response with the τ-NEAT in different situations. They are precise temporal patterns, so it is supposed that the τ-NEAT will perform successfully on them. For each case, the position of the e-puck in each iteration can be viewed in the right images of Fig. 3. To setup this experiment in the MDB, two models must be considered and learned: one world model and one satisfaction model. The world model has three inputs (d, α and $action$) and two outputs (predicted d and α), while the satisfaction model has two inputs (predicted d and predicted α) and one output ($satisfaction$). These models are represented by means of the ANNs obtained by the τ-NEAT algorithm. The Episodic Buffer size was fixed to 15 episodes and a pure temporal replacement strategy was used.

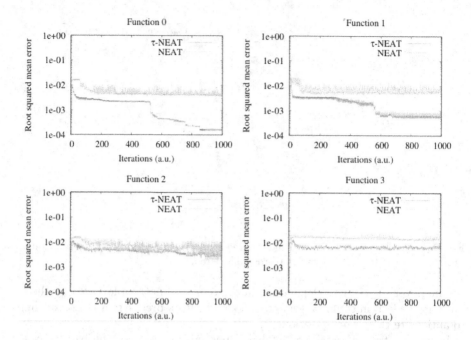

Fig. 4. Evolution of the error provided by NEAT and τ-NEAT

Fig. 4 displays the evolution of the root squared mean error averaged for the two outputs of the world model provided by the NEAT (with recurrent connections but no delays) and τ-NEAT algorithms when the e-puck robot follows the four dynamic patterns shown in Fig. 3. As it can be observed in the Figure, the NEAT version with synaptic delays outperforms the original one in all cases, which was the main objective of this experiment. In practical terms, it must be commented that with an error level below 1e-02, the task was successfully accomplished by the AIBO robot, while in any other cases the robot behavior was unstable (the input to the satisfaction model must be precise). As displayed in Fig. 4, such error level was obtained by the τ-NEAT in all the experiments.

4   Conclusions

In this work, the basic requirements for a cognitive architecture in order to deal with the learning scales of dynamic robotic problems are identified. Specific solutions for these requirements are proposed in a particular cognitive architecture, the Multilevel Darwinist Brain (MDB), which performs a gradual neuroevolutionary learning from episodes able to deal with dynamics in an intrinsic way. Adaptation to smooth changes is achieved through a gradual learning policy while adaptation to abrupt changes is obtained by means of a Long Term Memory element. Finally, to face time-dependent learning processes, the τ-NEAT algorithm is proposed and tested in a simple robotic experiment. With the inclusion of this algorithm in the MDB, the architecture contains all the identified requirements of dynamic learning in cognitive robotics, which must be validated in the near future in more realistic problems.

Acknowledgments. This work was partially funded by the Xunta de Galicia and European Regional Development Funds through projects 09DPI012166PR and 10DPI005CT.

References

1. Byrne, M.D.: Cognitive architecture. The Humancomputer Interaction Handbook 44(1), 97–117 (2003)
2. Asada, M., Hosoda, K., Kuniyoshi, Y., Ishiguro, H., Inui, T., Yoshikawa, Y., Ogino, M., Yoshida, C.: Cognitive Developmental Robotics: A Survey. IEEE Trans. on Autonomous Mental Development 1(1), 12–34 (2009)
3. Weng, J.: On developmental mental architectures. Neurocomputing 70(13-15), 2303–2323 (2007)
4. Weng, J., Hwang, W.S., Zhang, Y., Evans, C.H.: Developmental Robots?: Theory, Method and Experimental Results. In: Proc. of the Int. Symposium on Humanoid Robots, pp. 57–64 (1999)
5. Morse, A.F., Greeff, J.D., Belpeame, T., Cangelosi, A.: Epigenetic Robotics Architecture (ERA). IEEE Trans. on Autonomous Mental Development 2(4), 325–339 (2010)
6. Vernon, D.: Enaction as a conceptual framework for developmental cognitive robotics. Paladyn 1(2), 89–98 (2010)
7. Baranes, A., Oudeyer, P.Y.: R-IAC: Robust intrinsically motivated exploration and active learning. IEEE Trans. on Autonomous Mental Development 1(3), 155–169 (2009)
8. Bellas, F., Duro, R.J., Faina, A., Souto, D.: Multilevel Darwinist Brain (MDB): Artificial Evolution in a Cognitive Architecture for Real Robots. IEEE Trans. on Autonomous Mental Development 2(4), 340–354 (2010)
9. Bellas, F., Becerra, J.A., Duro, R.J.: Construction of a Memory Management System in an On-line Learning Mechanism. In: Proceedings ESANN 2006, pp. 26–28 (2006)
10. Stanley, K.O., Miikkulainen, R.: Evolving neural networks through augmenting topologies. Evolutionary Computation 10(2), 99–127 (2002)
11. Stanley, K.O.: Efficient Reinforcement Learning through Evolving Neural Network Topologies. In: Proceedings of the GECCO 2002 Conference, pp. 569–577 (2002)
12. Stanley, K.O., Bryant, B.D., Miikkulainen, R.: Real-time neuroevolution in the NERO video game. IEEE Trans. on Evolutionary Computation 9(6), 653–668 (2005)
13. Duro, R.J., Reyes, J.S.: Discrete-time backpropagation for training synaptic delay-based artificial neural networks. IEEE Trans. on Neural Networks 10(4), 779–789 (1999)
14. Salgado, R., Bellas, F., Santos-Diez, B., Caamaño, P., Duro, R.J.: A Procedural Long Term Memory for Cognitive Robotics. Optimizing Adaptive Learning in Dynamic Environments. In: Proceedings of the EAIS 2012 Coference, pp. 1–8 (2012)

A Vision-Based Dual Anticipatory/Reactive Control Architecture for Indoor Navigation of an Unmanned Aerial Vehicle Using Visual Topological Maps

Darío Maravall, Javier de Lope, and Juan Pablo Fuentes Brea

Department of Artificial Intelligence, Faculty of Computer Science, Universidad
Politécnica de Madrid, Madrid, Spain - Centro de Automática y Robótica
(UPM-CSIC) Universidad Politécnica de Madrid, Madrid, Spain
dmaravall@fi.upm.es, javier.delope@upm.es,
juanpablo.fuentes.brea@alumnos.upm.es

Abstract. Indoor navigation of an unmanned aerial vehicle is the topic
of this article. A dual feedforward/feedback architecture has been used
as the UAV´s controller and the K-NN classifier using the gray level image histogram as discriminant variables has been applied for landmarks
recognition. After a brief description of the aerial vehicle we identify the
two main components of its autonomous navigation, namely, the landmark recognition and the controller. Afterwards, the paper describes
the experimental setup and discusses the experimental results centered
mainly on the basic UAV´s behavior of landmark approximation which
in topological navigation is known as the beaconing or homing problem.

Keywords: Unmanned Aerial Vehicles, Vision-based dual anticipatory
reactive controllers, Nearest Neighbors Methods.

1 Introduction

For the autonomous navigation of the UAV we have used a visual topological
map in which the landmarks or relevant places are modeled as the vertices of a
labeled graph and the edges correspond to specific UAV´s maneuvers.

As the landmarks are visual references, a fundamental problem in visual topological navigation is landmark recognition, so that we devote a complete section
of the paper to this topic (see paragraph 2 below "Automatic Recognition of
Visual Landmarks").

For the UAV´s controller we have applied a vision-based dual feedforward
/ feedback control architecture. In the sequel we describe both components
of the UAV´s navigation: first, landmark recognition and afterwards, the dual
controller.

The paper ends with the experimental work in our laboratory and the final
conclusions.

J.M. Ferrández Vicente et al. (Eds.): IWINAC 2013, Part II, LNCS 7931, pp. 66–72, 2013.
© Springer-Verlag Berlin Heidelberg 2013

1.1 Unmanned Aerial Vehicles (UAV)

An UAV can be regarded as a autonomous robot and it has the capacity to fly within an environment, in this paper through indoor environment. We are using the quadrotor Parrot AR.Drone 2.0 [8] as robotics research platform, available to the general public.

All commands and images are exchanged with controller via a WiFi ad-hoc connection. The AR.Drone has a vision sensor implemented as a HD camera, and has four motors to fly through the environment.

Fig. 1. Parrot AR.Drone 2.0

This UAV support four types of movement along its axes $(roll, pitch, gaz, yaw)$ allowing you to move on the three coordinates of the space 3D (x, y, z): sideways, forward/back, vertical speed and rotation about its vertical axis.

The AR.Drone can be used for visual autonomous navigation in environments using machine learning approaches, and it's used in many applications as surveillance tasks, rescue tasks, and can perform human-machine interactions.

2 Automatic Recognition of Visual Landmarks

The UAV´s navigation system utilizes the onboard camera to capture the environment images. These images are classified and used by the controller in order to generate the control command s in real time. More specifically, as the navigation system is based on a topological map it is vital to have an efficient classification of the landmarks images to guarantee a correct guidance of the UAV.

For landmark recognition we have used the gray levels standard histogram as the discriminant variables and the k-NN algorithm as the classifier. As it is well-known, the k-NN algorithm is an efficient memory-based classifier based on a stored data base of labeled exemplars and any new case to be recognized is assigned the most frequent class among the k nearest neighbors in the training data base.

dataset of landmark 1

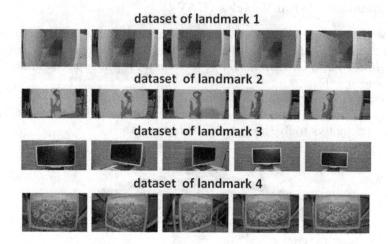

dataset of landmark 2

dataset of landmark 3

dataset of landmark 4

Fig. 2. Dataset of several visual landmarks

To evaluate the performance in landmark recognition we have generated a training data set formed by four different landmarks [7] displayed in Fig. 2. Notice that in the data set are actually stored the gray level histograms of the corresponding landmark images.

In our experiments we have considered four landmarks: a picture, a TV set, a sculpture and a door. For the classifier evaluation we have applied the leaving-one-out crossvalidation technique.

During the experimental validation of the k-NN algorithm we have tried several values of the design parameter: k = 1, 2 and 3. For all cases we have obtained excellent results with a classification error close to 0%. Fig. 3 displays a typical confusion matrix.

	predicted class 1	predicted class 2	predicted class 3	predicted class 4
actual class = 1	10	0	0	0
actual class = 2	0	10	0	0
actual class = 3	0	0	10	0
actual class = 4	0	0	0	10

Fig. 3. Confusion matrix

Apart from the efficiency of the k-NN algorithm itself we believe that the excellent recognition results obtained are mainly due to the discriminant variables provided by the graylevel histogram of the landmarks images.

3   The Feedforward/Feedback Controller

Fig. 4 displays the block-diagram of the dual feedforward / feedback controller [4][6]. Notice that both the feedforward or anticipatory controller and the feedback or reactive controller [2][3] receive as input the same image error, which is the difference between the target or desired image (i.e. the image corresponding to the current identified landmark) and the current image captured by the UAV´s on board camera.

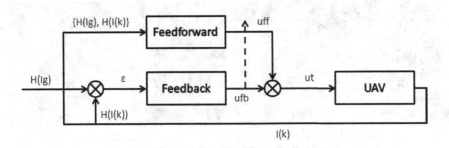

Fig. 4. The feedforward/feedback controller: Notice that the error signal of both controllers is obtained as the difference between the histogram of the recognized landmark or histogram of the goal image $H[Ig]$ and the histogram of the current image $H[I(k)]$

More specifically, this vision-based error signal is obtained as the histogram of the identified landmark or histogram of the goal image $H[Ig]$ minus the histogram of the current UAV´s captured image $H[I(k)]$ during the k iteration of the controller. The feedback controller is implemented as a conventional PD control [5] and the feedforward controller is based on a inverse model [1] using a conventional neural network based algorithm.

4   Experimental Work: UAV´s Navigation through Doors

To test experimentally the proposed vision-based dual controller we have chosen the basic UAV´s navigation skill of "door approaching and crossing". The basic idea to test this UAV´s navigation skill is to get the UAV to fly towards a door as a target landmark in its visual topological map. Once the UAV is approaching its target landmark and after its correct recognition it can activate its dual vision-based controller in order to safely transverse the door by monitoring and controlling the visual error.

Fig. 5 display a sequence of the images captured by the onboard UAV´s camera while performing a door navigation maneuver. We have also displayed in Fig. 6 the visual error signal that converges to zero as expected during this door navigation maneuver.

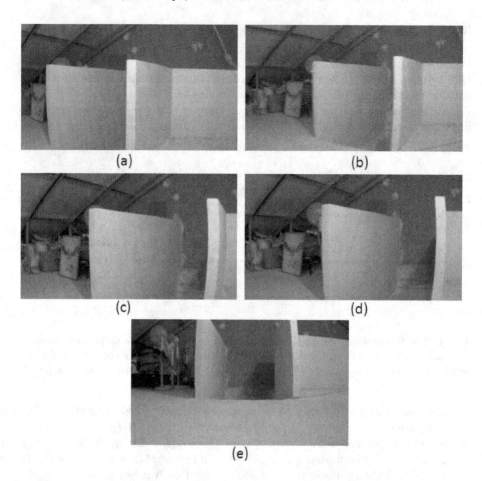

Fig. 5. Notice that the sequence (a)-(e) includes the successive images captured by the UAV while performing the maneuver: (a) is the initial state and (e) is the goal image

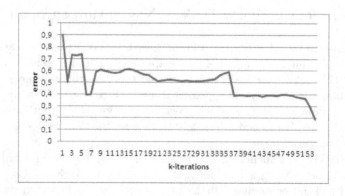

Fig. 6. The visual error signal during the door-landmark approximation and crossing maneuver

Notice also the control curves in Fig. 7 giving an idea of the UAV´s control efforts applied during this maneuver. In both cases, k denotes the number of iterations of the controller during the maneuver (sequence (a)-(e)): (a) and (b) have been caught between k=1 and k=9, (c) and (d) between k=10 and k=39, and finally (e) corresponds from k=40 to the end of the maneuver. At each iteration, the controller sends a control signal to the UAV; the time between two consecutive control signals is 30 ms.

During this approaching experiment, the UAV has used the pitch actuator (forward / back) and yaw actuator (rotation on its axis z), which were obtained by the controller based on the error signals received by the classifier k-NN.

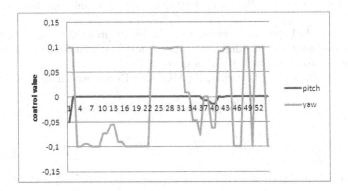

Fig. 7. The control signals during the door approximation and crossing maneuver

From the experimental results obtained in our laboratory and shown in Figures 6 and 7 we can conclude that the UAV is able to successfully perform in real time the fundamental skill of landmark door approximation and crossing by mean of the proposed vision-based dual feedforward/feedback controller.

5   Conclusions and Future Work

This paper has presented a vision-based dual anticipatory/reactive controller for indoor navigation of an UAV that uses a visual topological map to autonomously navigate in the environment. We have also described the basic problem of visual landmark recognition for which we have implemented a k-NN classifier using the standard graylevel histogram values as discriminant variables, giving excellent recognition accuracy. The proposed navigation system has been experimentally tested on the basic skill of door-landmark approximation and crossing maneuver. Future work is planned towards the UAV´s autonomous navigation in a whole building by means of a visual topological map of a complete building.

References

1. Kawato, M.: Feedback-Error-Learning Neural Network for Supervised Motor Learning. In: Advanced Neural Computers (1990)
2. Wolpert, D.M., Kawato, M.: Multiple paired forward and inverse models for motor control. Neural Network 11, 1317–1329 (1998)
3. Kawato, M.: Internal models for motor control and trajectory planning. Neurobiology 9, 718–727 (1999)
4. Imamizu, H., Kawato, M., et al.: Human cerebellar activity reflecting an acquired internal model of a new tool. Nature 403, 192–195 (2000)
5. Maravall, D., de Lope, J.: Multi-objective dynamic optimization with genetic algorithms for automatic parking. Soft Computing 11(3), 249–257 (2007)
6. Barlow, J.S.: The Cerebellum and Adaptive Control. Cambridge University Press (2002)
7. Maravall, D., de Lope, J., Fuentes, J.P.: Fusion of probabilistic knowledge-based classification rules and learning automata for automatic recognition of digital images. Pattern Recognition Letters (in press, 2013)
8. Piskorski, S., Brulez, N., D'Haeyer, F.: AR.Drone Developer Guide SDK 2.0, Parrot (2012)

Alignment in Vision-Oriented Syntactic Language Games for Teams of Robots Using Stochastic Regular Grammars and Reinforcement Learning

Jack Mario Mingo, Darío Maravall, and Javier de Lope

Departamento de Inteligencia Artificial,
Universidad Politécnica de Madrid, Spain

Abstract. This paper approaches the syntactic alignment of a robot team by means of dialogic language games by applying online probabilistic reinforcement learning algorithms. The main contribution of the paper is the application of stochastic regular grammars, with learning capability, to generate the robots' language. First, the paper describes the syntactic language games, in particular the type of grammar and syntactic rules of the robots' language and the dynamic process of the language games which are based on dialogic communicative acts and a reinforcement learning policy that allows the robot team to converge to a common language. Afterwards, the experimental results are presented and discussed. The experimental work has been organized around the linguistic description of visual scenes of the blocks world type.

Keywords: Stochastic Grammars, Reinforcement Learning, Dynamics of Artificial Languages, Language Games, Multi-robot Systems, Self-Collective Coordination and Computational Semiotics.

1 Introduction

Language, at both lexical and syntactic levels, is one of the fundamental cognitive skills necessary for the development of advanced and intelligent multi-robot systems as it allows communication and cooperation among the individuals of a robotic group. In previous work, using the so called language games concept [1] which is partly inspired in the ideas of Wittgenstein and de Saussure about the public and conventional dimensions of linguistic meaning, we have applied online reinforcement learning algorithms to the self-emergence of a common lexicon in robot teams [2]. In that work we modelled the lexicon or vocabulary of each robot as a look-up-table mapping the **referential** meanings (i.e. the objects or situations and states of the environment) into symbols. In this paper we extend our previous work on multi-robot lexical alignment through language games into multi-robot syntactic alignment also through dialogic language games.

J.M. Ferrández Vicente et al. (Eds.): IWINAC 2013, Part II, LNCS 7931, pp. 73–80, 2013.
© Springer-Verlag Berlin Heidelberg 2013

2 Multi-robot Syntactic Alignment

Although we acknowledge the fundamental relevance of lexical competence concerning language use and meaning [3] we do also believe that compositional, structural or simply syntactic competence is vital for an agent to efficiently describe reality in a symbolic, linguistic way, so that in this paper we approach the syntactic alignment of a robot team by means of dialogic language games applying also on-line reinforcement learning algorithms. The remaining of the paper is organized as follows. First, we describe the syntactic language games, in particular the type of grammar and syntactic rules of the robots' language and the dynamic process of the language games which are based on dialogic communicative acts and a reinforcement learning policy that allows the robot team to converge to a common language. Afterwards, the experimental results are presented and discussed. The experiments have been organized around the linguistic description of simple visual scenes of the block world type. A paragraph on conclusions and future work closes the paper.

3 Description of the Syntactic Language Game

In a first step, alignment language games must be applied so that the robot team get a common lexicon for the objects present in the environment as well as a common lexicon for the spatial relationships (*right, left, in front, behind*). The acquisition of the objects lexicon is performed through a fully autonomous interactive process in which the robots are able to converge to common names for the different objects perceived as sensory discriminant variables, typical of Pattern Recognition methods. This coordination process is fully autonomous and based on a reinforcement learning algorithm that control the robot team's alignment of the mapping of names and objects besides an unsupervised clustering algorithm applied for the objects categorization. As for the acquisition of the common spatial lexicon we have applied a supervised process based on injecting to each robot the corresponding spatial concepts: right, left, front and behind. In this paper we do not describe the concrete processes behind these lexical alignments as we rather concentrate on the syntactic alignment.

After acquiring both the objects lexicon and the spatial relationships lexicon, the robot team engage in a dialogic language game aimed at converging to a common grammar as explained in the sequel.

3.1 The Environment's Visual Description

The robots are situated in an environment that they perceive trough visual sensors or video cameras. As in this paper we are mainly interested in the lexical and syntactic alignment of the robot team, we depart from linguistic sentences obtained as linguistic descriptions of the scenes captured by the robots' video cameras and after their corresponding segmentation and analysis by means of standard computer vision techniques. Under this linguistic-centred set-up we have worked with digital scenes as the ones displayed in Figures 1 and 2.

Fig. 1. Digital scene: book on the right of the ball

Fig. 2. Digital scene: pencil on the left of the glasses

3.2   The Robots' Grammar

For this kind of scenes the robots' language is formed by sentences like "*object such* is on the *right* of *object such*" or "the *book* is on the *right* of the *ball*", which can be formalized as a string aRb where **a** and **b** are objects names and **R** is a spatial relationship. As we are interested in using a language formalism that allows the robots to learn their language by means of an interactive process, we propose to use stochastic grammars with learning capability for visual scenes description as explained in the sequel. In this regard, the probabilities associated to the production rules provide the plasticity of the robots' language by changing these probabilities by means of a reinforcement learning algorithm.

3.3   Stochastic Learning-Grammars for Visual Scene Description

As commented above, in order to allow the robots to develop a suitable language for the description of the type of visual scenes displayed in figures 1 and 2 that are composed of two objects and in order to allow also the syntactic alignment of a group of robots, we propose to use stochastic grammars in which the probabilities of the production rules can be learnt by the robots through reinforcement as they engage in language games. The terminal vocabulary of the robots' language is formed by the objects appearing in the visual scenes and by the corresponding spatial relationships (*right, left, above, below* ...). The probabilistic production rules of the proposed grammar are the following:

$$
\begin{aligned}
\text{<sentence>} &::= \text{<object> <relation> <object>} & p_1 \\
&\mid \text{<relation> <object> <object>} & p_2 \\
&\mid \text{<object> <object> <relation>} & p_3 \\
\text{<object>} &::= \text{anObject} \\
\text{<relation>} &::= \text{aSpatialRelation}
\end{aligned}
$$

Where the probabilities of the first rewriting rules p_1, p_2 and p_3 are initialized at the indifference position $p = 1/3$ and by means of a reinforcement-learning are changed to allow the robots to converge to the common optimum rewriting rules. To this end we have applied the so called Linear Reward-Inaction algorithm, L_{RI} [4] in which the probabilities are updated as follows:

$$P_i(k+1) = P_i(k) + \lambda\beta(k)[1 - P_i(k)] \tag{1}$$

$$0 \le \lambda \le 1$$

$$P_j(k+1) = P_j(k) + \lambda\beta(k)P_j(k); j <> i \tag{2}$$

Where λ stands for the learning rate and $\beta(k)$ stands for the reinforcement signal got from the environment with a value "1" associated to success and a value "0" associated to failure. To balance the exploration-exploitation dilemma we have added the following decision rule for the selection of the probabilistic rewriting rules:

If [*random* $\le \delta$] **then**
        [pure random rule selection or roulette-like selection]
else
        [select the rule with the maximum probability]
end if

Where δ controls the balance between exploration and exploitation in the learning process.

3.4   The Dialogic Syntactic Language Game

For a team of N robots, each of them with its own learning stochastic grammar, the way to converge to syntactic consensus by means of reinforcement learning is based on the following procedure [see pseudo code in Figure 3. First, a sequence of what we call language games rounds is performed until the team converges to an optimal communication system or syntactic consensus in which all the robots use the same optimal stochastic grammar. In each language game round all the possible pairwise communicative acts or dialogs are performed. Each communicative act or dialog taking place between two robots proceeds as displayed in Figure 4. Thus the scenes of the training data set $(I_1, I_2, ...I_p)$ are sequentially and simultaneously presented to both robots that utter their corresponding sentences, and depending on the success or failure of each communicative act the probabilities associated to the production rules applied by the robots are incremented or decremented, respectively. Obviously, a success in a communicative act happens whenever the sentences uttered by both robots coincide.

```
for k=1,2, .... Max rounds do
        Execute all the possible communication acts
        Compute the Communicative Efficiency of the robot team CE(k)
        If CE(k) = 100% in three consecutive rounds then
            Break
        end if
end for
```

Fig. 3. Pseudocode of the reinforcement learning-based syntactical coordination procedure

In this work, we define the Communicative Efficiency (CE) as follows:

$$CE(k) = (NSD/ND)100 \tag{3}$$

where NSD stands for *number of successful dialogs* and ND stands for *number of dialogs*.

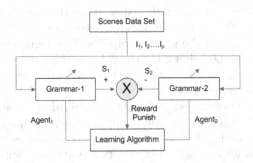

Fig. 4. Dialogic Syntactic Language Game. Notice that for each image, each robot utters its own sentence according to its private grammar and a reward or penalty signal is sent to the learning algorithm for the updating of the probabilities of the production rules

4   Experimental Results

The experimental work has been phocused on moderate robot teams sizes typical of realistic multi-robot systems in the order of ten or fewer robots and and we have also experimented with scenes involving a few specific objects and their spatial relationships. More specifically, the scenes of the training data set contain four specific objects (*book, pencil, glasses* and *ball*) and two relations (*left* and *right*). It is important to observe that the introduction of more objects and spatial relations does not imply any significant change or problem in the robots' syntactic alignment.

We conducted a series of 100 experiments with different groups of robots. We started with a group of 10 robots and we added 10 robots each time until a group of 60 robots was formed. In all the experiments we use the standard reinforcement learning algorithm we explained above. In order to make up their sentences starting from the stochastic learning grammar, the robots choose the rule with the greater probability in most cases but occasionally they can choose a random rule. In the first round, all the robots choose a rule randomly. During each language game round all the possible pairwise communicative acts or dialogs are performed and all the rules which allow the robot to make up the sentence are rewarded or punished depending on the success of fail in the communicative acts. In the case of success, both robots share the same sentence to describe a spatial situation and the syntactical consensus is attained when all the team use the same sentences to describe whatever spatial situation. For N robots the maximum number of communicative acts in a language game round is $\binom{N}{2} = N(N-1)/2$.

In Figure 5 we show some cases of learning curves displaying the results when syntactical consensus is attained with a similar number of rounds. Horizontal axis shows the rounds while vertical axis shows the communicative efficiency (CE) value expressed as percentage.

As we can see in the figure, in all the experiments the communicative efficiency rises up as rounds go by. According to this figure the more robots are in the group, the more steady-increase is the learning process. This behaviour is observed for groups with 40, 50 and 60 robots. On the other hand, smaller groups involving 10, 20 and 30 robots show a first stage with an almost-constant communicative efficiency before this value rises up.

Finally, Figure 6 shows success probabilities attained in the series of 100 experiments. Syntactical consensus was achieved in all the experiments we conducted in less than 4000 rounds. However, significant consequences can be extracted from the figure. As we can see, best results are attained with 30, 40 and 50 robots in the group while the worst result is associated to 60 robots. Results for groups with 10 and 20 robots are located among those bigger groups. The more robots are added to the group, the faster the consensus is attained because we only need around 400 rounds for groups with 30, 40 and 50 robots. However, the improvement is limited to a number of robots. Adding more robots beyond this limit do not improve the process as the results for 60 robots show.

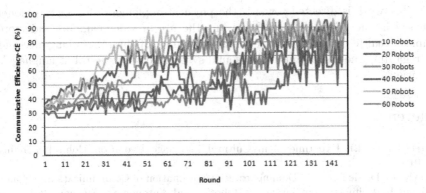

Fig. 5. Learning curves associated to different team sizes

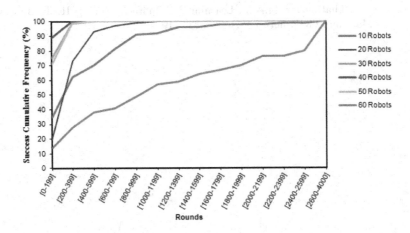

Fig. 6. Success Probabilities

5   Conclusions and Further Work

It has been proposed the use of stochastic regular grammars with learning capability for the syntactic alignment of a robot team. The probabilities associated to each rewriting rule of the regular grammar provide the robots' language with a plasticity that allows the robots to change their sentences aimed at converging to a common language with optimal communicative efficiency. The experimental scenario contemplated in the paper is based on digital images of scenes of the blocks world type. After a previous lexical alignment of the robot team in both the objects lexicon and the spatial relationships lexicon, the robots engage in dynamic syntactic games aimed at efficiently communicate the scenes linguistic description by applying a probabilistic reinforcement learning algorithm. The simulation-based experimental results show that the robot team converges to the optimal communicative efficiency in just a few dialogic rounds.

Further work is directed towards the physical implementation of the model proposed in the paper, in both regarding the use of video-cameras and more complex "real-life" images for the visual grounding of the robots language and also regarding the use of musical or sound synthesizers for the robot language's phonation.

References

1. Steels, L. (ed.): Experiments in Cultural Language Evolution. John Benjamins (2012)
2. Maravall, D., de Lope, J., Domínguez, R.: Coordination of Communication in Robot Teams by Reinforcement Learning. Robotics and Autonomous Systems (2012) (in press), doi:10.1016/j. robot 201207007
3. Marconi, D.: Lexical Competence. MIT Press (1997)
4. Narendra, K., Thathachar, M.A.L.: Learning Automata- A Survey. IEEE Trans. on Systems, Man, and Cybernetics 4(4), 323–334 (1974)

A Preliminary Auditory Subsystem Based on a Growing Functional Modules Controller

Jérôme Leboeuf-Pasquier, Gabriel Fatíh Gómez Ávila,
and José Eduardo González Pacheco Oceguera

DIP, CUCEI, University of Guadalajara, Mexico
jerome.lep@gmail.com

Abstract. In the present paper, a learning based controller that performs as an auditory subsystem is described. Based on the Growing Functional Modules (GFM) paradigm, the auditory subsystem is the result of the interconnection of four kinds of components: Global Goals, Acting Modules, Sensations and Sensing Modules. The resulting controller is radically different from conventional speech recognition due to its ability to gradually learn in context a set of vocal commands while performing in a virtual environment. Recognition is considered as satisfactory when the robot's behavior is in accordance with the commands' meaning. This learning process may be compared to how a dog learns to obey its master's orders. The experiment described in this paper illustrates the design process of a GFM controller. It exhibits the behavior of the control process and principally, exposes the inherent philosophy of the GFM approach.

Keywords: Machine learning, knowledge acquisition, learning based control, connectionism, artificial brain.

1 Introduction

In computer sciences, building applications that are able to interact with the real world always represents a huge challenge due to the complexity, the illegibility and the dynamics of the environment. To face this challenge, programmers must drastically reduce the intricacy of the modeled environment; otherwise, the number of lines of code would increase drastically to be able to produce adequate answers to any possible situation. Classical Artificial Intelligence does not offer much help because it mainly consists in replacing the usual sequential-imperative paradigm by another one, for example functional or logical. Anyhow, the precise representation of a natural environment requires a huge coding effort. In contrast, biological entities have the ability to gradually face the complexity of the real world and this, thanks to the learning abilities of their brains. Conceptually, these brains perform as controllers providing real time responses to the events perceived by their sensory system. Hence, a biological brain may be assimilated to a learning based controller. In that sense, the GFM approach slightly modifies the classic control loop concept to fit the inherent requirements of an artificial

J.M. Ferrández Vicente et al. (Eds.): IWINAC 2013, Part II, LNCS 7931, pp. 81–91, 2013.

brain. The GFM controller schematic is shown in figure 1; the three main differences include:

- First, the classical "input reference value" is integrated to the controller under the concept of "Global Goals". These components are in charge of specifying the intrinsic motivations and thus, of inducing a proficient behavior.
- Second, the controller's architecture is divided in two areas, the "Sensing" and the "Acting" ones. The Sensing Area interprets the feedback sent by the controlled system, commonly a real or a virtual robot. This feedback is referred to as Sensations and it is processed by a set of interconnected sensing modules. These sensing modules produce Perceptions as output that are transferred to the Acting Area and processed by a set of Acting Modules. These acting modules, according to these Perceptions and the input values of the Global Goal, trigger a new command to the system. This whole process is performed during each cycle of the control loop.
- Third, Acting and Sensing Areas are composed of respectively interconnected Sensing and Acting Modules (illustrated in figure 3).

The creation of a specific GFM controller is facilitated by a graphic editor, thus the design process mainly consists in interconnecting four kinds of components, Sensation, Acting Module, Sensing Module and Global Goal as illustrated in figure 3. A step by step example of a former design process is fully described in [1] meanwhile a previous application to a virtual robot was proposed in [2]. Until now, several simple GFM controllers have been designed but recently our challenge has been to implement three basic subsystems: the visual subsystem, the humanoid's equilibrium subsystem and the auditory subsystem. The two last ones have been completed; the implementation of the auditory one is described in the present paper.

As mentioned in the previous section, our approach is learning based though distant from supervised learning. The controller gradually improves its behavior by taking into account its previous performance; no tag needs to be associated to a newly memorized word.

A virtual trainer facilitates the learning process repeating words until achieving a satisfactory learning level. In that sense, the trainer's role is conceptually similar to the dog training process : Vocal instructions are emitted in a specific context but, there is no way to explain their meaning to the dog or to make explicit the expected behavior. Taking the context into account, the dog will produce attempts of response that the trainer's reaction (feedback) may then, either strengthen or weaken. This kind of training process is naturally embodied in the control loop: each time the controller decides to trigger a command to the system, it will wait for the corresponding feedback in order to actualize the internal structure of the acting and sensing modules that were involved in this selection.

Historically, GFM controllers have not been conceived to perform speech recognition; but their architecture appears herein to be sufficiently generic to

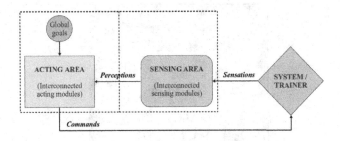

Fig. 1. The GFM controller and its control loop

perform speech recognition though this subsystem introduces a higher challenge due to the temporal dimension of the auditory signal.

2   The Virtual Environment and the Trainer

As mentioned in the previous section, the GFM controller performs in context. Consequently, an environment, at least a virtual one, must be provided to allow this process. The virtual environment is in charge of first, ensuring the consistency of the robot's behavior and second, generating the required feedback to the controller. In such experiment, only a subset of the simulation functionalities is employed. The environment is composed of a little maze (see figure 2) where the robot may perform a few actions like grasping or moving in any direction. Each of these actions is performed when the corresponding command is triggered by the controller. Some behavior rules are integrated to the environment, including:

- The robot is allowed to grasp an object only if its gripper is free and there is an object at its position.
- The robot is allowed to release an object only if it has previously grasped one.
- The order to move in one direction is given only if there is no wall that prevents it.
- A move over an object is prohibited if the robot has already grasped one.

The presence of these rules is important because the controller is expected to assimilate them in order to facilitate its recognition process.

Each command sent by the controller to the robot will be performed according to its close environment. Next the feedback aimed at the controller is generated as a sequence of sensations that reflects this environment. Each sensation is coded as an int32 value and the feedback sequence includes 13 values: position in X, position in Y, East contact, West contact, North contact, South contact, gripper state, object ident, object color, object size, satisfaction of the trainer and finally the audio event codified with two values. This feedback sequence is specified during the controller's design as a list of Sensations that are represented by green rectangles in figure 3.

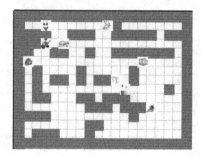

Fig. 2. The virtual environment of the simulator

In order to teach a few words to the controller, the trainer associated to the simulation elaborates a step by step challenge based on vocal orders. For example, a challenge may consist on the following sequence of vocal commands: "South, South, East, Grab, East, Right, Up, Release". The performance of the controller is evaluated according to its response to each order; that is if it triggers each one of the expected commands. Moreover, the trainer applies a pedagogy that consists in repeating more often the less understood orders.

3 Audio Signal Processing

The signal processing algorithm incorporates signal pre-processing and post-processing. Pre-processing consists in various steps commonly employed in Speech Recognition. First, the signal is treated with a pre-emphasis filter to increase the amplitude of certain frequencies relatively to others in order override noise. Then the Cepstrum [6] and the Linear Predictive Code (LPC) [7] feature extraction are applied on the signal as well. Over the years, techniques have been developed to improve LPC or Cepstrum analysis. These procedures generally apply any number of filters to the signal before the feature extraction to improve the recognition rates. In some other cases, as described in this paper, the technique consists in combining different feature extraction procedures to improve the recognition accuracy rate [8].

During LPC feature extraction analysis, we observed that the most relevant information of the signal appeared on the firsts formants. Increasing the order of the extraction would only cause the resulting formants to converge at some point, therefore increasing the difficulty to separate the information without adding much. Consequently, only the first formants were considered. On the other hand, the Cepstrum analysis results exhibit that the odd formants are those which carry the most valuable vocal information from the signal. For this reason, these formants were used from the extracted Cepstrum features.

When applying pre-processing to syllables, the extraction of LPC and Cepstrum features showed different and valuable representations of the signal. This leads to apply both methods. Further experimentation brought us to combine them: this results in a more complete representation of the vocal signal. Pre-processing's parameters are as follows: LPC window size is 1024; LPC order is 1; Cepstrum order is 12; LPC harmonics kept is 1; Cepstrum harmonics kept is 7; Combination order is 1(lpc) 1(cep) 3(cep) 5(cep) 7(cep) 9(cep) 11(cep) 13(cep).

The purpose of the post-processing is to clean the signal and to discretize it to obtain sequences of events compatible with the GFM protocol. First, the chosen harmonics from the LPC and Cepstrum feature extraction process are brought together so they can be jointly performed. Then they are filtered by a cleaning module in two steps: first, silences at the beginning and at the end of the signal are eliminated and second, any noise found at the beginning of the sequence of events is filtered out. This filtering is achieved by looking for frequencies known to be out of the voice range and by searching for small pieces of information that can be considered as silence because of their lack of useful information. Once having a clean signal, the pre-processing searches for the maximum and minimal frequencies of each formant, this with the purpose of finding atypical behavior that can be filtered out. At the signal discretization step, these atypical frequencies only occupy bands with very few events. They make small contributions and can be discarded. When vocal events are completely clear, silence are inserted between each vocal event to increase the separation and allow the SER module to easily encounter the end of each sequence. Finally, the signal is discretized through bit shifts and each formant is converted into two 32-bits integers, which on binary base represent the events' positions on each frequency band used. Post-processing parameters are as follows: signal inhibition is true; silence inserted between syllables is 22 events; frequency bands is 64; bits per event is 64.

The corpus's selection is entirely determined by the simulation-trainer. A total of eighteen words pronounced 10 times each by a single speaker, that is a total of 180 audio files. Each word was spoken slowly marking very short silences between every vowel. The words used consisted on three categories:

- Cardinal Points: Spanish words for North, South, East and West (respectively: Norte, Sur, Este and Oeste).
- Instructions of movement: Spanish words for Down, Forward, Up, Back, Right, Left, Up head, Downhill (respectively: Abajo, Adelante, Arriba, Atrás, Derecha, Izquierda, Sube, Baja).
- Actions: Spanish words for Open, Grab, Close, Release, Drop, Take (respectively: Abre, Agarra, Cierra, Deja, Suelta, Toma).

During the recording phase, the hardware and environment conditions were as follows: a unclean speech environment (CS), a quality set microphone (Sennheiser e-835) and a high-end sound card (Creative Sound Blaster X-FI Platinum); the signal to noise ratio (SNR) of CS was around 30 dB.

4 The Controller

The GFM control loop has been described in section 1; the specific controller designed to perform as an auditory subsystem is shown in figure3 as represented in the GFM graphic editor [1]. The editor facilitates the creation, configuration and interconnection of the four kinds of components. This controller results as the interconnection of one Global Goal, thirteen Sensations, three Sensing Modules and one Acting Module. These components' roles are given below.

4.1 Sensations

First kind of component, the Sensations are encapsulated in green rectangles and localized in the Sensing Area. They correspond to the sequence of values that the system sends as feedback. For example, Sensations #12 and #13 represent the occurrence of a speech signal event.

4.2 Sensing Modules

Second kind of component, the Sensing modules are encapsulated in blue rectangles and localized in the Sensing Area. According to figure 3, both previously mentioned sensation values are given in input to the SER sensing module. This module has an internal structure that performs a comparison between the current input sequence of speech signal events and the memorized ones. The dynamic internal representation memorize the input sequence when the comparison process does not trigger any response. Otherwise, the SER module may trigger in output a maximum of three[1] responses that corresponds to the best recognized syllables.

Then, these responses are given in input to a DMC sensing module that dynamically builds and compares Markov chains. Hence, this module is in charge of determining the most probable time sequence of input syllables that produces a "valid" word. Hence, the resulting sequence of syllables represents a word or an expression whose value is triggered in output. The degree of validity gradually results from the accuracy of this response on the environment. Once again, if no word could be produced then, a learning process occurs to memorize the new potential input sequences.

Finally, the best response of the DMC module is sent to the FBC sensing module jointly with four sensations identified in figure 3 as contE, contW, contN, contS that indicate the contact sensors activation of the robot in each of the four directions. The FBC module generates a response that corresponds to a combination of the different inputs and that takes place in a specific context. For example, the response enhances the fact that the words "move left" occurs quite always when there is no "west contact". Learning for this module consists basically in computing frequencies over input values.

The third kind of component, a single Global Goal, is identified by a red circle. Its associated type and the value are respectively, "Cst" and "1". This means that this component constantly sends the constant request value 1 to the acting module connected in output.

---

[1] This value is one of the controller's parameters.

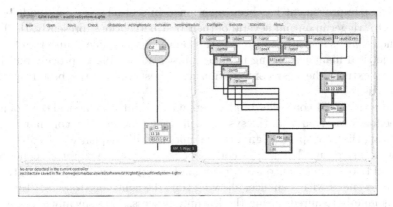

Fig. 3. The GFM implemented auditory subsystem. Sensations are associated to the green color, Sensing Modules to the blue, Global Goals to the red and Acting Modules to the orange

4.3 Acting Modules

The fourth kind of component is a CI module, identified by an orange rectangle. The output value of the Global Goal #1 is given in input to the CI acting module. The engine and dynamic internal structure of the CI acting module has been described in details in [3]. Its internal structure is based on a graph of states, the goal state is indicated by the first value "11" of the middle field that refers to the Sensation #11 that represents the "satisfaction" of the trainer that is, a value that indicates if the previous challenge has been or not overcome. This first feedback value "11" jointly with the input value "1" from the Global Goal signifies that the causal inference is always aimed at satisfying the trainer. In that sense, the CI module will trigger to the system one of the specified commands, presently: move to a specific direction, grab/release...

4.4 Global Goals

The module's decision of triggering a specific command take into account its actual state, the input request, the "satisfaction" state but also the second feedback value #18 that refers to the output of the FBC sensing module and indicates the actual context. That is a vocal command is actuated according to the environment. Therefore, the word "move left" may trigger different commands, either "step up" or "negate" according to the occurrence of a contact in the west direction.

5 Improvement to the GFM Controller

As mentioned in introduction, the GFM controller has not been specifically designed to perform speech recognition. To face this challenge, the original GFM controller architecture has required some improvements.

The first improvement has been mentioned in section4, sensing modules are allowed to trigger in output several of their best perceptions; presently, the three best ones. This is specified in the editor's canvas by a double connection line. In a symmetrical manner, sensing modules must also be able to process multiples inputs to extract the best combination and thus produce the best responses-perceptions.

Besides the GFM controller implements until recently a classical control loop: each feedback received from the system will trigger a corrective command. Now, considering that the speech signal is dynamic, a whole sequence of feedback values commonly occurs before the controller is ready to interpret a sequence of events and to produce an output command. In the actual implementation, a sequence is ended with a silence, allowing thus the controller to detect their end which is mainly required during the learning process. Thus, meanwhile the controller requires more sensations to perform, a Do-Nothing command is triggered to the controlled system.

6   Experimental Results

The main experiment was focused on the evaluation of four primary functions of the robot: mobility, sensor reading, action and perception. This evaluation was performed by the trainer giving instructions to the robot; that is, asking it to move to a specific direction using synonyms as "forward" and "right", to report the status on its sensors using cardinal point words, to pick up or drop an object, and to determine whether the object it is currently holding is a toy or a meal. Results are monitored on the trainer's challenge window that describes the acquisition level of each challenge. The experiment was considered successful when the robot learns every challenge sent by the trainer showing its ability to correctly perform it from that point forward.

To evaluate the behavior of the described auditory subsystem, three experiments have been performed; all of them include the CI acting module and its associated CST Global Goal. The first experiment only considers the SER sensing module. The experiment consists in giving in input a single sequence of eighteen words/orders during the training cycle and to manually decrease the

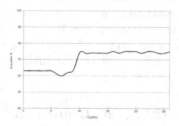

Fig. 4. Success rate evolution resulting from the first experiment

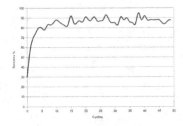

Fig. 5. Success rate evolution resulting from the second experiment

module's threshold to prohibit any new learning phase. Then, during each cycle, the four last sequences are repeatedly given in input allowing module's adaptation. The evaluation is performed considering the three best syllable-candidates triggered by the SER module. Figure 4 shows the success rate obtained during the first 25 cycles: the recognition rate increases from approximately 60% to 75% and then oscillates around this last value.

In order to improve the recognition process, the DMC sensing module is connected in output of the SER sensing module. The DMC module learns gradually to trigger the most probable word corresponding to a combination of the three best syllables. The results presented in figure 5 shows that the recognition rate reaches a maximum of 95% after 25 cycles then the success rate oscillates with prorated value of 84.3%. Compared with the previous architecture, a maximum improvement rate close to 20%.

The third experiment should include the two previous sensing modules plus the third one, the FBC module, in order to improve recognition using the robot's context. Thus, for example, in this experiment, the recognized order "left" would be taken into account only when there is no left contact (according to the third rule presented in section 2). Nevertheless the frequency of the occurrences depends of the context; for example the robot may perceive much more left contacts than right ones. To avoid this dependency between the success rate and the robot's environment, a specific virtual application has been developed to randomly generate equiprobable contexts. Thus, the FBC module is evaluated alone to measure its adaptation ability. Figure6 shows how the number of errors first grows quickly to then stay constant. During each cycle, a valid perception, composed of an order of motion and the robot's side contacts, is given in input. After 215 of such cycles, learning is completed and no more error happens in this artificial context.

In conclusion to these experiments, first adding a DMC module to the design of the controller and then a FBC module contribute to improve the recognition rate. It also appears that the adaptation mechanism of the SER module requires improvements to reach at least 85% of success rate on such small vocabulary. Undoubtedly, better experimental results may easily be obtained using classical techniques. Nevertheless, the purpose of this project was not to improve or compete with this class of systems. The purpose of this project was to show

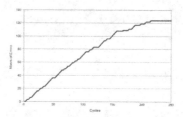

Fig. 6. Evolution of the accumulated number of errors during the third experiment

the feasibility of learning from the environment in order to satisfy the trainer, a motivation provided by the CST global goal. In fact, we show that the proposed auditory subsystem is able to learn in context that is, in absence of any explicit knowledge given by a human entity - typically the manual association of semantic tags. Consequently, the modifications implemented in this new version of the GFM controller allow processing temporal sequences of events like those of speech signal.

7   Conclusions

This preliminary auditory subsystem developed in [5] shows the well founded of the proposed learning based approach. During the experimental phase, vocal commands have been learned in context and executed by a virtual robot in a similar process to dog training. The GFM controller employed was not originally conceived for this kind of processing; for example, in a recent project, such a controller has been implemented as the - learning based - equilibrium system of a bipedal robot[4]. To adapt the controller architecture to this class of challenge, two major improvements have been implemented: first, the processing of sequences of events as part of the feedback loop; second, the triggering of multiple answer-candidates in output of sensing modules.

Such result clearly reinforces the feasibility of the GFM proposal to generate "artificial brains" able to learn from their environment in order to satisfy some internal motivations. Moreover this artificial brain is obtained from a graphical design and thus, does not require a programming phase.

Future work is focused on the development of more complex controllers whose behavior and learning abilities should be equivalent to those of a pet. Interacting with the real world indeed constitutes a very complex challenge. Overcoming such challenge requires the implementation of learning based controllers.

Acknowledgment. This paper is based upon work supported by the Jalisco State Council of Science and Technology of Mexico: projects FOMIXJAL-2008-05-98486 and COECYTJAL-UdeG-5-2010-1-821.

References

1. Leboeuf-Pasquier, J., González Pacheco Oceguera, E.: Designing a Growing Functional Modules "Artificial Brain". Broad Research in Artificial Intelligence and Neuroscience 3(2), 5–15 (2012)
2. Leboeuf-Pasquier, J.: Applying the GFM Prospective Paradigm to the Autonomous and Adaptive Control of a Virtual Robot. In: Gelbukh, A., de Albornoz, Á., Terashima-Marín, H. (eds.) MICAI 2005. LNCS (LNAI), vol. 3789, pp. 959–969. Springer, Heidelberg (2005)
3. Leboeuf-Pasquier, J.: A Growing Functional Module Designed to Trigger Causal Inference. In: Proceedings of the 4th Intl Conference on Informatics in Control, Automation & Robotics, ICINCO, pp. 456–463 (2007)
4. Galvan, R.: The Equilibrium Subsystem of a Humanoid Robot based on the GFM Paradigm. Master Thesis in Information Technology, CUCEA, University of Guadalajara (2012)
5. Gómez-Avila, G.: An Auditory Subsystem based on the GFM Paradigm. Master Thesis in Information Technology, CUCEA, University of Guadalajara (2011)
6. Oppenheim, A.V., Schafer, R.W.: From Frequency to Quefrency: A History of the Cepstrum. IEEE Signal Processing Magazine (September 2004)
7. Markel, J.D., Gray Jr., A.H.: Linear Prediction of Speech. Springer, New York (1976)
8. Antoniol, G., Rollo, F., Venturi, G.: Linear predictive coding and cepstrum coefficients for mining time variant information from software repositories. In: MSR International Workshop on Mining Software Repositories (2005)

Robust Multi-sensor System
for Mobile Robot Localization

A. Canedo-Rodriguez[1], V. Alvarez-Santos[1], D. Santos-Saavedra[1], C. Gamallo[1],
M. Fernandez-Delgado[1], Roberto Iglesias[1], and C.V. Regueiro[2]

[1] CITIUS (Centro Singular de Investigación en Tecnoloxías da Información),
Universidade de Santiago de Compostela, Santiago de Compostela, Spain
[2] Department of Electronics and Systems, Universidade da Coruña, A Coruña, Spain
adrian.canedo@usc.es

Abstract. In this paper, we propose a localization system that can combine data supplied by different sensors, even if they are not synchronized, or if they do not provide data at all times. Particularly, we have used the following sensors: a 2D laser range finder, a Wi-Fi positioning system (designed by us), and a magnetic compass. Real world experiments have shown that our algorithm is accurate, robust, and fast, and that it can take advantage of the strengths of each sensor, and minimise its weaknesses.

Keywords: Sensor fusion, robot localization, Wi-Fi positioning, particle filter.

1 Introduction

Mobile robot localization is the problem of determining the pose (position and orientation) of a robot relative to a map. This problem has been largely considered as one of the most important in mobile robotics, because knowledge of the robot pose is required by almost any robotic task. For instance, we are currently involved in the development of a tour-guide robot able to learn and reproduce routes of interest to the visitors of social events [1]. This tour-guide robot requires accurate localization in order to work properly.

The core issue in robot localization arises from the fact that the robot pose has to be estimated from noisy sensor measurements, being lasers [2], sonars [3], and cameras [4] the most widely used. Most localization systems only use one sensor, therefore they are not robust [5] when: a) the data rate is irregular, b) the data is highly noisy, c) the sensor fails to provide data, d) different areas look alike to the sensor, etc. These problems get worse in social environments, because there are people moving around, environment changes, etc.

In order to solve these issues, we propose a localization system that combines the evidence supplied by several sensors. Our system is based on the Augmented Montecarlo Localization Algorithm (particle filter localization) [6], and unlike other systems (e.g. Kalman Filters), it can handle naturally: a) sensors with non-Gaussian noise, b) non-synchronized sensors, c) sensors with different data

J.M. Ferrández Vicente et al. (Eds.): IWINAC 2013, Part II, LNCS 7931, pp. 92–101, 2013.
© Springer-Verlag Berlin Heidelberg 2013

rates, or d) sensors that do not provide data at all times (e.g. sensor failures). Moreover, our system is scalable, in the sense that it can handle a variable number of sensors. In this regard, we demonstrate the use of the system with three sensors: 2D laser range finder, Wi-Fi, and magnetic compass. To use Wi-Fi, we have developed a Wi-Fi positioning system, which we describe in this paper as well.

Real world experiments have proven that our system is more robust than those which rely on one sensor only, because we use complementary sensors that do not have correlated errors [5]. For instance, the Wi-Fi alone is not accurate, and the laser alone can not discriminate among geometrically similar rooms, but the combination of the two achieves both accuracy and discrimination power.

The remainder of this paper is organized as follows. Section 2 provides a general description of our system. Section 3 describes the sensor fusion process, and the three sensors mentioned. Section 4 contains the experimental results, and Section 5 the main conclusions of our work.

2   General Description

Figure 1-(a) shows a representation of our system. Its general goal is to estimate, at any time t, the robot pose s_t from: 1) perceptual information Z_t (sensor measurements), and 2) control data u_t (robot movement provided by odometry encoders). The "pose probability estimation" step (Fig. 1-(a)) computes the likelihood $bel(s_t)$ of each pose from u_t, and Z_t. Then, the "pose estimation" step calculates s_t, a state vector that contains the most probable robot pose:

$$s_t = (x_t, y_t, \theta_t) \tag{1}$$

where x_t and y_t are the position components, and θ_t is the orientation one. In this paper, the term s_t will be called the robot state or pose interchangeably.

2.1   Pose Probability Estimation

Following a Bayesian Filtering approach, the likelihood assigned to each robot pose $bel(s_t)$ will be the posterior probability over the robot state space conditioned on the control data u_t and the sensor measurements Z_t [7]:

$$bel(s_t) = p(s_t | Z_t, u_t, Z_{t-1}, u_t, ..., Z_0, u_0) \tag{2}$$

Assuming that the current state s_t suffices to explain the current u_t and Z_t (Markov assumption), we can estimate $bel(s_t)$ recursively [7]:

$$bel(s_t) \propto \left[\int p(s_t | s_{t-1}, u_t) bel(s_{t-1}) ds_{t-1} \right] p(Z_t | s_t) \tag{3}$$

In this Equation, the term $\int p(s_t | s_{t-1}, u_t) bel(s_{t-1}) ds_{t-1}$ represents the *prediction* step of Fig. 1-(a). This step infers the new $bel(s_t)$ from $bel(s_{t-1})$ (previously

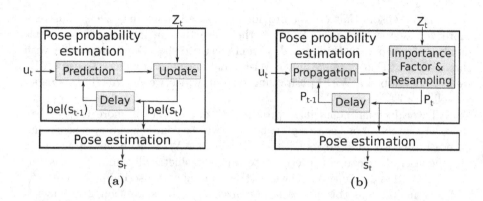

Fig. 1. a) Block diagram of the localization system from a recursive Bayesian estimation perspective. b) Block diagram of the system implemented with particle filters. At each time t, the system estimates the pose of the robot based on the robot movement (control information u_t and the sensor measurements Z_t).

stored in the *delay* buffer) and u_t. On the other hand, the term $p(Z_t|s_t)$ corresponds to the *update* step in Fig. 1-(a). This term performs the fusion of the data provided all the sensors. Note that $Z_t = \{z_t^1, z_t^2, ..., z_t^{N_s(t)}\}$ is the set of all sensor measurements at time t ($N_s(t)$ is the number of sensors available at time t). Therefore, $p(Z_t|s_t) = p(z_t^1, z_t^2, ..., z_t^{N_s(t)}|s_t)$ represents the probability that, at time t, the system receives the sensor measurements $\{z_t^1, z_t^2, ..., z_t^{N_s(t)}\}$ conditioned on the state s_t. This joint probability function may be very hard to estimate in practice, specially if we have non-synchronized sensors that work at different rates. Instead, we assume that the sensor measurements are conditionally independent given the state of the robot, therefore:

$$bel(s_t) \propto \left[\int p(s_t|s_{t-1}, u_t)bel(s_{t-1})ds_{t-1} \right] \prod_{k=1}^{N_s(t)} p(z_t^k|s_t) \qquad (4)$$

In order to be able to apply Equation 4, we must know: 1) the initial belief distribution $bel(s_0)$ (it can be chosen randomly), 2) the *motion model* of the robot $p(s_t|s_{t-1}, u_t)$, and 3) the *measurement model* $p(z_t^k|s_t)$ of each sensor k. The motion model represents the probability of transition from the state s_{t-1} to the state s_t, provided u_t. This model depends on the odometry of the robot, but it is common to assume that it follows a multivariate Normal distribution [7]:

$$p(s_t|s_{t-1}, u_t) \sim \mathcal{N}(f_{mov}(s_{t-1}, u_t), \Sigma_s) \qquad (5)$$

where f_{mov} is a function that models the movement of the robot, and Σ_s represents the noise of the model. On the other hand, the measurement model $p(z_t^k|s_t)$ depends on the nature of each specific sensor. This will be described in detail in Sec. 3.

Implementation with Particle Filters. Equation 4 can be approximated very efficiently using particle filters [7]. These filters approximate $bel(s_t)$ as a set of M random weighted samples or particles:

$$bel(s_t) \approx P_t = \{p_t^1, p_t^2, ..., p_t^M\} = \{\{s_t^1, \omega_t^1\}, \{s_t^2, \omega_t^2\}, ..., \{s_t^M, \omega_t^M\}\} \quad (6)$$

where each particle p_t^i consists, at time t, of a possible robot state s_t^i and a weight ω_t^i (likelihood) assigned to it. We calculate Equation 4 by means of the Augmented Montecarlo Localization algorithm (Augmented-MCL) with Low Variance Resampling [7]. In essence, this algorithm proceeds as follows. Initially, all particles are distributed randomly over the state space, and assigned equal weights $(\frac{1}{M})$. Then, the algorithm performs the following steps (Fig. 1-(b)):

1. Propagation. Each particle is assigned a new state according to the motion model (Equation 5).

$$s_t^i \sim p(s_t^i | s_{t-1}^i, u_t) \quad \forall\, i \in \{1, 2, ..., M\} \quad (7)$$

2. Importance factor. The weight of each particle is re-computed taking into account the measurements of the available sensors:

$$\omega_t^i \propto \prod_{k=1}^{N_s(t)} p(z_t^k | s_t^i) \quad \forall\, i \in \{1, 2, ..., M\} \quad (8)$$

3. Resampling. The algorithm constructs a new particle set P_t from P_{t-1}. First, it draws M_{NR} particles from P_{t-1} with a probability proportional to their weight ω_t^i. We use the Low Variance Resampling technique [7] to accomplish this task. Second, it generates a variable number of M_R random particles, to increase the robustness against problems such as the "robot kidnapping problem" [7]. Finally, the union of the M_{NR} non-random and the M_R random particles ($M_{NR} + M_R = M$) will form the new particle set P_t.

2.2 Pose Estimation

In this stage, represented both in Figs. 1-(a) and 1-(b), we choose the most likely robot pose from the particle set P_t. To do this, we perform the following operations:

1. Clustering. We perform an agglomerative clustering on the particle set P_t. Each particle will be assigned to the cluster with the closest centroid, provided that it is closer than a maximum distance min_{CD}. Otherwise, the particle will create its own cluster. In this context, each cluster represents an hypothesis of the robot pose.
2. Hypothesis selection. We select the most likely hypothesis, provided that it exceeds a minimum likelihood min_{HL}. We calculate the likelihood of each hypothesis as the sum of the weights of its particles, as calculated in Eq. 8.

3   Sensor Fusion

In this paper, we analyse the use of our localization algorithm with the data supplied by three different sensors: 2D laser range finder, Wi-Fi, and magnetic compass. Equation 8 provides us with an elegant way to fuse this data, only by specifying the "model" of each sensor. This Equation confers great scalability to our system, because it allows us to work with a variable number of sensors. On the one hand, this enables us to increase the number of sensors if required. On the other, this ensures that our system can work even if only a subset of the sensors is available. This is very important, because our sensors will not be synchronized, and they might have different data rates. In addition, this ensures that the system will work even if some sensors fail.

3.1   2D Laser Range Finder

At any time instant t, a 2D laser range finder provides a vector $\boldsymbol{l_t} = \{l_t^1, l_t^2, ..., l_t^{N_L}\}$ of N_L range measurements between the robot and the nearby objects. This vector is usually called the laser signature. Using an occupancy map [7] of the environment, we could pre-compute the laser signature $\boldsymbol{l_e(s)}$ expected for any possible pose in this map. Thus, we can approximate the laser "sensor model" by the similarity between $\boldsymbol{l_e(s)}$ and $\boldsymbol{l_t}$:

$$p(z^l|\boldsymbol{s}) = \left[\sqrt{1 - \frac{\sqrt{\sum_{i=1}^{N_L} l_t^i \cdot l_e^i(\boldsymbol{s})}}{N\sqrt{\sum_{i=1}^{N_L} l_e^i(\boldsymbol{s}) \sum_{i=1}^{N_L} l_t^i}}}\left[\frac{1}{N_L}\sum_{i=1}^{N_L} max\left(1 - \frac{|l_e^i(\boldsymbol{s}) - l_t^i|}{max_{LD}}, 0\right)\right]\right] \quad (9)$$

The first term of the previous equation calculates the Hellinger distance among both laser scans [8], to estimate the shape similarity among them. To take scale into account, the second term calculates the average difference among each pair of range measurements $(l_t^i, l_e^i(\boldsymbol{s}))$, normalized in the range $[0, 1]$ using a triangular function. The parameter max_{LD} (maximum laser difference) indicates the maximum allowed difference among each pair of laser ranges. Note that this "sensor model" is what we need in order to incorporate the information of this sensor into our system using Equation 8.

3.2   Wi-Fi

Our Wi-Fi positioning system provides an estimate $z^w = (x^w, y^w)$ of the robot position using the signals received from Wi-Fi landmarks such as Wi-Fi Access Points (APs). More concretely, our system estimates x^w and y^w using two regression functions f_x and f_y:

$$x^w = f_x(POW_1, POW_2, ..., POW_{N_w}) \quad (10)$$

$$y^w = f_y(POW_1, POW_2, ..., POW_{N_w}) \tag{11}$$

where N_w is the number of APs in the environment, and POW_i is the power in dBms of the ith AP. In order to learn f_x and f_y, we have chosen the ϵ-Support Vector Regression technique with Gaussian Radial Basis Function kernels (ϵ-SVR-RBF) [9]. The prediction error of the ϵ-SVR-RBF can be approximated by a zero mean Laplace distribution [9], and therefore the "sensor model" of our Wi-Fi positioning system will be:

$$p(z^w|s) = \left[\frac{1}{2\sigma_x^w} e^{-\frac{|x^w - x|}{\sigma_x^w}} \right] \left[\frac{1}{2\sigma_y^w} e^{-\frac{|x^y - y|}{\sigma_y^w}} \right] \tag{12}$$

We will see in Section 4 that the noise parameters σ_x^w and σ_y^w can be estimated from the training data that has been used to obtain f_x and f_y.

3.3 Magnetic Compass

A magnetic compass provides the orientation θ^c of the robot with respect to a fixed frame. We will assume that the measurement noise is Gaussian:

$$p(z^c|s) = \left[\frac{1}{\sigma^c \sqrt{2\pi}} e^{-\frac{(\theta - \theta_c)}{2(\sigma^c)^2}} \right] \tag{13}$$

where θ is the orientation component of the robot pose s, and σ^c is a noise parameter of this sensor, which can be estimated experimentally. This model is reasonable in absence of strong magnetic interferences. In other case, a more complex alternative might be needed. We will tackle this task in future works.

4 Experimental Results

We have performed two different experiments. The first one was intended to validate our Wi-Fi positioning system. The second, to evaluate the performance of our localization system, and the contribution of every sensor to it. These experiments took place at the CITIUS building (Centro Singular de Investigación en Tecnoloxías da Información da Universidade de Santiago de Compostela, Spain), in an area of $750\,m^2$ approximately. We have used a Pioneer P3DX robot equipped with a SICK-LMS100 laser, a Devantech CMPS10 magnetic compass, and a processing unit with Wi-Fi connectivity (Intel Core i7-3610QM @ 2.3GHz, 6GB RAM). Prior to the experiments, we have built an occupancy map of the environment using the GMapping SLAM algorithm [10].

4.1 Wi-Fi Positioning System

In order to train and test our Wi-Fi positioning system, we collected three data sets: one for training with $N_{train} = 1891$ patterns, one for validation with $N_{val} = 722$ patterns, and one for testing with $N_{test} = 644$ patterns. Each pattern contains 31 inputs (signal power of 31 APs) and 2 outputs (position where the measurement was taken).

We have trained two ϵ-SVR-RBF using the training set with $\epsilon = 0.001$. We have evaluated different combinations of the remaining parameters [9], C and γ, with the validation set. With the best combination ($C = 2^3$ and $\gamma = 2^{-16}$), we achieved an average error of 4.87 *meters* with the test set. Finally, we estimated the Wi-Fi sensor model (Sec. 3.2) through a *5-fold cross validation* with the training set [9]. This yielded the values $\sigma_x^w = 2.59$ and $\sigma_y^w = 2.42$.

4.2   Particle Filter Localization

We have tested our algorithm along two different robot trajectories: a) a trajectory of 100 meters which the robot took 161 seconds to follow, and b) a trajectory where the robot was just spinning for 51 seconds. Both trajectories started in the room marked as A in Fig. 2-(a). To collect the ground truth for the experiments, we have executed our algorithm at $8\,Hz$ with the data of each trajectory, providing it with the initial true pose of the robot. Then, each calculated pose was either accepted or rejected by an expert (comparing the real laser signature with the signature expected from that pose). We are aware that the use of a high precision localization system would have been a better option to build the ground truth. However, we did not have access to a system able to work on such big areas. Anyway, the expert estimated a maximum error of $30\,cm$ with respect to the real ground truth.

Once the ground truth had been collected, we evaluated the algorithm with different sensor combinations: 1) laser only (L), 2) laser and compass ($L + C$), 3) Wi-Fi only (W), 4) Wi-Fi and compass ($W + C$), 5) Wi-Fi and laser ($W + L$), and 6) Wi-Fi, laser and compass ($W + L + C$). We executed our algorithm 10 times with each combination, without providing it with knowledge of an initial estimation of the robot pose. Each localization step took always less than $100\,ms$. In all the experiments, we used the following parameters: 1) $M = 1000$ particles (Sec. 2.1), 2) $max_{CD} = 3\,m$ (Sec. 2.2), 3) $min_{HW} = 0.75$ (Sec. 2.2), 4) $max_{LD} = 2\,m$ (Sec. 3.1), 5) $\sigma_x^w = \sigma_y^w = 5m$ (Sec. 3.2), 6) $\sigma^c = 1\,rad$ (Sec. 3.3).

Figure 2 shows two results of our algorithm with the first trajectory. The ground truth of this trajectory starts in the place signaled as "Start", and follows the path of the squared marks. The estimated trajectory is indicated with a continuous line. First of all, Fig. 2-(a) shows an example of poor performance when using only laser data. Note that the first pose estimation calculated by the algorithm was wrong: while it calculated the one marked as "Wrong Localization", the correct one was the one marked as "Initial True Position" (Fig. 2-(a)). This happened because the laser can not distinguish among rooms A, B, and C, so the algorithm only converged (wrongly) when the hypotheses reached the corridor. However, the generation of random particles allowed to recover from this error (marked as "Recovery" in Fig. 2-(a)). On the other hand, Fig. 2-(b) shows a successful example when using Wi-Fi, laser, and compass data. Thanks to the Wi-Fi, the algorithm was able to estimate from the very beginning that the robot started in room A. Then, the algorithm tracked the robot pose correctly, even when the robot traversed areas of strong magnetic interference.

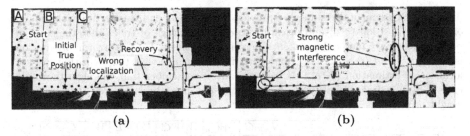

Fig. 2. Sample trajectories generated by our localization algorithm. We represent samples of the ground truth with squares, and the trajectory estimated by the algorithm with a line. a) Incorrect trajectory when using laser-only localization. b) Correct trajectory when using laser, Wi-Fi localization, and compass.

Other than these qualitative results, Table 1 provides quantitative results of these experiments:

1. *Laser-only (L)*. The low $\%t_{loc}$ indicates that the algorithm does not converge properly. The big percentages of $e_{xy} > 5m$ and $e_\theta > \pi/4$ demonstrates a big probability of incorrect convergence. This happens because the laser can not distinguish among rooms A, B or C (Fig. 2-(a)), so it either does not choose among the hypotheses, or chooses one randomly.
2. *Laser and compass (L+C)*. The compass deteriorates the performance of the laser: it tends to enhance the $\%t_{loc}$, at the cost of increasing e_{xy}. Again, none of the sensors can discriminate among rooms A, B, and C, but the compass information tends to reduce the diversity of particles. This increases the probability of convergence towards a random hypothesis.
3. *Wi-Fi only (W)*. It has high e_{xy} values, but the percentage of $e_{xy} > 5m$ is very low. This indicates that Wi-Fi is very adequate to discriminate among spatially distant hypotheses.
4. *Wi-Fi and compass (W+C)*. The compass improves slightly the performance of the Wi-Fi. Noticeably, it reduces the percentage of $e_\theta > \pi/4$ to a 0%.
5. *Wi-Fi and laser (W+L)*. The Wi-Fi helps the laser to discard hypotheses. This improves $\%t_{loc}$, and reduces the percentage of $e_{xy} > 5$. However, the Wi-Fi adds noise to the system, which tends to increase the values of e_{xy} and e_θ when the algorithm is tracking a correct pose.
6. *Wi-Fi, laser, and compass (W+L+C)*. This is the best combination. It shows a percentage of $e_{xy} < 5$ above the 99%, and a percentage of $e_\theta < \pi/4$ of 100%. Moreover, the high value of $\%t_{loc}$ (above 80.59%), and its low variability indicate a fast and stable convergence (convergence time below 11 seconds).

To summarize the results, we have seen that the combination of information sources with non-correlated errors tends to perform better that single sensors alone. Moreover, we have seen that our system can handle a variable number of sensors (scalability), even if they have different data rates. In the future, we plan to test our system extensively in dynamic environments crowded with people.

Table 1. Experimental results. We provide averages (μ) and standard deviations (σ) of: error in position (e_{xy}) and orientation (e_θ), percentage of times that the error in position is below 5 meters ($e_{xy} < 5m$), percentage of times that the error in orientation is below $\pi/4$ rad ($\%e_\theta < \pi/4$), and percentage of times that the algorithm gives a localization estimation ($\%t_{loc}$). It is important to remember that the algorithm only gives a pose estimate if at least the weight of one hypothesis exceeds the minimum weight min_{HW}.

		L	$L+C$	W	$W+C$	$W+L$	$W+L+C$
Traj. 1	$\mu(e_{xy})$	1.81	2.67	2.72	2.54	0.77	0.52
	$\sigma(e_{xy})$	2.25	2.88	0.30	0.36	0.38	0.15
	$\mu(e_\theta)$	0.14	0.07	0.25	0.12	0.15	0.08
	$\sigma(e_\theta)$	0.18	0.02	0.05	0.03	0.12	0.02
	$\mu(\%e_{xy} < 5m)$	87.49	76.55	93.36	98.58	97.60	99.32
	$\sigma(\%e_{xy} < 5m)$	18.18	22.99	2.37	0.07	3.45	0.002
	$\mu(\%e_\theta < \pi/4)$	97.76	100	95.02	98.30	97.16	99.79
	$\sigma(\%e_\theta < \pi/4)$	5.80	0	3.50	0.67	0.45	0.06
	$\mu(\%t_{loc})$	74.40	79.87	76.65	86.98	84.16	93.27
	$\sigma(\%t_{loc})$	15.62	7.70	6.41	3.68	2.63	2.18
Traj. 2	$\mu(e_{xy})$	0.42	5.65	1.22	1.83	0.92	0.59
	$\sigma(e_{xy})$	0.44	5.77	0.42	0.71	0.61	0.30
	$\mu(e_\theta)$	1.04	0.07	1.80	0.17	1.62	0.18
	$\sigma(e_\theta)$	1.68	0.05	0.57	0.06	1.37	0.11
	$\mu(\%e_{xy} < 5m)$	100	44.44	100	99.03	100	100
	$\sigma(\%e_{xy} < 5m)$	0	52.71	0	2.23	0	0
	$\mu(\%e_\theta < \pi/4)$	66.67	100	13.39	100	48.93	100
	$\sigma(\%e_\theta < \pi/4)$	57.73	0	21.46	0	48.45	0
	$\mu(\%t_{loc})$	13.92	40.98	45.10	70.20	60.39	80.59
	$\sigma(\%t_{loc})$	25.24	25.30	15.95	18.64	13.58	4.59

5   Conclusions

We have presented a localization algorithm for mobile robotics based on particle filters. This algorithm is able to fuse data provided by different information sources, even if they are not synchronized, or if they have different data rates. We have shown the use of our algorithm with: a 2D laser range finder, a Wi-Fi positioning system, and a magnetic compass. Real world experiments have demonstrated that our algorithm is accurate, robust, and fast. Moreover, we have shown that it can work with a variable number of information sources (scalability). Finally, we have proven that the performance of a combination of information sources with uncorrelated errors is better than that of a single one.

In addition, we have designed and developed the Wi-Fi positioning system mentioned above. This system uses ϵ-Support Vector Regression with Gaussian Radial Basis Function kernels to predict the position of the robot from the strength of the signal of Wi-Fi Access Points. Real world experiments showed that our Wi-Fi positioning system committed an error below 5 m.

In the future, we plan to test our system extensively in different environments, such as dynamic environments crowded with people. Additionally, we will use more information sources, such as a camera mounted on the robot, or external CCTV-like cameras which can locate the robot in the environment [11]. Finally, we plan to enhance our algorithm with more advanced resampling techniques.

Acknowledgements. This work was supported by the research projects TIN2009-07737, and TIN2012-32262, the grant BES-2010-040813 FPI-MICINN, and by the grant "Consolidation of Competitive Research Groups, Xunta de Galicia ref. 2010/6".

References

1. Alvarez-Santos, V., Canedo-Rodriguez, A., Iglesias, R., Pardo, X.M., Regueiro, C.V.: Route learning and reproduction in a tour-guide robot. In: Ferrández, J.M., Álvarez, J.R., de la Paz, F., Javier Toledo, F. (eds.) IWINAC 2013, Part II. LNCS, vol. 7931, pp. 112–121. Springer, Heidelberg (2013)
2. Thrun, S., Beetz, M., Bennewitz, M., Burgard, W., Cremers, A.B., Dellaert, F., Fox, D., Haehnel, D., Rosenberg, C., Roy, N., et al.: Probabilistic algorithms and the interactive museum tour-guide robot minerva. International Journal of Robotics Research 19(11), 972–999 (2000)
3. Tardós, J.D., Neira, J., Newman, P.M., Leonard, J.J.: Robust mapping and localization in indoor environments using sonar data. The International Journal of Robotics Research 21(4), 311–330 (2002)
4. Gamallo, C., Regueiro, C., Quintía, P., Mucientes, M.: Omnivision-based kld-monte carlo localization. Robotics and Autonomous Systems 58(3), 295–305 (2010)
5. Canedo-Rodriguez, A., Santos-Saavedra, D., Alvarez-Santos, V., Regueiro, C.V., Iglesias, R., Pardo, X.M.: Analysis of different localization systems suitable for a fast and easy deployment of robots in diverse environments. In: Workshop of Physical Agents, pp. 39–46 (2012)
6. Gutmann, J.-S., Fox, D.: An experimental comparison of localization methods continued. In: IEEE/RSJ International Conference on Intelligent Robots and Systems, vol. 1, pp. 454–459. IEEE (2002)
7. Thrun, S., Burgard, W., Fox, D., et al.: Probabilistic robotics, vol. 1. MIT Press, Cambridge (2005)
8. Pollard, D.: A user's guide to measure theoretic probability, vol. 8. Cambridge University Press (2001)
9. Chang, C.-C., Lin, C.-J.: Libsvm: a library for support vector machines. ACM Transactions on Intelligent Systems and Technology 2(3), 27 (2011)
10. Grisetti, G., Stachniss, C., Burgard, W.: Improved techniques for grid mapping with rao-blackwellized particle filters. IEEE Transactions on Robotics 23(1), 34–46 (2007)
11. Canedo-Rodriguez, A., Iglesias, R., Regueiro, C.V., Alvarez-Santos, V., Pardo, X.M.: Self-organized multi-camera network for a fast and easy deployment of ubiquitous robots in unknown environments. Sensors 13(1), 426–454 (2012)

Implicit and Robust Evaluation Methodology for the Evolutionary Design of Feasible Robots

Andrés Faíña, Felix Orjales, Francisco Bellas, and Richard J. Duro

Integrated Group for Engineering Research,
Universidade da Coruña, 15403, Ferrol, Spain
{afaina,felix.orjales,fran,richard}@udc.es
http://www.gii.udc.es

Abstract. This paper deals with the evolutionary design of feasible and manufacturable robots. Specifically, here we address the problem of defining a methodology for the evaluation of candidate robots that guides the evolution of morphology and control towards a valid design when transferred to reality. We aim to minimize the explicit knowledge introduced by the designer in the fitness function. As a consequence of this higher flexibility, we must include elements to ensure that the obtained robots are feasible. To do it, we propose an extension of the principles proposed by classical authors from traditional evolutionary robotics to brain-body evolution. In this paper we describe this methodology and show its application in a benchmark example of evolutionary robot design. To this end, previously presented elements like the structural definition of the robotic units, the encoding of the morphology and control and the specific evolutionary algorithm applied are also briefly described.

1 Evolving Feasible Robots

Several authors have researched the coevolution of morphology and control to obtain virtual creatures specifically designed for a task [1][2]. Nevertheless, for practical reasons and to get results in a reasonable amount of time, most of these works have employed inaccurate simulators which haven´t generated realistic robot behaviors. Also, the lack of an easy method or technique to translate these creatures to real robots have led to a slowdown in the use of evolutionary approaches in the design of new robots for several years.

Nowadays modular robotics allows building new robots with different morphologies in seconds as an aggregation of several homogeneous or heterogeneous modules [3]. As a consequence, now search is being carried out in order to obtain good modular configurations that, after building the robot, can be applied to different tasks. However, applying evolution to these types of systems still presents some serious challenges. Even though we now have set of standardized building blocks which should simplify the construction of the robots and limit the dimensionality of the search space, this search space becomes discontinuous and its deceptiveness is increased.

J.M. Ferrández Vicente et al. (Eds.): IWINAC 2013, Part II, LNCS 7931, pp. 102–111, 2013.

Researchers employ two main avenues to address this problem. Some authors impose some knowledge on the morphology and control to obtain good solutions easily, thus reducing the search space and consequently the diversity of solutions [4][5]. Other authors introduce several parameters within the fitness function in order to lead evolution towards their goal. This method facilitates that robots can perform their tasks successfully, but it generates a very difficult tuning process in order to produce the correct fitness function for the combination of requirements that arise [6].

In this paper we deal with how to obtain robust and feasible robots without imposing knowledge in order to obtain new designs, and using non-complex fitness functions to simplify the evolutionary process. We address the problem employing methodologies similar to those that classical authors in traditional evolutionary robotics (that is, when they were evolving just the control system for the robots) [7] proposed and extendign them to the simultaneous generation of the morphologies and behaviors of robots that can be transferred to the reality. Here these principles are merged with a procedure for simplifying the fitness definition and extended to be applied to modular robotics. The resulting methodology is tested using an evolutionary robot design benchmark task.

The rest of the paper is structured as follows. Section II is devoted to a brief description of the modular architecture that will be used in evolution. Section III summarizes the evolutionary design system, called EDMoR, which includes a previously presented encoding and algorithm, and the evaluation definition methodology, which is the focus of this work. In section IV we present a typical design experiment in modular robots using EDMoR and, finally, in section V the main contributions of this work are summarized.

2   Modular System Definition

In [8] we presented a modular robotic architecture based on a set of simple heterogeneous modules to maximize flexibility and simplicity in automatic design processes. This architecture has been used in the present work to analyze the evaluation methodology, so the robots designed in this paper will be made up of the modules defined by it.

Currently, four types of actuator modules have been fully designed and laboratory prototypes haven been fabricated. Two of them produce linear motions (slider and telescopic modules) and the other two produce rotational motions (rotational and hinge modules). Additionally, there is a passive module that acts as a structural base. These prototypes are displayed in Fig. 1. The slider, telescopic and rotational modules contain cube shaped structures called nodes. These nodes act as connection bays and their size is 54x54x54 mm. The free sides of the nodes correspond to connection mechanisms.

Details about the mechanical design, connector, communications, energy, sensors and control of the modular architecture can be found in [8].

3 Evolutionary Design System

To address the problem of evolving complete robots to obtain feasible structures we have developed a system called EDMoR (Evolutionary Designer of Modular Robots) that includes the main elements involved in this process: encoding, algorithm and evaluation. They will be described in this section.

Encoding. The EDMoR system uses a direct encoding of the morphology of the individuals as trees, where every robotic module depends on a parent module, except for the root module. The encoding is composed of n nodes and n-1 links between nodes. Each node represents one module and it contains the information about the module type, its control parameters and the c connections with the child nodes. Moreover, each node stores the number of connected sides of the module. The links between nodes contain information about the child node and all the variables required to connect it in a unique position and orientation with respect to the parent module. It must be highlighted here that the flexibility provided by the large number of connection bays present in the modules (see Fig. 1) implies a huge number of possible structures depending on the selected side and the orientation of the child modules.

Algorithm. The search spaces defined by modular structures when evolving morphology and control are very badly configured. They are deceptive, discontinuous, extremely multimodal, etc. This is determined by the encoding of the individuals and, as we will discuss later in detail, by how they are evaluated. In [9] we presented a constructive evolutionary strategy to be used in EDMoR with the aim of softening the search space by avoiding abrupt changes in the morphology. Moreover, the algorithm takes into consideration that the evolutionary temporal scales of the morphology and the control system are quite different.

Summarizing its operation, the constructive evolutionary strategy used in EDMoR starts with a random population with a few modules per robot. Then it executes five different stages which generate mutations in the robot until a stop

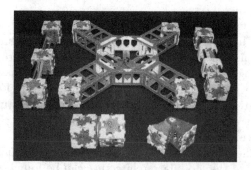

Fig. 1. Modular System: Slider module (left), passive base module (center), telescopic module (right), rotational module (bottom left) and hinge module (bottom right)

criterion is met. These stages are a *Growing phase* (adds modules to the robot), a *Morphological adaptation phase* (changes the position or orientation of some modules), a *Control adaptation phase,* (changes the control parameters of some modules), a *Pruning phase* (deletes the modules that do not increase the fitness) and a *Replacement phase* (some individuals of the population are replaced by mutations of the best robots, typically employing a symmetry operator).

Evaluation. A key aspect in the evolutionary design of feasible and manufacturable systems is the evaluation of individuals. As commented in the introduction, this evaluation is mainly performed in simulation because it is not affordable to build all the candidate solutions in reality. Typically, the designer creates the simulated environment and explicitly defines a fitness function that guides evolution towards an objective while satisfying different real-world constraints. This procedure involves a high degree of designer intervention that is undesirable because it is limited by his/her capacities (consequently non-generalizable) and because it restricts the search space, preventing the emergence of more flexible and original solutions. This idea of allowing a more flexible evolution is not novel, but it has been widely rejected when trying to design manufacturable systems due to the unfeasibility of most of the emergent solutions.

Here we are interested in addressing this approach in a different way. We aim to develop a methodology to evaluate individuals in a more implicit way, allowing emergent solutions while maintaining their feasibility by means of a realistic simulator where the physical constraints can be easily incorporated and where the main features of the environment can be properly varied during evaluation. The problem of obtaining simulated results that can be transferred to reality is a classical one in the evolutionary robotics literature. It was considered by Jakobi in [7] for the case of evolving robot controllers in simulation with fixed morphologies. He proposed a procedure whereby a set of basic features of the real environment and basic interactions with it must be carefully defined and randomly varied during the evaluation phase, leading to controllers that are highly robust in real world operation. Some of the ideas introduced by Jakobi will be used here.

Specifically, we propose a methodology for the evaluation of individuals in the evolutionary design of modular robots that implies the following steps:

1. The objective of the robot behavior must be clearly specified.
2. According to the established objective, a very simple fitness function must be defined that can be easily implemented by the designer using high-level measurements with the aim of preserving a search space as general as possible to ease the search process. This fitness function should not include physical constraints explicitly.
3. Physical constraints should be included in the simulation as a part of the world instead of being terms added to the fitness functions. Soft constraints are preferred rather than hard constraints to make the search easier.
4. A basic set of features of the real environment that is relevant for the behavior must be defined.

5. All the features of the simulated environment that are out of the basic set and that could modify the robot's design must be randomly varied from evaluation to evaluation to make the robot tolerant to them.
6. All the features of the simulated environment that are in the basic set must be varied in a predefined operational range (related to the desired behavior) from evaluation to evaluation.

It seems clear that this methodology implies a detailed definition of the problem by the designer and one could think that the work is practically the same as in the typical procedure of defining an explicit fitness function that contains all the constraints and elements to be evaluated. This is generally not so as it is very difficult to predict how very complex explicit fitness functions may be exploited by an Evolutionary Algorithm and, at the same tiem, these complex fitness functions lead to very difficult search spaces. As we will show in the next section with a practical example, following the methodology proposed here simplifies the process notably, because most of the constraints and features that must be considered are implicitly introduced through the simulator, maintaining a very simple explicit fitness function.

4   Incremental Design of a Modular Robot

To show the robotic structures that can be obtained using the EDMoR system with the evaluation methodology presented above, we consider here the problem of automatically designing a feasible robot capable of moving through rugged surfaces carrying a payload.

First, a simulation model of the problem has been created. We have used the Gazebo dynamic simulator [10] as the basic tool for evaluating the candidate robots. To this end, all the modules presented in section II were carefully modeled, so the transfer of the simulated robots to the prototypes is direct.

When an individual is evaluated in Gazebo, it is decoded and simulated for a fixed period of time until a reliable fitness value is obtained. To this end, a specific control system must be defined. In this first study, and with the aim of using a very simple control policy to promote the emergence of more adapted morphologies, all the modules are controlled using a sinusoidal function that provides the module position pos (displacement distance or angle between the two parts of the module) using:

$$pos = \alpha \cdot A_{max} \cdot sin(\omega \cdot V_{max} \cdot t + \varphi) \tag{1}$$

being t the simulation time, α the amplitude control parameter in the range $[0,1]$, ω the angular velocity control parameter in the range $[0,1]$, φ the phase shift between $[0,360)$, A_{max} is the maximum amplitude and V_{max} is the maximum angular velocity. These two final constants depend on the type of module. As a consequence, the control parameters that are included in the chromosome and must be adjusted are α, ω and φ.

For the sake of clarity, we have organized the practical example in two parts of incremental complexity. In the first one, we do not consider the payload and focus the design in obtaining a robot capable of moving in rugged surfaces. According to the proposed methodology, we must define the following evaluation details:

1. The objective of the robot to be designed is: moving forward as far as possible in a general rugged terrain and minimizing the energy consumption.
2. The fitness function that has been defined to reach such objective is:

$$f = \begin{cases} distance, & if\ f < distance_{th} \\ distance + \frac{f_{max}}{N} \cdot (N - n), & if\ f > distance_{th} \end{cases} \quad (2)$$

   where $distance$ is the distance travelled by the payload, $distance_{th}$ is a distance threshold value, f_{max} is the fitness of a robot with zero modules, N is the maximum number of modules allowed and n is the current number of modules of the individual that is being evaluated. The idea behind this function is rewarding those individuals that cover a minimum distance $distance_{th}$ using a low number of modules, which is directly related with low energy consumption. The control of the number of modules is not considered until the individual reaches a minimum fitness to make the search space easier.
3. In this case, we define one physical constraint: the robot has to walk without large deviations to avoid solutions that cover high distances but not forward. The constraint is implemented in a soft way and introduced as a modification of the simulation environment. Specifically, if the robot moves outside a virtual corridor defined in the simulator, the evaluation is stopped. The last position of the robot is employed to calculate the travelled distance. Therefore, robots that do not break this constraint can make use of the whole the simulation time and, potentially, can produce better fitness.
4. The basic set of features in this case is made up of three parameters related with the obstacles of the floor, height h_o, width w_o, and length l_o, which constitute the features of the environment where the robots must be robust.
5. In this case, we do not consider any feature out of the basic set that is relevant for the design (for example, we could have included the initial position of the robot and varied it randomly between evaluations, but in this case this feature is considered irrelevant).
6. We have defined five different rugged floors, constructed with boxes to work as small obstacles. The sizes of these boxes are randomly selected between 0 to 15 cm in height and from 0.6 to 2 m of width and length. The robots are evaluated five times, one in each floor, and the resulting fitness is the minimum obtained by the robot in these five evaluations.

Table 1. Configuration Parameters of the EDMoR System in the Experiments

	Population	Maximum evaluations	Evaluation time	f_{th}	f_{max}	N
Value	80	40.000	44s	0.3	0.8	16

Once the evaluation procedure has been established, we can start with the evolutionary process with the parameters displayed in Table 1. We have run the algorithm four times obtaining the evolution curves displayed in Fig. 2 (left). The blue line represents the average fitness of these four runs while the red one corresponds to the mean fitness of the population. As can be seen, we obtain robots that cover distances of over 1.3 meters while preserving a clear diversity in the population. Moreover, in this figure we have represented the evolution of the average number of modules in the population (green line). The maximum allowed number of modules was 16, and with the selected fitness function the algorithm is able to obtain successful robots using 12 modules on average. It must be pointed out that the best individual obtained in each execution was made up of a number of modules below this average value in all cases.

As an example, the three top images of Fig. 3 show one of the best-evaluated robots obtained in this case. It is made up of seven modules, the base, four sliders and two telescopic modules, symmetrically placed with respect to the base (a consequence of the symmetry mutation operator commented in section 3). As can be observed, the robot drags the base, but it is very efficient in rugged surfaces like those used in the evaluation because the modules act as propulsion legs and the base is a kind of appendix that is dragged and provides stability.

Another interesting solution the EDMoR system found in this case is the one presented in the bottom images of Fig. 3 where motion over the rugged terrain is achieved by rotating the platform, much like a wheel. This solution reinforces the idea of how a more flexible search space leads to more original solutions.

We face now the complete design problem considering that the robots must carry a payload in addition to moving over rugged terrains. In this case, according to the proposed methodology, the evaluation steps are specified as follows:

1. The objective of the robot to be designed is: moving forward as far as possible in a general rugged terrain carrying a payload of different weight and minimizing the energy consumption.
2. The fitness remains unchanged.
3. We add a new physical constraint: the robot cannot drop the payload during its movement. Again, this constraint is implemented in a soft way and introduced modifying the simulation environment. Specifically, if the payload falls, the simulation is stopped. Considering both restrictions, the last position of the payload is employed to calculate the travelled distance.
4. The basic set of features in this case is made up of the previous parameters related with the obstacles of the floor, height (h_o), width (w_o), and length (l_o), and now we include the payload parameters weight (w_p), size (s_p) and payload friction coefficient (μ_p).
5. Again, we do not consider any feature out of the basic set that is relevant for the design.
6. In addition to the five rugged floors used in the previous setup, we include five different cubic payloads of different weights (from 0.4 to 0.8 Kg), sizes (from 0.3 to 0.6 m) and friction coefficients (from 0.8 to 10). The robots are evaluated five times, one in each floor and with a different payload. The resulting fitness is the minimum obtained in these five evaluations.

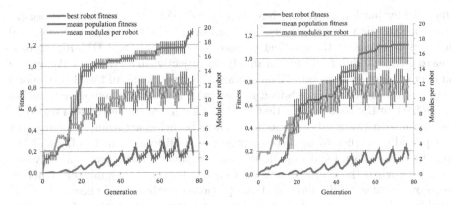

Fig. 2. Evolution of the fitness and average number of modules for the experiments over ground with obstacles (left) and with a payload (right)

Fig. 3. Examples of simulated robots obtained in the experiment without payload

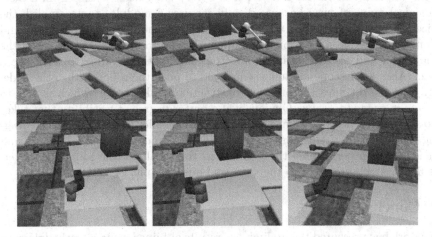

Fig. 4. Examples of simulated robots obtained in the second part of the experiment

Once the evaluation procedure has been established, we start the evolutionary process. Evolution results are displayed in Fig. 2 (right). As expected, the total distance decreases as the problem is more complex, but the EDMoR system is able to design modular robots that satisfy the objective in all the runs. Again, the number of modules is around 12 on average. The top images of Fig. 4 correspond to one of the best robots obtained in this case. It has two symmetric branches in the back of the base module that starts with a hinge connected to a telescope and finish with two rotational modules. These branches act again as contracting legs providing thrust to the whole structure. Additionally, in the front of the base there is a rotational module joined to a slider that generate an advance movement when the center of mass is frontally balanced, helping the structure to go over obstacles. To show that the proposed methodology allows obtaining feasible structures, Fig. 5 (left) contains a prototype construction of this robot, which is able to operate in rugged terrains successfully. The morphology of the robot guarantees that the base inclination is not too large and, as a result, the payload can be transported even with small obstacles on the floor.

Fig. 5. Prototypes of two final robots obtained in the experiments

Finally, in the bottom images of Fig. 4 we display the best solution obtained in this experiment. As we can see, it is a very efficient structure from a morphological point of view because it only requires 6 modules to accomplish the objective. It is composed by a branch of two hinge modules to obtain pitch and roll movements and a rotational module in the end to increase the length of the serial chain and also to rotate the bottom part of this module to work like a wheel. Also, a telescope module is employed to push the robot in the other side. Fig. 5 (right) displays a prototype construction of this robot, which is able to operate in rugged terrains. Obviously, if a better degree of feasibility that takes into account the possible damages the robot could suffer from friction when dragging the base is desired, more features should be included in the base set and outside it when performing the simulations that evaluate the robots. However, the fitness structure remains the same and quite simple.

5   Conclusions

This work has presented the preliminary results obtained using a methodology to define the evaluation of candidate solutions in the evolutionary design of feasible robots. The methodology is based on the establishment of an implicit fitness function that must include the minimal knowledge imposed by the designer in order to simplify its work and to allow the emergence of original solutions. Moreover, with the aim of obtaining manufacturable robots that can be directly transferred to reality, the methodology establishes a set of steps to be performed in order to ensure that the solutions are robust in their real operation. These steps are an adaptation of the principles developed by Jakobi and imply the definition of a basic set of features where the desired robot must be robust during its operation, which are varied during the evaluation process following a predefined pattern. This methodology has been included in EDMoR, a modular robot evolutionary design system. We have analyzed here EDMoR's response with the proposed methodology in a benchmark task. These results are very promising and, consequently, we are now improving EDMoR to apply it in more complex tasks.

Acknowledgments. This work was partially funded by the Xunta de Galicia and European Regional Development Funds through projects 09DPI012166PR and 10DPI005CT.

References

1. Sims, K.: Evolving virtual creatures. In: Proceedings of the 21st Annual Conference on Computer Graphics and Interactive Techniques, pp. 15–22. ACM (1994)
2. Komosinski, M.: The framsticks system: versatile simulator of 3d agents and their evolution. Kybernetes 32, 156–173 (2003)
3. Yim, M., Shen, W., Salemi, B., Rus, D., Moll, M., Lipson, H., Klavins, E., Chirikjian, G.: Modular self-reconfigurable robot systems. IEEE Robotics & Automation Magazine 14, 43–52 (2007)
4. Marbach, D., Ijspeert, A.: Online optimization of modular robot locomotion. In: IEEE International Conference on Mechatronics and Automation, vol. 1, pp. 248–253. IEEE (2005)
5. Rommerman, M., Kuhn, D., Kirchner, F.: Robot design for space missions using evolutionary computation. In: IEEE Congress on Evolutionary Computation, pp. 2098–2105 (2009)
6. Farritor, S., Dubowsky, S.: On modular design of field robotic systems. Autonomous Robots 10(1), 57–65 (2001)
7. Jakobi, N.: Evolutionary robotics and the radical envelope-of-noise hypothesis. Adaptive Behavior 6(2), 325–368 (1997)
8. Faiña, A., Orjales, F., Bellas, F., Duro, R.J.: First steps towards a heterogeneous modular robotic architecture for intelligent industrial operation. In: Workshop on Reconfigurable Modular Robotics, IROS, San Francisco (2011)
9. Faíña, A., Bellas, F., Souto, D., Duro, R.J.: Towards an evolutionary design of modular robots for industry. In: Ferrández, J.M., Álvarez Sánchez, J.R., de la Paz, F., Toledo, F.J. (eds.) IWINAC 2011, Part I. LNCS, vol. 6686, pp. 50–59. Springer, Heidelberg (2011)
10. Koenig, N., Howard, A.: Design and use paradigms for gazebo, an open-source multi-robot simulator. In: IEEE/RSJ International Conference on Intelligent Robots and Systems, vol. 3, pp. 2149–2154 (2004)

Route Learning and Reproduction in a Tour-Guide Robot

Víctor Alvarez-Santos[1], A. Canedo-Rodriguez[1], Roberto Iglesias[1],
Xosé M. Pardo[1], and C.V. Regueiro[2]

[1] Centro Singular de Investigacion en Tecnoloxias da Informacion
Universidade de Santiago de Compostela, Spain
victor.alvarez@usc.es
[2] Department of Electronics and Systems
Universidade da Coruña, Spain

Abstract. Route learning and reproduction in tour-guide robots is usually performed with the help of an expert in robotics. In this paper we describe a novel approach to these tasks, which reduces the intervention of an expert to a minimum. First, the robot is able to learn routes while following a human acting as a route instructor. Then, anyone can easily ask the robot to reproduce a route using various hand gestures. In order to achieve an accurate route learning and reproduction we use a novel localization algorithm, which is able to combine various sources of information to obtain the robot's pose. Moreover, the path planning and obstacle avoidance used to navigate while reproducing routes are also described in this article. Finally, we show through several trajectories how the robot is able to learn and reproduce routes.

Keywords: tour-guide robot, route learning, route reproduction, human following.

1   Introduction

In the coming years, personal service robots are expected to become a common element in most homes or offices, playing an important role as appliances, servants and assistants; they will be our helpers and elder-care companions. These robots will need to be capable of acquiring a sufficient understanding of the environment, being aware of different situations, as well as establishing a successful communication with humans in order to be able to cooperate with them.

We are currently building a general purpose tour-guide robot, which learns routes from an instructor, and then shows these routes to the visitors of the event where the robot operates. The development of the control software for this robot involves many open challenges such as robot localization, human identification, and robot navigation in crowded and challenging environments.

In this work, we describe the route learning and reproduction processes in our robot. The route learning takes place while following an instructor, who only needs to move in the environment showing the route to the robot. The reproduction of

J.M. Ferrández Vicente et al. (Eds.): IWINAC 2013, Part II, LNCS 7931, pp. 112–121, 2013.

a route by the robot is performed on user's demand. During the reproduction, the robot travels to the origin of the route, and starts to mimic the route learnt from the instructor in the past. We want to remark that other service robots like a robotic wheelchair will also benefit from this ability: the wheelchair could learn a route in a hospital while following a nurse, and then be able to travel back to a patient's room.

2   Related Work

In the late nineties, the first well-known tour-guide robots (*Rhino* [1] and *Minerva* [2]) had no online route learning abilities. In fact, points of interest and the rest of information which make up a route were introduced in the robot by an expert. In these robots, the users could select the exhibit that they wanted to visit using a touch-screen located at the robot. This strategy for route recording and route learning, with minor variations, became a common element in the tour-guide robots that were developed afterwards. One of them was *RoboX* [3], which was designed for long time operation in a public exposition where the routes were also precoded in the robot by experts. These routes were shown to visitors as soon as anyone approached the robot. Other tour-guide robots [4] randomly choose a visitor and guide him to an exhibit, which was selected based on information from RFID tags, which are used to detect which exhibit has not been visited yet by the human. Urbano [5] is another tour-guide robot, which requires a manual creation of a navigation graph, and a database of objects and locations. This, allows the robot to dynamically generate routes that will be shown to a visitor depending on which category is the user classified on. Robotinho [6] was one of the first humanoid tour-guide robots, and it could give tours to people who show interest in the robot. These tours were also manually introduced in a pre-operational stage of the robot.

A recent tour-guide robot [7] allows the user to select which exhibit he wants to visit using speech recognition. It can also identify which part of the exhibit needs further explanation by recognizing pointing gestures. Despite of these improvements in route management, this tour-guide robot still requires a pre-operational stage in which an expert introduces location and other information about the exhibits available. Another recent tour-guide robot [8] states that new routes can be created on the fly, although no details about the process are given. Finally, one of the latest tour-guide robots that has been developed [9] explores different ways of reproducing routes: it can keep track of the human which is following him and wait for him while reproducing the route. However, the definition of the routes is still done by an expert, and the same ahppens with the selection of which route should be reproduced.

3   System Overview

The latest version of the robot that we use as prototype for our tour-guide robot can be seen in Fig 1 (left). It consists of a Pioneer P3DX robot base, a laser

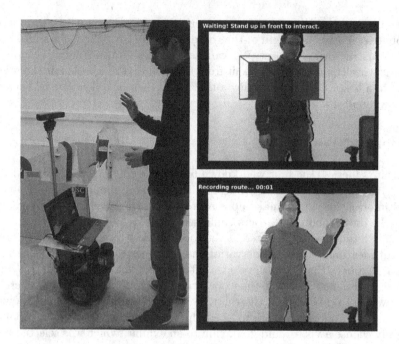

Fig. 1. Left: picture of our prototype of tour-guide robot interacting with a human. Right: captures of the robot's user interface with augmented reality.

range finder, and a range camera which is located at the top of the robot. Its main processing unit is a laptop.

An instructor can teach routes to the robot by commanding the robot to follow him. The instructor can be anyone from the staff of the event where the robot will operate, and this person will not need to be an expert in robotic. The instructor can also inform the robot about the *points of interest* along the route. Thus, every time that the instructor reaches one of those *points of interest* he will be able to record an explanatory *voice messages*.

On the other hand, the visitors of an event can ask the robot to show them those routes. The robot will search and travel to the first point of the route, and start to mimic the full route. When the robot arrives at any *point of interest*, it will reproduce the corresponding *voice message*, which has been previously recorded by the instructor.

We have developed an interface which provides visual feedback to the users so that they know whether they are being properly detected by the robot. On the other hand, this interface also uses augmented reality (virtual buttons) so that either the instructor or the visitor will be able to use hand gestures to start recording a route or voice messages along it (instructor), or select and reproduce a route (visitor).

In order to successfully accomplish the tasks of route recording and route reproduction, we need a robust robot localization that we have developed. As we will see in the next section, the use of this localization strategy makes us consider two different stages: the *deployment stage* and the *operational stage*.

4   The Deployment Stage

In the location system that we use in our robot [12] we combine information from different sources, which have to be previously set up when the robot arrives to a new environment. This process is performed by an expert in a short period of time.

Therefore, the goal of this stage is to generate what we call the *model* of each sensor. By *model* we understand a function that is able to determine the probability of a sensor reading provided a robot position. These models will be used to localize the robot in the environment considering the sensor readings at each instant (section 5.1). For this reason, we need to move the robot around the environment collecting data: odometry, laser signatures and signal strength of each Wi-Fi access point (AP).

Firstly, we use this data to build a laser-based map of the environment. Using this map, we can pre-compute the expected laser signature l_e (with N_L readings) from each possible pose s in this map and compare it with the actual laser signature l_t. This allows us to approximate the laser *sensor model* by Eq. 1:

$$p(z^l|s) = \left[\sqrt{1 - \frac{\sqrt{\sum_{i=1}^{N_L} l_t^i \cdot l_e^i(s)}}{N\sqrt{\sum_{i=1}^{N_L} l_e^i(s) \sum_{i=1}^{N_L} l_t^i}}} \right] \left[\frac{1}{N_L} \sum_{i=1}^{N_L} max\left(1 - \frac{|l_e^i(s) - l_t^i|}{max_{LD}}, 0\right)\right] \quad (1)$$

The first term of the previous equation calculates the Hellinger distance which estimates shape similarity. To take scale into account, the second term computes the average difference among each pair of range measurements $(l_t^i, l_e^i(s))$. The parameter max_{LD} (maximum laser difference) indicates the maximum allowed difference among each pair of laser ranges.

Secondly, we use [12] the signal strength of the Wi-Fi AP and the ϵ-Support Vector Regression technique with Radial Basis Function kernels (ϵ-SVR-RBF [16]) to provide an estimate $z^w = (x^w, y^w)$ of the robot position from the signal strength of the audible Wi-Fi APs. The prediction error of the ϵ-SVR-RBF can be approximated by a zero mean Laplace distribution [16], and therefore the *sensor model* of our Wi-Fi location system can be approximated by Eq. 2:

$$p(z^w|s) = \left[\frac{1}{2\sigma_x^w} e^{-\frac{|x^w - x|}{\sigma_x^w}} \right] \left[\frac{1}{2\sigma_y^w} e^{-\frac{|x^y - y|}{\sigma_y^w}} \right] \quad (2)$$

where the x^w and y^w are the coordinates corresponding to the position of the robot that has been estimated using the signal strength of the Wi-Fi APs.

5 The Operational Stage

Using the information recorded during the deployment stage, we will be able to localize the robot at any instant, because of this, during the operational stage any instructor can teach routes to the robot, or any visitor can demand the reproduction of these routes. Both route learning and reproduction involve several critical tasks which need to be properly addressed: robot location in the environment, the detection, identification and tracking of the instructor, the path planning of the routes and a safe navigation of the robot in a crowded environment. In the next sections, we will describe the most relevant asoects of each of these tasks.

5.1 Robot Localization

As we have already introduced in section 4, our localization algorithm [12] is able to combine information from different sources: a laser scanner, a compass, and a Wi-Fi positioning system.

To this extent, we fuse the information from all the data sources by means of a particle filter [17]. The most relevant contribution of this filter lays on the weighting of the particles p^i using the N sensors available. Each particle p_i consists of a estimated position of the robot s_i, and a weight $\hat{\omega}^i$. The weight of each particle is re-computed taking into account the meassurements of the available sensors:

$$\hat{\omega}^i = \prod_{k=1}^{N} p(z^k|s^i) \quad \forall\, i \in P \tag{3}$$

It is important to notice that in order to compute the weight of the particles, we need to use the models $p(z|s)$ of each sensor obtained in the deployment stage (Eq. 1 and Eq. 2).

This solution [12] is a indoor localization robust to sensor failures or sensor with different data acquisition rates, because it allows us to keep updating the particle filter even if not all the sensors are available.

5.2 Route Learning

Route learning in our tour-guide robot is performed under the command of an instructor who moves along the desired route while being followed by the robot. That instructor does not have to be an expert in robotics, thus, anyone can teach new routes to our robot by simply letting the robot follow him.

Therefore, the robot will follow the instructor, logging its pose every metre (*way-points* from now on). Every time that the robot logs a new *way-point*, it will make a sound to let the user know that the route is being recorded. In addition, during the route teaching, the instructor can stop at any *point of interest* and record a voice message.

It is obvious that the performance of the person following behaviour is critical for this task, that is why in the past [11] we have presented a study of several colour and texture features, which are used to distinguish the instructor from the rest of the people which moving around the robot. In order to select which ones would perform best, we evaluated 27 visual characteristics, and selected eight of them based on which features shared the lowest mutual information. The features that we have selected are: the HSV colour space, the second derivatives in both image axis, the Canny edge image, the Centre-symmetric LBP, and the MPEG-7 edge histogram. In addition, we use an online feature weighting procedure [11], which is able to increase the weight of those features which are more discriminant at each moment. The weight of each feature f is dynamically updated based on the feature's discrimination power calculated with the Hellinger distance d_f:

$$d_f = \sqrt{1 - \sum_{i=1}^{B} \sqrt{h_1(i)h_2(i)}} \qquad (4)$$

where h_1 is the normalized distribution of the feature in the instructor, and h_2 the normalized distribution of the same feature in the rest of the people present in the scene where the robot operates, and B the number of bins in which we have discretised the distributions. This allows us to obtain a visual distance between humans detected in the image and a visual model of the instructor, this visual distance combined with the physical distance (predicted position vs actual position) can be used to obtain a probability of a human to be the instructor that the robot is actually following.

Moreover, we have developed our own human detector [10] which is able to perform pixel-level segmentation of humans from the background. This point is quite important because it has significantly improved the feature extraction, and thus, the accuracy of the visual distance between humans and the instructor.

We have evaluated the person following ability in many real world scenarios with people of all ages. One of those scenarios was the Domus museum (A Coruña, Spain), which presents many challenges: strong illumination changes, an uneven floor and a crowded environment.

The advantage of this route learning procedure is that it is not performed on the deployment stage, which would increase the robot's deployment time, and also that anyone can be followed regardless of his knowledge about the robot.

5.3   Route Reproduction

Route reproduction is a critical behaviour of the robot.

When the robot is about to reproduce a route, it picks the first point from the route and tries to reach it. For this task, we use a global planner based on Dijkstra's algorithm [13] in order to find the shortest path between the current robot's pose and the first point of the route. Once the robot has calculated its global path towards the goal, the local planner is responsible for safely moving the robot towards the environment. Thus, local navigation is performed using the

Dynamic Window Approach (DWA) [14], which consists in forward-simulating various trajectories which were generated by sampling the linear and angular velocities of the robot. Then, each simulated trajectory is scored based on some parameters, such as distance to the goal and proximity to obstacles. The trajectory with the highest score is chosen, and the velocities which generated that trajectory are executed in the robot's base. This algorithm, allows us to continuously evaluate the trajectory, and thus avoid obstacles such as persons walking in the robot's path.

Once the robot has reached the first point in the route, the robot will pick the next *way-point* and navigate towards it. This is done until there last point is reached or the visitor stops the robot. Moreover, the *points of interest* are treated in the same way, with the difference that the robot stops in that point for a while, until the voice message is played back.

In order to speed up route reproduction, we are flexible when it comes to decide whether the robot has reached a *way-point* or not. Therefore, we have set a distance of 0.3m as acceptable, as well as a difference in the yaw goal of 0.4 radians. Moreover, whenever there is a point within the route that is unreachable the robot can skip it and travel to the next one. The robot will be making bell sounds every time it reaches a point and a error sound if it skips a point.

Fig. 2. $100m^2$ robotics laboratory where the experiments were conducted. We prepared an environment with 5 rooms, and recorded and reproduced several routes.

In order to test the route learning and reproduction of our robot, we have recorded two routes in our robotics laboratory (Fig. 2), where we have settled an indoor environment with five rooms. We have chosen this laboratory as the first test-bed for route-management in our robot because it is similar in size ($100m^2$) and space arrangement (five rooms) with most home or offices nowadays. The routes that we used for testing are illustrated in Fig. 3 where we can see the robot's trajectory during route recording (drawn in a dark-grey and dotted line), and the trajectory (drawn in a light-grey and continuous line) that the robot

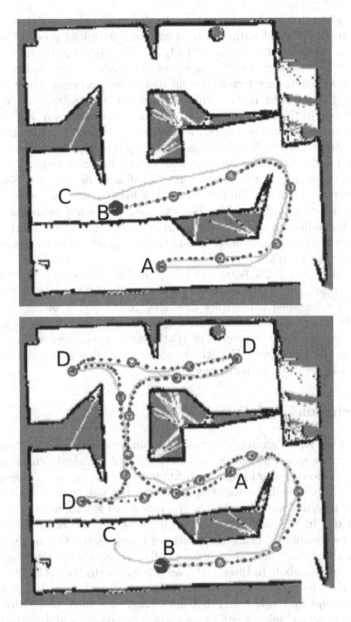

Fig. 3. Trajectories for recording and reproduction of two routes. The dotted trajectory represents the robot's trajectory while recording the route, and the circles are the recorded way-points. The light-grey lines are the robot's trajectories while reproducing these routes. The points marked with the letter A are the first points recorded in the routes, the points marked with the letter B are the end point of the routes, the points marked with the letter C are the location of the robot when it was asked to reproduce the route, and the points marked with the letter D are the locations of *points of interest* specified by the instructor while recording the route.

performed when it reproduced the route. In these examples we can see how the robot has to find a valid path and to travel from its initial pose (C in Fig. 3) to the initial point of the route (A in Fig. 3). This is not a trivial issue since the distance between both points can be several dozens of metres, and the path between those points might cross several rooms and corridors. Once the robot has travelled to the first point within the route, it moves from one *way-point* to the next one in the route (illustrated with dark-grey circles in Fig. 3). When the robot reaches a point of interest, it plays the voice message recorded by the instructor. Finally, when the robot finishes the reproduction of the route, it plays a bell sound and waited for new commands from its nearby users.

The first route (Fig. 3, top) was recorded in 56 seconds, and it is 15.3m long. In this route no points of interest were recorded by the instructor, but it is interesting in order to notice how fast can anyone teach a route path to the robot in a home-like environment with narrow corridors. In this experiment, the robot had to travel 29.2m in order to find the first point of the route and mimic the complete route path.

The second route (Fig. 3, bottom) was recorded in 155 seconds, and it is 40.8m long. When reproducing this route the robot travelled 52.8metres. In this route, the instructor recorded three voice messages at the points of interest. These messages were very short: around 5 seconds each one. In this way we can isolate the time used for route management (path teaching and gesture interaction) from that of the message itself: with our tour-guide robot a complex route can be recorded in few time by non experts.

6   Conclusions and Future Work

In this work we have described the route learning and reproduction behaviours in our tour-guide robot. Moreover, we have presented a critical element for those behaviours: a multi-sensor fusion algorithm for robot localization. Moreover, we have successfully tested the route management through two routes that were successfully recorded and played back in a real world environment.

Like we did in the past [11] [10], we plan to test the route management and robot localization in challenging scenarios such as the Domus museum (A Coruña, Spain), with which we have been collaborating during the development of this tour-guide robot. In those tests, we will adhere to the ISO/IEC-9126 for evaluating the quality of the route management with real users. In addition, we also plan to improve the human-robot interaction that takes place while reproducing a route by including a pan camera which can rotate and monitor the user to whom the robot is showing the route.

Acknowledgements. This work was funded by the research projects TIN2009-07737, TIN2012-32262, the grant BES-2010-040813 FPI-MICINN, and by the Galician Government (Consolidation of Competitive Research Groups, Xunta de Galicia ref. 2010/6).

References

1. Buhmann, J., Burgard, W., Cremers, A.B., Fox, D., Hofmann, T., Schneider, F.E., Strikos, J., Thrun, S.: The mobile robot Rhino. AI Magazine 16(2), 31 (1995)
2. Thrun, S., Bennewitz, M., Burgard, W., Cremers, A.B., Dellaert, F., Fox, D., Hahnel, D., Rosenberg, C., Roy, N., Schulte, J., et al.: MINERVA: A second-generation museum tour-guide robot. In: Proceedings of IEEE International Conference on Robotics and Automation, vol. 3 (1999)
3. Jensen, B., Froidevaux, G., Greppin, X., Lorotte, A., Mayor, L., Meisser, M., et al.: The interactive autonomous mobile system RoboX. In: IEEE/RSJ International Conference on Intelligent Robots and Systems, vol. 2, pp. 1221–1227 (2002)
4. Shiomi, M., Kanda, T., Ishiguro, H., Hagita, N.: Interactive humanoid robots for a science museum. In: Proceedings of the 1st ACM SIGCHI/SIGART Conference on Human-Robot Interaction, pp. 305–312 (2006)
5. Rodriguez-Losada, D., Matia, F., Galan, R., Hernando, M., Montero, J.M., Lucas, J.M.: Urbano, an interactive mobile tour-guide robot. In: Seok, H. (ed.) Advances in Service Robotics, pp. 229–252. In-Teh (2008)
6. Faber, F., Bennewitz, M., Eppner, C., Gorog, A., et al.: The humanoid museum tour guide Robotinho. In: The 18th IEEE International Symposium on Robot and Human Interactive Communication, RO-MAN 2009, pp. 891–896 (2009)
7. Avilés, H., Alvarado-González, M., Venegas, E., Rascón, C., Meza, I.V., Pineda, L.: Development of a Tour–Guide Robot Using Dialogue Models and a Cognitive Architecture. In: Kuri-Morales, A., Simari, G.R. (eds.) IBERAMIA 2010. LNCS, vol. 6433, pp. 512–521. Springer, Heidelberg (2010)
8. Yelamarthi, K., Sherbrook, S., Beckwith, J., Williams, M., Lefief, R.: An RFID Based Autonomous Indoor Tour Guide Robot. In: IEEE 55th International Midwest Symposium on Circuits and Systems, MWSCAS, August 5-8, pp. 562–565 (2012)
9. Bueno, D.R., Viruete, E., Montano, L.: An autonomous tour guide robot in a next generation smart museum. In: 5th International Symposium on Ubiquitous Computing and Ambient Intelligence (2011)
10. Alvarez-Santos, V., Iglesias, R., Pardo, X.M., Regueiro, C.V., Canedo-Rodriguez, A.: Gesture-based interaction with voice feedback for a tour-guide robot. Submitted to Journal of Visual Communication and Image Representations (2013)
11. Alvarez-Santos, V., Pardo, X.M., Iglesias, R., Canedo-Rodriguez, A., Regueiro, C.V.: Feature analysis for human recognition and discrimination: Application to a person following behaviour in a mobile robot. Robotics and Autonomous Systems 60(8), 1021–1036 (2012)
12. Canedo-Rodriguez, A., Alvarez-Santos, V., Santos-Saavedra, D., Gamallo, C., Fernandez-Delgado, M., Iglesias, R., Regueiro, C.V.: Robust multi-sensor system for mobile robot localization. In: Ferrández, J.M., Álvarez, J.R., de la Paz, F., Javier Toledo, F. (eds.) IWINAC 2013, Part II. LNCS, vol. 7931, pp. 92–101. Springer, Heidelberg (2013)
13. Dijkstra, E.W.: A note on two problems in connexion with graphs. Numerische Mathematik 1(1), 269–271 (1959)
14. Fox, D., Burgard, W., Thrun, S.: The dynamic window approach to collision avoidance. Robotics & Automation Magazine 4(1), 23–33 (1997)
15. Grisetti, G., Stachniss, C., Burgard, W.: Improved Techniques for Grid Mapping with Rao-Blackwellized Particle Filters. IEEE Transactions on Robotics 23(1), 34–46 (2007)
16. Chang, C.-C., Lin, C.-J.: LIBSVM: A library for support vector machines. ACM Trans. Intell. Syst. Technol. 2(3) (2011)
17. Thrun, S., Burgard, W., Fox, D.: Probabilistic Robotics. MIT Press (2003)

Meta-ensembles of Classifiers
for Alzheimer's Disease Detection
Using Independent ROI Features

Borja Ayerdi, Alexandre Savio*, and Manuel Graña

Grupo de Inteligencia Computacional (GIC), Universidad del País Vasco
(UPV/EHU), San Sebastian, Spain

Abstract. Due to its growing social impact, prodromal detection of
Alzheimer's disease is of paramount importance. Biomarkers based on
Magnetic Resonance Imaging (MRI) are one of the most sought results
in the neuroscience community. In this paper we evaluate several ensembles of classifiers trained and tested in a two level ensemble scheme
as follows: the 116 regions of interest (ROI) of the Anatomical Automatic Labeling (AAL) brain atlas are used to compute disjoint feature
sets from the Grey-matter probability maps from the segmentation of
the T1 weighted MRI of each subject; ROI features are the summary
statistics inside this ROI; one ensemble of classifiers is trained on each
independent ROI feature data set; the final classification of each subject is given by the combination of the classifications of each ROI, as
meta-ensemble classifier. Experiments are performed on the 416 subjects
(316 controls and 100 patients) of the OASIS database. We perform a
hold-out of the 20% of the data for model selection, computing a leave-
one-out validation on the 80% remaining data. Results are computed
without circularity. Tested classifiers are the Extreme Learning Machines
(ELM), Bootstrapped Dendritic Computing (BDC), Hybrid Extreme
Random Forest (HERF) and Random Forest (RF). The best performance
achieved is 80.8% accuracy, 77.1% specificity and 92.5% specificity with
BDC. We also report the most discriminant ROIs obtained in the model
selection phase.

1 Introduction

Alzheimer's Disease (AD) is one of the most important causes of disability in
the elderly and with the increasing proportion of elderly in many populations,
the number of dementia patients will rise also. Due to the socioeconomic importance of the disease in occidental countries there is a strong international
effort focus in AD. In the early stages of AD brain atrophy may be subtle and
spatially distributed over many brain regions, including the entorhinal cortex,
the hippocampus, lateral and inferior temporal structures, as well as the anterior
and posterior cingulate.

---

* This work has been partially supported by the "Ayudas para la Formación de
Personal Investigador" fellowship from the Gobierno del Pais Vasco.

J.M. Ferrández Vicente et al. (Eds.): IWINAC 2013, Part II, LNCS 7931, pp. 122–130, 2013.
© Springer-Verlag Berlin Heidelberg 2013

Machine learning methods have become very popular to classify functional or structural brain images to discriminate them into two classes: normal or a specific neurodegenerative disorder [1]. The development of automated detection procedures based in Magnetic Resonance Imaging (MRI) and other medical imaging techniques [2] is of high interest in clinical medicine. It is important to note that these techniques are aimed to help clinicians with more statistical evidence for the diagnosis, it is not intended to substitute any other existing diagnosis procedure.

Most published classification methods working on MRI data train a single classifier. However, it is challenging to train only a global classifier that can be robust enough to achieve good classification performance, mostly due to noise and small sample size of neuroimaging data. Previous works propose a classification method via aggregation of regression algorithms fed with histograms of deformations from the volume registrations [3]. Another study shows a local patch-based subspace ensemble method which builds multiple individual classifiers based on different subsets of local patches with the sparse representation-based classifier [4]. In [5] subsets of ranked features from neuroimaging data are used to in an ensemble of linear Support Vector Machine (SVM) classifiers.

In this paper we use modulated Grey-Matter (GM) maps computed from the OASIS database [6]. This data is then partitioned according to the Automatic Anatomical Labeling (AAL) atlas [7] regions of interest (ROIs) and we create datasets of statistical features from each ROI for each subject. The overall classification system is a meta-ensemble combining the results of the independent classifications of each ROI. A model selection phase is performed to find the ROIs which give best classification performance and the majority threshold needed for positive classification in the majority voting criterion, which is intended to alleviate the problems due to class unbalance (316 controls vs. 100 patients) of the datasets. The remaining data from the model selection hold-out is used for a leave-one-out validation. Four classification algorithms are compared: Extreme Learning Machines (ELM)[8], Bootstrapped Dendritic Computing (BDC)[9], Hybrid Extreme Random Forest (HERF)[10,11] and Random Forest (RF)[12]. We also report the ROIs chosen in the model selection step for each classifier, allowing to compare discriminant brain locations against the findings of other methods [13,14].

Section 2 gives a description of the subjects selected for the study, the image processing, feature extraction details, cross-validation and classifier algorithms. Section 3 gives the classification performance results and in section 4 we provide conclusions of this work and further research suggestions.

2 Materials and Methods

2.1 Data

In this study we use data of all the subjects from a public available brain MRI database, the first Open Access Series of Imaging Studies (OASIS) [6]. These

subjects were selected from a larger database of individuals who had participated in MRI studies at Washington University, they were all right-handed and
older adults had a recent clinical evaluation. Older subjects with and without
dementia were obtained from the longitudinal pool of the Washington University
Alzheimer Disease Research Center (ADRC). This release of OASIS consists of
a cross-sectional collection of 416 male (119 controls and 41 patients) and female
(197 controls and 59 patients) subjects aged 18 to 96 years (218 aged 18 to 59
years and 198 subjects aged 60 to 96 years). Further demographic and image acquisition details can be found in [6]. The data we are using are the skull-stripped
and corrected for intensity inhomogeneity volumes.

2.2 Preprocessing

The spatial normalization of each subject of the database is performed with the
FMRIB Software Library (FSL) FNIRT [15]. A four step registration process
with increasing resolution and a scaled conjugate gradient minimization method
has been performed using the default parameters, nearest neighbour interpolation and the standard MNI brain template. The Jacobian matrix at each voxel
site describes the speed of change of the deformation in the neighboring area
of each voxel. The determinant of the Jacobian matrix $\mathbf{J}_i$ (aka Jacobian) is
commonly used scalar measure of the amount of distortion necessary to register
the images. A value $\det(\mathbf{J}_i) > 1$ implies that the neighborhood adjacent to the
displacement vector in voxel i was stretched to match the template (i.e., local
volumetric expansion), while $\det(\mathbf{J}_i) < 1$ is associated with local shrinkage.

Apart, we segment the subjects with FSL FAST [16] into 3 volume brain
tissue probability maps: grey (GM), white (WM) matter, and cerebral-spinal
fluid (CSF). In this study we are interested in the GM maps, which we multiply
by the Jacobians from the non-linear registration in order to get a modulated
GM map in the standard MNI space. Subsequently, these maps, after smoothing
with a 2mm Full-Width Half-Maximum (FWHM) Gaussian filter, are the basis
for feature extraction. A visual check has been performed for all images in every
processing step carried out in this experiment.

2.3 Feature Extraction

The Automatic Anatomical Labeling (AAL) atlas [7] is used to partition the
GM maps into 116 brain anatomical regions. In this study we compute for each
AAL anatomical region from each subject GM map 7 statistical measures: the
maximum voxel value, the minimum, the mean, the variance, the median, the
kurtosis and the skewness. Resulting in 116 sub-datasets of 416 subjects with 7
features each.

2.4 Classifiers

Random Forests Decision trees [17,18] are built by recursive partitioning of the
data space. A univariate (single attribute) split is recursively defined for each

tree node, from the root to the leaves, of the tree using some criterion (e.g., mutual information, gain-ratio, gini index). The data space and data samples are then partitioned according to the univariate test. Tree leaves correspond to the actual assignment of data samples to classes. Random forests are ensembles of decision trees where each individual decision tree is built on a bootstrapped training data subset and on a random subset of the input variables. The majority voting rule decides the class assignment to the input data.

ELM The Extreme Learning Machine (ELM) [8] is a very fast training algorithm for single-layer feed-forward neural networks (SLFN). The key idea of ELM is the random initialization of the SLFN hidden layer node weights. Consider a set of M data samples $(\mathbf{x}_i, y_i)$ with $\mathbf{x}_i \in \mathbb{R}^d$ and $y_i \in \Omega$. Then, a SLFN with N hidden neurons is modeled as the following expression $\mathbf{y} = \sum_{i=1}^{N} \beta_i f(\mathbf{w}_i \cdot \mathbf{x} + b_i)$, $j \in [1, M]$, where $f(x)$ is the activation function, $\mathbf{w}_i$ the input weights to the i-th neuron in the hidden layer, b_i the hidden layer unit bias and β_i are the output weights. The application of this equation to all available data samples can be written in matrix form as $\mathbf{H}\beta = \mathbf{Y}$, where $\mathbf{H}$ is the hidden layer output matrix defined as the output of the hidden layer for each input sample vector, $\beta = (\beta_1 \ldots \beta_N)^T$ and $\mathbf{Y} = (\mathbf{y}_1, \ldots, \mathbf{y}_M)^T$. The output weights β are estimated computing the least squares solution $\hat{\beta} = \mathbf{H}^\dagger \mathbf{Y}$, where $\mathbf{H}^\dagger$ is the Moore–Penrose generalized inverse of the matrix $\mathbf{H}$.

Hybrid Extreme Rotation Forest. The Hybrid Extreme Rotation Forest (HERF) algorithm [10,11] is summarized as follows: Let $\mathbf{x} = [x_1, \ldots, x_n]^T$ be a random vector composed of n feature variables, F is the feature variable set and X is the data set containing N data samples in a matrix of size $N \times n$. The i-th row of X, denoted by $X_i = [x_i^1, x_i^2, \ldots, x_i^N]$ is the sample vector of the i-th feature variable F_i. Let Y be a vector containing the class labels of the data samples, $Y = [y_1, \ldots, y_N]^T$, $y_i \in \Omega$. Denote by $D_1, \ldots, D_L$ the classifiers in the ensemble. As with most ensemble methods, we need to specify L in advance. All classifiers can be trained in parallel. To construct the training set for classifier D_i, we carry out the following steps: (1) Partition the set of feature variables F into K subsets of variables. (2) For each subset of feature variables, extract the data from the training data set, and compute the rotation matrix performing the Principal Component Analysis (PCA) transformation. (3) Compose the global rotation matrix reordering columns according to the original data. (4) Transform the train and test data applying the same rotation matrix.

Bootstrapped Dendritic Classifiers. The Bootstrapped Dendritic Classifiers (BDC) is a collection of dendritic classifiers [9], $C(x; \psi_j)$, $j = 1, \ldots, N$, where ψ_j are independent identically distributed random vectors whose nature depends on their use in the classifier construction. Each DC classifier casts a unit vote for the most popular class of input x. Given a dataset of n samples, a bootstrapped training dataset is used to train $C(x; \psi_j)$. The independent identically distributed random vectors ψ_j determine the result of bootstrapping. In conventional RF they also determine the subset of data dimensions $\hat{d}$ such that $\hat{d} \ll d$ on which each tree is grown, in this paper we are not dealing with this kind of DC randomization, which will be studied elsewhere. The main parameters for the experimental evaluation of the BDC are the number of trees and the maximum depth of each DC given by the maximum number of dendrites allowed. Limiting the number of dendrites is a kind of regularization that weakens the classifier.

2.5   Meta-ensemble Classification

The overall classification is a meta-ensemble process. For each ROI we train a separate classifier, which can be a single or ensemble classifier. The meta-ensemble classification performs a majority voting on the results of the ROI classifiers. To care for the effect of class unbalance we assume a majority threshold, that is the minimum number of ROIs that must agree in order to decide on the most frequent class. This majority threshold acts as a correction for the *a priori* probability (which is unknown to the model selection algorithm).

2.6   Model Selection and Validation

Algorithm 1 specifies the experimental setting including the model selection and validation phases. Model selection is performed on a 20% of the data holding out 80% for validation. Model selection aims to select the subset of ROIs giving best classification performance and the majority threshold value. Model selection proceeds as follows, first the classifier performance on each ROI features is estimated by a leave-one-out (LOO) procedure. Second, the ROIs are ordered according to their independent performances. Third, a greedy search for the optimal subset of ROIs is performed, adding them according to the previous order and testing their LOO performance for varying majority threshold values. The maximum performance gives the optimal ROI subset and majority threshold value.

Classifier parameter settings. Each ensemble (i.e. BDC, RF and HERF) is composed of 5 classifiers. The number of ELMs in HERF ensembles is either one third or two thirds, this decision is taken on the model selection phase. For ELMs, the number of hidden nodes is fixed to 14 nodes. In the case of BDC the depth given by the number of dendrites D is fixed to 31 and the box size to 0.8.

3   Results

We report accuracy $((TP + TN)/N)$, sensitivity $(TP/(TP + FN))$, specificity $(TN/(FP + TN))$, for each ensemble classifier. In table 1 we show the classification performance of the ensembles. Best results correspond to the BDC, but in general results are below other conventional approaches in the literature. The presented results do use very few ROIs for the classification, so that the

---

Algorithm 1. Experiment cross-validation procedure

Let be $X = \{x_1, \ldots, x_n\}$ input data $x_i \in \mathbb{R}^d$, and $Y = \{y_1, \ldots, y_n\}$ the input data class labels $y_i \in \{0, 1\}$.

N is the number of samples.

K is the number of classifiers in the ensemble.

R is the number of ROIs.

1. Perform dataset partition $X = X^m \cup X^v$, $Y = Y^m \cup Y^v$, where (X^m, Y^m) contains the 20% of the dataset.

model selection using (X^m, Y^m)

1. for $i = 1 : R$
   (a) Select from X^m the i-th ROI data, denoted X_i^m .
   (b) Compute by LOO, the ROI accuracy, sensitivity and specificity.(a_i, ss_i, sp_i) of the ensemble classifier on X_i^m
2. end for
3. sort the ROIs according to decreasing value of $ss_i + sp_i$
4. for $i = 1 : R$
   (a) Select from X^m the i-th ROI data, denoted X_i^m , adding it to the incremental model selection feature set $X_i^{inc} \leftarrow X_{i-1}^{inc} \cup X_i^m$, with $X_0^{inc} = \varnothing$.
   (b) for $\theta = 1 : i$
      i. Compute   by   LOO,   the   ROI   accuracy,   sensitivity   and specificity.$(a_i^\theta, ss_i^\theta, sp_i^\theta)$ of the *meta-ensemble* classifier on X_i^{inc} applying majority threshold θ.
   (c) endfor
5. Find the set of ROIs $(selectedROIs)$ and majority threshold θ^* giving the highest performance $ss_i^\theta + sp_i^\theta$

validation using (X^v, Y^v)

1. for i in *selectedROIs*
   (a) Select from X^v the i-th ROI data, denoted X_i^v , adding it to the final feature set $X^f \leftarrow \cup X_i^v$ used for testing.
2. endfor
3. Compute final accuracy, sensitivity and specificity by LOO on X^f using θ^* as the majority threshold.

---

number of features is very small compared with other methods in the literature. Figures 1 and 2 show the ROIs selected from the model selection with each of the classifiers. The RF has selected the largest number of ROI (49), while the BDC selected the smallest number (10), meaning that BDC achieves its results with only 70 features. The only ROI which has been selected for all the classifiers is the left Parahippocampal gyrus, which is known to show atrophies together with the entorhinal cortex in an early stage of visible anatomical degradation of the brain with Alzheimer's disease.

Table 1. Results

	Accuracy	Specificity	Sensitivity
RF	79.0	73.1	97.5
HERF	79.3	73.9	96.3
ELM	70.8	43.7	92.5
BDC	**80.8**	**77.1**	**92.5**

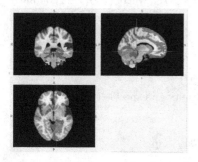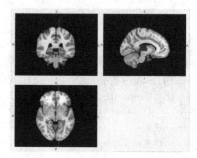

Fig. 1. Slices of the MNI standard template where the ROIs selected by ELM (left) and BDC (right) are colored

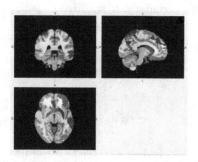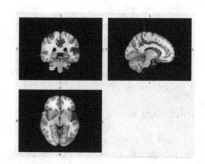

Fig. 2. Slices of the MNI standard template where the ROIs selected by RF (left) and HERF (right) are colored

4   Conclusions

In this paper we report classification results of an ensemble of different types of classifiers on features extracted from modulated GM probability maps partitioned with the AAL atlas. The sample is the complete cross-sectional OASIS database. For each subject, the modulated GM data is partitioned into 116 regions and 7 statistical values from each are used as feature vectors. The results are in agreement with most of our previous classification experiments [1,14,13].

It was of our aim to assess the performance of features built using *a priori* ROI maps against those experiments with supervised methods of feature selection. Unsupervised feature selection methods will most probably show worse classification performance than those which are supervised, but lead to more generalized systems and less fitted to the experimental database. We are aware that registration and segmentation errors can lead to biases in the accuracy of the classifiers. In addition, atrophy in brain structures can be interpreted as a late stage of AD and functional MRI could be used instead to detect previous stages of the disease. As future work we will be trying to successfully apply ensembles to deformation features, but we could also find more interesting approaches including functional MRI data.

Acknowledgments. We thank the Washington University ADRC for making MRI data available. This work has been supported by the MINECO grant TIN2011-23823, and UFI 11/07 of the UPV/EHU.

References

1. Chyzhyk, D., Graña, M., et al.: Hybrid dendritic computing with kernel-LICA applied to Alzheimer's disease detection in MRI. Neurocomputing 75(1), 72–77 (2012)
2. Davatzikos, C., Fan, Y., et al.: Detection of prodromal alzheimer's disease via pattern classification of MRI. Neurobiology of Aging 29(4), 514–523 (2008) PMID: 17174012 PMCID: 2323584
3. Chen, T., Rangarajan, A., et al.: CAVIAR: Classification via aggregated regression and its application in classifying OASIS brain database. In: Proceedings/IEEE International Symposium on Biomedical Imaging: From Nano to Macro, pp. 1337–1340 (2010)
4. Liu, M., Zhang, D., et al.: Ensemble sparse classification of alzheimer's disease. NeuroImage 60(2), 1106–1116 (2012) PMID: 22270352
5. Varol, E., Gaonkar, B., et al.: Feature ranking based nested support vector machine ensemble for medical image classification. In: 2012 9th IEEE International Symposium on Biomedical Imaging (ISBI), pp. 146–149 (May 2012)
6. Marcus, D.S., Wang, T.H., et al.: Open access series of imaging studies (OASIS): cross-sectional MRI data in young, middle aged, nondemented, and demented older adults. Journal of Cognitive Neuroscience 19(9), 1498–1507 (2007) PMID: 17714011
7. Tzourio-Mazoyer, N., Landeau, B., et al.: Automated anatomical labeling of activations in SPM using a macroscopic anatomical parcellation of the MNI MRI single-subject brain. NeuroImage 15(1), 273–289 (2002) PMID: 11771995
8. Huang, G.B., Zhu, Q.Y., Siew, C.K.: Extreme learning machine: Theory and applications. Neurocomputing 70(1-3), 489–501 (2006)
9. Chyzhyk, D.: Bootstrapped dendritic classifiers for alzheimer's disease classification on mri features. In: Graña, M., Toro, C., Posada, J., Howlett, R.J., Jain, L.C. (eds.) Advances in Knowledge-Based and Intelligent Information and Engineering Systems - 16th Annual KES Conference. Frontiers in Artificial Intelligence and Applications, vol. 243, pp. 2251–2258. IOS Press (2012)
10. Ayerdi, B., Graña, M.: Hybrid extreme rotation forest. Cognitive Computation (in press, 2013)

11. Chyzhyk, D., Ayerdi, B., Maiora, J.: Active learning with bootstrapped dendritic classifier applied to medical image segmentation. Pattern Recognition Letters (submitted, 2013)
12. Breiman, L.: Random forests. Machine Learning 45(1), 5–32 (2001)
13. Savio, A., García-Sebastián, M., et al.: Neurocognitive disorder detection based on feature vectors extracted from VBM analysis of structural MRI. Computers in Biology and Medicine 41(8), 600–610 (2011)
14. Savio, A., Graña, M.: Deformation based feature selection for computer aided diagnosis of alzheimer's disease. Expert Systems with Applications 40(5), 1619–1628 (2013)
15. Smith, S.M., Jenkinson, M., et al.: Advances in functional and structural MR image analysis and implementation as FSL. NeuroImage 23(suppl. 1) S208–S219 (2004) PMID: 15501092
16. Zhang, Y., Brady, M., et al.: Segmentation of brain MR images through a hidden markov random field model and the expectation-maximization algorithm. IEEE Transactions on Medical Imaging 20(1), 45–57 (2001) PMID: 11293691
17. Breiman, L., Friedman, J., Olshen, R., Stone, C.: Classification and Regression Trees. Wadsworth and Brooks, Monterey (1984)
18. Quinlan, J.R.: C4.5: Programs for Machine Learning. Morgan Kaufmann (1993)

Results on a Lattice Computing
Based Group Analysis
of Schizophrenic Patients on Resting State fMRI

Darya Chyzhyk* and Manuel Graña

Grupo de Inteligencia Computacional (GIC), Universidad del País Vasco
(UPV/EHU), San Sebastian, Spain

Abstract. We work on the definition of Lattice Computing approach to
identify functional networks in resting state fMRI data (rsfMRI) looking
for biomarkers of cognitive or neurodegenerative diseases. The approach
uses Lattice Auto-Associative Memories (LAAM) to compute a reduced
ordering h-function that can be thresholded or processed by morphological operators for network detection. Group analysis is performed on
the templates corresponding to each class of subjects computed by averaging their spatially normalized rsfMRI data. We inspect the Tanimoto coefficients computing the similarity between compared networks
to decide the appropriate threshold. Results on a dataset of healthy controls, schizophrenia patients with and without auditory hallucinations
show that the approach is able to find functionally connected cluster
differences discriminating the subjects suffering auditory hallucination.

1 Introduction

Resting state fMRI (rsfMRI) data has been used to study the functional connectivity in the brain [1–3], looking for temporal correlation of low frequency
oscillations in diverse areas of the brain. Because the subject is not performing
any explicit cognitive task, the functional network is assumed as some kind of
brain fingerprint, which can be used to detect biomarkers of cognitive or neurodegenerative diseases. Resting state fMRI experiments do not impose constraints
on the cognitive abilities of the subjects. For instance in pediatric applications,
such as the study of brain maturation [4], there is no single cognitive task which
is appropriate across the aging population. Several machine learning and data
mining approaches have been followed for the exploration of rsfMRI data: hierarchical clustering [5], independent component analysis (ICA) [6–8], fractional
amplitude of low frequency analysis [9], multivariate pattern analysis (MVPA)
[4, 10]. Graph analysis has been suggested [3] as a tool to study the connectivity
structure of the brain. Resting state fMRI has being found useful for performing
studies on brain evolution based on the variations in activity of the default mode
network [4], depression (using regional homogeneity measures) [11], Alzheimer's
Disease [12], and schizophrenia.

* This work has been partially supported by an FPU grant from Spain MEC.

J.M. Ferrández Vicente et al. (Eds.): IWINAC 2013, Part II, LNCS 7931, pp. 131–139, 2013.
© Springer-Verlag Berlin Heidelberg 2013

Schizophrenia is a severe psychiatric disease that is characterized by delusions and hallucinations, loss of emotion and disrupted thinking. Functional disconnection between brain regions is suspected to cause these symptoms, because of known aberrant effects on gray and white matter in brain regions that overlap with the default mode network. Resting state fMRI studies [13–15] have indicated aberrant default mode functional connectivity in schizophrenic patients. These studies suggest an important role for the default mode network in the pathophysiology of schizophrenia. Functional disconnectivity in schizophrenia could be expressed in altered connectivity of specific functional connections and/or functional networks, but it could also be related to a changed organization of the functional brain network. Resting state studies for schizophrenia patients with auditory hallucinations have also been performed [16] showing reduced connectivity.

This paper reports new results on the application of a Lattice Computing based multivariate morphology approach to the analysis of rsfMRI data. The general approach introduced in [17], consists in the application of Lattice Auto-Associative Memories (LAAMs) [18, 19] to the definition of a *LAAM-supervised ordering*, an specific kind of h-ordering [20], that allows the formal definition of morphological operators on multivariate data. In short, the a seed voxel BOLD time series is used to build a LAAM, which is then applied to the remaining voxels of the brain fMRI 4D data. The LAAM-supervised h-function consists in the LAAM recall error for each voxel. The supervised ordering is built on the h-function domain, and the map obtained from the whole brain volume is thresholded to detect functional connectivity. It can be also processed by morphological operators providing some specific features of the volume. The approach is a Lattice Computing correspondent to the correlation [21] or independent component analysis [22] based approaches to the functional connectivity discovery on rsfMRI data. Lack of space prevents to reproduce here the formalization of the approach, which can be found in [17].

Specifically, in this paper we perform experiments with background/foreground LAMM h-function on rsfMRI data from healthy controls (HC), Schizophrenia patients with and without auditory hallucinations (SZAH and SZnAH, respectively) looking for differences in network sites which may be useful as biomarkers or for feature extraction in classification experiments. The advances in this paper relative to [17] are the following: (a) we have built templates for each population by averaging the registered 4D data, (b) we process the whole brain volume, (c) we focus on the background/foreground h-function map, (d) we follow the work in [23] exploring the network induced by an specific localization in the brain, (e) threshold value is decided by inspection of the Tanimoto coefficients between the functional networks of each population class. Working on the population templates allows to assess group level effects that are visible on the average data.

Section 2 presents the data used for the experiments and the preprocessing performed. Section 3 presents some experimental results. Finally we give some conclusions and further work in section 4.

2   Materials and Preprocessing

The aim of the experimental work is to find systematic differences in networks identified from the resting state fMRI data that could be assumed as discriminating the underlying populations of healthy control subjects, schizophrenia patients with and without auditory hallucinations. The experiment shows that this approach detects quite different brain networks depending on the subject using the same h-function built from selected voxel seeds.

Materials. The results shown in this section are explorations over resting state fMRI data obtained from a 28 healthy control subjects (NC), and two groups of schizophrenia patients: 26 subjects with and 14 subjects without auditory hallucinations (SZAH and SznAH respectively), selected from an on-going study in the McLean Hospital, Boston, Ma. Details of image acquisition and demographic information will be given elsewhere. For each subject we have 240 BOLD volumes and one T1-weighted anatomical image.

Preprocessing pipeline. The data preprocessing begins with the skull extraction using the BET tool from FSL (http://www.fmrib.ox.ac.uk/fsl/). All the images was manually orient to AC-PC line. The functional images were coregistered to the T1-weighted anatomical image. Further preprocessing, including slice timing, head motion correction (a least squares approach and a 6-parameter spatial transformation), smoothing (FWHM=4mm) and spatial normalization to the Montreal Neurological Institute (MNI) template (resampling voxel size = 3 mm × 3 mm × 3 mm), temporal filtering (0.01-0.08 Hz) and linear trend removing, were conducted using the DPARSF (http://www.restfmri.net/forum/DPARSF) package. All the subjects have less than 3mm maximum displacement and less than $3^{\underline{o}}$ of angular motion. Z-scored images were used to calculate the time-average of BOLD across subjects for each group. Finally, we compute the 4D group-average of subjects from HC, SZAH and SZnAH groups to build corresponding templates for each population.

3   Experimental Results

According to the findings reported in [23], our experiments aim to obtain network localizations correlated with an specific voxel placed on the left Heschl's gyrus (LHG; MNI coordinates -42,-26,10) 1, preferentially from the auditory cortex in order to ascertain some effect related to the auditory hallucinations. The voxel time series used as the seed to build the LAAM corresponding to the foreground is extracted from this MNI coordinate from the template data of each group. We compute the map corresponding to the application of the background/foreground h-function map on each template using the foreground and background seed voxels identified in figures 1 and 2 respectively. The foreground voxel seed corresponds to the LHG, and the background voxel to the CSF in one of the ventricles, that is, to irrelevant or noisy time series.

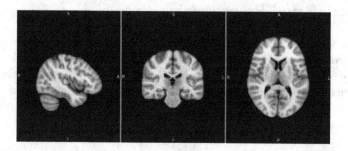

Fig. 1. Foreground voxel seed site from the left Heschl's gyrus (LHG; -42,-26,10)

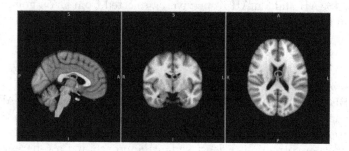

Fig. 2. Background voxel seed site from CSF of the ventricle

Computing the background/foreground h-function produces a real valued map over the brain volume, where functional networks are identified applying a threshold to this map. We would like to set this threshold so that the differences between the networks from each population are greatest while the size of the detected network sites (aka voxel clusters) are also greatest.

Given pairs of sets X and Y, the Tanimoto coefficient measures their similarity as the ration between the cardinalities of their intersection and union:

$$T(X,Y) = \frac{|X \cap Y|}{|X \cup Y|}$$

The Tonimoto coefficient $T \in [0,1]$. Complete dissimilarity corresponds to $T = 0$, identity corresponds to $T = 1$. Therefore, we look for small values of the Tanimoto coefficient, while the size of the sets $|X|$ and $|Y|$ remains significant. In fact, we deal with sets of voxels sites, which depend on the value of the threshold applied to the h-function map. Therefore, $X(\theta)$ is the set of voxel sites with h-function above the θ threshold. Consequently, we denote $T(\theta) = T(X(\theta), Y(\theta))$.

Figure 3 shows two plots. The first plot corresponds to the evolution of pairwise population network similarity measured by Tanimoto coefficients $T(\theta)$, increasing the threshold value. It can be appreciated that the similarities of the HC versus any of the two patient populations are very much the same. However, the Tanimoto coefficient of the two patient populations are significantly higher, confirming the intuition that they share many functional network traits. However, we are interested here in finding differences that may be useful for discrimination/classification. Therefore, we focus in the lower values: for $\theta > 0.7$, Tanimoto coefficients $T(\theta)$ comparing healthy and patients are close to zero , meaning that the networks are almost disjoint. The comparison between patient classes gives almost zero Tanimoto coefficients for $\theta > 0.8$. The second plot corresponds to the size of the networks, measured as the cardinality of the sets of voxel sites. It can be appreciated that for $\theta > 0.7$ the networks become very small, therefore, we have chosen $\theta = 0.7$ for the figures below.

Figure 4 shows the networks found related to the LHG seed voxel for the (a) healthy control (HC), (b) schizophrenic with auditory hallucinations (SZAH), and (c) schizophrenic without auditory hallucinations (SZnAH). Notice that the

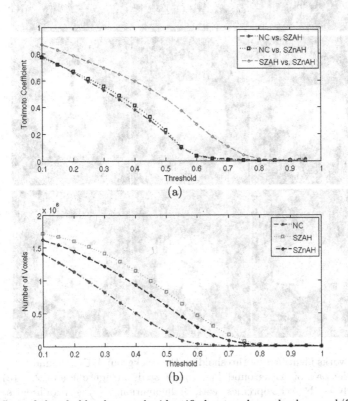

Fig. 3. Effect of threshold value on the identified networks on background/foreground h-function brain map. (a) Tanimoto Coefficient comparing networks from each pair of population, and (b) size of the detected clusters.

size of the SZAH network is bigger and they are more spread than the SZnAH network. There is a clear difference relative to the HC network. We think that these sites can be proposed as specific biomarkers. We have found that some of them are in agreement with previous reported findings [23], though the exhaustive listing of detection will be given elsewhere. Finally, we show in figure 5(a) the intersection between the SZnAH and SZAH networks, which is very small but significant, and in figure 5(b) the difference network corresponding to the clusters active only in SZnAH. There are many voxel sites which may be taken as biomarkers to discriminate between patient classes.

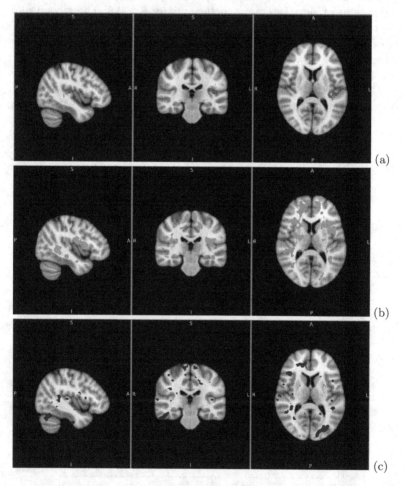

(a)

(b)

(c)

Fig. 4. Networks identified by thresholding the Background/Foreground h-function induced by the pair of background/foreground seeds in figure 1 and 2. . (a) healthy controls (HC), (b) schizoprenics with hallucinations, (c) schizophrenics without hallucinations.

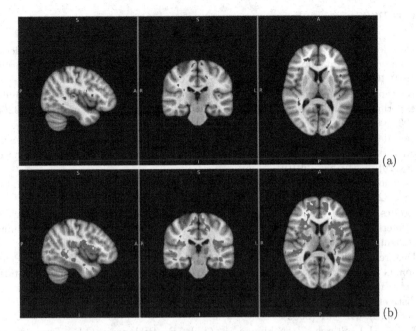

Fig. 5. Comparison of networks obtaining by thresholding background/foreground h-functions on the templates of the two types of schizophrenia patients (with and without auditory hallucinations): (a) the intersection network, (b) the network appearing only on the template of patients with hallucination (SZAH)

4 Conclusions

This paper gives further results on the discrimination of schizophrenic patients with and without auditory hallucinations from resting state fMRI data, using a lattice computing base technique for the computation of the functional networks. In this paper, the population study is performed on the average templates per population, i.e. we build average templates for healthy patients, schizophrenia patients with and without auditory hallucination. We compute the functional networks on these templates, using the central voxel of the LHG as the seed value, selecting an appropriate threshold for the h-function map obtained. The identified networks have clear differences between populations, which may allow to discriminate between them on an individual basis, using some classification approach. Further work is being directed to individual subject analysis, and future applications of classification approaches.

Acknowledgements. Ann K. Shinn from the McLean Hospital, Belmont, Massachusetts; Harvard Medical School, Boston, Massachusetts, US for providing experimental images. Support from project MINECO grant TIN2011-23823, and UFI 11/07 of the UPV/EHU. Darya Chyzhyk has a FPU grant from the spanish government.

References

1. Craddock, R., Holtzheimer III, P., Hu, X., Mayberg, H.: Disease state prediction from resting state functional connectivity. Magnetic Resonance in Medicine 62, 1619–1628 (2009)
2. Northoff, G., Duncan, N.W., Hayes, D.J.: The brain and its resting state activity–experimental and methodological implications. Progress in Neurobiology 92(4), 593–600 (2010)
3. van den Heuvel, M.P., Pol, H.E.H.: Exploring the brain network: A review on resting-state fmri functional connectivity. European Neuropsychopharmacology 20(8), 519–534 (2010)
4. Dosenbach, N.U.F., et al.: Prediction of individual brain maturity using fmri. Science 329, 1358–1361 (2010)
5. Cordes, D., Haughton, V., Carew, J.D., Arfanakis, K., Maravilla, K.: Hierarchical clustering to measure connectivity in fmri resting-state data. Magnetic Resonance Imaging 20(4), 305–317 (2002)
6. Demirci, O., Stevens, M.C., Andreasen, N.C., Michael, A., Liu, J., White, T., Pearlson, G.D., Clark, V.P., Calhoun, V.D.: Investigation of relationships between fMRI brain networks in the spectral domain using ICA and granger causality reveals distinct differences between schizophrenia patients and healthy controls. NeuroImage 46(2), 419–431 (2009)
7. Remes, J.J., Starck, T., Nikkinen, J., Ollila, E., Beckmann, C.F., Tervonen, O., Kiviniemi, V., Silven, O.: Effects of repeatability measures on results of fmri sica: A study on simulated and real resting-state effects. NeuroImage 56(2), 554–569 (2011)
8. Calhoun, V.D., Adali, T., Pearlson, G.D., Pekar, J.J.: A method for making group inferences from functional mri data using independent component analysis. Human Brain Mapping 14(3), 140–151 (2001)
9. Zou, Q.H., Zhu, C.Z., Yang, Y., Zuo, X.N., Long, X.Y., Cao, Q.J., Wang, Y.F., Zang, Y.F.: An improved approach to detection of amplitude of low-frequency fluctuation (alff) for resting-state fmri: Fractional alff. Journal of Neuroscience Methods 172(1), 137–141 (2008)
10. Pereira, F., Mitchell, T., Botvinick, M.: Machine learning classifiers and fMRI: A tutorial overview. NeuroImage 45(1, suppl. 1) S199–S209 (2009); Mathematics in Brain Imaging
11. Yao, Z., Wang, L., Lu, Q., Liu, H., Teng, G.: Regional homogeneity in depression and its relationship with separate depressive symptom clusters: A resting-state fmri study. Journal of Affective Disorders 115(3), 430–438 (2009)
12. Liu, Y., Wang, K., Yu, C., He, Y., Zhou, Y., Liang, M., Wang, L., Jiang, T.: Regional homogeneity, functional connectivity and imaging markers of alzheimer's disease: A review of resting-state fmri studies. Neuropsychologia 46(6), 1648–1656 (2008); Neuroimaging of Early Alzheimer's Disease
13. Mingoia, G., Wagner, G., Langbein, K., Scherpiet, S., Schloesser, R., Gaser, C., Sauer, H., Nenadic, I.: Altered default-mode network activity in schizophrenia: A resting state fmri study. Schizophrenia Research 117(2-3), 355–356 (2010); 2nd Biennial Schizophrenia International Research Conference
14. Zhou, Y., Liang, M., Jiang, T., Tian, L., Liu, Y., Liu, Z., Liu, H., Kuang, F.: Functional dysconnectivity of the dorsolateral prefrontal cortex in first-episode schizophrenia using resting-state fmri. Neuroscience Letters 417(3), 297–302 (2007)

15. Zhou, Y., Shu, N., Liu, Y., Song, M., Hao, Y., Liu, H., Yu, C., Liu, Z., Jiang, T.: Altered resting-state functional connectivity and anatomical connectivity of hippocampus in schizophrenia. Schizophrenia Research 100(1-3), 120–132 (2008)

16. Vercammen, A., Knegtering, H., den Boer, J., Liemburg, E.J., Aleman, A.: Auditory hallucinations in schizophrenia are associated with reduced functional connectivity of the temporo-parietal area. Biological Psychiatry 67(10), 912–918 (2010); Anhedonia in Schizophrenia

17. Graña, M., Chyzhyk, D.: Hybrid multivariate morphology using lattice auto-associative memories for resting-state fmri network discovery. In: IEEE 2012 12th International Conference on Hybrid Intelligent Systems (HIS), pp. 537–542 (2012)

18. Ritter, G.X., Sussner, P., Diaz-de-Leon, J.L.: Morphological associative memories. IEEE Transactions on Neural Networks 9(2), 281–293 (1998)

19. Ritter, G.X., Diaz-de-Leon, J.L., Sussner, P.: Morphological bidirectional associative memories. Neural Networks 12(6), 851–867 (1999)

20. Velasco-Forero, S., Angulo, J.: Supervised ordering in $\mathbb{r}^p$: Application to morphological processing of hyperspectral images. IEEE Transactions on Image Processing 20(11), 3301–3308 (2011)

21. Liu, D., Yan, C., Ren, J., Yao, L., Kiviniemi, V.J., Zang, Y.: Using coherence to measure regional homogeneity of resting-state fmri signal. Frontiers in Systems Neuroscience 4(24) (2010)

22. Beckmann, C.F., DeLuca, M., Devlin, J.T., Smith, S.M.: Investigations into resting-state connectivity using independent component analysis. Philosophical Transactions of the Royal Society of London - Series B: Biological Sciences 360(1457), 1001–1013 (2005)

23. Shinn, A.K., Baker, J.T., Cohen, B.M., Ongur, D.: Functional connectivity of left heschl's gyrus in vulnerability to auditory hallucinations in schizophrenia. Schizophrenia Research 143(2-3), 260–268 (2013)

Cocaine Dependent Classification on MRI Data Extracting Features from Voxel Based Morphometry

M. Termenon[1], Darya Chyzhyk[1], Manuel Graña[1],
A. Barros-Loscertales[2], and C. Avila[2]

[1] Grupo de Inteligencia Computacional (GIC), UPV/EHU
www.ehu.es/ccwintco
[2] Dpto. Psicologia Basica, Clinica y Psicobiologia, Universitat Jaume I,
Castellon de la Plana, Spain

Abstract. In this paper, we present a method to discriminate cocaine dependent patients and healthy subjects using features computed from structural magnetic resonance imaging (MRI). After image preprocessing, we compute voxel based morphometry (VBM) applying Gaussian smoothing with three different full width at half maximum (FWHM) kernel sizes. VBM clusters guide the feature extraction process used to classify subjects as cocaine dependent patients or healthy controls. We apply five well known classifiers from the WEKA platform. Classification results are good reaching accuracy, sensitivity and specificity values above 90%. It is possible to apreciate that as the smoothing kernel size grows, the features are less discriminative, but the VBM clusters identifying differences between both groups are bigger. We also obtain the location in the brain of the features selected and compare them with findings in the literature.

1 Introduction

This paper presents the application of feature extraction and classification algorithms for the automatic detection in T1-weigthed MRI of neuropsychiatric complications derived from the cocaine abuse. Cocaine is one of the most consumed illegal drugs and its chronic abuse may cause consequences such as ischemic, hemorrhagic strokes, depression and neuropsychological abnormalities [1]. Selected regions in the brain of cocaine consumers show functional, neurochemical and structural abnormalities that can be used to identify the differences between the brains of cocaine consumers and non-consumers [2].

Studies found structural differences in striatum (caudate and putamen), frontal gyrus, parahippocampus, posterior cingulate, amygdala, insula and cerebellum [3], ventromedial orbitofrontal, anterior cingulate, anteroventral insular and superior temporal cortices [2]. Functional MRI (fMRI) tests assert that chronic cocaine consumption may affect the attentional system in the right parietal lobe, making patients more prone to attention deficits [4] and detect

J.M. Ferrández Vicente et al. (Eds.): IWINAC 2013, Part II, LNCS 7931, pp. 140–148, 2013.

reductions in primary visual cortex and primary motor cortex after cocaine administration [5]. In white matter (WM), reduced fractional anisotropy (FA) in the genu and rostral body of the anterior corpus callosum in cocaine-dependent subjects compared to controls [6] and also lower FA and higher mean diffusivity in frontal and parietal WM regions [7]. Lim et al [8] suggested that duration of cocaine use was associated with decreased grey and white matter volumes. It has been shown [3] that cocaine dependence is associated with reduced gray matter (GM) volumes in the target structures of the dopaminergic system. They were able to demonstrate a GM reduction in the striatum by means of voxel-based morphometry in human users, thereby linking human results to animal models of addiction. A multimodal study [1] showed that cocaine dependent patients have generalized cerebral hypoperfusion.

In this paper, we experiment with VBM as feature selection pipeline. We select as features the voxel sites with p-value above a threshold. The p-values are obtained from a voxel based morphometry (VBM) [9] study. We found in the literature several studies that applied this methodology [10–14, 3] to study different neurodegenerative diseases. In this article, we want to analyze if VBM can be used as discriminative feature extraction method to differenciate cocaine dependent patients from halthy subjects. We also want to study how different smoothing kernel sizes affect the classification performance. Finally, a relevant research question is whether the location of the selected features are according to findings reported in the literature.

In Section 2, we introduce the database used in this experiment and we describe the image processing steps. Section 3 gives a description of the feature extraction process applied in this experiment. Section 4 details the different classifiers used and gives classification performance results obtained in the diverse computational experiments performed on the data. In Section 5, we report where in the brain are located the selected features. Finally, in Section 6, we give our conclusions of this work.

2   Materials

Thirty male cocaine-dependent patients (mean $= 34.41 \pm 6.62$) and thirty-five matched controls (mean $= 33.38 \pm 7.87$) participated in this study. The cocaine patients were recruited from the Addiction Treatment Service of San Agustín in Castellón, Spain. The inclusion criteria for cocaine dependence was based on the DSM-IV criteria. Control subjects were required to have no diagnosis of substance abuse or dependence. The exclusion criteria for all the participants included neurological illness, prior head trauma, positive HIV status, diabetes, Hepatitis C, or other medical illness and psychiatric disorders. Cocaine consumption was assessed with an urine toxicology test, which ensured a minimum period of abstinence of two to four days prior to MRI data acquisition. Groups were matched on the basis of age and level of education. All the participants were right-handed according to the Edinburgh Handedness Inventory [15]. They all signed an informed consent prior to participating in this study.

Images were acquired on a 1.5T Siemens Avanto (Erlangen, Germany) with a standard quadrature head coil. A high resolution 3D T1-weighted gradient echo pulse sequence was acquired (TE=4.9 ms; TR=11 ms; FOV=24 cm; matrix=256×224×176; voxel size=1×1×1).

2.1 Image Preprocessing

Preprocessing consisted of reorientation, tissue segmentation, bias correction and spatial normalization into a unified model [16] using Statistical Parametric Mapping (SPM8) running on Matlab. Reorientation was done according to the intercommissural line, AC-PC (anterior - posterior comissure) [17]. The parameter settings were: warp frequency cutoff to 25 mm, warping regularization light to 0.001, a thorough clean up of segmentations and a 1.5-mm3 voxel size resolution for normalization. Weighted hidden Markov Random Fields (HMRF) were applied to improve the accuracy of tissue segmentation.

Each brain was normalized to the tissue probability maps provided by the International Consortium for Brain Mapping (ICBM, http://www.loni.ucla.edu/Atlases/). First step consists of a linear transformation followed by nonlinear shape transformation. Segmented GM images were modulated to restore tissue volume changes after spatial normalization. Modulated GM segmented images were corrected only for nonlinear warping (http://dbm.neuro.uni-jena.de/vbm/segmentation/modulation/), making correcting for total intracranial volume of the individual unnecessary [18]. Thus, global brain volume effects were removed from the data to allow inferences on local GM volume changes.

3   VBM as Feature Extraction Process

Morphometry studies are a common tool to find brain structural differences between subjects. Voxel based morphometry [9] is a computational technique that solves the general lineal model (GLM) to obtain the effects of the categorical variable and other covariates and perform t-statistics over the GLM coefficients.

Different FWHM Gaussian kernels (σ) were applied to the segmented GM maps obtaining different results. We test this approach for smoothing kernels of FWHM set to 0, 3 and 6. Given the VBM results for each smooth kernel, we apply different p-values to select the voxels with most significative differences. We create a mask and apply it to all subjects after non linear registration (without smoothing). Afterwards, the features for each individual MRI data consists of a vector with the GM mask's voxel values. The selection of voxel sites for feature extraction depends on the smoothing kernel width.

In Table 1, we show the number of features that corresponds to each percentile. Each percentile corresponds to each mask created from the VBM results.

4   Classification Results

Waikato Environment for Knowledge Analysis (WEKA) is a set of machine learning algorithms for data mining analysis. WEKA is open source software

Table 1. Number of features corresponding to applied percentiles depending of the different FWHM Gaussian smoothing kernels

Percent.	# f($\sigma = 0$)	# f($\sigma = 3$)	# f ($\sigma = 6$)
0.01000	2,228	5,127	7,044
0.00750	1,667	3,854	5,153
0.00500	1,126	2,468	3,164
0.00250	557	1,132	1,268
0.00100	217	436	373
0.00075	168	322	322
0.00050	100	216	134

which contains tools for data pre-processing, classification, regression, clustering, association rules, and visualization [19, 20]. In our case WEKA permits to realize rapid experimentation with classification on MRI dataset of cocaine-dependent patients and healthy people. WEKA is available at http://www.cs.waikato.ac.nz/ml/weka/. Among the variety of classifiers included in WEKA, we chose groups of classifiers with different fundamental setup: support vector machines (SVM) [21] which is accepted as the standard classification tool, radial basis function (RBF) networks [22], Bayesian logistic regression (BLR) [23], naive Bayes (NB) [24] and random forest (RF) [25].

Due to the small number of data samples, we use leave-one-out cross-validation (LOO-CV) procedure. LOO-CV is a special case of k-fold cross-validation where k equals the number of instances in the data. In the case of cocaine dependent dataset, $k = 65$ for leave-one-out method. Classifiers parameters used for this experiment are presented in Table 2.

Table 2. Classifier's parameters

Classifier	Parameter	Value
SVM	c	1.0
	epsilon	1.0E-12
	filterType	Normalize training data
	Kernel	PolyKernel -C 250007 -E 1.0
	toleranceParameter	0.001
RBF	clusteringSeed	1
	minStdDev	0.1
	numClusters	2
	ridge	1.0E-8
Logistic	ridge	1.0E-8
NB	displayModelingOldFormat	False
	useKernelEstimator	False
	useSupervisedDiscretization	False
RF	maxDepth	0
	numTrees	10
	seed	1

To evaluate the correctness of classifiers and to compare them, we compute $Accuracy = \frac{TP+TN}{TP+TN+FP+FN}$, $Sensitivity = \frac{TP}{TP+FN}$ and $Specificity = \frac{TN}{TN+FP}$. As null hypothesis, we consider that the subject in study is a cocaine dependent patient, therefore, in these expressions, true positives (TP) are the number of diseased patient volumes correctly classified; true negatives (TN) are the number of control volumes correctly classified; false positives (FP) are the number of control volumes classified as diseased patients; false negatives (FN) are the number of diseased patient volumes classified as control subjects. These measures take values in the interval $[0, 100]$, where values close to to 100 indicate better performance.

The classification experiments based on VBM features with diverse smoothing parameters provide paradoxical results from the point of view of inference

Table 3. Classification results for the three different kernels and five classifiers. Acc(Sens)(Spec) means Accuracy, Sensitivity and Specificity. First table corresponds to results when no smoothing is applied ($\sigma = 0$). Middle table when $\sigma = 3$. Last table when $\sigma = 6$.

$\sigma = 0$	SVM	RBF	BLR	NB	RF
Percent.	Acc(Sens)(Spec)	Acc(Sens)(Spec)	Acc(Sens)(Spec)	Acc(Sens)(Spec)	Acc(Sens)(Spec)
0.01000	83.10(86.70)(80.00)	89.20(90.00)(88.60)	87.70(83.30)(91.40)	**90.80(90.00)(91.40)**	75.40(66.70)(82.90)
0.00750	84.60(80.00)(88.60)	**90.80(90.00)(91.40)**	81.50(76.70)(85.70)	89.20(90.00)(88.60)	80.00(70.00)(88.60)
0.00500	87.70(90.00)(85.70)	**92.30(96.70)(88.60)**	80.00(70.00)(88.60)	90.80(93.30)(88.60)	78.50(70.00)(85.70)
0.00250	83.10(80.00)(85.70)	**90.80(90.00)(91.40)**	73.80(66.70)(80.00)	86.20(86.70)(85.70)	78.50(66.70)(88.60)
0.00100	83.10(76.70)(88.60)	83.10(80.00)(85.70)	70.80(60.00)(80.00)	83.10(80.00)(85.70)	75.40(60.00)(88.60)
0.00075	81.50(73.30)(88.60)	81.50(76.70)(85.70)	67.70(60.00)(80.00)	84.60(83.30)(85.70)	67.70(50.00)(82.90)
0.00050	80.00(73.30)(85.70)	78.50(73.30)(82.90)	67.70(60.00)(80.00)	84.60(83.30)(85.70)	61.50(40.00)(80.00)

$\sigma = 3$	SVM	RBF	BLR	NB	RF
Percent.	Acc(Sens)(Spec)	Acc(Sens)(Spec)	Acc(Sens)(Spec)	Acc(Sens)(Spec)	Acc(Sens)(Spec)
0.01000	83.10(80.00)(85.70)	86.20(86.70)(85.70)	75.40(76.70)(74.30)	86.20(90.00)(82.90)	72.30(56.70)(85.70)
0.00750	86.20(86.70)(85.70)	**90.80(93.30)(88.60)**	78.50(80.00)(77.10)	89.20(93.30)(85.70)	81.50(80.00)(82.90)
0.00500	83.10(80.00)(85.70)	84.60(83.30)(85.70)	76.90(76.70)(77.10)	86.20(86.70)(85.70)	73.80(63.30)(82.90)
0.00250	83.10(80.00)(85.70)	83.10(80.00)(85.70)	73.80(66.70)(80.00)	84.60(83.30)(85.70)	75.40(66.70)(82.90)
0.00100	83.10(80.00)(85.70)	86.20(80.00)(91.40)	66.20(63.30)(68.60)	84.60(80.00)(88.60)	61.50(50.00)(71.40)
0.00075	83.10(80.00)(85.70)	86.20(80.00)(91.40)	67.70(63.30)(71.40)	86.20(76.70)(94.30)	64.60(50.00)(77.10)
0.00050	80.00(73.30)(85.70)	81.50(73.30)(88.60)	63.10(63.30)(62.90)	86.20(76.70)(94.30)	75.40(63.30)(85.70)

$\sigma = 6$	SVM	RBF	BLR	NB	RF
Percent.	Acc(Sens)(Spec)	Acc(Sens)(Spec)	Acc(Sens)(Spec)	Acc(Sens)(Spec)	Acc(Sens)(Spec)
0.01000	83.10(76.70)(88.60)	75.40(60.00)(88.60)	78.50(80.00)(77.10)	81.50(76.70)(85.70)	70.80(66.70)(74.30)
0.00750	84.60(80.00)(88.60)	83.10(73.30)(91.40)	76.90(76.70)(77.10)	81.50(76.70)(85.70)	70.80(56.70)(82.90)
0.00500	78.50(76.70)(80.00)	78.50(70.00)(85.70)	72.30(70.00)(74.30)	76.90(70.00)(82.90)	67.70(66.70)(68.60)
0.00250	83.10(80.00)(85.70)	75.40(53.30)(94.30)	66.20(56.70)(74.30)	78.50(56.70)(97.10)	70.80(60.00)(80.00)
0.00100	81.50(83.30)(80.00)	70.80(46.70)(91.40)	69.20(60.00)(77.10)	80.00(63.30)(94.30)	70.80(60.00)(80.00)
0.00075	80.00(80.00)(80.00)	67.70(40.00)(91.40)	66.20(60.00)(71.40)	75.40(53.30)(94.30)	64.60(40.00)(85.70)
0.00050	78.50(73.30)(82.90)	63.10(30.00)(91.40)	66.20(60.00)(71.40)	70.80(43.30)(94.30)	64.60(50.00)(77.10)

performed on statistical parametric maps. The classification accuracy decreases with the size of the smoothing kernel. In Table 3, we present the classification results achieved with VBM features at different values of the smoothing kernel size. The size of the feature set is determined by the p-value applied to obtain the VBM mask. It can be appreciated a systematic decrease in all accuracy values. These results exhibit one of the most salient differences between the inference and the machine learning approaches to the analysis of neuroimage data, classification accuracy is not dependent on the size of the effect. In fact, best classification results are obtained with the smaller sets of features and the smaller σ value.

5   Feature Location in the Brain

To obtain the location of features, we used the AtlasQuery tool of FSL, analyzing three different atlases: MNI structural atlas and Harvard-Oxford cortical and subcortical atlas. Most significant voxels for FWHM Gaussian kernels $\sigma = 0$, $\sigma = 3$ and $\sigma = 6$ and p-value=0.01 are shown in red on MNI template (Figure 1).

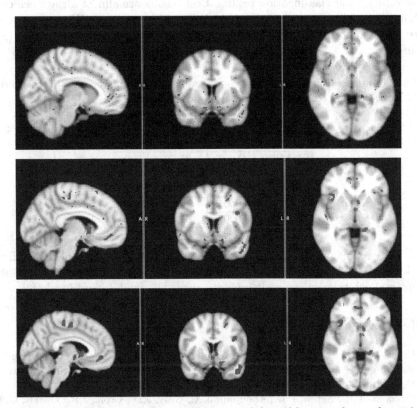

Fig. 1. Top: VBM Most significant voxels for smooth kernel $k = 0$ and $p-value = 0.01$. Middle: VBM Most significant voxels for smooth kernel $k = 3$ and $p-value = 0.01$. Bottom: VBM Most significant voxels for smooth kernel $k = 6$ and $p-value = 0.01$.

These significant voxels were located in the cerebral cortex (insular and frontal orbital), caudate, insula, precentral and paracingulate gyrus. These findings are in agreement with previous literature [2], [3].

6   Conclusions

In this paper, we present a method to discriminate cocaine dependent patients and healthy subjects using structural magnetic resonance imaging (MRI) volumes. Preprocessing the images ensures the anatomical correspondence between voxels across subject volumes. In this experiment, features are selected applying a p-value to results obtained from a VBM study. Different p-values and smooth kernels are considered in this process for posterior classification step. We test five well known classifiers in WEKA platform using LOO-CV validation methodology.

In Table 4, we show best accuracy results achieved for each procedure, showing also their associated area under ROC value. In case of tie, result shown in table is the one with less number of features used to classify. In general, as higher the σ value, worst classification results. Best results are almost always achieved without smoothing, best accuracy is 92.30% achieved with RBF network when there is no smooth Gaussian kernel applied and with a p-value of 0.001 (217 features).

Table 4. Best classification results for each classifier, indicating the smooth (sm) kernel applied, the p-value, accuracy and area under ROC

		VBM		
	sm	p-value	Accuracy	ROC
SVM	0	0.001	87.70	0.88
RBF	0	0.001	**92.30**	**0.93**
BLR	0	0.00050	87.70	0.93
NB	0	0.00050	**90.80**	**0.98**
RF	3	0.00075	81.50	0.87

We also report where in the brain are located the features obtained from VBM analysis. Brain regions are according to findings in the literature: insular and frontal orbital cortices and cingulate gyrus [2], caudate and insula [3].

References

1. Ernst, T., Chang, L., Oropilla, G., Gustavson, A., Speck, O.: Cerebral perfusion abnormalities in abstinent cocaine abusers: a perfusion MRI and SPECT study. Psychiatry Research: Neuroimaging 99(2), 63–74 (2000)
2. Franklin, T.R., Acton, P.D., Maldjian, J.A., Gray, J.D., Croft, J.R., Dackis, C.A., O'Brien, C.P., Childress, A.R.: Decreased gray matter concentration in the insular, orbitofrontal, cingulate, and temporal cortices of cocaine patients. Biological Psychiatry 51(2), 134–142 (2002) PMID: 11822992

3. Barrós-Loscertales, A., Garavan, H., Bustamante, J.C., Ventura-Campos, N., Llopis, J.J., Belloch, V., Parcet, M.A., Ávila, C.: Reduced striatal volume in cocaine-dependent patients. NeuroImage 56(3), 1021–1026 (2011)
4. Bustamante, J.C., Barrós-Loscertales, A., Ventura-Campos, N., Sanjún, A., Llopis, J.J., Parcet, M.A., Ávila, C.: Right parietal hypoactivation in a cocaine-dependent group during a verbal working memory task. Brain Research 1375, 111–119 (2011)
5. Li, S., Biswal, B., Li, Z., Risinger, R., Rainey, C., Cho, J., Salmeron, B.J., Stein, E.A.: Cocaine administration decreases functional connectivity in human primary visual and motor cortex as detected by functional MRI. Magnetic Resonance in Medicine 43(1), 45–51 (2000)
6. Moeller, F.G., Hasan, K.M., Steinberg, J.L., Kramer, L.A., Dougherty, D.M., Santos, R.M., Valdes, I., Swann, A.C., Barratt, E.S., Narayana, P.A.: Reduced anterior corpus callosum white matter integrity is related to increased impulsivity and reduced discriminability in Cocaine-Dependent subjects: Diffusion tensor imaging. Neuropsychopharmacology 30(3), 610–617 (2004)
7. Lane, S.D., Steinberg, J.L., Ma, L., Hasan, K.M., Kramer, L.A., Zuniga, E.A., Narayana, P.A., Moeller, F.G.: Diffusion tensor imaging and decision making in cocaine dependence. PloS One 5(7), e11591 (2010) PMID: 20661285
8. Lim, K.O., Wozniak, J.R., Mueller, B.A., Franc, D.T., Specker, S.M., Rodriguez, C.P., Silverman, A.B., Rotrosen, J.P.: Brain macrostructural and microstructural abnormalities in cocaine dependence. Drug and Alcohol Dependence 92(1-3), 164–172 (2008) PMID: 17904770 PMCID: 2693223
9. Ashburner, J., Friston, K.J.: Voxel-based Morphometry–The methods. NeuroImage 11(6), 805–821 (2000)
10. Hämäläinen, A., Tervo, S., Grau-Olivares, M., Niskanen, E., Pennanen, C., Huuskonen, J., Kivipelto, M., Hänninen, T., Tapiola, M., Vanhanen, M., Hallikainen, M., Helkala, E.L., Nissinen, A., Vanninen, R., Soininen, H.: Voxel-based morphometry to detect brain atrophy in progressive mild cognitive impairment. NeuroImage 37(4), 1122–1131 (2007)
11. Trivedi, M.A., Wichmann, A.K., Torgerson, B.M., Ward, M.A., Schmitz, T.W., Ries, M.L., Koscik, R.L., Asthana, S., Johnson, S.C.: Structural MRI discriminates individuals with mild cognitive impairment from age-matched controls: A combined neuropsychological and voxel based morphometry study. Alzheimer's & Dementia 2(4), 296–302 (2006)
12. Savio, A., García-Sebastián, M., Hernández, C., Graña, M., Villanúa, J.: Classification results of artificial neural networks for alzheimer's disease detection. In: Corchado, E., Yin, H. (eds.) IDEAL 2009. LNCS, vol. 5788, pp. 641–648. Springer, Heidelberg (2009)
13. Cousijn, J., Wiers, R.W., Ridderinkhof, K.R., van den Brink, W., Veltman, D.J., Goudriaan, A.E.: Grey matter alterations associated with cannabis use: Results of a VBM study in heavy cannabis users and healthy controls. NeuroImage 59(4), 3845–3851 (2012)
14. Geva, S., Baron, J.C., Jones, P.S., Price, C.J., Warburton, E.A.: A comparison of VLSM and VBM in a cohort of patients with post-stroke aphasia. NeuroImage: Clinical 1(1), 37–47 (2012)
15. Oldfield, R.C.: The assessment and analysis of handedness: the edinburgh inventory. Neuropsychologia 9(1), 97–113 (1971) PMID: 5146491
16. Ashburner, J., Friston, K.J.: Unified segmentation. NeuroImage 26(3), 839–851 (2005) PMID: 15955494
17. Talairach, J., Tournoux, P.: Co-Planar Stereotaxic Atlas of the Human Brain: 3-D Proportional System: An Approach to Cerebral Imaging. Thieme (January 1988)

18. Scorzin, J.E., Kaaden, S., Quesada, C.M., Müller, C., Fimmers, R., Urbach, H., Schramm, J.: Volume determination of amygdala and hippocampus at 1.5 and 3.0T MRI in temporal lobe epilepsy. Epilepsy Research 82(1), 29–37 (2008) PMID: 18691850
19. Hall, M., Frank, E., Holmes, G., Pfahringer, B., Reutemann, P., Witten, I.H.: The weka data mining software: an update. SIGKDD Explorations 11(1), 10–18 (2009)
20. Garner, S.R., Cunningham, S.J., Holmes, G., Nevill-Manning, C.G., Witten, I.H.: Applying a machine learning workbench: Experience with agricultural databases. In: Machine Learning in Practice Workshop, Machine Learning Conference, Tahoe City, CA, USA, pp. 14–21 (1995)
21. Vapnik, V.N.: Statistical Learning Theory. Wiley-Interscience (September 1998)
22. Buhmann, M.D.: Radial Basis Functions: Theory and Implementations. Cambridge University Press (July 2003)
23. Bishop, C.M.: Pattern Recognition and Machine Learning. Information Science and Statistics (2006)
24. John, G.H., Langley, P.: Estimation continuous distribution in bayesian classifiers. In: Proceedings of the Eleventh Conference on Uncertainty in Artificial Intelligence, pp. 338–345 (1995)
25. Breiman, L.: Random forests. Machine Learning 45(1), 5–32 (2001)

A Data Fusion Perspective
on Human Motion Analysis
Including Multiple Camera Applications

Rodrigo Cilla, Miguel A. Patricio, Antonio Berlanga, and José M. Molina

Computer Science Department. Universidad Carlos III de Madrid
Avda. de la Universidad Carlos III, 22
28270 Colmenarejo (Madrid). Spain
{rcilla,mpatrici}@inf.uc3m.es, {aberlan,molina}@ia.uc3m.es

Abstract. Human motion analysis methods have received increasing attention during the last two decades. In parallel, data fusion technologies have emerged as a powerful tool for the estimation of properties of objects in the real world. This papers presents a view of human motion analysis from the viewpoint of data fusion. JDL process model and Dasarathy's input-output hierarchy are employed to categorize the works in the area. A survey of the literature in human motion analysis from multiple cameras is included. Future research directions in the area are identified after this review.

Keywords: Human Action Recognition, Data Fusion, Computer Vision.

1 Introduction

The recognition of human movements [1] has been studied by the computer vision community for more than twenty years. The developments made during this period have enabled the creation of multiple systems. Automatic Surveillance [2], Ambient Intelligence [8] or Human Computer Interaction [5] are some of them. Abnormal behavior detection is employed in Video Surveillance Systems to detect suspicious behaviors that might be assessed as a thread. Smart home environments analyze actions and mood of the inhabitants to adapt the environment to their preferences, changing music or lighting conditions to make it more comfortable. Commercial gaming platforms employ advanced sensors to capture the real movements of the players, providing an enhanced and more realistic experience.

The aim of human movement analysis systems is to transform the pixel intensities in the input video sequences into a semantic intepretation of them. The interpretation might be defined at different knowledge levels. Aggarwal and Cai [1] propose a hierarchy of *gestures, actions, interactions* and *group activities*. *Gestures* are dened as elementary movements of a persons body part, and are the atomic components describing the meaningful motion of a person. *Actions* are dened as single-person activities that may be composed of multiple gestures

J.M. Ferrández Vicente et al. (Eds.): IWINAC 2013, Part II, LNCS 7931, pp. 149–158, 2013.

organized temporally. *Interactions* are human activities that involve two or more persons and/or objects. *Group activities* are dened as the activities performed by conceptual groups composed of multiple persons and/or objects. These levels should not be interpreted as closed sets, as many times it is not clear at what level operates a given system.

The first human motion analysis systems developed where limited to the usage of a single camera view. However, in recent years, with the aim of deploying human movement analysis systems in the real world, human movement analysis systems have incorporated multiple camera views, as they provide different advantages:

- Viewpoint invariance. The appearance of actions changes according to the orientation in the execution in the action with respect to the camera. Thus, employing multiple views provides complementary information to achieve a more robust recognition.
- Robustness towards occlusions. In real environments there is usually multiple furniture, walls or other objects that produce partial occlusions in the observed target. The way to overcome this limitation and not loss important motion information is to observe the scene from multiple viewpoints.
- Wider scene coverage. A single camera has a very limited coverage. Multiple cameras are needed to cover full scenes.

Data Fusion studies the efficient combination of measurements obtained from multiple sensors or, alternatively, the temporal measurements obtained from a single sensor, in order to achieve more specific inferences about the state of one or more entities than the ones that could be achieved by using a single, independent, sensor [14]. Human movement analysis systems are covered by this definition, independently of the number of cameras employed and the level of abstraction where the analysis is made. However, to the best of our knowledge, the recognition of human movement has not been studied from the viewpoint of data fusion. The purpose of this paper is to analyze human movement analysis applications from the viewpoint of data fusion.

1.1 Contributions

The contributions of this paper might be summarized as:

- A review of relevant data fusion concepts and frameworks.
- A characterization of Human Action Recognition systems from the viewpoint of the JDL process model.
- A survey of the literature of human action recognition from multiple cameras employing the taxonomy provided by Dasarathy's input-output framework.

1.2 Paper Organization

Paper is organized as follows. Section 2 presents the main concepts and frameworks developed by the data fusion community. Section 3 studies the relationship between the different data fusion levels and human action recognition. Section 4 surveys the area of human action recognition from multiple cameras. Section 5 concludes the paper discusing about hypothetical ways of colaboration between data fusion and human action recognition communities

2 Data Fusion

The Joint Directors of Laboratories Data Fusion Working Group currently defines Data Fusion as *The process of combining data or information to estimate or predict entity states* [23]. This definition is generic enough to cover a wide range of data association and combination problems appearing on different domains. Data fusion is not a discipline by itself, nor the combination of signal processing, artificial intelligence, estatistica estimation or systems enginering to solve state estimation problems.

Different frameworks have been developed to categorize data fusion systems: the JDL process model and Dasarathy's input-output model. These are complementary frameworks for the analysis of data fusion systems whose usage is widely extended. Both are introduced in next paragraphs an will be later employed for the analysis of human movement analysis systems.

2.1 The JDL Process Model

The JDL Data Fusion model [25] is the most widely used framework for the categorization of data fusion systems and algorithms. The first version was published in 1985 by the US Joint Directors of Laboratories (JDL) Data Fusion Working Group with the aim of providing a common framework to facilitate the communication between the communication between data fusion stakeholders and provide a conceptual framework for new developments. The JDL model is not an architectural paradigm nor a process model for the creation of data fusion system. Instead, it provides different levels of abstraction where the different algorithms employed in data fusion systems might be accommodated according to the kind of processing they perform.

The JDL data fusion model, after the 1998 revision [23], proposes five different levels of abstraction where the data fusion functions are accommodated (figure 1. These levels are:

- Level 0. *Signal/Feature Assessment.* This level includes the algorithms employed to enhance or combine the input signals of the fusion systems. The inferences made at this level do not make any assumption about the causes originating the signals. Typical operations at this level include spatial and temporal data alignment, data standardization and data preconditioning for bias removal.
- Level 1. *Entity Assessment.* Algorithms employed for the estimation of the current state of a individual entities are defined at this level. This includes target detection, classification, location, tracking and identity estimation.

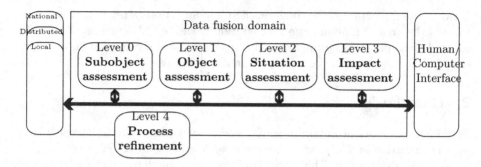

Fig. 1. The JDL data fusion model (1998 revision)

Processing at this level usually implies the association of observations to the corresponding responsible targets.

- Level 2. *Situation Assessment.* A situation is *a set of entities, their attributes, and relationships.* Thus, the task at this processing level is to infer the existent relationships between the analyzed entities employing the individual state estimations.
- Level 3. *Impact Assessment.* The purpose of the algorithms defined at this level is to predict future situations derived from the current and past inferred situations. This includes the computation of expected outcomes of actions executed to alter the current situation or the projection of the current situation to the future to predict the possible evolution.
- Level 4. *Process Assessment.* This level includes the algorithms employed to measure the real-time performance of the fusion system and improve it. This includes the reconfiguration of the sensors employed or the replacement of data fusion algorithms by others better adapted to the current or expected scenario.

2.2   Dasarathy's Input-Output Model

Dasarathy proposed an alternative categorization of Data Fusion systems according to the level of abstraction of the information at the input and output of the fusion system [6]. Three different levels of abstraction are defined: (1) *data*; (2) *features* and (3) *decisions*. Data is the lowest level of abstraction, corresponding to the raw measurements of the sensors, such pixel intensities or depth information. Features are transformations of the data to enhance some property such edges or curvature. Finally, decisions encode information about the certainty of a fact, in the form, among others, of probability estimates or fuzzy sets.

Data fusion systems are characterized according to this abstraction of their inputs and outputs as follows (figure 2):

- Data in-Data out (DAI-DAO) Fusion. At the lowest level of abstraction are systems processing *data* and generating *data.* An example of this kind of fusion systems are multispectral imaging devices: pixel intensities are captured

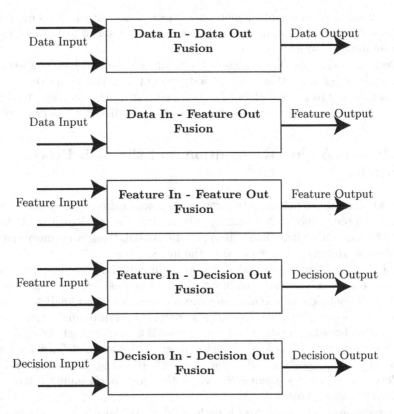

Fig. 2. Dasarathy Input-Output model

at different wavelengths to compose an image better describing the reality. High Dynamic Range (HDR) imaging is another example of a DAI-DAO fusion system, combining images taken with different exposition configurations to have a better representation of the details of dark and light regions of the scene.

- Data in- Feature Out (DAI-FEO) Fusion. At the next level of abstraction in the hierarchy are the systems processing *data* to generate *ig*features. Stereo vision systems are located at this level, as they compute disparity maps (*features*) from pixel intensities (*data*).
- Feature in-Feature Out (FEI-FEO) Fusion. At the mid level of the hierarchy are located systems processing *features*. The conceptually simpler are those generating *features* too. Due to the vague definition of what is a feature at this category lie a wide variety of systems. Fusion systems combining the measurements of the same state variable to provide a more robust estimation of the real value belong to this category.
- Feature In-Decision Out (FEI-DEO) fusion. The next abstraction level is related to pattern recognition systems, transforming *features* into *decisions*

about the class of the phenomena being recognized. At this level are defined those data fusion systems based on introducing a set of features computed from multiple sources into a classifier.

- Decision In-Decision Out (DEI-DEO) fusion. The highest level of abstraction includes the system that combine independent decisions about the phenomena to study to make a global decision about it. Decisions might be defined in different forms, such crisp values, probabilistic distributions or fuzzy sets.

3   Human Action Recognition and the JDL Process Model

This section analyzes human action recognition applications from the view point of the JDL process model. Next paragraphs analyze the relationship of JDL with different human movement analysis applications. JDL levels are confronted with the different abstractions presented at the introduction.

At JDL level 0 are image and video processing methods auxiliarly employed to enhance specific properties of input video sequences, but its definition does not allow to include any specificic method for human motion analysis.

Human movement analysis algorithms analyzing gestures and actions are defined at JDL level 1. The state variable to infer is a label characterizing the kind of action or gesture. This level contains most of the works defined for human motion analysis, as *gesture* and *action* are the better studied abstraction levels. JDL level 1 also includes *group activities*, as the group performing the movement is considered as a whole.

The recognition of *interactions* is performed at JDL level 2. Interactions might be human-human or human-object.

Level 3 in Human Action Recognition corresponds to the prediction of the future actions that person is going to do. However, to the best of our knowledge no applications at this level have been defined. The plan recognition problem [13], where the objective is to infer what is goal of an observed agent would be the closer sample to this level.

Levels 4 and 5 of the JDL process models have not been very exploited from human the human action recognition perspective. Level 4 studies how the information is presented to the system operator. Commercial video surveillance applications incorporate this capabilities, incorporating semantic information in the reports. Commercial gaming platforms with visual inputs represent the motions performed by the player with avatars. Fitness trainers represent with them how the player is performing a given exercise and how they should do it, in order to correct their performance and prevent hurts.

Level 5 would study the adaption of the algorithms employed to new conditions of the environment, such lighting or occlusions. However, to the best of our knowledge, no works have been reported proposing such applications.

4   Human Action Recognition from Multiple Cameras and Dasarathy's Input-Output Model

Dasarathy's input-output model introduced in section 3 provides a framework to categorize the works in Human Action Recognition employing multiple views of the scene being analyzed.

Human Action Recognition methods employing multiple cameras are defined at FEI-FEO, FEI-DEO and DEI-DEO levels. Although fusion at the data levels might be employed for human action recognition, they are not considered, as this kind of fusion is independent of the higher level task.

Diverse methods have been defined at the FEI-FEO data fusion level to combine the information obtained from multiple cameras. Different strategies have been defined at this level. It is possible to divide this works in three different categories: (1) methods projecting 2D features to 3D; (2) methods combining features in a subspace; (3) methods selecting the best available view.

Different 3D representation might be obtained from projecting 2D features to 3D. A popular approach is to recover the 3D shape projecting 2D silhouettes and recovering the visual hull[7,18,17]. Visual hull reconstruction requires accurate silhouette segmentation at the different available views. Recent works have proposed alternatives based on the projection of optical flow to 3D [9], or the projection of local interest points [10]. Other works recover the 3D star skeleton by the correspondence of the corresponding 2D skeletons [3]. The correspondence between action sketches might be computed from multiple views [27]. The main drawback of all these approaches is that they need from accurate camera calibration parameters to perform the projection of the features in 3D.

Alternative methods compute features for the 2D views available and combine them employing some simple scheme. The averaging of the multiple features representing pose, global and local motion has been proposed improving the results with respect to other alternatives [15]. A joint Bag-of-Words histogram might be constructed with the local feature descriptors obtained for each one of the views [26], but a higher performance is obtained with other fusion strategies. Projections maximizing the cross-covariance between the $\mathcal{R}$-transform derivatives computed at each view have been defined to learn a joint subspace where the action recognition is performed [12]. Two level Linear Discriminant Analysis is employed to learn silhouette projections maximizing the separability of the action classes [11]. Cilla et al. proposed variations of Canonical Correlation Analysis to perform the fusion of the diferent motion descriptors computed from the different views [4]. All this methods provide more flexible solutions for the combination of the features obtained from multiple cameras. However, the experimental results show a lower performance than the methods based on 3D reconstruction.

The last class of methods is based on computing a measurement of the quality of each view available, in order to select the best and perform the recognition with the data from that view. A first approach to the selection of the best view is made estimating the orientation of the human with respect to the camera [21]. A measurement based on the properties of the silhouette has been proposed [15].

Other proposed measure in the case of employing local features is to choose the camera with the highest number of detections [26]. Different utility measures have been proposed studying the saliency, concavity or variations of silhouette stacks [20]. The main drawback of this approaches is that they do not exploit the complementary information that might be present at each view.

The next category of works examined employing multiple views of the scene for the recognition of human actions are those defined at the FEI-DEO level. This works model the existing correlations among the multiple observations in the structure of the classifier employed for the prediction of the actions. The concatenation of the input features is the most straightforward procedure to perform the fusion [26,15]. The Fused HMM [24] proposes to model correlations among observations coupling the values of the hidden state chains of parallel HMMs defined for each view. Histograms of local features have been fused rotating the ordering of the inputs to account for the variations in the orientation of the inputs [22]. The main drawback of this works is their lack of flexibility, assuming that the camera configurations remain unchanged between train and test steps. A procedure for the alignment of camera views where the configuration changes from train to test steps is defined in [19], but requiring the knowledge of relative camera placement.

The last category of works employing multiple views performs the fusion at the DEI-DEO level, combining the outputs of action classifiers applied to each one of the camera views. Majority voting has been the most common technique for the fusion of decisions [15,16]. A weighted voting strategy has been proposed in [28], correcting each vote according to the value of the observed feature. Cilla et al. [5] have proposed to learn an error model to weight the predictions made from the different cameras, improving the overall result.

5   Conclusions

This work has analyzed human movement understanding applications employing data fusion concepts and frameworks. The different levels of the JDL process model have been compared to the different steps needed to perform human action recognition. It has been shown that most of the human action recognition algorithms are defined at JDL level 1. At level 2 are defined algorithms studying interactions. Other levels have not been really exploited and they should be targets of future research.

Dasarathy's Input-Ouput hierarchy has been employed to categorize multi-camera human action recognition applications. Existing works have been categorized under three conceptual classes according to the data abstractions employed.

It is clear from this work the existing relationships between data fusion and human movement analysis. However, human movement analysis applications have not been developed according to data fusion practices. Future works will have to exploit these potential sinergies to improve human movement analysis systems.

References

1. Aggarwal, J.K., Ryoo, M.S.: Human activity analysis. ACM Computing Surveys 43(3), 1–43 (2011)
2. Castanedo, F., Gomez-Romero, J., Patricio, M.A., Garcia, J., Molina, J.M.: Distributed data and information fusion in visual sensor networks. In: Distributed Data Fusion for Network-Centric Operations, p. 435 (2012)
3. Chen, D., Chou, P.C., Fookes, C.B.: Multi-view human pose estimation using modified five-point skeleton model, pp. 17–19 (2008)
4. Cilla, R., Patricio, M.A., Berlanga, A., Molina, J.M.: Multicamera action recognition with canonical correlation analysis and discriminative sequence classification. In: Ferrández, J.M., Álvarez Sánchez, J.R., de la Paz, F., Toledo, F.J. (eds.) IWINAC 2011, Part I. LNCS, vol. 6686, pp. 491–500. Springer, Heidelberg (2011)
5. Cilla, R., Patricio, M.A., Berlanga, A., Molina, J.M.: A probabilistic, discriminative and distributed system for the recognition of human actions from multiple views. Neurocomputing 75(1), 78–87 (2012)
6. Dasarathy, B.V.: Sensor fusion potential exploitation-innovative architectures and illustrative applications. Proceedings of the IEEE 85(1), 24–38 (1997)
7. Gkalelis, N., Kim, H., Hilton, A., Nikolaidis, N., Pitas, I.: The i3DPost Multi-View and 3D Human Action/Interaction Database. In: 2009 Conference for Visual Media Production, pp. 159–168 (November 2009)
8. Gómez-Romero, J., Serrano, M.A., Patricio, M.A., García, J., Molina, J.M.: Context-based scene recognition from visual data in smart homes: an information fusion approach. Personal and Ubiquitous Computing, 1–23 (2011)
9. Holte, M.B., Chakraborty, B.: A Local 3D Motion Descriptor for Multi-View Human Action Recognition from 4D Spatio-Temporal Interest Points, vol. (c), pp. 1–13 (2011)
10. Holte, M.B., Moeslund, T.B., Nikolaidis, N., Pitas, I.: 3D Human Action Recognition for Multi-view Camera Systems. In: 2011 International Conference on 3D Imaging, Modeling, Processing, Visualization and Transmission, pp. 342–349 (May 2011)
11. Iosifidis, A., Tefas, A., Nikolaidis, N., Pitas, I.: Multi-view human movement recognition based on fuzzy distances and linear discriminant analysis. Computer Vision and Image Understanding 116(3), 347–360 (2012)
12. Karthikeyan, S., Gaur, U., Manjunath, B.S., Grafton, S.: Probabilistic subspace-based learning of shape dynamics modes for multi-view action recognition. In: 2011 IEEE International Conference on Computer Vision Workshops, ICCV Workshops, pp. 1282–1286 (November 2011)
13. Kautz, H., Allen, J.F.: Generalized plan recognition. In: Proceedings of the Fifth National Conference on Artificial Intelligence, Philadelphia, PA, vol. 19, p. 86 (1986)
14. Liggins, M.E., Hall, D.L., Llinas, J.: Handbook of multisensor data fusion: theory and practice, vol. 22. CRC (2008)
15. Määttä, T., Aghajan, H.: On efficient use of multi-view data for activity recognition, pp. 158–165 (2010)
16. Naiel, M.A., Abdelwahab, M.M.: Multi-view Human Action Recognition System Employing 2DPCA Motaz El-Saban, pp. 270–275 (2010)
17. Pehlivan, S., Duygulu, P.: A new pose-based representation for recognizing actions from multiple cameras. Computer Vision and Image Understanding 115(2), 140–151 (2011)

18. Peng, B., Qian, G., Rajko, S.: View-invariant full-body gesture recognition via multilinear analysis of voxel data. In: 2009 Third ACM/IEEE International Conference on Distributed Smart Cameras, ICDSC, pp. 1–8 (August 2009)
19. Ramagiri, S., Kavi, R., Kulathumani, V.: Real-time multi-view human action recognition using a wireless camera network. In: 2011 Fifth ACM/IEEE International Conference on Distributed Smart Cameras, pp. 1–6 (August 2011)
20. Rudoy, D., Zelnik-Manor, L.: Viewpoint Selection for Human Actions. International Journal of Computer Vision 97(3), 243–254 (2011)
21. Shen, C., Zhang, C., Fels, S.: A Multi-Camera Surveillance System that Estimates Quality-of-View Measurement. In: 2007 IEEE International Conference on Image Processing, pp. III-193–III-196 (2007)
22. Srivastava, G., Iwaki, H., Park, J., Kak, A.C.: Distributed and lightweight multi-camera human activity classification. In: 2009 Third ACM/IEEE International Conference on Distributed Smart Cameras, ICDSC, pp. 1–8 (August 2009)
23. Steinberg, A.N., Bowman, C.L., White, F.E.: Revisions to the JDL data fusion model. American Inst. of Aeronautics and Astronautics, New York (1998)
24. Wang, Y., Huang, K., Tan, T.: Multi-view Gymnastic Activity Recognition with Fused HMM, pp. 667–677 (2007)
25. White, F., et al.: A model for data fusion. In: Proc. 1st National Symposium on Sensor Fusion, vol. 2, pp. 149–158 (1988)
26. Wu, C., Khalili, A.H., Aghajan, H.: Multiview activity recognition in smart homes with spatio-temporal features. In: Proceedings of the Fourth ACM/IEEE International Conference on Distributed Smart Cameras, ICDSC 2010, p. 142 (2010)
27. Yan, P., Khan, S.M., Shah, M.: Learning 4D action feature models for arbitrary view action recognition. In: 2008 IEEE Conference on Computer Vision and Pattern Recognition, pp. 1–7 (June 2008)
28. Zhu, F., Shao, L., Lin, M.: Multi-View Action Recognition Using Local Similarity Random Forests and Sensor Fusion. Pattern Recognition Letters (May 2012)

Evaluation of a 3D Video Conference System Based on Multi-camera Motion Parallax

Miguel A. Muñoz[1], Jonatan Martínez[1], José Pascual Molina[1,2],
Pascual González[1,2], and Antonio Fernández-Caballero[1,2]

[1] Instituto de Investigación en Informática de Albacete (I3A), 02071-Albacete, Spain
[2] Universidad de Castilla-La Mancha, Departamento de Sistemas Informáticos,
02071-Albacete, Spain
Pascual.Gonzalez@uclm.es

Abstract. Video conference systems have evolved little regarding 3D vision. An exception is where it is necessary to use special glasses for viewing 3D video. This work is based primarily on the signal of vision motion parallax. Motion parallax consists in harnessing the motion of the observer, and offering a different view of the observed environment depending on his/her position to get some 3D feeling. Based on this idea, a client-server system has been developed to create a video conference system. On the client side, a camera that sends images to the server is used. The server processes the images to capture user movement from detecting the position of the face. Depending on this position, an image is composed using multiple cameras available on the server side. Thanks to this image composition, and depending on the user standpoint, 3D feeling is achieved. Importantly, the 3D effect is experienced without the use of glasses or special screens. Further, various composition models or change modes between cameras have been included to analyze which of them achieves a greater improvement of the 3D effect.

Keywords: 3D video conference, Multi-camera, Motion parallax.

1 Introduction

Video conferencing is a video (and in some cases, audio) simultaneous communication system which allows two or more physically distant people to see each other. Video conferencing has evolved a lot over the years and services like Skype or Messenger have become popular. The approach has allowed even to send files between computers and even multiple videos at a time. However, technology has advanced very little in terms of 3D video visualization [1]. Currently, viewing 3D video often has the disadvantage of requiring color filter glasses. These glasses hinder communication between partners because their lenses, whether color filters, polarized or LCD shutter hide the look of the person, so important in non-verbal language that people use. There are alternatives free of glasses, called auto-stereoscopic systems, such as the Nintendo 3DS console, but at the cost of the loss of resolution and still the 3D effect seems limited to a few inches in front and behind the screen.

J.M. Ferrández Vicente et al. (Eds.): IWINAC 2013, Part II, LNCS 7931, pp. 159–168, 2013.

In this paper, the limitation to be saved is the need to use special glasses to perceive 3D video. The idea underlying this work is to use multiple cameras and to change from one image to another as if users were face to face. So, if one of them moves his/her head to the side, the point of view is changed and thus the image perceived by the user. Advantages of our proposal are that the price of the (web) cameras is very low and that no additional devices such as screens with 3D technology are required. To achieve the desired effect, face movement is detected to change from one view to another. This process is intended to operate the visual motion parallax. Indeed, many animals, including humans, have two eyes with overlapping visual fields that use parallax to gain depth perception [2], [3]; this process is known as stereopsis [4]. In computer vision the effect is used for computer stereo vision [5].

2   Motion Parallax for 3D Video Conferencing

Two physical eyes horizontally separated produce in our retinas two different images of the world. However, this difference between images or retinal disparity is not the only cue that our brain uses to extract the depths of the objects for building a 3D image of the world. In fact, in order to focus on an object, the eyes jointly with convergence or movement of the eye muscles form the so-called binocular cues. Another group is the monocular cues that the brain extracts from an image projected on a single eye, such as linear perspective, relative sizes, lights and shadows, texture gradient, atmospheric attenuation, and motion parallax of the lens accommodation. Motion parallax has not received as much attention as retinal disparity, but the final result can be as good or better than using stereoscopy [6], [7].

Motion parallax, the change of position of an object in the image, may be due to movement of the object so as to change of the viewer's point of view. The object image travels through the retina and through motion parallax it is possible to determine the depth and distance of objects. Due to the effect of motion parallax, the nearest objects give the impression of moving faster than the farthest. Objects closer to the viewer's point of view give the impression of moving in the opposite direction to their motion, while distant objects move in the same direction. Our proposal mainly develops on this monocular motion parallax signal. When the user (in front of his/her computer) moves his/her head, he/she receives a different image of the remote user, thus changing the point of view as in reality. The user also gets a greater sense of depth.

There have been two pseudo-3D effects for video conferencing using motion parallax from a single camera [8]. The first effect called frame box consists in creating a virtual box where the remote video is stuck in a plane behind the virtual box. Depending on the user's viewpoint, the virtual box moves according to his/her position, creating the impression that the remote user is within the screen. This technique does not require the difficult task of segmentation of foreground and background. The second effect called layered video performs segmentation of foreground and background layers. The layers which are placed at different depths are displaced according to the user's viewpoint. However, the algorithms used to create this effect are not robust to small changes in the background. Another vision system using pseudo 3D segmentation of layers with a generic camera has been described [9]. To separate the background layer from the foreground, first a picture of the background is taken. Once the two layers are positioned at different depths, they are turned so that they are perpendicular to the user's viewpoint. To this end, the center position of the user's face is detected. Thanks to motion parallax some 3D illusion is generated.

In another article multiple cameras are used to track face movements [10]. Algorithms that detect facial features, even when the user rotates the head, are applied. The authors discuss the importance of tracking the movement of the user to exploit motion parallax. Also, it is shown that the effects achieved by separating layers are not in line with reality, because objects are not flat in real life. To avoid this problem, they work with multiview images using multiple cameras to capture the scene. Various rendering models are introduced to represent the scene from a particular viewpoint. This task is not trivial and has attracted much research over the past two decades.

3   A Proposal for 3D Video Conferencing

With a single camera motion parallax is provided by moving objects. However, in a video conference partners often stand still in front of the camera or move very little. To use this cue, each partner should be able to change his/her point of view on the other with the simple movement of the head, as you do in your daily life. This requires a remote robotic camera to reproduce the movement, or multiple cameras to be deployed on the remote computer, automatically selecting the one that best approximates the viewpoint at all times. This work focuses on the latter solution.

With multiple cameras on the remote computer or server, a first option is to transmit the camera image that best fits the standpoint of local user or client. To follow the movement of the local user, images from his/her camera and software that processes head-tracking can be used. Other options include using the many cameras on the remote computer to compose and send a panoramic image composed of all cameras' overlapping images. This creates a panoramic image that best matches the standpoint of the local user. Not all regions of the images that make up the panoramic image fit perfectly with each other, because of the different perspectives provided by each camera. This effect may embarrass some users. One option to reduce the visibility of the images out of focus is to update them at a slower rate than the image in focus. Another option is that only the image in the focus is displayed in color, leaving the others in gray. In both cases, we rely on what comprises the human retina itself, consisting of color-sensitive cones in the central region, the fovea, and motion sensitive rods on its periphery.

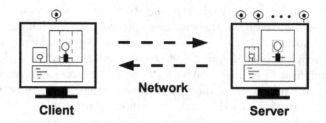

Fig. 1. Architecture of the 3D video conference system

Fig. 1 shows a schematic view of the proposed 3D video conferencing system. On a computer is the client part of the system which uses a single camera. In another computer is the server using multiple cameras. The client gets the user's camera image and sends it to the server through a network. Then, the server detects the user's face in the image received from the client and, depending on its position, generates an image composed by the multiple cameras focusing on the user. The cameras are aligned in parallel. This process is repeated until the video conference ends.

3.1   Camera Change Modes

On the server, the resulting image to be transmitted to the client consists of a composition of images from the cameras connected to the server. There are three ways to compose an image.

Full Camera Image. In the first mode, the image size is the size of a single camera image, and, whichever is the camera selected each time, the full image obtained from the camera is shown. This mode is the simplest, yet provides poorer results since a major impact is appreciated in the change between images.

Panoramic Image. In this mode, the height of the resulting image is the height of the camera images. Its width is the total width resulting from a prior adjustment of the camera images. If no prior camera adjustments are made, the chosen width is the sum of the widths of all the camera images selected by the server. In this case, a composition of images placed by their adjustment settings is obtained. The active camera, chosen according to the detected position of the user, shows its proper image above the rest. In addition, we have the possibility for pieces of images for unselected cameras at each instant of time to be presented in grayscale to give less importance to the user when viewing the composite image. We can also set the period of updating this less important section every given number of frames. The upside of this mode is that you have a complete view of all the space that can be see on the server side. By contrast, when there are many cameras in the selected server the composite image is very wide with respect to height in this mode.

Adjusted Aspect Ratio Image. This latter mode is an enhancement of the former. We said that when the number of cameras increases, the resulting image is too wide. For example, if we want to see the image in full screen, the image is not optimal viewing. Indeed, it would look like a movie with black bars above and below, which wastes the screen size. This can be solved by adding the power to select a portion of the resulting image to display. This selection is called aspect ratio, as one can select an aspect ratio proportional to the height of the image to be displayed by the client. In this proposal an aspect ratio of 4:3, 16:9 and 16:10 can be chosen. This section may be fixed, showing only the previously selected setting in the camera, or this section can be transferable across the entire image. That is, when the user moves his/her head to a side, the aspect ratio proportionally displays the image portion required at all times.

3.2 Camera Adjustments

When the server communicates with the client, the image sent is a composition of images from the server. In order to compose images that make sense to the client user, they must have been calibrated on the server before starting communication. The calibration of the selected cameras, consisting in overlapping by matching their areas, is precisely what we call camera adjustment. This gives the user the possibility to adjust the server camera images previously selected on his/her computer. Initially, the adjustment space fits the number of cameras.

The image part, both above and below, is lost in the final composition after tuning in the adjustment space. The preset height for the final composite image is the height of one of the cameras. It takes the value of the first image which will be used entirely. The final width of the composite image is the final part where the last adjusted camera image ends. The regions between the adjusted images which are not filled by any image are completed in the final composition of images with color black. This decision is taken to best adjust vertical images (in addition to horizontal ones), and to least possible notice a change between cameras.

4   Steps for 3D Video Conferencing

4.1   Face Detection

The detection of the client user's face is performed from the received image on the server. To perform the face detection, we used OpenCV [11] cvHaarDetectObject function that returns a structure CvSeqfrom the image where the face has to be detected. There we can find the positions where the faces are found in the image.

4.2   Filtering

Hysteresis is the tendency of a material to retain one of its properties in the absence of the stimulus that generated it. When composing the resulting image,

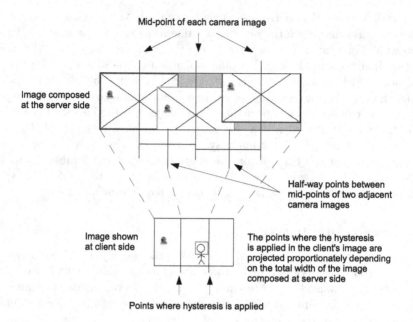

Fig. 2. Diagram showing the hysteresis

there are some visible space areas that are shared by two cameras. When choosing which is the main camera in the area, the position of the client user's face is taken into account to compose the resultant image. So that, when the user moves his/her head from left to right, one or the other camera is either chosen. To select a camera, for each of the points where the user's face can be positioned, firstly, sections are limited through the midpoint between the centers of the cameras. These points are proportionally projected to the width of the image of the client. These are the points where hysteresis is applied (see Fig. 2). When exceeding each of these points, either to the left or right, the main image will be one or the other, depending on whether the center of the user's face is in one or other section.

Up to this point, in principle, it is not necessary to do anything more. But the problem is that the detection of the user's face is not always perfect, and between consecutive frames the center point of the face can vary by several pixels. That is why, when the center of the face is just a midpoint between cameras, there is the flickering effect. This effect appears in consecutive (or very close) frames, when the user is still the image may be chosen in one camera or the other in very small time slots, resulting in a rapid change between cameras. To avoid this effect, the concept of hysteresis is applied allowing changes between cameras only if hysteresis points exceed a certain margin.

The flicker effect problem when switching between cameras on the server is saved with the implementation of the hysteresis. But there is another annoying effect that due to the instability of the face detection algorithm.

This disturbance is detected on the server side, where the position of the face is marked with a square face. It also detected on the client side, being this more evident when the adjusted aspect ratio mode is selected. To correct this disturbance a comparison in face detection between current and previous frame is performed. If the difference between frames of the absolute sum of the four corners of the face detection square is lower than the value established for the filter, the new position is discarded and the previous position is retained.

4.3 Composition of the Server Image

On the server, the resulting image to be transmitted to the client shall consist of a composition of images from the cameras connected to the server. At the time of making the composite image on the server, two aspects previously discussed have to be addressed: camera adjustment and camera change mode. For full camera mode, the adjustment between the cameras makes no sense. For the other two camera change mode, the previous setting of the camera images is relevant.

5   Data and Results

An experiment was carried out with users to analyze the various ideas for implementing techniques of motion parallax. The goal of the experiment is to test different aspects of the settings, especially the aspects linked to 3D sensation produced. For this, we design a test scenario consisting of two computers, one that makes the client function (see Fig. 3a) and another the server function (see Fig. 3b). On the client side a single Logitec QuickCam Pro camera observes the user's movements. On the server another computer is available which connects five cameras of the same model Logitec QuickCam Sphere AF. Using one single type of camera enables the characteristics of resolution, contrast and lighting, etc., to be the same. This facilitates the integration and adjustment of the resulting image and produces a more homogeneous final effect in the composite image. Within the server, which is the one to get the images observed by the client, the cameras are located at a distance of approximately 15-20 cm, all directed toward a central point of the object to be observed. Since in our case the observed element is the person who is in front of the server platform, his nose is taken as reference. Thereafter, the necessary adjustments are made on the cameras to adjust the generation of the final image. Similarly, the other parameters related to tracking the user's face in the client platform are configured.

After configuration we test the system with the help of 8 users of similar ages (25 to 30 years). Two women and six men are placed in front of the client platform. The users are asked to make small head movements in order to see different views of the object. In this case the object is the person who is in front of the server platform. In turn, on the server platform the way of representing the motion parallax is randomly changed. Likewise, each user performs independent tests in order to avoid being influenced by the experience of other users. After various tests in which each user explores the different display options, each of

Fig. 3. Test setup. (a) Client user. (b) Server user.

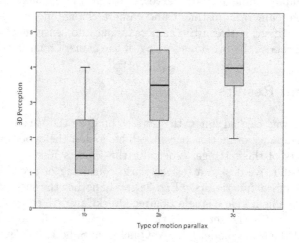

Fig. 4. 3D sensation of the three types of motion parallax

them is passed a test to analyze the user experience. The questionnaire is shown in Table 1. As can be seen, there is a common goal to analyze the 3D perception for each type of representation offered. In turn, we analyze some specific aspects of each of the systems.

The users were asked to assess the questions on a scale of 1-5: 1 (none), 2 (very little), 3 (normal), 4 (a lot) and 5 (very much). The results are shown in Table 2. The table reflects the average of the values given by each user for each question. If we analyze the results, it can be seen that the best option is the option valued three, corresponding to adjusting the aspect ratio when the image is formed by the composition of several images. Furthermore, said composition matches the monitor screen. This choice can be justified if we analyze the results of the responses to the question about 3D sensation or perception of each one of the options. As shown in Fig. 4, the answer to question 3c gets the best results, and therefore, this is the the motion parallax management mode that better 3D feeling offers.

Table 1. User experience questionnaire

Rate to 1 (none), 2 (very little), 3 (normal), 4 (a lot) to 5 (very much) the following questions.

1. Full camera image, when there is no image composition.
   (a) Do you like the change between cameras?
   (b) Do you feel 3D sensation?
2. Panoramic image, when the image is formed by the composition of multiple images.
   (a) Do you like the panoramic image to be so wide? In full screen black bars appear above and below.
   (b) Do you feel 3D sensation?
3. Adjusted aspect ratio image, when the image is formed by the composition of multiple images, and the composition fits the monitor screen.
   (a) Do you like the image to fit your monitor? In full screen you can see this effect better.
   (b) Do you like the aspect ratio to be translatable? That is, when the composite image also moved from side to side.
   (c) Do you feel 3D sensation?
4. Common issues for points 2 and 3 above. These relate to the least significant section (hereinafter LSS) of the image, ie in the composite image these are laterally out of focus.
   (a) Do you like the LSS in color?
   (b) Do you like the LSS in grayscale?
   (c) Do you like the LSS to be updated every 0 frames? That is, the LSS is left as is.
   (d) Do you like the LSS to be updated every 1 frames? That is, the LSS is always updated.
   (e) Do you like the LSS to be updated every 5 frames? That is, when it seems that the LSS stutters.
5. General considerations.
   (a) Which option do you like best? (1st, 2nd or 3rd)
   (b) Do you prefer the previous option versus traditional video conferencing? That is, both ends of the communication with a single camera.
   (c) Do you find practical the necessary interaction to move your head sideways?

Table 2. Results of the user experience questionnaire

Question	1a	1b	2a	2b	3a	3b	3c	4a	4b	4c	4d	4e	5a	5b	5c
Mean response	2.0	1.9	2.9	3.4	4.1	3.9	4.0	3.3	3.7	3.3	4.0	1.9	3	4.3	3.9

In addition, we analyze user perception of the least significant section (questions grouped in paragraph 4 of the questionnaire). The result of putting that image in gray or leaving it in color does not affect the sensation perceived by the users. The rate the use of color with a value of 3.3 and the use of gray with a value of 3.4. This may be because in reality we spend some resemblance to the non-central focus of our image. Finally, it is interesting to note in this section that users prefer all areas of the image to be updated at the same time, so that no gaps occur between parts of the image. This may be because, if there are gaps, the user may perceive and produce some distraction. Moreover, the process of moving the head is well received, 3.9, very close to value "a lot". This is because when performing this movement a flat image can be perceived with a certain three-dimensional aspect.

6   Conclusions

The first aim of this study was the use of multiple cameras and their combined images to create a system simulating a 3D visualization without the use of special glasses, exploiting the motion parallax signal in 3D vision. Hysteresis is introduced to control the flicker effect and a filter is included to prevent vibration effect of the resulting image on the server.

The second objective was to conduct and evaluate the proposed client-server system. This objective has been fully achieved, enabling optimal visualization and synchronized imaging. As for the simulation of motion parallax based 3D visualization, as tested users have noticed quite 3D effect in some of the implemented representation options.

Acknowledgements. This work was partially supported by Spanish Ministerio de Economía y Competitividad / FEDER under TIN2012-34003 and TIN2010-20845-C03-01 grants.

References

1. Azari, H., Cheng, I., Daniilidis, K., Basu, A.: Optimal pixel aspect ratio for enhanced 3D TV visualization. Computer Vision and Image Understanding 116(1), 38–53 (2012)
2. Yoonessi, A., Baker Jr., C.L.: Contribution of motion parallax to segmentation and depth perception. Journal of Vision 11(9), 1–21 (2011)
3. Fernández-Caballero, A., López, M.T., Saiz-Valverde, S.: Dynamic Stereoscopic Selective Visual Attention (DSSVA): Integrating motion and shape with depth in video segmentation. Expert Systems with Applications 34(2), 1394–1402 (2008)
4. López-Valles, J.M., Fernández, M.A., Fernández-Caballero, A.: Stereovision depth analysis by two-dimensional motion charge memories. Pattern Recognition Letters 28(1), 20–30 (2007)
5. Fernández-Caballero, A., López, M.T., Mira, J., Delgado, A.E., López-Valles, J.M., Fernández, M.A.: Modelling the stereovision-correspondence-analysis task by lateral inhibition in accumulative computation problem-solving method. Expert Systems with Applications 33(4), 955–967 (2007)
6. Li, I.K.Y., Peek, E.M., Wünsche, B.C., Lutteroth, C.: Enhancing 3D applications using stereoscopic 3D and motion parallax. In: Proceedings of the Thirteenth Auatralasian User Interface Conference, pp. 59–68 (2012)
7. Fernandez, J.M., Farell, B.: A neural model for the integration of stereopsis and motion parallax in structure from motion. Neurocomputing 71(7-9), 1629–1641 (2008)
8. Zhang, C., Yin, Z., Florêncio, D.: Improving depth perception with motion parallax and its application in teleconferencing. In: IEEE International Workshop on Multimedia Signal Processing, MMSP 2009, pp. 1–6 (2009)
9. Harrison, C., Hudson, S.E.: Pseudo-3d video conferencing with a generic webcam. In: Tenth IEEE International Symposium on Multimedia, ISM 2008, pp. 236–241 (2008)
10. Zhang, C., Florêncio, D., ZHang, Z.: Improving immersive experiences in telecommunication with motion parallax. IEEE Signal Processing Magazine 28(1), 139–144 (2010)
11. OpenCV. Open source computer vision (2013), http://opencv.org/

Abandoned Object Detection
on Controlled Scenes Using Kinect

Antonio Collazos, David Fernández-López, Antonio S. Montemayor,
Juan José Pantrigo, and María Luisa Delgado

Universidad Rey Juan Carlos, C/Tulipán s/n, 28933 Móstoles, Spain

Abstract. This paper presents a new approach for the detection of abandoned
objects in partially controlled environments using images from a low cost depth
sensor. To reach this goal, we propose a systems which involves: (i) a two phase
object segmentation based on 3D points clustering, (ii) an object selection based
on permanence and object dimensions and, finally, (iii) a state machine
monitoring capable to deal with occlusions. The proposed system exclusively
considers depth images, which makes it independent of lighting. In order to obtain evidences of the system performance in real conditions, we have conducted
an experimental study on video sequences captured into a public bus, obtaining
successful results.

Keywords: abandoned object detection, depth sensor, real conditions.

1 Introduction

The growing demand for video surveillance systems in public areas involves an increase
in the amount of information to analyse. This increase in the amount of information has
become intractable for human supervisors. For this reason, it is particularly interesting
to design and implement autonomous systems capable of extracting information from
these environments and generating alerts automatically. It is expected that displaying
these alerts to a human operator in an understandable language instead of the whole
visual captured information reduces their stress and workload. As a consequence, it is
also expected that it improves their performance.

The problem of automatic detection of abandoned objects in public areas is currently
a research topic of great interest due to the large number of people concentrated in these
areas and the potential security risks that entails the abandonment of objects. This topic
has gained importance in recent years, even entering as an important topic of some international conferences in video surveillance. As a result, there have been many authors
who have made contributions in this area using many different techniques. For example,
in [1], authors use a trans-dimensional Markov Chain Monte Carlo tracking approach
to separate moving and static blobs. This system can also separate static blobs in at-
tended and unattended objects by the analysis of the behaviour of moving blobs. An
important and remarkable issue regarding the abandoned object detection topic is that
most of the works are heavily based on background subtraction methods, using adaptive background models, and the analysis of changes of the foreground and background

J.M. Ferrández Vicente et al. (Eds.): IWINAC 2013, Part II, LNCS 7931, pp. 169–178, 2013.
© Springer-Verlag Berlin Heidelberg 2013

along the time. Background models are usually supported by computationally demanding statistic techniques, where the *Mixture of Gaussians* model (MoG) [2] is the most extended, but some authors prefer to use simpler approaches to enhance performance [3]. Also, some works go beyond this type of techniques using a two layer background method [4][3] where one of the background model is updated more frequently than the other one. This background model allows to improve the background subtraction and enables differentiation between moving, inactive and removed objects. Moreover, the background model can be refined by using some techniques to clean the result of the subtraction, or to determine the model update frequency using optical flow [5].

Nevertheless, realistic sequences are complex due to variations in lighting conditions. For this reason, although the adaptive background subtraction can alleviate this problem, in most cases this is not enough. To deal with this issue, some authors use different color models like HSV where most of the illumination information is separated in the V channel [5]. Other approaches try to segment shadows in the image [6][7], which are the most problematic illumination issues, so they can remove them from the background subtraction. Once the objects present in the image are segmented from the background, it is needed to determine which of them are actually abandoned objects. The first step to get this consists of selecting the objects that remain in the same position along the recent time, for example selecting pixels that remain active the 70% of frames in the last ones [7]. Note that objects detected can be persons who are not moving. Due to this issue, a lot of works try to differentiate between them, usually by analysing mathematical properties and shape of the blob [5][7].

Microsoft's Kinect's recent arrival on the market has enabled the people, researchers or not, to have a depth sensor at low cost. This type of technology can be a qualitative leap forward in vision applications, enabling researchers to create more robust applications. So at this time, there are many works in progress using this technology. Most of these works focuses on gesture recognition and biometrics [8][11], areas for which this sensor was developed. However, some works focus on different areas like 3D reconstruction, robotics [9] and navigation [10]. In this paper we propose to take advantage that depth sensors provides to deal with video surveillance tasks. Specifically we propose an abandoned object detector based exclusively on depth sensor capable to work inside a public bus. This detector has also been tested under real conditions.

The remainder of the paper is organized as follows. Section 2 describes the algorithm implemented for the detection of abandoned objects. Section 3 analyzes the obtained results. Finally Section 4 depicts the conclusions extracted from this work.

2   Algorithm

The abandoned object detection algorithm deployed can be divided into five diferent blocks. The first one deals with the capture stage, the peculiarities of the data obtained from the depth sensor and how the system works with it. The second block depicts the background model used to perform a background subtraction adapted to the problem. The third one associates pixel in the image with different objects in the scene.

The fourth block classifies the objects identified from the previous stage in 'people' or 'non people'. Finally, the last block classified objects in function to their permanence in the scene and their position with respect to the people in the scene.

2.1 Capture and Preprocessing

Although not using RGB images may mean the loss of relevant information, we have decided to use only the data from the depth sensor for detection purposes, so RGB will only be used for visualization. This decision is due to the need to be as independent as possible to lighting conditions. Data obtained from the depth sensor is coded as an image in which a pixel value represents the distance from the lens plane to its corresponding point displayed in the image. There are two artifacts that must be taken into account when working with this images:

– **Black holes.** They usually appear in zones where the infra-red sensor is unable to capture the light emitted because the point where is projected is too far, too close, or occluded. This is also produced when light is projected on materials with a high reflexivity, like glass. In some cases, the angle at which the focused area reflects the light emitted by the device can also produce this effect in the image. These black holes may come and go during the capture.
– Pixels in the image can behave unstable. This **pixel instability** varies in function of the material where the light is projected and usually increases with the distance.

To avoid black holes, these pixels are filled with values in the same position of the background model. This process prevents error detections in the next stages of the algorithm. To improve the pixel stability, a temporal average of the recent frames is performed and used instead of the present frame. An example of the final result of this step of the algorithm can be seen on Figure 1.

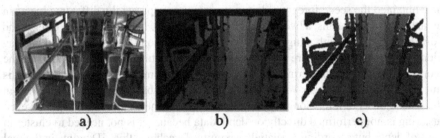

Fig. 1. Image depth preprocessing: (a) RGB image, (b) captured depth image containing black holes and unstable pixels, (c) depth image preprocessed

2.2 Background Subtraction

Background subtraction becomes an important part of video surveillance systems. Also, due to lighting conditions it is usually needed an adaptive background model. However depth sensor abstracts lighting from us, so it is possible to use a model as simple as a static background model. Nevertheless, depth image defects explained the previous subsection point make necessary to perform a little background training during some frames, before start applying the proper background subtraction algorithm.

$$I_{out}(x,y) = \begin{cases} 1 \text{ if } I_{depth}(x,y) < B(x,y) + Th \\ 0 \text{ otherwise.} \end{cases} \tag{1}$$

where $B(x,y)$ is the pixel in the (x,y) position of the background and Th is a threshold value. To avoid image defects, the background training would set each pixel to the minimum value found in the image during the training, except in the case that this value equals zero (black hole). Finally, at the end of the training, black holes in the background are filled with the maximum distance value of the sensor. This prevents the miss detection of objects placed before black holes in the background. Figure 2 shows an example of the background subtraction process.

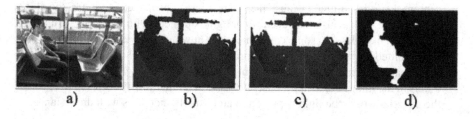

Fig. 2. Background subtraction: (a) RGB image of the current frame, (b) depth image of the frame at time t, (c) background model, (d) background subtraction

2.3   Clustering

Once obtained the scene foreground, it is necessary to separate the different objects in the image. To achieve this, a two-phase strategy is applied. In the first phase, foreground is divided in regions with a high cohesion and in the second one the regions that may belong to the same object are merged. To perform the first phase of this part of the algorithm, a k-means clustering on the depth image is made. K-means clustering is a method of cluster analysis which aims to partition n observations into k clusters in which each observation belongs to the cluster with the nearest mean [12]. However, clustering is not performed directly on depth data because it is not needed to cluster in levels of depth but according to spatial proximity. To achieve this, 3D points in the real world corresponding to the pixels of the image are obtained by applying the calibration model of the camera to depth data. In Figure 3, an example of the difference between applying k-means to depth image and applying k-means to 3D points is shown.

In the second phase, to merge the regions belonging to the same object, the algorithm is focused on the boundaries between them. In order to do so, it is determined that two regions belong to the same object if the minimum distance between border points

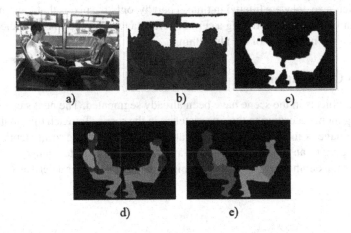

Fig. 3. K-means clustering: (a) RGB image of the current frame, (b) depth image of the current frame, (c) background subtraction, (d) k-means clustering based on depth image, (e) k-means clustering based on 3D points

in both regions does not exceed a certain threshold. Note that the distance computed is a Euclidean distance between the real world 3D points corresponding to the border pixels. In Figure 4 the result of this process is shown.

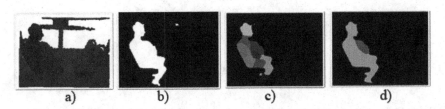

Fig. 4. Region merge: (a) depth image of the current frame, (b) background subtraction, (c) k-means clustering based on depth image, (d) regions after merging clustered image

2.4 Object Detection

Once the objects in the scene are detected, we need to decide which of them may correspond to people and which may correspond to objects to be monitored. To determine this, an approach based on the real size and area of the object has been chosen. Although this approach is simpler than others in the state of the art [7][5], the possibility of getting true dimensions offers a good enough robustness without compromising performance.

Nevertheless, the clustering process performed before may not separate the objects in all frames and in some of them can separate parts of people that can have the size of an object. therefore to provide a stable input to the next stage, in addition to filtering objects by their size, they are filtered in time. Thereby, only the pixels that remain active as objects (and not as people) during a certain percentage of recent time, are selected to be analysed in the next stage of the algorithm.

2.5   Object Classification

At this point, objects in the scene have been already segmented. The next step consists of classifying them according to their persistence in the scene. To reach this goal, every pixel in the image is modelled as a finite state machine according to its depth values and its belonging to an object. However, to avoid undesirable state changes due to irregularities in the depth images, pixels with high spatial and temporal derivative are not updated.

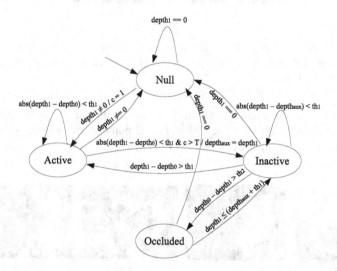

Fig. 5. Finite state machine associated to each pixel in the image for classifying objects

Figure 5 shows the finite state machine that models the pixel history. Its possible states are *null*, *active*, *inactive* and *occluded*. In this diagram, $depth_0$ corresponds to the pixel depth in the previous frame, $depth_1$ is the pixel depth in the current frame and $depth_{aux}$ is the depth stored when the pixel rises the state *inactive*. This is done to avoid the loss of objects when they are occluded. The model also provdes a counter c used to determine when the pixel should take *inactive* state (from the *active* state, when the counter reaches T value). On the other hand, th_1 and th_2 are threshold values. The former is related to the permanence of the object while the later is related to the occlusion of the object. Then, objects the *inactive* state, are maintained although if they look occluded (see Figure 6 for an example). Finally, objects are then labelled as *inactive* or *abandoned*. The latter label produce alerts in the system.

It is assumed that an object has been abandoned when it's owner has left their area of influence. For simplicity, we will establish that an object is abandoned when there is

Fig. 6. Object occlusion: (a) Bag detected as *abandoned* object, (b) the bag is occluded, but it is maintained in memory

no person at a certain distance from it. In this case, we will set the threshold distance to 50 cm. In this way, for each inactive object, if the people in the scene are separated a distance from the object larger than a predefined threshold, then it is labelled as abandoned object. This strategy also makes that people limbs are not detected as abandoned objects.

3   Experimental Results

The experiments reported in this section were performed on an Intel Core i7 920 2.67Ghz with 8 GB of RAM and Windows Vista 64 bits OS. To test the proposed algorithm, we captured sequences from different camera positions into a real public bus (see Figure 7). These sequences represent one of the situations displayed in Figure 8 and described as follows:

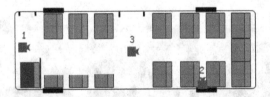

Fig. 7. Sensor configurations

A Person #1 enters in the scene carrying a backpack and sits on a window seat leaving his backpack on the next seat. After a while ($\simeq$ 30s) Person #1 leaves the scene without the backpack.

B Person #1 enters in the scene carrying a backpack and sits on an aisle seat leaving his backpack on the next seat. At the same time Person #2 sits behind Person #1. After a while ($\simeq$ 30s) Person #1 leaves the scene without the backpack. After a while ($\simeq$ 30s) Person #2 leaves his seat standing momentarily in front of the abandoned backpack.

Fig. 8. Possible scenarios in test sequences: (a) situation A, (b) situation B, (c) situation C, (d) situation D

C Person #1 enters in the scene carrying a backpack and sits on an aisle seat leaving his backpack between his legs. After a while ($\simeq 30s$) Person #1 leaves the scene without the backpack. After a while ($\simeq 30s$) Person #2 enters the scene and sits in the same seat that Person #1 was taking previously.

D Person #1 enters in the scene carrying a backpack, leaves it on an aisle seat and leaves the scene. After a while ($\simeq 30s$) Person #2 enters the scene standing momentarily in front of the abandoned backpack and sits behind it.

Table 1 shows the results obtained by the algorithm on the considered scenes. As it can be seen, a high abandoned object detection rate is achieved and, in contrast, the false positive rate is low. As far as the false positives is concerned, most of them are caused by people limbs. They appear separated from the rest of the body and occluded by other elements in the scene. Sequence 2 offers a view more free of obstacles, and the algorithm detects a lower number of false positives. In other cases false positives may occur due to pixel instability not assimilated by the preprocessing stage. Figure 9 shows the results obtained in sequence 2B, where people, *active*, *inactive*, *occluded* and *abandoned* objects are highlighted.

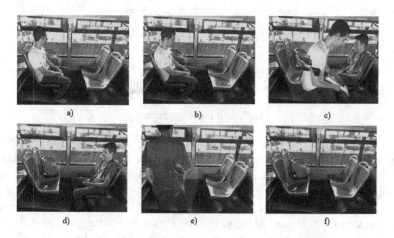

Fig. 9. Selected representative frames from sequence 2B

Table 1. Analysis of the abandoned objects detected on sequences recorded. Rows correspond to true positives (TP), false negatives (FN) and false positives (FP), respectively.

Sequence	1A	1B	1C	1D	2A	2B	2C	2D	3A	3B	3D
TP	1	1	1	1	1	1	1	1	1	1	1
FN	0	0	0	0	0	0	0	0	0	0	0
FP	0	2	0	2	0	0	0	0	1	0	1

4 Conclusions

In this paper, we have presented an abandoned object detector exclusively based on the depth image provided by a low cost depth sensor like Microsoft Kinect. An important advantage over traditional camera-based detectors is that it manages 3D information and actual distances without the requirement of complex stereo systems. Another great advantage is the independence from lighting that provides this approach that may even work in the absence of light. On that basis, an algorithm capable of segmenting objects using this 3D information and determining when these objects can match unattended ones has been developed. Applying this algorithm we have also obtained good results on sequences recorded in a real public bus.

As future works, we aim to improve the efficiency of the algorithm and combine information from multiple depth devices working simultaneously. It would provide a broader perspective of the scene which could achieve more complete results and more accuracy in the detection.

Acknowledgements. This work has been supported by the Cátedra de Ecotransporte, Tecnología y Movilidad between University Rey Juan Carlos and the Empresa Municipal de Transportes de Madrid (EMT) through the BusSeguro project, by the Spanish Ministry of Economy and Competitiveness grant TIN2011-28151 and by the Government of the Community of Madrid grant ref S2009/TIC-1542.

References

1. Smith, K.C., Quelhas, P., Gatica-Perez, D.: Detecting Abandoned Luggage Items in a Public Space. PETS, 06–39 (2006)
2. Stauffer, C., Grimson, W.E.L.: Adaptive background mixture models for real-time tracking. In: Computer Vision and Pattern Recognition, vol. 2, pp. 637–663 (1999)
3. Singh, A., Sawan, S., Hanmandlu, M., Madasu, V.K., Lovell, B.C.: An Abandoned Object Detection System Based on Dual Background Segmentation. In: Advanced Video and Signal Based Surveillance, pp. 352–357 (2009)
4. Porikli, F., Ivanov, Y., Haga, T.: Robust Abandoned Object Detection Using Dual Foregrounds. Eurasip Journal on Advances in Signal Processing 2008, 1–12 (2008)
5. Bhargava, M., Chia-Chih, C., Ryoo, M.S., Aggarwal, J.K.: Detection of abandoned objects in crowded environments. In: Advanced Video and Signal Based Surveillance, pp. 271–276 (2007)

6. Spagnolo, P., Caroppo, A., Leo, M., Martiriggiano, T., D'Orazio, T.: An Abandoned/Removed Objects Detection Algorithm and Its Evaluation on PETS Datasets. In: IEEE International Conference on Video and Signal Based Surveillance (2006)
7. Wen, J., Gong, H., Zhang, X., Hu, W.: Generative model for abandoned object detection. In: IEEE International Conference on Image Processing, pp. 853–856 (2009)
8. Breidt, M., Biilthoff, H., Curio, C.: Robust semantic analysis by synthesis of 3D facial motion. In: IEEE International Conference on Automatic Face & Gesture Recognition and Workshops, pp. 713–719 (2011)
9. Khandelwal, P., Stone, P.: A Low Cost Ground Truth Detection System for RoboCup Using the Kinect. In: Röfer, T., Mayer, N.M., Savage, J., Saranlı, U. (eds.) RoboCup 2011. LNCS, vol. 7416, pp. 515–527. Springer, Heidelberg (2012)
10. Zöllner, M., Huber, S., Jetter, H.-C., Reiterer, H.: NAVI - A Proof-of-Concept of a Mobile Navigational Aid for Visually Impaired Based on the Microsoft Kinect. In: Campos, P., Graham, N., Jorge, J., Nunes, N., Palanque, P., Winckler, M. (eds.) INTERACT 2011, Part IV. LNCS, vol. 6949, pp. 584–587. Springer, Heidelberg (2011)
11. Reyes, M., Ramírez-Moreno, J., Revilla, J.R., Radeva, P., Escalera, S.: ADiBAS: Sistema Multisensor de Adquisición Automática de Datos Corporales Objetivos, Robustos y Fiables para el Análisis de la Postura y el Movimiento. VI Congreso Iberoamericano de Tecnologíad de Apoyo a la Discapacidad (2011)
12. Hartigan, J.A., Wong, M.A.: A K-Means Clustering Algorithm. Journal of the Royal Statistical Society. Journal of the Royal Statistical Society. Series C (Applied Statistics) 28, 100–108 (1979)

People Detection in Color and Infrared Video Using HOG and Linear SVM

Pablo Tribaldos[1], Juan Serrano-Cuerda[1], María T. López[1,2],
Antonio Fernández-Caballero[1,2], and Roberto J. López-Sastre[3]

[1] Instituto de Investigación en Informática de Albacete (I3A), 02071-Albacete, Spain
[2] Universidad de Castilla-La Mancha, Departamento de Sistemas Informáticos,
02071-Albacete, Spain
Antonio.Fdez@uclm.es
[3] Universidad de Alcalá, Dpto. de Teoría de la señal y Comunicaciones,
28805-Alcalá de Henares (Madrid), Spain

Abstract. This paper introduces a solution for detecting humans in smart spaces through computer vision. The approach is valid both for images in visible and infrared spectra. Histogram of oriented gradients (HOG) is used for feature extraction in the human detection process, whilst linear support vector machines (SVM) are used for human classification. A set of tests is conducted to find the classifiers which optimize recall in the detection of persons in visible video sequences. Then, the same classifiers are used to detect people in infrared video sequences obtaining excellent results.

Keywords: Human classification, Color video, Infrared video, HOG, Linear SVM.

1 Introduction

In smart spaces visual surveillance, real-time detection of people (e.g. [1], [2]) and their activities [3] is performed both in visible (e.g. [4], [5], [6]) and infrared spectrum (e.g. [7], [8]). Therefore, it seems interesting to find a single solution to detect people in both types of videos. Most methods described for the detection of people are divided into two steps, namely extraction of image features and classification of the images according to these features.

In this sense, histogram of oriented gradients (HOG) is a feature extraction technique for the detection of objects [9]. Its essence is that the shape of an object in an image can be described by means of the intensity distribution of the gradients. The great advantage of a detector obtained using HOG descriptors is that it is invariant to rotation, translation, scaling and illumination changes. Therefore, it has been applied successfully in both visible spectrum images (e.g. [10], [11], [12], [13]) and infrared images (e.g. [14], [15], [16], [17]). In our approach, we are firstly interested in discovering if there are HOG descriptors for extracting human features that are equally valid for color and infrared images.

After using HOG descriptors, support vector machines (SVM) are usually used in the classification stage. SVM are a set of supervised learning algorithms which

J.M. Ferrández Vicente et al. (Eds.): IWINAC 2013, Part II, LNCS 7931, pp. 179–189, 2013.

were introduced for linearly separable [18] and linearly non-separable [19] data. SVM have been used in classification and regression problems in many fields such as text recognition, bioinformatics and object recognition, among others. They have also been used successfully in the detection of persons (e.g. [20], [21]). Here, we are also interested in knowing if linear SVM trained with color images provide good results in classifying infrared images without re-training. Should this be true, we could overcome the lack of large enough datasets in the infrared spectrum.

2   Detection of Humans in Color and Infrared Video

2.1   HOG for Feature Extraction

Histogram of oriented gradients (HOG) consists of a series of steps that provide an array of image features representing the objects contained in an image in a schematic manner. The image features are later used to detect the same objects in other images. In our particular case, we are interested in obtaining strong features for human detection.

Global Normalization of the Gamma/Color Image. This first step is undertaken to reduce the influence of the effects of image lightning changes. In order to normalize the color of an image, histogram equalization is applied. The $\sqrt{RGB}$ function is used for gamma normalization. Each pixel is obtained from the square root of its channel values.

Gradient Computation. A first derivative edge detection operator is launched to estimate the image gradients. Specifically, filter kernels $G_x = [-1\ 0\ 1]$ and $G_y = [-1\ 0\ 1]^T$ are applied to x and y axes, respectively, as well as a smoothing value $\sigma = 0$. This way the image contours, shape and texture information are obtained. Furthermore, resistance to illumination changes is achieved. The gradient is calculated for each color channel, and the locally dominant gradient is used to achieve invariance against color.

Orientation Binning. This step generates the HOG descriptors. Local information on the direction of the gradient is used in the way SIFT [22] does. It aims to produce an encoding that is sensitive to the local image content, while being resistant to small changes in attitude or appearance. Orientation binning divides the image into regions called "cells" of $n \times n$ pixels. Gradients or orientations of the edges at each cell pixels are accumulated in a 1-D histogram. The combined histograms form the orientation histogram. Each orientation histogram divides the range of angles of the gradient in a fixed number of bins. The gradient value of each pixel of the cell is used in the orientation histogram for voting.

Local Normalization. Now, the cells are grouped into sets called "blocks", and each cell block is normalized. A cell can belong to several overlapping blocks. Therefore, it appears several times in the final vector, but with different normalization. Indeed, the normalization of each block depends on the cell which it belongs to. Normalization provides better invariance against lightning, shadows and contrast of the edges. The descriptors of the normalized blocks are precisely the HOG descriptors. Dalal and Triggs [9] explore four different methods for block normalization: L1-norm (see equation (1)), L1-sqrt (2), L2-norm (3) and L2-hys. Let ν be the non-normalized vector containing all histograms in a given block, $\|\nu\|_k$, its k-norm for $k = 1, 2$ and e be some small constant (the exact value, hopefully, is unimportant). Finally L2-hys is L2-norm followed by clipping (limiting the maximum values of ν to 0.2) and re-normalizing. The normalization factor can be one of the following:

- L1-norm

$$f = \frac{\nu}{(\|\nu\|_1 + e)} \tag{1}$$

- L1-sqrt

$$f = \sqrt{\frac{\nu}{(\|\nu\|_1 + e)}} \tag{2}$$

- L2-norm

$$f = \frac{\nu}{\sqrt{\|\nu\|_2^2 + e^2}} \tag{3}$$

HOG Descriptors Combination. In the last stage of the process all blocks are combined into a dense grid of overlapping blocks, covering the detection window to obtain the final feature vector.

2.2   Linear SVM for Classification

Given the features of two objects, an SVM seeks a hyperplane optimally separating the features of an object from the other. An SVM maximizes the margin of separation between the two classes, so that one side of the hyperplane contains all objects of a class, and the other one the other objects. The vectors closest to the margin of separation are called support vectors and are used for classification. The accuracy of an SVM may be degraded in the case that data are not normalized. Normalization can be performed at the level of input features or at kernel level (in the feature space).

The classification task involves separating data into training and testing. Each instance of the training set contains a target value, which is the class label, and a series of attributes such as the observed features. The goal of SVMs is to create a model based on training data to predict the target values of the test dataset by only knowing their attributes. Given a training set with instance-label pairs $(x_i, y_i), i = 1, \ldots, l$, where $x_i \in R^n$ e $y \in \{0, -1\}^l$, an SVM requires the solution of the following optimization problem:

$$\min_{w,b,\xi} \quad \frac{1}{2}w^T w + C\sum_{i=1}^{l}\xi_i \tag{4}$$

subject to

$$y_i(w^T\phi(x_i) + b) \geq 1 - \xi_i, \quad \xi_i \geq 0 \tag{5}$$

Here training vectors x_i are mapped into a large or even infinite dimensional space by function ϕ. SVMs seek a linear hyperplane with the maximum margin separator in this dimensional space. $C > 0$ is the error penalty parameter. Function $K(x_i, x_j) \equiv \phi(x_i)^T\phi(x_j)$ is called the kernel function. LibSVM [23] offers the following four main kernel types:

- linear: $K(x_i, x_j) = x_i^T x_j$.
- polynomial: $K(x_i, x_j) = (\gamma x_i^T x_j + r)^d, \gamma > 0$.
- radial basis function (RBF): $K(x_i, x_j) = \exp(-\gamma||x_i - x_j||^2), \gamma > 0$. Variable γ can be expressed as $\gamma = 1/(2\sigma^2)$.
- sigmoidal: $K(x_i, x_j) = \tanh(\gamma x_i^T x_j + r)$.

In this work it was decided to use a linear kernel, $K(x_i, x_j) = x_i \times x_j$, where $x_i, x_j \in N$ are the feature vectors. A linear kernel uses to work fine when handling only two classes and is quite easier to refine, as it only has one parameter affecting performance, namely C, the soft margin constant.

3   Data and Results

3.1   Parameters for Performance Evaluation

Let us define *positive image* as an image containing one or more persons and *negative image* as an image where no person appears. The parameters used to validate the goodness of the proposed classifier are:

- FP (false positives): number of images that are negative but have been classified as positive.
- FN (false negatives): number of images that are positive but have been classified as negative.
- TP (true positives): number of positive images that are correctly classified, that is, number of hits.
- P: number of positive test images.
- N: number of negative test images.
- T: number of test images:

$$T = P + N \tag{6}$$

- *accuracy*: percentage of the number of correctly classified test images:

$$accuracy = \frac{TP}{P} \cdot 100 \tag{7}$$

– *precision*: percentage of true positives among all positives detected:

$$precision = \frac{TP}{TP + FP} \cdot 100 \qquad (8)$$

– *recall*: percentage of true positives among all positives:

$$recall = \frac{TP}{TP + FN} \cdot 100 \qquad (9)$$

3.2 Parameters for HOG Feature Extraction

The recommended parameters used for extracting HOG descriptors [9] are provided in Table 1. These have been used without modifications in our approach.

Table 1. Recommended values for the extraction of HOG features

Parameter	Value
Window size	64 × 128 pixels
Block size	2 × 2 cells
Cell size	8 × 8 pixels
Number of angle divisions	9 (no sign, 180°)
Overlap	8 × 8 pixels (stride = 8)
Gaussian smoothing	No
Histogram normalization	L2-hys
Gamma correction	Yes
Max number of detection window scalings	64

3.3 People Detection in Color Video

Description of Training and Test Databases. Two people image databases widely addressed in the scientific community have been used to train and test the proposal in the visible spectrum. These are INRIA (Institut National de Recherche en Informatique et en Automatique) "Person Dataset" (available at http://pascal.inrialpes.fr/data/human/) and MIT (Massachusetts Institute of Technology) "Pedestrian Data" (available at http://cbcl.mit.edu/software-datasets/PedestrianData.html). The MIT training database of people was generated from color images and video sequences taken in a variety of seasons using several different digital cameras and video recorders. The pose of the people in this dataset is limited to frontal and rear views. The MIT pedestrian database contains 923 positive images; each image was extracted from raw data and was scaled to the size 64×128 and aligned so that the person's body was in the center of the image. The data is presented without any normalization.

The INRIA person database also contains images of people in different positions, backgrounds and with different lightning (see Table 2). There are also people partially occluded. This dataset was collected as part of research work on detection of upright people in images and video. The dataset contains images from several different sources. Only upright persons (with person height > 100)

Table 2. Description of INRIA person database

	# of images	Size (pixels)
Positive training images	2,416	96 × 160
Negative training images	1,218	Not normalized
Positive test images	1,126	70 × 134
Negative test images	453	Not normalized

Table 3. Description of final test dataset

	# of images	Size (pixels)
Positive training images	4262	64 × 128
Negative training images	12180	64 × 128
Positive test images	1126	64 × 128
Negative test images	4530	64 × 128

are marked in each image, and annotations may not be right; in particular at times portions of annotated bounding boxes may be outside or inside the object.

Before using images from this database to extract their features, it is recommended to normalize them to 64 × 128 pixels, and to get sub-images of their negative images. In our case, we have added:

- 10 sub-images of size 64 × 128 pixels are randomly extracted from each negative image.
- A centered window of size 64 × 128 pixels is extracted from each positive image.
- A mirror image of each positive image (reflection on the vertical axis) is obtained.

Therefore, $2,416$ positive training images, $12,180$ negative training images, $1,126$ positive test images and $4,530$ negative test images are extracted. Also, in order to increase the number of positive images to train the classifier, the mirror images of the MIT dataset are obtained. Table 3 shows the final set of images used for training and testing.

Description of the Training Process. During the training process the SVMs supplied by LibSVM and LibLINEAR [25] are used. The models generated will be used later for human detection. LibLINEAR offers several SVMs for linear classification; we use L2-regularized L1-loss (dual), L2-regularized L2-loss (primal) and L2-regularized L2-loss (dual). The influence of the soft margin constant C on the three kernels is studied. Table 4, Table 5 and Table 6 show the results for each kernel, respectively.

The aim is to find the best kernels to classify the color training input images to apply them to the detection of people in new images. The more accurate the results of the kernel are, the better the future detection results. In this case, in order to assess the goodness of a kernel we will use the *recall* evaluation parameter to obtain the minimum possible number of false negatives, although some more false positives may appear. From the previous study, we conclude to use kernels L2-regularized L2-loss (dual) with $C = 10$ and L2-regularized L2-loss (dual) with $C = 0.001$, as their respective *recall* values are very close (96.36% and $96,63\%$).

Table 4. Influence of parameter C in the L2-regularized L2-loss (dual) linear kernel

C	Hits	FP	FN	precision (%)	recall (%)	accuracy (%)
0.0001	5582	16	58	98.57904085	94.849023	98.69165488
0.001	5590	24	42	97.86856128	96.269982	98.8330976
0.1	5580	37	39	96.71403197	96.536412	98,6562942
1	5574	41	41	96.35879218	96.358792	98.55021216
10	5577	41	38	96.35879218	96.625222	98.60325318
100	5577	41	38	96.35879218	96.625222	98.60325318

Table 5. Influence of parameter C in the L2-regularized L2-loss (primal) linear kernel

C	Hits	FP	FN	precision (%)	recall (%)	accuracy (%)
0.0001	5581	16	59	98.57904085	94.760213	98.67397454
0.001	5591	24	41	97.86856128	96.358792	98.85077793
0.1	5577	36	43	96.80284192	96.181172	98.60325318
1	5578	35	43	96.89165187	96.181172	98.62093352
10	5578	36	42	96.80284192	96.269982	98.62093352
100	5578	36	42	96.80284192	96.269982	98.62093352

Table 6. Influence of parameter C in the L2-regularized L1-loss (dual) linear kernel

C	Hits	FP	FN	precision (%)	recall (%)	accuracy (%)
0.0001	5565	21	70	98.13499112	93.783304	98.39108911
0.001	5592	16	48	98.57904085	95.737123	98.86845827
0.1	5578	38	40	96.62522202	96.447602	98.62093352
1	5575	43	38	96.18117229	96.625222	98.5678925
10	5576	42	38	96.26998224	96.625222	98.58557284
100	5577	41	38	96.35879218	96.625222	98.60325318

Description of the Results. The classification in the visible spectrum is performed on the test images by using both selected kernels. Here, the best *accuracy* is 90.33% using the classification model L2-regularized L2-loss (dual) [24] with $C = 10$. The performance results are offered in Table 7. Also, some result images in the visible spectrum are shown in Fig. 1.

Table 7. Performance results in color video

Kernel	L2-regularized L2-loss (primal) C= 0.001	L2-regularized L2-loss (dual) C = 10
Mean detection time (ms)	336	297
Hits	896	908
False positives	63	32
False negatives	222	229
Accuracy (%)	89.41	90.33
Precision (%)	93.43	96.60
Recall (%)	80.14	79.86

Fig. 1. Some results in color video

3.4 People Classification in Infrared Video

In order to test the proposal in infrared spectrum, we manually labeled 112 infrared images recorded by our research team. Of course, as stated previously, we use the parameters and kernels obtained during the training color images detection and classification phases. Here, the best performance results obtained for human detection in infrared are offered in Table 8. These come from the use of model L2-regularized L2-loss (primal) with $C = 0.001$. Now, *accuracy* is 94.64, which is astonishingly very high compared to the *accuracy* in color images. The reason for this increment is probably the fact the infrared images have been annotated manually and very carefully. Lastly, some resulting images in infrared are shown in Fig. 2.

Table 8. Performance results in infrared video

Kernel	L2-regularized L2-loss (primal) $C = 0.001$	L2-regularized L2-loss (dual) $C = 10$
Mean detection time (ms)	870	924
Hits	188	186
False positives	3	6
False negatives	8	6
Accuracy (%)	94.64	94.54
Precision (%)	97.92	96.88
Recall (%)	96.91	96.88

Fig. 2. Some results in infrared video

4   Conclusions

The initial objective of this work was to efficiently detect humans in color and infrared video. For this, we use the HOG algorithm for extracting image features, and a linear SVM for classification of the features. This combination allows detecting humans in images with high accuracy, both in visible and infrared spectrum. The HOG algorithm obtains the feature vectors from the training color images of the proposed databases. Then, a linear SVM seeks a hyperplane capable of separating the feature vectors in two classes in the most optimal way.

As it has been demonstrated, after using the recommended parameters for the feature detector and selecting a couple of kernels suited for SVM in the color spectrum, the approach works well for in the visible and infrared spectra, providing an *accuracy* of 90.33% and a *recall* of 79.86% for automatically annotated images in the visible spectrum, and an *accuracy* of 94.64% and *recall* of 96.91% for manually annotated infrared images.

Acknowledgements. This work was partially supported by Spanish Ministerio de Economía y Competitividad / FEDER under TIN2010-20845-C03-01, TIN2010-20845-C03-03,   IPT-2011-1366-390000   and   IPT-2012-0808-370000 grants.

References

1. Delgado, A.E., López, M.T., Fernández-Caballero, A.: Real-time motion detection by lateral inhibition in accumulative computation. Engineering Applications of Artificial Intelligence 23(1), 129–139 (2010)
2. Fernández-Caballero, A., López, M.T., Castillo, J.C., Maldonado-Bascón, S.: Real-time accumulative computation motion detectors. Sensors 9(12), 10044–10065 (2009)

3. Chaquet, J.M., Carmona, E.J., Fernández-Caballero, A.: A survey of video datasets for human action and activity recognition. Computer Vision and Image Understanding (2013), http://dx.doi.org/10.1016/j.cviu.2013.01.013

4. Moreno-Garcia, J., Rodriguez-Benitez, L., Fernández-Caballero, A., López, M.T.: Video sequence motion tracking by fuzzification techniques. Applied Soft Computing 10(1), 318–331 (2010)

5. López, M.T., Fernández-Caballero, A., Fernández, M.A., Mira, J., Delgado, A.E.: Visual surveillance by dynamic visual attention method. Pattern Recognition 39(11), 2194–2211 (2006)

6. Fernández-Caballero, A., Mira, J., Fernández, M.A., López, M.T.: Segmentation from motion of non-rigid objects by neuronal lateral interaction. Pattern Recognition Letters 22(14), 1517–1524 (2001)

7. Fernández-Caballero, A., Castillo, J.C., Serrano-Cuerda, J., Maldonado-Bascón, S.: Real-time human segmentation in infrared videos. Expert Systems with Applications 38(3), 2577–2584 (2011)

8. Fernández-Caballero, A., Castillo, J.C., Martínez-Cantos, J., Martínez-Tomás, R.: Optical flow or image subtraction in human detection from infrared camera on mobile robot. Robotics and Autonomous Systems 58(12), 1273–1281 (2010)

9. Dalal, N., Triggs, B.: Histograms of oriented gradients for human detection. In: IEEE Computer Society Conference on Computer Vision and Pattern Recognition, CVPR 2005, vol. 1, pp. 886–893 (2005)

10. Meysam, S., Farsi, H.: A robust method applied to human detection. International Journal of Computer Theory and Engineering 2(5), 692–694 (2010)

11. Wang, X., Han, T.X., Yan, S.: An HOG-LBP human detector with partial occlusion handling. In: IEEE International Conference on Computer Vision, ICCV 2009, pp. 32–39 (2009)

12. Zhu, Q., Yeh, M.C., Cheng, K.T., Avidan, S.: Fast human detection using a cascade of histograms of oriented gradients. In: IEEE Conference on Computer Vision and Pattern Recognition, CVPR 2006, vol. 2, pp. 1491–1498 (2006)

13. Marin, J., Vazquez, D., Geronimo, D., Lopez, A.M.: Learning appearance in virtual scenarios for pedestrian detection. In: IEEE Conference on Computer Vision and Pattern Recognition, CVPR 2010, pp. 137–144 (2010)

14. Zhang, L., Wu, B., Nevatia, R.: Pedestrian detection in infrared images based on local shape features. In: IEEE Conference on Computer Vision and Pattern Recognition, CVPR 2007, pp. 1–8 (2007)

15. Suard, F., Rakotomamonjy, A., Bensrhair, A., Broggi, A.: Pedestrian detection using infrared images and histograms of oriented gradients. In: IEEE Intelligent Vehicles Symposium, IVS 2006, pp. 206–212 (2006)

16. Bertozzi, M., Broggi, A., Grisleri, P., Graf, T., Meinecke, M.: Pedestrian detection in infrared images. In: IEEE Intelligent Vehicles Symposium, IV 2003, vol. 3, pp. 662–667 (2003)

17. Dong, J., Ge, J., Luo, Y.: Nighttime pedestrian detection with near infrared using cascaded classifiers. In: IEEE International Conference on Image Processing, ICIP 2007, vol. 6, pp. 185–188 (2007)

18. Boser, B.E., Guyon, I.M., Vapnik, V.N.: A training algorithm for optimal margin classiers. In: Fifth Annual Workshop on Computational Learning Theory, COLT 1992, pp. 144–152 (1992)

19. Cortes, C., Vapnik, V.N.: Support-vector networks. Machine Learning 10(3), 273–297 (1995)

20. Papageorgiou, C., Poggio, T.: A trainable system for object detection. International Journal of Computer Vision 38(1), 15–33 (2000)

21. Ronfard, R., Schmid, C., Triggs, B.: Learning to parse pictures of people. In: Heyden, A., Sparr, G., Nielsen, M., Johansen, P. (eds.) ECCV 2002, Part IV. LNCS, vol. 2353, pp. 700–714. Springer, Heidelberg (2002)
22. Lowe, D.G.: Object recognition from local scale-invariant features. In: IEEE International Conference on Computer Vision, ICCV 1999, vol. 2, pp. 1150–1157 (1999)
23. Chang, C.C., Lin, C.J.: LibSVM: a library for support vector machines. ACM Transactions on Intelligent Systems and Technology 2(3), 1–27 (2011)
24. Keerthi, S.S., Sundararajan, S., Chang, K.W., Hsieh, C., Lin, C.J.: A sequential dual method for large scale multi-class linear SVMs. In: Proceeding of the 14th ACM SIGKDD International Conference on Knowledge Discovery and Data Mining, pp. 408–416 (2008)
25. Fan, R.E., Chang, K.W., Hsieh, C.J., Wang, X.R., Lin, C.J.: LibLINEAR: a library for large linear classification. Journal of Machine Learning Research 9, 1871–1874 (2008)

Smart Spaces and Monitoring Simulation

Coral García-Rodríguez, Rafael Martínez-Tomás, and José Manuel Cuadra-Troncoso

Dpto. Inteligencia Artificial. Escuela Técnica Superior de Ingeniería Informática
Universidad Nacional de Educación a Distancia, Juan del Rosal 16, 28040 Madrid, Spain

Abstract. This piece of work is framed in our group research and is about behaviors identification of the smart spaces monitoring. Particularly, if this monitoring uses cameras, some habitual problems related to lights and shadows in the scene because they make the task of recognition difficult, the identification of objects and even the images captured to analyze the correct functioning of these systems appear. However, the experimentation is difficult in any case because the installation of cameras and other sensors are laborious and it can entail a high cost. Thus, to solve these problems we are developing a simulator in order to create a scene with actors, objects and sensors, which the user requires and can reproduce a defined plot. Thereby we can get two objectives: a) Check the correct location and appropriate characteristics of the equipment installed, virtually, and b) The system can emulate the event generations of the real system and with this it can confirm the utility of the installation. In the global frame of our research, besides, it allows to unhook of the physical level and focus on the behavior interpretation from that virtual smart space. Thus, the problems of generating images and events, which probe the correct functioning of the systems and other possible failures derived directly from the images captured by the devices, are solved.

The proposal is showed in this article and is also exemplified with a possible real scene, developed indoors. The objective consists of elaborating a preventive diagnosis about the activity monitorized and detecting what is happening in a particularly moment. In all, it simulates smart spaces and monitoring behaviors. This piece of work emphasizes: (1) The capacity of generating a lot of scenarios with innumerable characteristics, reproducing scenes which exist in real life, which implies that simulated sensors generate the correct events, events obtained from the simulated scene observe. (2) From the simulated events or from the virtual images directly, the interpretation of the actions performed by the actors in the plot and the interaction with objects placed on stage, getting all these events associated to the activities observed.

1 Introduction

At present it seems necessary the human behavior monitoring as a vital element to prevent the health and the safety of people. To ensure that we use many different sensors and videos. Nowadays there are a lot of work dedicate to the low level activity identification (as enter or go out in a perimeter) which can interpret as a simple group of sensors. With this line of work we want to be able to identify automatically high level activities or human behaviors.

J.M. Ferrández Vicente et al. (Eds.): IWINAC 2013, Part II, LNCS 7931, pp. 190–199, 2013.
© Springer-Verlag Berlin Heidelberg 2013

There are some precedents which have being detecting low level activities; one of these examples is Frédéric Cupillard, François Brémond and Monique Thonnat[6] with their studies about it or M. A. Ali, S. Indupalli and B. Boufama[2].

Our group is developing a project at the Instituto de Investigación en Informática de Albacete in collaboration with the Universidad de Castilla La Mancha called *Horus*[5][12] which is dedicated to multisensory recognition, specially with videos. This tool has a set of sensors data as the input which develops a specific plot, and through it we get the interpretation of what is occurring in the input sequence. *Horus* is a system composed of plenty of internal modules with different functions: *Data Acquisition, Sensor Fusion, Pixel-level Segmentation, Pixel-level Segmentation Fusion, Blob Detection, Object Identification, Object Classification, Object Tracking, Event Detection, Event Fusion* and *Activity Detection*. During these phases the system analyzes the input pictures and gets two outputs for this analysis: objects and events of the plot obtained.

Because of this situation described previously, we have the necessity of developing a simulator which facilitates the creation of scenarios that show context and situation.

This project continues with the line of research which has been carried out with the project Avisa, Avisa-2 and INT-3. The proposed simulator influences the multisensory recognition directly. An approximation of the role of the tool appears in Figure 1:

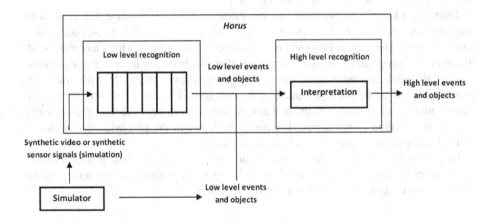

Fig. 1. Function scheme

As the figure shows, the simulator proposed is responsible for providing synthetic videos (or images) extracted from the simulation after the creation of the scenario and defining, executing the plot, and generating the set of objects and events that occurred during the simulation created and simulated therein.

It is noteworthy that the outputs of the simulator and multisensory recognition *Horus* project are the same type, these are: the objects in the plot and the events produced by either the simulator or *Horus*. These outputs should be equivalent because this shows that multisensory recognition is right and works correctly. As the simulator creates virtual scenes and generates events, it is easy to integrate it in *Horus* and compare the multisensory recognition results of *Horus* and the multisensory recognition results of our simulator. Besides to work in this high semantic level, the simulator avoids the data collection problems or other vision problems (like lights, shadows, positioning sensors in the environment and so on) because the sensors can be configured as the user wants.

To develop the simulator many possibilities were considered. At present, a lot of videogames and online games based on simulators of virtual reality exist. Videogames could be classified as one gender or another depending on their graphical representation, the type of interaction between the player and the machine, the settings and game system and so on. These are summarized in three genres: *Action, Adventure* or *Simulation*. The latter includes two subtypes that have been a crucial model for the development of the application that is proposed here: *building simulation* and *life simulation games*. *Building simulation games* are based on the program that provides the user with all the tools to build a project, which should be as realistic as possible. This kind of games have one peculiarity: the user can experiment, make decisions and affect the performance of the simulation created. On the other hand, *life simulation games* exist. In these the simulator focuses on controlling a character in all aspects of his life. One of the most famous examples of this are *The Sims*[7] and *Second Life*[9]. The realism with which these games are made and the timeline which allows the actor to evolve, eat, sleep, live and interact with other people, growing old and even dying. In both cases, the feeling and the closeness to reality are what make them so popular.

There are a lot of examples about games based on virtual simulation, but at present there are not developed projects that use similar simulators to develop tasks out of the gaming world. The project proposed is one of these few projects which has innovated this area, as a proposal to develop a simulator that solves the creation and design tasks of scenarios and development of user-model plots. Thanks to this tool the user can produce a scene as he likes, recreating real or imaginary scenarios in which you can place a number of objects, sensors and actors. The sensors introduced in the scenario capture information of the scene. The actors included in the plot will develop user-defined actions and interact with the objects in the scene.

In section 2, we describe the proposed simulator as well as the event generation that simulator produces. In section 3, we show and explain an experimentation stage, and in section 4, we describe the conclusions of this research.

2 Proposal

2.1 Simulator Description

The simulator which is going to be created has to be able to modeling a 3D scenario and simulate a staging using it. To perform this, the simulator is going to be divided into two different parts:

- **Components**: this part consists in modeling a scenario to suit the user, imitating a real scenario or inventing an imaginary one. This will create and define:
  - Scenario: the stage can be created by the user step by step introducing walls, floor, doors, windows and so on. If the user does not need a specific scenario, he can also choose one from a predefined scenarios provided by the simulator.

- Objects in the scene with which the characters can interact: furniture and other objects can also introduce in the scene. Characters can interact with some of these objects (doors, windows, bags and so on).
- Actors: characters will be introduced on the stage. They are who are going to perform the different actions in the plot, interacting with the environment modeling. The objective is simulating most of the interesting movements that a human can perform, but actually, these actors are be able to execute several actions like walking, running, jumping, crouching down, sitting and so on.
- Sensors: many sensors can be introduced in the scenario, their function is to collect information on the plot. The function of cameras will be recording the scene and what is happening in it.
- The plot: after creating the scenario, objects and sensor. The user must define the sequence of actions that actors will perform. To do that, the user select the position where he wants the actor to move or the object which he wants the character to interact. The ideal concept is the user can introduce a narrative description of the plot in the simulator and automatically the simulator reproduces it, but actually the user has to point a trajectory with the positions and actions which actor has to represent.
- **Simulation**: it consists in represent the actions defined. In this simulation, the actors represent the actions that are specified in the previously mentioned plot, on the *Creation and definition*. This simulation has two different parts:
  - Simulation of the plot: the characters perform the behavior previously defined. This part is corresponds to the evolution of the actors of the plot.
  - Simulation of the monitoring and sensing: depending on what is captured from the sensors some images are recorded and some events are generated. These two outputs are produced in parallel, at the same time. These are:
    * The signal: sensor data. In the case of cameras the signal will be images, and in the case of another sensor the signal will be, for example, the activation of the sensor.
    * The eventualization: the simulator converts this kind of signals in medium level events, not engaged with sensors.

The following subsection shows the event types.

Eventualized Scenario. There are three levels in the scenario:

- The first level is engaged with sensors. It consists in information captured directly by the sensor. This information is composed by events with information like the sensor identifier, the sensor location, the identifier of the person who has just activated de sensor or the instant when this action occurred. It generates low level events, for example, this level would produced an event like *RFID(identifier-sensor, localization-sensor, identifier-person, instant)*.
- The medium level is not engaged with sensors but it depends on the context of the scene. From the low level events generate in the first level, the simulator extracts the medium level events. For example, from the information of the low level the simulator extracts in which room the sensor with the identifier registered are, or

which person had entered in this room from the person identifier registered on the low level. A example of it would be the event *Enters(actor, room, instant)* or *Begins-to-walk(actor, position, instant)*.

– The high level is the level in which events not engaged with sensors are abstracted and interpreted resulting high level events (composed events). The vision of these events is extracted from the multisensory processing [3] [10]. An example of this could be the event *Leaves(actor, object, (position, instant))* [11].

One of the main sensors which exist in the monitoring are cameras, because of that we emphasize the video events. There are two types of events[11]: output events at an activity level, according to the result of continuous monitoring of the level or activity request, for example *Appears, Begins-to-walk, Detected-man-object, Disappears, Is-close-to, Is-going-to* or *Not-detected-man-object*; and instantaneous events used in the monitoring of the scene, for example *Holding, Not-holding, Standing, Stopped* or *Walking*. These events have some variables values like a which corresponds to the actor who perform the action, x which corresponds to the position in the scenario, t which corresponds to the instant when the action is performed, o which corresponds to the object which is involved in the scene and d which corresponds to the door which the actor is near to. The format of these events are *Begins-to-walk(a Actor, (x (PosX, PosY, PosZ)), t Instant)*, and an example of this could be *Begins-to-walk(a Actor 1, (x (586.4, 0.0, 994.6)), t 46.87)*. To obtain more information about the events you can consult the references [10] and [11].

Instead of the previous simple events, composed events exist that can be inferred to the previous simple events. An example of this is the *Leaves* event, which could be inferred from the following events and situation: *if the actor is carrying an object, after that the object appears at one position and the actor is not carrying the object, it means that the actor has abandoned the object*. These events have the same variables that the previous events. The format of these events are *Carrying(a Actor, o Object, ((x (PosX-Initial, PosY-Initial, PosZ-Initial)), t Initial-Instant), (x' (PosX-Final, PosY-Final, PosZ-Final)), t' Final-Instant))*, and an example of this could be *Carrying(a Actor 1, o Bag 1, ((x (996.6, 0.0, 995.7)), t 35.13), (x' (1000.8, 0.0, 994.2)), t' 43.09))*. As previously, to obtain more information about the events you can consult the references [10] and [11].

2.2   Tools Used

Different tools were used to create the simulator proposed, the main tools were Unity 3D in first place and 3DS Max in second place.

3DS Max[1] is used in character modeling, structures and objects that make up the plot, the scenario and the animations of each of them (characters moving, doors opening and closing and so on).

Unity 3D[13] was used to create the rest. Characters, objects and structures modeled in 3DS Max are imported into Unity 3D. After this, the interface is defined in the program, and thanks to this all the possibilities of simulating plots and scenarios may be defined. These are: the creation, adaptation and modification of scenarios, characters, objects and other sensors involved in the plot and the definition of the actions which actors will perform. After defining the plot, Unity 3D will also be the program which permits its implementation and simulation, producing the events associated with each action, object, sensor and actor and extracting image frames (or other data) which allow, if necessary, the creation of a video with the sequence of images. Besides this, this tool was chosen because it is one of the game creation programs more suitable due to the powerful rendering engine which it has, like the performance and frame rate, which is vital in capturing images, one of the objectives that the simulator pursues. Is necessary to stress the animations that characters and objects in the scene can also have, as well as the details of the plot that can be added or deleted, such as the lights and shadows, something vital for multisensory recognition. It is also noteworthy that Unity 3D is a technological multiplatform, which means that making small adjustments to the code, the application could run the same simulation on computers with different operating systems (Linux or Mac) in addition to that currently used (Windows).

Programming Languages. The language used in the development of the simulator was JavaScript, the language used by the Unity 3D tool. JavaScript is an interpreted programming language. It is defined as object-oriented, prototype-based, dynamic and weakly typed.

3   Application Case

Figure 2 shows an example of the use of the application in which the possibilities of the tool and the outputs that it produces is detailed.The aim of this application case is to reproduce a situation in which a person leaves an object in a strange place as a symptom of disorientation.

As is explained above, the tool allows the creation and definition of the scenario. There are two possibilities: the user creates and mounts the whole scenario, step by step, or chooses a scenario that the simulator gives among the examples. In this case it has created a home and a woman (the actor), and she has been introduced into the living-room (the room located in the centre of the house). There is also a bag (an object) at one side of the living-room. Note that there is a menu in the upper left side which can be used to modify, create or delete elements of the stage, actors and objects.

Another aspect to consider is the sensors added to the scene. In this case it has introduced two video cameras, one in the living-room and another in the kitchen.

The simulator operates in four different sections where four different cameras can be associated, as shown in Figure 3. In this case, as there are only two cameras, only two are associated. The allocation of these cameras to the sections is marked at the bottom right of the menu, while the view from each of the cameras is shown at the bottom left.

After the creation, definition of the stage, its objects, sensors and actors are finished, the plot is defined. In this example the plot consists of the following steps:

1. The actor starts standing on the left of the living-room.
2. Walks around the living-room to get to the bag, where she stops.
3. Picks up the bag.
4. Leaves the room.
5. Enters the kitchen.

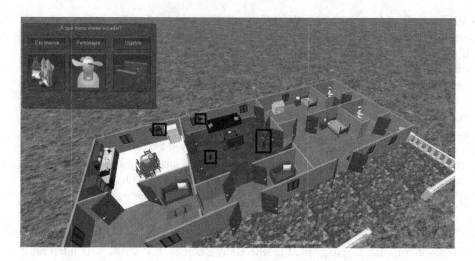

Fig. 2. Model creation

Fig. 3. Configuring the sensors

6. Walks up to a given point of the kitchen, where she stops.
7. Leaves the bag as a sign of disorientation or confusion.
8. Goes away from the bag.

After that some images belonging to each step may be seen at Figure 4. Note that two rooms exist in this figure, the living-room and the kitchen, and the plot is performed in both of them. The first four images are about the plot development in the living-room and the last four images are about the plot development in the kitchen. As logical, there are two cameras involved in this plot which follow the movements of the actor and generates de events [4].

This simulation generates the following events [10] [11].As we saw before, the action was divided in eight steps. These steps encompass the following events:

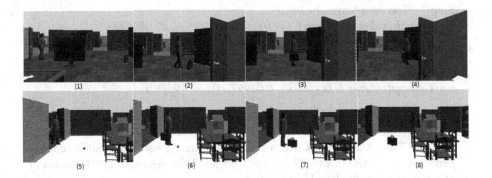

Fig. 4. Camera vision

- **Step 1** - The actor is stopped in the living-room:
  - *Not-detected-man-object(a Actor 1, (x (996.6, 0.0, 995.5), t 35.1312))*

- **Step 2** - The actor walks towards the bag and stops near to it:
  - *Begins-to-walk(a Actor 1, (x (996.6,0.0, 995.5), t 35.1312))*
  - *Stops(a Actor 1, (x (1000.6, 0.0, 994.6), t 43.06987))*
  - *Walking(a Actor 1, (x (996.6, 0.0, 995.5), t 35.1312), (x' (1000.6, 0.0,994.6),*
    t' 43.06987))

- **Step 3** - The actor picks up the bag:
  - *Crouches-down(a Actor 1, (x (1000.6,0.0, 994.6), t 43.06987))*
  - *Detected-man-object(a Actor 1, o Bag 1,(x (1000.6, 0.0, 994.6), t 43.06987))*
  - *Holding(a Actor 1, o Bag 1, (t 35.1312, t' 44.60263))*
  - *Disappears(o Bag 1, (x (1001.1,0.0, 994.6), t 44.60263))*
  - *Gets-up(a Actor 1, (x (1000.6, 0.0, 994.6), t 44.60263))*

- **Step 4** - The actor leaves the living-room:
  - *Begins-to-walk(a Actor 1, (x (1000.6, 0.0, 994.6), t 46.11248)*
  - *Is-going-to(a Actor 1, d Door 11, t 48.36957)*
  - *Is-close-to(a Actor 1, d Door 11, t 48.41647)*
  From this set of events the event *Goes-out* of the living-room is inferred.

- **Step 5** - The actor enters the kitchen: there are no events involved in this step, but
  thanks to step 5 knowing that the actor left the living-room and knowing that this
  specific door is between the living-room and the kitchen, it can be inferred that the
  actor has just entered in the kitchen, provoking the event *Enters*.

- **Step 6** - The actor walks up to a given point of the kitchen:
  - *Stops(a Actor 1, (x (1003.9, 0.0, 993.1), t 51.96685))*
  - *Walking(a Actor 1, (x (1000.6, 0.0, 994.6), t 46.11248), (x' (1003.9, 0.0, 993.1),*
    t' 51.96685))

- **Step 7** - The actor leaves the bag:
  - *Crouches-down(a Actor 1, (x (1003.9, 0.0, 993.1), t 51.96685))*
  - *Not-detected-man-object(a Actor 1, o Bag 1, (x (1003.9, 0.0, 993.1), t 51.96685))*
  - *Not-holding(a Actor 1, o Bag 1, (t 44.60263, t' 53.49518))*
  - *Appears(o Bag 1, (x (1004.4, 0.0, 992.9), t 53.49518))*
  - *Gets-up(a Actor 1, (x (1003.9, 0.0, 993.1), t 53.49518))*

- **Step 8** - The actor goes away from the bag:
  - *Begins-to-walk(a Actor 1, (x (1003.9, 0.0, 993.1), t 53.77924))*

  The step 7 and 8 are actions which could implicate the person is disoriented, depends on the context in which he is involved, and from this set of events the alarm of a patient disoriented could be inferred.

These events extracted between step 1 and 8 are the inputs of *Horus*, and thanks to them it can infer the high level event which implies "The person *p* has abandoned the object *o* in the *kitchen*" according to the article [11]. It is also important to emphasize that events generated by the simulation of the proposed application are the same as *Horus* multisensory recognition of would provide, so that we could extract the same conclusion in both cases. Note, therefore, that it would be really easy to check the reliability of multisensory recognition, so that it would be enough to compare the two outputs generated by the simulator and the multisensory recognition. Besides, thanks to the simulator, it can check if there are enough sensors in specific allocations to obtain the conclusion what the user is looking for, and if not, the user can add more sensors of modified the sensors which are using to obtain it.

4   Conclusions

This paper introduces a new type of 3D simulator that can create and model scenarios step by step, objects, actors and sensors which will be present in the scene. With all these components the user can define the actions which the characters will perform and represent the plot based on those actions defined. Images and other sensor data are obtained of each of cameras or other sensors and the information relating to the sensors embedded in the scene are obtained from this plot. A file events extracts from the actions that the actor is performing on stage. The sequence of images obtained from the simulator are used for input for other projects dedicated to multisensory recognition. From this recognition a list of events is generated, these events will compare with the events generated by the simulator and thanks to them is possible to verify the correct functioning of the project dedicated to multisensory recognition knowing if it is necessary to add more sensors to solve these problems.

As an added benefit, the simulator can also be used to create realistic scenarios and test the sensors placements and orientations, which allows transfer this result to a real scenario saving the time and the money which would be spent on testing them.

What is presented here is a tool with many possibilities. This tool reduces a lot of time and costs eliminating the need for real videos and the time that takes to record a video which serves as input to the multisensory recognition project. It also saves time in the analysis of the results obtained from the multisensory recognition, because instead of looking at events obtained result by result they can be used the events generated by the simulator and compare the two output files. It also emphasizes the easily and useful of the tool, because it has a simple and basic interface where the user can define their specific scenarios and plots quickly and with no charge.

A future application of this piece of work includes the analysis of the visual problems, like lights and shadows which distort the images that cameras record, or the noise which influences in the acceptance of the sensors, anticipating it.

Acknowledgements. The authors are grateful to the Spanish CiCYT for financial aid for the project TIN-2010- 20845-C03-02.

References

1. Autodesk, Inc. 3ds max (2001)
2. Boufama, B., Ali, M.A.: Tracking multiple people in the context of video surveillance. In: Kamel, M., Campilho, A. (eds.) ICIAR 2007. LNCS, vol. 4633, pp. 581–592. Springer, Heidelberg (2007)
3. Rivas-Casado, Á., Martínez-Tomás, R.: Event detection and fusion model for semantic interpretation of monitored scenarios within ASIMS architecture. In: Ferrández, J.M., Álvarez Sánchez, J.R., de la Paz, F., Toledo, F.J. (eds.) IWINAC 2011, Part I. LNCS, vol. 6686, pp. 521–530. Springer, Heidelberg (2011)
4. Casado, A.R., Martínez-Tomás, R., Fernández-Caballero, A.: Multi-agent system for knowledge-based event recognition and composition. Expert Systems 28(5), 488–501 (2011)
5. Castillo, J.C., Fernández-Caballero, A., Serrano-Cuerda, J., Sokolova, M.V.: Intelligent monitoring and activity interpretation framework - int^3-horus general description. In: Graña, et al. (eds.) [8], pp. 970–979
6. Cupillard, F., Brémond, F., Thonnat, M.: Group behavior recognition with multiple cameras. In: WACV, pp. 177–183. IEEE Computer Society (2002)
7. Electronic Arts, Inc. The sims (2001)
8. Graña, M., Toro, C., Posada, J., Howlett, R.J., Jain, L.C. (eds.): Advances in Knowledge-Based and Intelligent Information and Engineering Systems - 16th Annual KES Conference, San Sebastian, Spain, September 10-12. Frontiers in Artificial Intelligence and Applications, vol. 243. IOS Press (2012)
9. Linden Research, Inc. Second life (2003)
10. Tomás, R.M., Casado, A.R.: Knowledge and event-based system for video-surveillance tasks. In: Mira, J., Ferrández, J.M., Álvarez, J.R., de la Paz, F., Toledo, F.J. (eds.) IWINAC 2009, Part I. LNCS, vol. 5601, pp. 386–394. Springer, Heidelberg (2009)
11. Martínez-Tomás, R., Rincón, M., Bachiller, M., Mira, J.: On the correspondence between objects and events for the diagnosis of situations in visual surveillance tasks. Pattern Recognition Letters 29(8), 1117–1135 (2008)
12. Sokolova, M.V., Castillo, J.C., Fernández-Caballero, A., Serrano-Cuerda, J.: Intelligent monitoring and activity interpretation framework - int^3-horus ontological model. In: Graña, et al. (eds.) [8], pp. 980–989
13. Unity Technologies. Unity 3d (2009)

Experimental Study of the Stress Level at the Workplace Using an Smart Testbed of Wireless Sensor Networks and Ambient Intelligence Techniques

F. Silva[1], Teresa Olivares[2], F. Royo[2], M.A. Vergara[2], and C. Analide[1]

[1] University of Minho, Braga, Portugal
[2] Albacete Research Institute of Informatics
University of Castilla-La Mancha, 02071 Albacete, Spain
{fabiosilva,analide}@di.uminho.pt, {teresa,froyo,vergara}@dsi.uclm.es

Abstract. This paper combines techniques of ambient intelligence and wireless sensor networks with the objective of obtain important conclusions to increase the quality of life of people. In particular, we oriented our study to the stress at the workplace, because stress is a leading cause of illness and disease. This article presents a wireless sensor network obtaining information of the environment, a pulse sensor obtaining hear rate values and a complete data analysis applying techniques of ambient intelligence to predict stress from these environment variables and people attributes. Results show promise on the identification of stressful situations as well as stress inference through the use of predictive algorithms.

Keywords: Ambient Intelligence, Intelligent Environments, Wireless Sensor networks, Body Area Networks, Environmental Monitoring, Stress Detection.

1 Introduction

Ambient Intelligence (AmI) is considered one of three emergent technologies: Ubiquitous Computing, Ubiquitous Communication and Intelligent User Interfaces. AmI systems aim to change how people interact with technology and the environment integrating concepts from psychology, social sciences and artificial intelligence to increase the quality of life. AmI makes this possible by anticipating and predicting future needs and desires while taking in consideration aspects like safety, economy and comfort. One concept usually linked to AmI is ubiquitous computing, a concept proposed in [1] by Mark Weiser. In this environment, computational units are embedded in its surroundings functioning and hidden from view.

Directly connected with ambient intelligence is also how the sensing of the environment is done. In order to deliberate which actions an intelligent system may do in an environment it is necessary to obtain the information of the

J.M. Ferrández Vicente et al. (Eds.): IWINAC 2013, Part II, LNCS 7931, pp. 200–209, 2013.

environment needs. Significant levels of data must be constantly and ubiquitously collected to provide the data and information needed.

Body Area Networks (BANs) allow the integration of miniaturized and low power sensor nodes in, on, or around the human body to monitor body functions and the surrounding environment [2][3]. BANs provide long term health monitoring of patients under natural physiological states without constraining their normal activities. The typical architecture of a BAN includes sensor nodes deployed in, on or around the body and a BAN coordinator that gathers data coming from all the sensor nodes. This BAN coordinator can store the information for further analysis or can forward it to an intermediate network that serves as a gateway to send the data through the Internet to the expert user (i.e., nurses, doctors, ...) placed in a smart space. Our experiments will use a wireless sensor testbed to emulate this intermediate network. A testbed of wireless sensor nodes provides a reliable platform where to test new protocols and applications in a controlled way.

BANs enable continuous measurement of physiological parameters, such as heart rate, muscular tension, skin conductivity, breathing rate and volume, during the daily life of a user. Those parameters can be combined with contextual information extracted from the environment through wireless sensor networks (temperature, humidity, light, etc.), and all together could be used to infer emotions, mood, depression, and levels of stress and anxiety. Using BANs for emotion, mood, stress or depression recognition is an open research problem and requires development of novel signal processing techniques to interpret and fuse the data collected by multiple sensors [4].

In this work we are going to use a testbed of wireless sensor networks and a BAN. The testbed of 43 fixed nodes is continuously gathering different environment parameters (temperature, humidity, light and CO2). The BAN, composed of a pulse sensor node, is continuously gathering heart rate values. This two networks work together with the objective of detecting possible situations of stress at the workplace using ambient intelligence techniques. This work join two interesting research lines (sensor networks and ambient intelligence) with the purpose of achieve innovative conclusions.

According the World Health Organization a healthy job is likely to be one where the pressures on employees are appropriate in relation to their abilities and resources, to the amount of control they have over their work, and to the support they receive from people who matter to them. As health is not merely the absence of disease or infirmity but a positive state of complete physical, mental and social well-being, a healthy working environment is one in which there is not only an absence of harmful conditions but an abundance of health-promoting ones [5].

Environment and physical working conditions are important organizational risk factors [6]. Our research takes the evaluation of different important parameters related with environment and physical working conditions, using a testbed of wireless sensor network and the extension of the testbed, with sensors installed in mobile elements (BANs). The whole wireless network transmits the data to the right base station to store data for a following data analysis. This paper focuses on both, the evaluation with the tesbed and the data analysis.

The rest of this paper is organized as follows: Section 2 summarizes some related work. The testbed used is detailed in Section 3. Section 4 describes the analysis of environment and user data. Finally, Section 5 gives some concluding remarks and interesting future works.

2   Previous Work

2.1   Stress Monitoring Using Wireless Sensor Networks

There are some interesting works about emotion recognition and stress monitoring with wireless sensors. In [7] we can see a wearable system for ambulatory stress monitoring recording a number of physiological variables known to be influenced by stress. This system is not ZigBee based, it employs a radio module from TI (Texas instruments Inc.), a connection-based low-power lightweight sensor network protocol for small RF networks supporting two basic topologies: strictly peer-to-peer and a star topology. The authors use a star topology where all sensor (leaf nodes) are connected to a sensor hub (the root). If this root node fails the system would support permanent disconnections. Regarding this issue, our mobile nodes send data to the testbed nodes actuating as parents. If a failure is produced in one node, any other node can act like a parent node. In [8], the author gathers values from a skin temperature sensor, a heart rate sensor and a skin conductance sensor. The signals from the sensors are input into a microcontroller where all the processing takes place and carried out though ZigBee technology. Data are stored in a computer. It is stored for data analysis and feature extraction for emotion recognition. The four basic emotions observed in this project are happy, sad, angry and neutral. The system can be used by anyone sitting within a range of 30m of a computer. The sensors used on this system are not too much "wearable". There is an special surface with sensors where a person has to put his hand. Subjects chosen to take part in the experiments have are sitting for neutral state.

2.2   Ambient Intelligence Techniques

Monitor people well being and increases in stress are with the problems covered by ambient intelligence and ubiquitous monitoring. As such, there are model proposals to explain and potentially predict any undesirable effects considering a series of factors believed to influence user/occupant behaviour [9]. Other approaches, relate environment attributes to people attributes such as heart rate which may be used to infer thermal comfort levels and detect stress induced by environment attributes and unoptimised values [10].

In Paola et al. [11], ambient intelligence is used within a sensor network testbed to derive useful from sensor spread across the environment. The data retrieved from sensors is used to derive rich information such as user attendance, perceptible temperature, user activity and user preferences. All this information is learned unobtrusively by the system fusing sensor data without requiring user attention or direct user input.

Learning in ambient intelligence is often made using machine learning algorithms such as time series algorithms, evolutionary algorithms and statistical inference. These methods sanction the acquiring of past and current trends and predict future results. From information assembled from different environments, machine learning techniques may derive models of behaviour and interaction based on specialized backgrounds (e.g., users, environment, social interaction or consumption). For such analysis the use of statistical programs such as R [12] and machine learning workbenches such as Weka [13] is common. Such knowledge about environment and user attributes is usually stored in knowledge databases which are used in deliberative actions or to derive reactive rules to be applied in the system.

3   The Testbed as a Powerful and Continuous Smart Tool

The Albacete Research Institute of Informatics (I3A) has its own testbed of wireless sensor network deployed in its facilities, known as I3ASensorBed [14]. This allows researchers to test in controlled environments. The testbed consists of 43 fixed nodes including temperature, humidity and light sensors, opened/closed doors and windows, movements, air condition (CO, CO2 and dust) and power consumption. All these sensors allow I3ASensorBed monitor weather conditions and the possibility to add mobile nodes with additional sensor, as hearth rate [15]. The distributed nodes are TelosB from Crossbow, TmoteSky from Sentilla and MTM-CM5000-MSP from Maxfor. It also uses nT-A3500 computers from Foxconn, known as supernodes, to redirect data. To facilitate re-program of nodes and debugging of tests, nodes are connecting through USB cable to the nearest supernode. This supernode is the responsible to redirect information to the central server. Along with the existing wiring, the nodes have the standard 802.15.4 (ZigBee) for wireless communications emulating realistic environments. The use of the testbed that covers a large work area with multiple workers allows the collection and study of data that is of interest in several areas, especially in the creation of an appropriate working environment for the workers. This is an ideal smart space to contribute to the measurement and experimentation of people health, safety and well-being.

4   Analysis of Environment and User Data

In this section, the use of ambient intelligence to unobtrusively detect and assess stress situations is made using information gathered by the sensor network detailed in section 3. The data was collected under real work conditions from an wireless pulse sensor, and embedded temperature, luminosity and humidity sensor for periods of time ranging from 10 minutes to 2 hours. Using these values, an heart rate change algorithm was validated and used to learn about stressful conditions in the environment.

4.1   Preliminary Study

It is possible to observe that values for environment attributes values are constantly changing throughout the day, see figure 1(a), and it is even possible to detect human actions to correct values such as luminosity during some periods of the day, see figure 1 (b). Those corrections are made with artificial lightning to increase luminosity values during work hours and reach comfortable values for people. Such action are taken in order to decrease stress caused by poor environments and facilitate the activities being made.

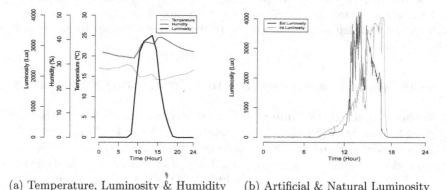

(a) Temperature, Luminosity & Humidity   (b) Artificial & Natural Luminosity

Fig. 1. Environment Monitoring

In the literature review, it is possible to infer a range of ideal values for attributes that while not being directly responsible for health risks, they should increase stress on people. Table 1 demonstrates the use of information about the activity are performing and the values for luminosity deemed acceptable.

Table 1. Luminosity values required by activity

Activity	Luminosity Required
Short Visit	[50, 100]
Working areas with few visual tasks	[100, 150]
Office Work	[250, 500]
Mechanical Workshops, Office Landscapes	750
Normal Detailed Work	1000
Detailed Work	[1500, 2000]

Likewise, thermal comfort is an area of study that has seen important contributions to assess thermal discomfort and thermal stress in people. The equation of human heat balance is one theory commonly used to derive algorithms that capture the thermal sensation from people, rather than just environment temperature. One theory based on these principles is the Predictive Mean Vote (PMV), a scale from -3 to 3 that classifies thermal sensation and has a correspondence to thermal discomfort [16]. Other approach uses an index that translates temperature readings in an environment to a set of stable conditions inside a room environment obtaining what is known as Physiological Equivalent Temperature (PET) [17]. Relationships from both algorithm are explored in Table 2, as well as, indication of thermal stress. The relationship with real temperature values in the environment is dependent on variables such as air speed, clothing, temperature and humidity which change human thermal perception. Studies confirming an influence of thermal conditions and variations in heart rate in people are found in Liu et al, where a medical exam ECG to confirm such relationship between these attributes [10].

Table 2. Temperature Perception

PMV	PET	Thermal Sensation	Stress Sensation
< -3.5	< 4	Very Cold	Extremely cold stress
$[-3.5, -2.5]$	$[4, 8]$	Cold	Strong cold stress
$[-2.5, -1.5]$	$[8, 12]$	Cool	Moderate cold stress
$[-1.5, -0.5]$	$[12, 16]$	Slightly cool	Slight cool stress
$[-0.5, 0.5]$	$[16, 24]$	Neutral	No thermal stress
$[0.5, 1.5]$	$[24, 28]$	Slightly warm	Slight warm stress
$[1.5, 2.5]$	$[28, 32]$	Warm	Moderate warm stress
$[2.5, 3.5]$	$[32, 36]$	Very Hot	Strong warm stress
> 3.5	> 36	Very Hot	Extremely warm stress

In order to validate stress levels with thermal and luminosity values, values from heart rate of people can be used to evaluate if significant changes are occurring.

The use of different sensors enables the application of sensor fusion projects empowered by sensor networks that produce information contextualized to the environment and the people on it. As such, a number of preliminary tests conducted in order to assess the viability of the analysis being undertaken. For these analysis, a test subject was monitored in different thermal situations, in comfortable situation and feeling cold. The statistical t-tests were conducted assuming a normal distribution for the values of the heart rate, a 95% significance value and hypothesis concerning an increase, decrease and no change in the average heart rate of the test subject. These tests found statistical evidence that heart rate increased when on the situation where the test subject felt cold with a $p < 0.05$, thus accepting the hypothesis that the heart rate average increased. It is important to notice that in these tests, the test subject did not change its emotional state or any other factor that could affect heart rate values and as such these changes in heart rate averages are concluded to have been caused from environment induced stress that raises the test subject heart rate. Such analysis corroborates other studies that point the same developments in similar conditions and validates the ability for the testbed to produce significant values that can be used in the analysis of stress using values from existing sensors [10]. The same approach with similar sensors was also developed to infer people emotional states from skin temperature sensors and heart rate sensors [18], adding evidence that people react to changes in values of attributes such as temperature.

4.2   Detection of Change in Attribute Values

In order to detect changes in monitored attributes it is necessary to identify points in time that show increases or decreases that have statistical significance for the analysis. Taking heart rate of people as an example, the statistical t-test assuming a normal distribution for heart rate values can be use to create a time series algorithm to detect changes in the average of heart rate. The approach undertaken uses windows of data gathered by a testbed and continuously assesses whether or not the average heart rate is increasing, decreasing or neutral over windows of data with t-test with a significance of 95%, see figure 2.

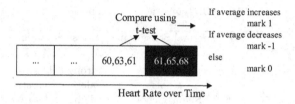

Fig. 2. Window Analysis

Using this strategy it is possible to create a counter variable, adding all the marks created over time, that provides a contextual information on the balance between increases and decreases in heart rate. An illustration of these counters can be observed in figure 3(a). Substituting the counter value for the average heart rate being used we can produce the same graph but in relationship to real heart rate values, figure 3(b).

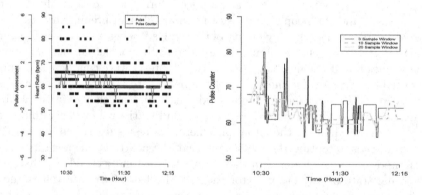

(a) Heart Rate vs Change Counter   (b) Heart Rate Average Detection over Different Windows of Data

Fig. 3. Heart Rate Monitoring and Assessment

The size of the windows of data is important as it influences the number of detections. Big windows detect less changes while smaller windows detect more changes. Moreover, it is also observed that the bigger the window of data the more stable

Table 3. Number of change detection by windows size and attribute

Size of Window	Heart Rate	Temperature	Luminosity	Humidity
3	41	306	104	281
10	22	79	49	77
20	15	36	24	36

change detections are. Table 3 demonstrates the number of detections using different windows of data for different attributes monitored in the environment.

4.3   Predicting Stress through Environment and People Attributes

A person's heart rate is influenced by a number of factors such as emotional state, stress or physical activity. In the study presented, evidence suggests that environmental variables affect heart rate. As it was presented, comfort values for attributes such as temperature, luminosity or humidity affect the heart rate of people inside an environment. Furthermore, it is possible to link changes in temperature, luminosity and humidity with changes in people's heart rate thus validating the use of comfort values for human well being and stress reduction.

Using datasets created using the attributes monitored from the environment and change markers derived from the attribute change detection algorithm it is possible to train predictive classifiers to predict according to current environment condition how people heart rate will respond. For instance using classifier algorithm from the Weka workbench [13] it is possible to assess with relative precision whether the average heart rate will increase, decrease or maintain. Aside the use of a counter variable to the balance between changes, indicator marking the moment an increase and decrease on the average of each attribute, temperature, humidity luminosity and heart rate as well as the l heart rate value. With the dataset generated, some experiments were used on machine learning algorithms, namely the Naive Bayes and J48 algorithms to predict the increases and decreases in heart rate average using. The first algorithm is probabilistic and it builds conditional probability structure to classify instances of data. On the other hand, the J48 is an implementation of the C4.5 decision tree algorithm which uses an decision tree to classify instances. Both algorithm provide an internal structure that is understandable an may be used to see how changes in attribute values change the output attribute.

Table 4. Heart rate change prediction from environment attributes

Data window	Algorithm	Correct	Correct Up	Correct No Change	Correct Down
3	Naives Bayes	88.7%	0%	100%	0%
3	J48	88.5%	0%	99.7%	0%
10	Naives Bayes	63.25%	0%	77.5%	36.3%
10	J48	72.5%	0%	100%	0%
20	Naives Bayes	62.2%	37.5	77.2%	37.5%
20	J48	70.3%	57.1%	86.3%	37.5%

Table 4 demonstrates the results from a cross-validation test on the trained classifiers. It is possible to observe that dataset with lower values for the window of data yield poor predictive values, however at higher window improves the predictive ability. These results indicate that there may exist a balance between the number of detected changes and the predictability of the heart rate changes. Smaller windows identify more changes but are not easy to predict, larger windows give fewer change detentions but improve predictive tasks.

4.4 Analysis

Analysing the results presented, it is possible to correlate heart rate of people with environment conditions, as non-comfortable conditions result in higher heart rate suggesting discomfort to the user. The window analysis used with tests with statistical significance has shown to identify changes in the heart rate condition of people. However, the results also evidence that there are some balances that must be taken into consideration when building systems for monitoring and identification of stress induced by the environment. Smaller windows of data result in less stable analysis with more change identifications and less predictive accuracy in the algorithms tests. On the other hand, bigger windows of data decrease the number of change identifications and increase the predictive accuracy of the system. This balance between the size of the window of data, number of change identifications and predictive accuracy determines the overall sensitivity and predictive accuracy of the system. Computational analysis show that it is possible for the sensor node inside the sensor network to handle the workload from detecting change in attribute averages and performing predictions using trained classifiers. However, the initial classifier training should be handled by a central server with more computational power as it performs intensive computational operations.

5   Conclusion and Future Work

In this paper, it is presented a study that correlates thermal comfort to environment induced stress through the use of ambient intelligence techniques and sensor networks. Results show promise on the identification of stressful situations as well as stress inference through the use of predictive algorithms. Though for predictive tasks and hypothesis making it is needed a learning phase for each person while stress identification can be made in real time without a learning context as it can be inferred by the context provided by other sensors. Future work should extend the capabilities of the purposed algorithms or take into consideration, user activities, emotion detection and emotion on stress assessment.

Acknowledgement. This work was supported by the Spanish MEC and MICINN, as well as European Commission FEDER funds, under Grants TIN2009-14475-C04 and TIN2012-38341-C04-04. The work is also partially supported by a portuguese doctoral grant, SFRH/BD/78713/2011, issued by the Fundação da Ciência e Tecnologia (FCT) in Portugal.

References

1. Weiser, M.: The Computer for the Twenty-First Century. Scientific American 265(3), 94–104 (1991)
2. Ullah, S., Higgins, H., Braem, B., Latre, B., Blondia, C., Moerman, I., Saleem, S., Rahman, Z., Kwak, K.S.: A comprehensive survey of Wireless Body Area Networks: on PHY, MAC, and Network layers solutions. Journal of Medical Systems 36(3), 1065–1094 (2012)
3. Chen, M., Gonzalez, S., Vasilakos, A., Cao, H., Leung, V.C.M.: Body Area Networks: A Survey. Mobile Networks and Applications 16(2), 171–193 (2010)
4. Cooperating Objects NETwork of Excellence: Recognizing Emotions using Wireless Sensor Networks (2011)
5. World Health Organization: Stress at the workplace (2013),
   http://www.who.int/occupational_health/topics/stressatwp/en/
6. Brun, J.-P.: Work-related stress: scientific evidence-base of risk factors, prevention and costs
7. Choi, J., Ahmed, B., Gutierrez-Osuna, R.: Ambulatory Stress Monitoring with Minimally-Invasive Wearable Sensors. Comput. Sci. and Eng., Texas A&M (2010)
8. Tauseef, M.: Human Emotion Recognition Using Smart Sensors. Ph.D. dissertation, Massey University (2012)
9. Acampora, G., Loia, V.: A proposal of ubiquitous fuzzy computing for Ambient Intelligence. Inf. Sci. 178(3), 631–646 (2008)
10. Liu, W., Lian, Z., Liu, Y.: Heart rate variability at different thermal comfort levels. European Journal of Applied Physiology 103(3), 361–366 (2008)
11. Paola, A.D., Gaglio, S., Re, G.L., Ortolani, M.: Sensor 9 k: A testbed for designing and experimenting with WSN-based ambient intelligence applications. Pervasive and Mobile Computing 8(3), 448–466 (2012)
12. R Core Team: R: A Language and Environment for Statistical Computing, R Foundation for Statistical Computing, Vienna, Austria (2012),
    http://www.r-project.org/
13. Hall, M., National, H., Frank, E., Holmes, G., Pfahringer, B., Reutemann, P., Witten, I.H.: The WEKA Data Mining Software: An Update. SIGKDD Explor. Newsl. 11(1), 10–18 (2009)
14. Ortiz, A., Royo, F., Galindo, R., Olivares, T.: I3ASensorBed: a testbed for wireless sensor networks. Tech. Rep. (2011)
15. PulseSensor (2013), http://pulsesensor.myshopify.com/
16. Fanger, P.O.: Thermal comfort: Analysis and applications in environmental engineering. Danish Technical Press (1970)
17. Höppe, P.: The physiological equivalent temperature - a universal index for the biometeorological assessment of the thermal environment. International Journal of Biometeorology 43(2), 71–75 (1999)
18. Quazi, M., Mukhopadhyay, S.: Continuous monitoring of physiological parameters using smart sensors. In: 2011 Fifth International Conference on Sensing Technology, pp. 464–469 (November 2011)

Towards Usability Evaluation of Multimodal Assistive Technologies Using RGB-D Sensors

José Alberto Fuentes[1], Miguel Oliver[1], Francisco Montero[1,2],
Antonio Fernández-Caballero[1,2], and Miguel Angel Fernández[1,2]

[1] Instituto de Investigación en Informática de Albacete (I3A), 02071-Albacete, Spain
[2] Departamento de Sistemas Informáticos, Universidad de Castilla-La Mancha,
02071-Albacete, Spain
{JoseA.Fuentes,Miguel.Oliver}@gmail.com,
{Francisco.MSimarro,Antonio.Fdez,Miguel.FGraciani}@uclm.es

Abstract. To date there are many solutions in the field of assistive technologies addressing different kinds of disabilities. Each solution has opted for very specific (and incompatible) hardware and software technologies. Recently, new devices initially destined to electronic entertainment are appearing. They have joined in a single sensor various types of technologies typical for assistance. In this paper, we show and evaluate how RGB-D sensors are capable of replacing traditional heterogeneous technologies and a single device covers several products in the field of multimodal human-computer interaction and assistive technologies. Furthermore, a prototype of a software equivalent to a traditional assistive technology product is shown.

Keywords: Assistive technology, RGB-D sensor, Human-computer interaction, Multimodal interface.

1 Introduction

In the field of assistive technologies there are many solutions for the elderly, disabled and impaired people [4], [3], [6], [7], [9]. Each solution has opted for very specific and incompatible hardware and software technologies. Thus, a person who suffers from several types of disabilities needs to manage multiple products. Besides, these solutions provide different peripherals that the user has to adapt to. In some cases, the user wears peripherals on his/her body to interact with the system. Certainly, the experience between the user and the system is improved by providing new forms of human-computer interaction (HCI). Recently, new devices initially destined to electronic entertainment have appeared. These have joined various types of products typical of assistive technologies in a single peripheral. There are now some devices of this kind on the market. The best known so-called RGB-D sensors are Microsoft Kinect and ASUS Xtion Pro / Pro Live (see Fig. 1).

This paper shows how new RGB-D sensors can replace heterogeneous technologies and how a single RGB-D sensor can cover several areas in the field

J.M. Ferrández Vicente et al. (Eds.): IWINAC 2013, Part II, LNCS 7931, pp. 210–219, 2013.

Fig. 1. RGB-D sensors. Left: ASUS Xtion Pro. Right: Microsoft Kinect.

of usability in HCI [14], [15] and assistive technologies. We demonstrate their multimodal character by describing the general features and showing how they perform many tasks previously dominated by other products. Furthermore, a prototype of a software equivalent to a more traditional product is shown. The prototype demonstrates the user/system interaction using the user's head, but without using any specially dedicated additional peripheral.

2   Assistive Technologies at a Glance

During the last years, assistive technologies have evolved to satisfy the needs of people with disabilities. The solutions in this area cover a multitude of disabilities. People with reduced mobility, impaired hearing and vision, have improved their quality of life thanks to a number of hardware and software solutions. Some solutions require additional hardware for the user to interact with the system, such as small peripherals that the user holds in his/her body (e.g. special glasses, webcams, etc.). The most important aspects addressed by assistive technologies are the problems of mobility, vision and auditory perception. Therefore, we next present three solutions, one for each disability mentioned.

2.1   Gesture-Based Interaction

Firstly, let us introduce a tool for reduced mobility people who can only interact with the system using their head. A solution consists of a small device that is put on the user's head using a rubber hair. The included software allows users to interact with the system by only performing head movements. This product makes computer access simpler for people with reduced or null mobility in the upper extremities. The peripheral is an adaptation that enables the user to interact just as easily as with a conventional keyboard and mouse, but he/she uses the computer without his/her hands. The software has been developed by volunteers of several pathologies such as quadriplegia, multiple sclerosis and brain paralysis. There are equivalent solutions for people with limited mobility [5]. As these can use their hands, the systems provide suitable control devices such as joysticks or special keyboards.

2.2   Voice-Based Interaction

Other solutions in the field of assistive technologies allow the user to interact through speech [10]. Voice recognition technologies are not new and commercial software speech recognition has been available since 1990. Despite the apparent success of such technologies, there are very few people that use them to interact with computers. The first reason for it is related to user privacy. Perhaps the user does not want anyone to hear what he/she is writing. Another reason is that it is usually faster to write than to talk. Additionally, the user might disturb others when he/she is talking and inaccuracies may appear due to environmental noise. However, in the field of high degree of disability these technologies are particularly useful. The user interacts with the system through a microphone, and information is sent to the user via the display or speakers. This software allows the user to communicate with the system through simple voice commands. Surfing the Internet is an action the user performs through his/her voice. Another typical feature of this type of software is that it allows the user to dictate text.

2.3   Iris Motion-Based Interaction

This kind of solution is relatively new. For many years, solutions in this field have proposed the use of special glasses or a helmet. Subsequently, many developments use webcams to perform digital image processing to determine the position of the iris [20]. The results are reasonably good. The latest proposal in this area is the eyeCan project by Samsung. The goal is to design a system to help people who suffer some kind of paralysis or physical disability that prevents them from handling a mouse. Thus, a small device that is put on glasses has been designed, which allows users to control the mouse pointer by moving their eyes. The glasses incorporate a small camera and an infrared detector to track the movement of the eyes.

3   RGB-D Sensors for Multimodal Interaction

The three solutions discussed above include various forms of HCI. But it is necessary to use different devices to achieve the complete objective. On one hand, an additional peripheral to the head is required, while a microphone and speakers are needed on the other. Finally, for the third case, special glasses are required. Is it plausible to use a single RGB-D sensor to face the overall challenge?

Current RGB-D sensors on the market are quite similar, and all of them share a set of features. Such sensors have become popular in recent years thanks to video game consoles [18]. The first system to incorporate a motion device was Nintendo Wii. Since 2006 many initiatives have been performed using this device. The project goal was the identification and study of a new interaction scenario, where communication problems existed between the user and computer. It was also intended to develop a hardware and software system to reduce the problems encountered and to raise the potential users' satisfaction. Wii completely transformed the concept of HCI. However, the revolution in the field of RGB-D sensors came with the Xbox 360. In 2010, Microsoft introduced the Kinect RGB-D sensor, and then other similar products hit the market (e.g. ASUS Xtion Pro and Pro Live or SoftKinetic DetphSense). Microsoft proposes a completely new way to interact with the system [12]. Indeed, Kinect allows interacting only with the body (without additional devices). Although the device was designed to run on Xbox 360, the sensor was well received among software developers. So Microsoft decided to release a PC version. The features of the new version are similar to the initial version of the sensor. To date, these features are usually present in most RGB-D sensors on the market. The most remarkable Kinect features described next are the color web camera, the depth camera and the audio system consisting of four directional microphones.

3.1   The Kinect RGB Camera

A color webcam is a device especially useful in the field of assistive technologies. Until recently, its main function was to enable video conferences. Since the very beginning, this device incorporates a microphone to efficiently conduct video conferences. Thanks to powerful artificial vision algorithms, webcams allow more specific tasks related to the detection of patterns. Through pattern recognition it is possible to detect the movement of arms, eyes and head. Thus, a user interacts more efficiently with the system using different body parts. Also, face recognition is being used in a variety of solutions in the field of assistive technologies. Face recognition allows to infer specific states of the user (or patient). The RGB camera together with the depth camera (described next) performs a more efficient facial recognition [1].

RGB-D sensors can also work as webcams, replacing conventional color cameras. The resolution currently provided by color cameras is quite good. The color camera resolution of the sensors discussed is typically 640×480 pixels at a rate of 30 frames per second (fps). The latest version of Microsoft's SDK allows a resolution of 1280×960 at 12 fps [16]. Obviously, it is foreseen that the resolution increases in the near future. Thus, the image processing capacities of the sensor will also enhance.

3.2   The Kinect Depth Camera

One of the most remarkable features of this new device is the depth camera. The depth camera is the most important element of the sensor because it is responsible for tracking the human skeleton through algorithms provided by the device's software (e.g. [19], [22]). The depth sensor consists of an infrared laser projector combined with a CMOS sensor, which captures video data in 3D under any ambient lightning conditions. The sensing range of the depth sensor is tunable. The Kinect software is capable of automatically calibrating the device based on the user's physical environment, accommodating under the presence of furniture or obstacles. The infrared laser projector, combined with the RGB camera and the Kinect software algorithms, provides 20 individually treatable points of the user's skeleton. Thus, a developer captures the motion of a given point and associates it to a predefined interaction event. Another remarkable aspect is that the software detects more than one person. Kinect specifically allows monitoring two users simultaneously.

Again, conventional webcams do not provide this kind of features. In addition, the depth camera can be used for face recognition because small changes are determined quite accurately. The reconstruction of 3D environments is another task that is usually performed with traditional webcams [11]. A depth camera significantly helps in this task. This is another important reason to bet on this type of sensor.

3.3   The Kinect Directional Microphones

Speech recognition has also been very important in the field of assistive technologies. In fact, early studies and tests in speech recognition are now over 60 years old. Some of the problems that this technology has faced are the existence of homophonic words, the disparity between speakers and too noisy environments. Devices like Kinect have overcome some of these problems. This particular device (thanks to its software) identifies voices and reduces noise in inadequate rooms. Typically, conventional microphones or webcams with microphone are used for speech-related solutions. These microphones, with a reasonably good quality, have covered the needs of users. However, problems related to poor quality of sound are found in certain environments. The devices discussed in this paper include a set of microphones for the implementation of speech recognition. Only the first version of the ASUS sensor incorporates no microphones. In the case of the Microsoft sensor, it has four directional microphones for the talking people. The inclusion of microphones in these devices is very important and shows the intention to replace conventional webcams.

4   Usability Evaluation of Kinect-Based Head Movement Interaction

The ISO/IEC 25062 standard [13] has been used as reference to assess and reporting the usability during planning and developing of the performed test. Also, SUS (System Usability Scale) questionnaire [2] is used to assess the user satisfaction.

4.1   Goals

This experiment has been performed to demonstrate the usability (*effectiveness*, *efficiency* and *satisfaction*) of the Kinect RGB-D sensor in the field of assistive technologies.

4.2 Hypotheses

Our experiment should to contribute to show how to interact with the mouse pointer with an RGB-D sensor, but the user has not to wear a device on his/her head in this case. The main problem is that the user with reduced or null mobility in his/her upper extremities can not put the enPathia device in his/her head, it requires human support. In Fig. 2 the enPathia device is shown and in Fig. 3 the same example with Microsoft Kinect is also provided.

This experiment demonstrate how MS Kinect can replace two types of assistive technology. On the one hand, a traditional component to interact with the mouse pointer is replaced by a Kinect sensor. On the other hand, voice is used to perform simple actions. And, in both cases, MS Kinect is used.

4.3 Method

Participants. For this usability testing, eight adults were recruited. The usability testing was done by a group of eight users as specified in [13], although other references such as [17] suggest that by providing at least five users 80% of the problems related to usability can be identified. The participants were between the ages of 23 and 31, with an average of 26 years; there were five males and three females. None of them reported significant usability or accessibility problems.

Apparatus. Two devices were used in this usability testing: enPathia [8] and MS Kinect [16]. The device used to interact with the system through the mouse is a small device that the user must carry on his/her head. This is a device commercialized by Eneso called enPathia. enPathia is a product specially designed for people with reduced or null mobility in their upper extremities. It is a peripheral that enables the user to work with a computer just as easily as with a conventional mouse and keyboard. But, there is no need to use hands.

For this experiment Microsoft Kinect for Xbox 360 has been used under the Beta 2 SDK version. To move the mouse, the user uses his/her head and to do left click, right click, double left click and calibrate the system, the user must utter the words "left", "right", "double" and "calibrate", respectively. With only four words and the movement of the head, the user performs lots of actions. The application runs automatically when the user logs into the computer.

Procedure. During the experiment each participant completes three tasks. The difficulty of the tasks is incremental:

- Task 1. Move the cursor around the screen. To perform this task, the user uses the movement of his/her head to move the mouse. This task was used as a first system approximation and training to determine the ease of movement of the cursor.
- Task 2. The user has to position the cursor on specific points of the screen. The user uses the movement of his/her head to move widgets and the mouse into the indicated screen areas. This task was also used for training purposes and to determine the accuracy of each device.

Fig. 2. Interaction with enPathia

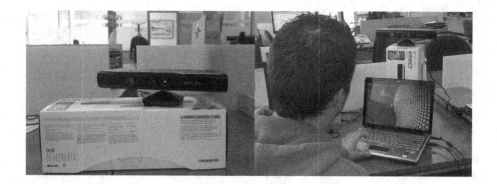

Fig. 3. Interaction with Microsoft Kinect

- Task 3. Finally, the user has to open a word processor, write his/her name
  and save the file. This actions requires the mouse movement and the but-
  ton interaction emulation. Users use the movement of his/her head to move
  the mouse (enPathia and Kinect) and the voice (Kinect) to press buttons
  on the mouse emulation. This task determines the full functionality of the
  application and the main task of our experiment was to measure the goal
  achievement, task time and completion rate/Task time to this task (see
  Table 1).

4.4   Results

Outcomes from this usability testing are shown in Table 1 and Table 2. A t-test is
used to determine if the differences in means are statistically significant. The first
thing we notice is the difference in means between enPathia and Kinect. enPathia
is faster (37.7 seconds) compared to Kinect (39.5 seconds). However, the other
relevant data is the p-value. Because we are not making any assumptions ahead
of time about what might be faster (enPathia or Kinect), we look at the p-value

Table 1. Results for *Task 3*

Task 3	enPathia / MS Kinect					
Participant #	Completion rate (%)		Task time (seg)		$\frac{\text{Completion rate}}{\text{Task time}}$ (%)	
1	100%	100%	29.95	40.00	3%	2%
2	100%	100%	25.05	29.76	4%	3%
3	100%	100%	56.68	59.86	2%	2%
4	100%	100%	30.72	32.07	3%	3%
5	100%	90%	37.18	37.18	3%	2%
6	90%	100%	50.25	45.27	2%	2%
7	100%	100%	32.22	35.91	3%	3%
8	100%	100%	39.65	35.85	3%	3%
mean	99%	98%	37.7	39.5	3%	3%
max	100%	100%	56.7	59.9	4%	3%
min	90%	90%	25.1	29.8	2%	2%

Table 2. SUS questionnaire for enPathia and MS Kinect

Participant #	enPathia SUS	MS Kinect SUS
1	85	92.5
2	95	87.5
3	80	92.5
4	92.5	77.5
5	82.5	85
6	85	90
7	75	80
8	82.5	92.5
mean	84.6	87.2
max.	95	92.5
min.	75	77.5

for the two-tailed distribution. The p-value is about 0.33, which is above the 0.05 threshold. Therefore, it is possible to conclude that there is no statistically significant difference in task times between the use of enPathia and MS Kinect, at the 0.05 level.

Table 2 shows the results obtained through the SUS questionnaire for enPathia and MS Kinect. With the data presented, it may be concluded that enPathia has a level of satisfaction close to 85%, although Kinect is better in this aspect (87%). Also the collected data shows that some users prefer the enPathia system, but these users are the least. A paired samples t-test is used and the p-value of 0.46 indicates that there is no significant difference between the two devices, since this number is considerably bigger than 0.05.

5   Discussion and Conclusions

We conclude that Kinect is a good alternative for many existing devices in the field of assistive technologies. Compared with the enPathia device, it has been demonstrated that the time to perform the same tasks are similar and there is not a significant difference between devices. Taking into account that the user satisfaction is higher with Kinect, the use of such a sensor in the area of assistive technologies is highly recommended.

Let us emphasize that assistive technologies are increasingly present in our lives. Many hardware manufacturers have made transparent this type of technology to bring technology to everyone. In recent years, assistive technologies have converged to facilitate their use and development of solutions in this area. The emergence of RGB-D sensors is another reason to think that convergence will be better in the future. In addition, major hardware manufacturers such as Microsoft, ASUS, Samsung, SoftKinetic or Sony [21] have chosen this type of technology for their new devices. The use of this technology is becoming popular in other areas such as health and tourism and provide a great opportunity to bring technology to people with disabilities. And, further enhancement of the features and increased performance of the sensors will lead to more accurate systems.

Acknowledgements. This work was partially supported by Spanish Ministerio de Economía y Competitividad / FEDER under projects TIN2012-34003 and TIN2010-20845-C03-01.

References

1. Arumugam, D., Purushothaman, S.: Emotion classification using facial expression. International Journal of Advanced Computer Science and Applications 2(7), 92–98 (2011)
2. Brooke, J.: SUS: a "quick and dirty" usability scale. Usability Evaluation in Industry. Taylor & Francis (1986)
3. Carneiro, D., Castillo, J.C., Novais, P., Fernández-Caballero, A.: Multimodal behavioural analysis for non-invasive stress detection. Expert Systems with Applications 39(18), 13376–13389 (2012)
4. Castillo, J.C., Carneiro, D., Serrano-Cuerda, J., Novais, P., Fernández-Caballero, A., Neves, J.: A multi-modal approach for activity classification and fall detection. International Journal of Systems Science (in press, 2013)
5. Chang, Y.J., Chen, S.F., Chuang, A.F.: A gesture recognition system to transition autonomously through vocational tasks for individuals with cognitive impairments. Research in Developmental Disabilities 32(6), 2064–2068 (2011)
6. Costa, A., Castillo, J.C., Novais, P., Fernández-Caballero, A., Simoes, R.: Sensor-driven agenda for intelligent home care of the elderly. Expert Systems with Applications 39(15), 12192–12204 (2012)
7. Doukas, C., Metsis, V., Becker, E., Le, Z., Makedon, F., Maglogiannis, I.: Digital cities of the future: extending home assistive technologies for the elderly and the disabled. Telematics and Informatics 28(3), 176–190 (2011)

8. enPathia - Eneso - Tecnología para personas con discapacidad,
   http://www.eneso.es/producto/enpathia

9. Fernández-Caballero, A., Castillo, J.C., Rodríguez-Sánchez, J.M.: Human activity
   monitoring by local and global finite state machines. Expert Systems with Applications 39(8), 6982–6993 (2012)

10. Harada, S., Wobbrock, J.O., Landay, J.A.: Voicedraw: a hands-free voice-driven
    drawing application for people with motor impairments. In: Ninth Annual ACM
    Conference on Assistive Technologies 2007, pp. 27–34 (2007)

11. Henry, P., Krainin, M., Herbst, E., Ren, X., Fox, D.: RGB-D mapping: using
    Kinect-style depth cameras for dense 3D modeling of indoor environments. The
    International Journal of Robotics Research 31(5), 647–663 (2012)

12. Hernández-López, J.J., Quintanilla-Olvera, A.L., López-Ramírez, J.L., Rangel-
    Butanda, F.J., Ibarra-Manzano, M.A., Almanza-Ojeda, D.L.: Detecting objects
    using color and depth segmentation with Kinect sensor. Procedia Technology 3,
    196–204 (2012)

13. ISO/IEC 25062: Software engineering – Software product Quality Requirements
    and Evaluation (SQuaRE) – Common Industry Format (CIF) for usability test
    reports (2006)

14. López-Jaquero, V., Montero, F., Molina, J.P., González, P., Fernández-Caballero,
    A.: A seamless development process of adaptive user interfaces explicitly based
    on usability properties. In: Feige, U., Roth, J. (eds.) EHCI-DSVIS 2004. LNCS,
    vol. 3425, pp. 289–291. Springer, Heidelberg (2005)

15. López-Jaquero, V., Montero, F., Molina, J.P., Fernández-Caballero, A., González,
    P.: Model-based design of adaptive user interfaces through connectors. In: Jorge,
    J.A., Jardim Nunes, N., Falcão e Cunha, J. (eds.) DSV-IS 2003. LNCS, vol. 2844,
    pp. 245–257. Springer, Heidelberg (2003)

16. Microsoft Corporation: Microsoft Kinect for Windows SDK - V1.0 Release
    notes (2012), http://www.microsoft.com/en-us/kinectforwindows/develop/
    release-notes.aspx

17. Nielsen, J.: Why You Only Need to Test with 5 Users (2000),
    http://www.useit.com/alertbox/20000319.html

18. Roccetti, M., Marfia, G., Semeraro, A.: Playing into the wild: a gesture-based
    interface for gaming in public spaces. Journal of Visual Communication and Image
    Representation 23(3), 426–440 (2012)

19. Schwarz, L.A., Mkhitaryan, A., Mateus, D., Navab, N.: Human skeleton track-
    ing from depth data using geodesic distances and optical flow. Image and Vision
    Computing 30(3), 217–226 (2012)

20. Sirohey, S., Rosenfeld, A., Duric, Z.: A method of detecting and tracking irises and
    eyelids in video. Pattern Recognition 35(6), 1389–1401 (2002)

21. Sony Computer Entertainment Inc.: User-driven three-dimensional interactive
    gaming environment (2011), http://www.google.com/patents/US20120038637

22. Zhang, Q., Song, X., Shao, X., Shibasaki, R., Zhao, H.: Unsupervised skeleton ex-
    traction and motion capture from 3D deformable matching. Neurocomputing 100,
    170–182 (2012)

Fusion of Overhead and Lateral View Video for Enhanced People Counting

Juan Serrano-Cuerda[1], Marina V. Sokolova[1], Antonio Fernández-Caballero[1,2], María T. López[1,2], and José Carlos Castillo[3]

[1] Instituto de Investigación en Informática de Albacete (I3A), 02071-Albacete, Spain
[2] Universidad de Castilla-La Mancha, Departamento de Sistemas Informáticos, 02071-Albacete, Spain
Antonio.Fdez@uclm.es
[3] Instituto Superior Técnico, Instituto de Sistemas e Robótica, 1049-001 Lisbon, Portugal

Abstract. This article introduces a multi-camera system for real-time people counting. The proposed system is built from INT^3-Horus, a framework for intelligent monitoring and activity interpretation. The system uses an indoor overhead video camera and a lateral view video camera to detect people moving freely in smart spaces. The segmentation is performed from both synchronized input videos. Then, information is fused to enhance the overall efficiency. The people counting system is flexible in detecting individuals as well as groups. Also, people counting is independent of the trajectories and possible occlusions of the humans present in the smart space. The initial results offered are very promising.

Keywords: People counter, Multi-camera, INT^3-Horus framework.

1 Introduction

People counting has been widely addressed during the last few years, mainly for surveillance applications [1], [4]. This paper is focused on a system that calculates in real-time the number of people that are present in for use in smart spaces such as monitored halls or rooms. The people counting system described in this paper has been developed from INT^3-Horus, a framework for intelligent monitoring and activity interpretation. People are monitored by a couple of indoor color video cameras. There is an overhead video camera overlooking the scenario [16], as well as a lateral video camera monitoring the same scenario. People move freely, as there is no clear entrance/exit at the monitored smart space. Also, there is no initial limitation in the number of people to be detected; single humans appear in the scenario, and also groups of people are allowed. The overhead camera is used to avoid the majority of occlusions present in lateral video camera installations. Recently, people detection from a single overhead view camera has demonstrated good results [19]. In this paper, a lateral video camera is added to enhance the accuracy of the proposed people counting system.

J.M. Ferrández Vicente et al. (Eds.): IWINAC 2013, Part II, LNCS 7931, pp. 220–229, 2013.
© Springer-Verlag Berlin Heidelberg 2013

2   The INT³-Horus Framework

INT³-Horus is conceived as a framework to carry out monitoring and activity interpretation tasks [10], [11], [12]. This is an ambitious goal given the huge variety of scenarios and activities that can be faced [14], [3], [20]. The framework establishes a set of operation levels where clearly defined input/output interfaces are defined. Inside each level, a developer places his/her code, encapsulated in a module in accordance with the operation performed. Although a set of levels are proposed in INT³-Horus to cover all the steps of a generic multi-sensor and activity interpretation system in smart spaces [15], [18], [7], the philosophy underlying the framework allows a flexible set of levels to be adapted to a given final system [2], [5].

The framework infrastructure as well as the modules layout is based on the Model-View-Controller (MVC) paradigm [17], which allows to isolate the user interface from the logical domain for an independent development, testing and management. The MVC paradigm divides an application into three main entities, defining their main roles as well as the connections among them. *Model* manages the application data, initializes objects and provides information about the application status. In event-driven systems, the *Model* informs the *View* and the *Controller* about information changes. *View* provides a representation of the *Model* information (performing just simple operations) to fit user requirements. Finally, *Controller* receives the inputs to the application and interacts with the *Model* to update its objects, and with the *View* to represent the new information.

The INT³-Horus framework allows easy code integration, providing users with module templates to put their code into them. These templates already have the necessary connections to access the rest of INT³-Horus components, not only the data model or the user interface, but also the controller to trigger each module's execution. Together with the easy addition of new functionalities, a state of the art framework for monitoring and activity detection must take into account several information sources (sensors) and INT³-Horus is not an exception. These sources are mainly related to image sensors since they are the most widespread for monitoring tasks; but other sensor technologies, like commercial sensors and wireless sensor networks (WSNs), are also integrated to show the generic purpose of INT³-Horus.

3   The INT³-Horus Levels for People Counting

The main goal of an efficient people counting system in smart spaces is to obtain the number of humans inside the field of view of a camera. In this particular case, we describe a system based on an indoor overhead camera coupled with a lateral view one with the highest possible accuracy when counting people. For this purpose, four processing levels are selected from INT³-Horus framework (see Fig. 1). The lower one, *data acquisition*, is in charge of collecting data from both cameras. The next level, *segmentation*, uses an approach based on background subtraction to isolate the humans in the scenes captured from the

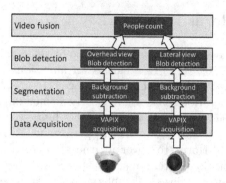

Fig. 1. INT^3-Horus levels and modules to develop the people counting system

cameras (e.g. [8], [6], [9]). Some filtering and heuristics are applied in the next level, *blob detection*, to enhance segmented humans, dealing with false positives and splitting groups of people into individuals. The two modules running at this level are different since they deal with different camera perspectives, leading to different heuristics. After the *blob detection* process it is already possible to obtain the number of humans contained on each camera view with good results from both image streams. Nevertheless, the aim of this work is to enhance the human counting process through the fusion of the information from the two cameras. For this reason, another level, namely *video fusion*, matches the humans detected from the previous level to generate a unified output.

3.1   The Data Acquisition Level

At *data acquisition* level, a specific module is in charge of capturing images from Axis cameras. The module uses VAPIX, an HTTP-based application programming interface which provides functionality for requesting images, controlling network camera functions (pan-tilt-zoom, relays, etc.) and setting/retrieving internal parameter values (see http://www.axis.com/techsup/cam_servers/ dev/index.htm). The proposed people counting system only needs images obtained from two networked cameras. For this purpose, two module instances run in parallel.

3.2   The Segmentation Level

The main objective of the *segmentation* level is to perform the initial detection of the humans present in the scene from both image streams. The processing of both streams is common at this level, having two instances of the same module running simultaneously. An adaptive Gaussian background subtraction is performed on input image I_Z, obtained from the indoor overhead camera, as shown in Fig. 2a. The subtraction is based on the OpenCV implementation of a well-known algorithm [13]. The algorithm builds an adaptive model of the scene background based on the probabilities of a pixel to have a given color level. An example

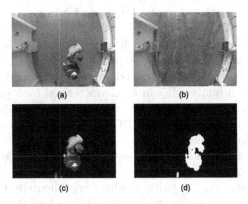

Fig. 2. Images generated at segmentation level. (a) Current frame. (b) Background model. (c) Foreground. (d) Segmented image.

of this model is shown in Fig. 2b. A shadow detection algorithm, based on the computational color space used in the background model, is also used. After the background segmentation is performed, initial background segmentation images (I_{BZ} and I_{BL} for overhead and lateral view, respectively) are obtained as shown in Fig. 2c.

However, the resulting image contains some noise that has to be eliminated. For this, an initial threshold θ_0 (experimentally fixed as 1/16 of the number of possible gray levels in the image) is applied as shown in equation (1).

$$I_{ThZ}(x,y) = \begin{cases} min, \text{if } I_{BZ}(x,y) \leq \theta_0, \\ max, \text{otherwise} \end{cases} \tag{1}$$

and

$$\theta_0 = \frac{max}{16} \tag{2}$$

where min is fixed to 0 (since we are obtaining binary images) and max is the maximum gray level value for a pixel in I_{BZ} (e.g. 255 for an 8-bit image). Notice that, although only the equation for the overhead view is shown, the process for the lateral view is totally equal to this one, with the only difference of using I_{BL} and I_{ThL} instead of I_{BZ} and I_{ThZ}.

After the previous operation, two morphological operations, namely opening and closing, are performed to eliminate the remaining noise of the image, obtaining I_{SZ} in the overview view and I_{SL} in the lateral view, as shown in Fig. 2d. After the first noise reduction, the number of white pixels (corresponding to possible humans) is counted in the image. If the value is greater than a 50% of the area of the image during a predefined time Δt (usually one second), it is estimated that a big lightning change has occurred in the scene (e.g. a light switch turned on/off or a door was opened/closed). In this case, the algorithm is initialized to build a new background based on the new lightning conditions of the smart space.

3.3 The Blob Detection Level - Overhead View

Now, human candidates are extracted from binary image I_{SZ}, paying special attention to the existence of groups of people. For this purpose, the concept of region of interest (ROI) is introduced. A ROI is defined as the minimum rectangle containing a human. It can be characterized by a pair of coordinates (x_{min}, y_{min}),(x_{max}, y_{max}), corresponding to the upper-left and lower-right limits of the ROI, respectively. All detected ROIs are used to annotate the humans detected in the scene in a list L_Z.

In first place, human candidates are extracted from the scene. With this objective, connected components (blobs) are extracted from I_{SZ}. Next, blobs with a ROI area lower than A_{minZ} (with a value experimentally fixed according to scene features such as the scene area or the height where the camera is placed) are discarded. A new area threshold A_{GZ} is also established based on similar factors. Blobs with a ROI area lower than A_{GZ} are considered to contain a single human and the ROI containing it is enlisted in L_Z. Otherwise, blobs B_{GZ} with a ROI area greater than A_{GZ} are analyzed separately, since they are considered to possibly contain a group of humans. Now, each human belonging to these groups is extracted individually. In this case, a new subimage I_{GZ} is created containing the ROI that delimits B_{GZ}, as shown in Fig. 3a. Then, a new series of morphological openings are performed, since occlusions are less frequent in an overhead view than in a lateral view, obtaining a new image I_{GZ}. An example of the result of these operations is offered in Fig. 3b. Next, blobs are searched in the new image. Blobs with an area greater than A_{minZ} are annotated in a list of group blobs L_{Z_G}, whilst the others are discarded. If L_{Z_G} is empty at the end of the search, the original ROI with blob B_{GZ} is enlisted as a single human; but, it will be marked as a possible group that could not be separated. Finally, blobs from L_{Z_G} are enlisted in L_Z, where the number of humans in the space (people counting) is the number of blobs contained in L_Z. The detected humans in the smart space are shown in Fig. 3c for this running example.

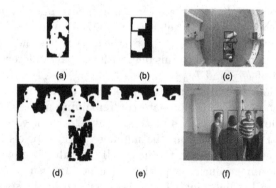

Fig. 3. Results of blob detection level. Overhead view: (a) Original ROI. (b) Separated ROIs. (c) Final Result. Lateral view: (d) Original ROI. (e) Separated ROIs. (f) Final Result.

3.4   The Blob Detection Level - Lateral view

This stage is also performed after an initial background subtraction, in parallel with the human detection from the overhead view. The input image to this stage is the background subtraction image I_{SL}. The ROIs are used to insert the humans detected in the scene in a list L_L. Similarly to the segmentation on the overhead view, the blobs from binary image I_{SL}, obtained from the background subtraction, are extracted, discarding those with a ROI area lower than A_{minL} (fixed similarly to A_{minZ} according to the scene features). Then, the remaining blobs are separated according to their area. Those with an area lower than an area threshold A_{GL} (based on the scene features) are enlisted as single humans in L_L.

Blobs B_{GL} (as shown in Fig. 3d), considered to contain more than a single human, are analyzed separately. Here, the objective is to find the heads of the humans in a group. Firstly, a subimage I_{GL} is gotten in the upper fifth area of B_{GL}. The next step is to perform morphological opening and closing operations to eliminate small artifacts not belonging to human candidates. The result of this operation is shown in Fig. 3e. Now, a new area threshold A_H is fixed according to the scene features. Blobs B_H with an area greater than A_H are considered to belong to a human. This way new humans are enlisted in L_{L_G} with the x_{min}, x_{max} and y_min coordinates of B_H and the y_{max} coordinate of B_{GL}. In a similar way to the human detection in overhead view, if no human heads are found and L_{G_L} is empty at the end of the process, the blob is enlisted as a single human. In other case, blobs from L_{G_L} are enlisted in L_L, and the number of humans present in the lateral view is the number of elements in the list. The final result is shown in Fig. 3f.

3.5   The Video Fusion Level

The *video fusion* level uses a module which is launched when the area of a blob of interest (BOI), detected in the overhead view, is greater than a predefined value. The area of a BOI from the overhead view is studied when two or more persons are very close to each other. In this case, the module requests information from the lateral view camera. The decision on how many people are present in the scene is generated as a joint solution based on the responses provided by the segmentation levels on overview and lateral views. The decision generated by the module is based on the analysis of four variables:

1. Area of the overhead view BOI.
2. Number of blobs segmented in the current overhead view frame, $N_{b,o,(t)}$.
3. Number of blobs segmented in the previous overhead view frame, $N_{b,o,(t-1)}$.
4. Number of blobs segmented in the current lateral view frame, $N_{b,l,(t)}$.

When a BOI is detected, an approximate number of humans, N_{approx}, is calculated. The simplest way to obtain this approximate number of people consists in dividing the area by a value corresponding to the average area of a human.

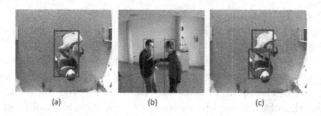

Fig. 4. (a) A too great area of an overhead view BOI is detected. (b) Segmentation result in the current lateral camera view frame. (c) Segmentation result in the previous overhead view frame.

The rest of variables are used to assess this issue. In this sense, $N_{b,o,(t)}$ is corrected starting from the newly calculated N_{approx}. Then, the corrected value $N_{b,o,(t)}$ is compared with the number of people segmented in the current lateral view frame, $N_{b,l,(t)}$ and in the previous overhead view frame, $N_{b,o,(t-1)}$. Variable $N_{b,o,(t-1)}$ ensures that there is no possibility of gross changes in the number of humans detected between successive frames. Fig. 4 illustrates the functioning of the fusion module. Here, a BOI has been detected with the overhead camera (see Fig. 4a). Fig 4b and Fig 4c show the blobs detected by the lateral view camera in the current frame and the overhead camera at the previous frame, respectively.

So, the number of humans N_h, in agreement with the proposed algorithm, is calculated as shown in equation (3):

$$N_h = \begin{cases} N_{b,l,(t)}, \text{if } N_{b,o,(t-1)} + N_{approx} \leq N_{b,l,(t)}, \\ N_{b,o,(t-1)} + N_{approx}, \text{otherwise} \end{cases} \tag{3}$$

4 Data and Results

Two different video sequences are recorded from a couple of Axis cameras to test our proposal. The first camera records an overhead view from the roof of a hallway and the second camera is placed in a sidewall of the same hallway. A first sequence shows different people walking along the hallway (individually, and in groups of two or three individuals). Generally, people do not stop and do not cross their paths in this sequence, except for the final frames of the video, where two people meet and talk for a while in the center of the smart space and another person approaches them. Fig. 5a shows that people counting from the overhead view marks the ROI as a potential group (depicted as a rectangle). Nevertheless, the fusion approach uses lateral view people counting (see Fig. 5b) to discern that there are two people in that area, with the final total count of people displayed as a number in the overhead view result at the bottom-right corner of the space.

The second sequence is more complex. In this video sequence, up to five people appear in the scene, crossing and intersecting their paths, which results in a great amount of occlusions. People also meet for two minutes without being added to the background, and partially occluding themselves. Once again, Fig.

Table 1. Comparison of the results of overhead view and fusion-based approach for people counting

Sequence	Humans present	Humans detected in overhead view	Humans detected after fusion	TP_O	TP_F	FP_O	FP_F	FN_O	FN_F	Sn_O	Sn_F	Sp_O	Sp_F	$F\text{-}S_O$	$F\text{-}S_F$
1	952	825	900	811	881	14	19	141	71	0,85	0,93	0,98	0,97	0,91	0,95
2	2963	2711	2751	2629	2664	82	87	334	299	0,89	0,90	0,97	0,96	0,92	0,93
Total	3915	3536	3651	3440	3545	96	106	475	370	0,88	0,91	0,97	0,97	0,92	0,94

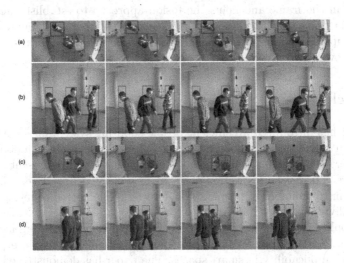

Fig. 5. Qualitative results. First sequence: (a) Overhead view. (b) Lateral view. Second sequence: (c) Overhead view. (d) Lateral view.

5d demonstrates how in lateral view the system is able to separate a group of two people that are too close to be separated in the overhead view (see Fig. 5c).

Table 1 shows a comparison of the results obtained from the overhead view and from the proposed fusion approach. In order to evaluate the performance, measures of specificity (Sp), sensitivity (Sn) and F-Score (F-S) are provided. These statistics are calculated as shown in equations (4), (5) and (6), respectively.

$$\text{specificity} = \frac{TP}{TP + FP} \tag{4}$$

$$\text{sensitivity} = \frac{TP}{TP + FN} \tag{5}$$

$$\text{F-Score} = \frac{2 \times \text{specificity} \times \text{sensitivity}}{\text{specificity} + \text{sensitivity}} \tag{6}$$

where TP (true positives) is the number of correct detections in the smart space, FP (false positives) is the number of detections of humans which are not actually present, and FN (false negatives) is the number of humans present in the scene which have not been detected by our algorithm. Since the table shows the result of both overhead view and fusion-based counting, the statistics of the first approach are shown with subindex O, while the results from the fusion approach are displayed with the subindex F.

Notice that the first sequence shows a greater improvement with the proposed fusion technique that the second one. This is due to the fact that that the first sequence has a great amount of groups of two people walking alongside each other and partially occluding themselves. This is why people counting in the overhead view is not able to separate them, while the lateral algorithm finds the two heads in the frame and helps the fusion approach to establish the number of people in the scene. The second sequence shows a greater amount of people in the groups but with a bigger separation between themselves, resulting in a lower improvement ratio.

5   Conclusions

This paper has introduced an efficient people counting system for smart spaces. The system based on an indoor overhead video camera and an indoor lateral video camera counts in real-time the number of people that are present in a given smart space. There is no restriction in the motion of the people. Even, there is no limitation in the number of people to be detected. The people counting systems accepts individuals as well as groups of people.

The people counting system described in this paper has been developed from INT³-Horus, a framework for intelligent monitoring and activity interpretation with special applicability in smart spaces. The paper has demonstrated the usefulness of the framework and the accuracy of the developed system.

Acknowledgements. This work was partially supported by Spanish Ministerio de Economía y Competitividad / FEDER under TIN2010-20845-C03-01 grant. This work was also partially sponsored by the project CMU-PT/SIA/0023/2009 under the Carnegie Mellon Portugal Program and its Information and Communications Technologies Institute.

References

1. Boltes, M., Seyfried, A.: Collecting pedestrian trajectories. Neurocomputing 100, 127–133 (2013)
2. Carneiro, D., Castillo, J.C., Novais, P., Fernández-Caballero, A.: Multimodal behavioural analysis for non-invasive stress detection. Expert Systems with Applications 39(18), 13376–13389 (2012)
3. Castillo, J.C., Fernández-Caballero, A., Serrano-Cuerda, J., Sokolova, M.V.: Intelligent monitoring and activity interpretation framework - INT3-Horus ontological model. In: Advances in Knowledge-Based and Intelligent Information and Engineering Systems, pp. 980–989 (2012)
4. Chaquet, J.M., Carmona, E.J., Fernández-Caballero, A.: A survey of video datasets for human action and activity recognition. Computer Vision and Image Understanding (2013), http://dx.doi.org/10.1016/j.cviu.2013.01.013

5. Costa, A., Castillo, J.C., Novais, P., Fernández-Caballero, A., Simoes, R.: Sensor-driven agenda for intelligent home care of the elderly. Expert Systems with Applications 39(15), 12192–12204 (2012)
6. Delgado, A.E., López, M.T., Fernández-Caballero, A.: Real-time motion detection by lateral inhibition in accumulative computation. Engineering Applications of Artificial Intelligence 23(1), 129–139 (2010)
7. Fernández-Caballero, A., Castillo, J.C., Rodríguez-Sánchez, J.M.: Human activity monitoring by local and global finite state machines. Expert Systems with Applications 39(8), 6982–6993 (2012)
8. Fernández-Caballero, A., Castillo, J.C., Serrano-Cuerda, J., Maldonado-Bascón, S.: Real-time human segmentation in infrared videos. Expert Systems with Applications 38(3), 2577–2584 (2011)
9. Fernández-Caballero, A., Castillo, J.C., Martínez-Cantos, J., Martínez-Tomás, R.: Optical flow or image subtraction in human detection from infrared camera on mobile robot. Robotics and Autonomous Systems 58(12), 1273–1281 (2010)
10. Gascueña, J.M., Fernández-Caballero, A.: On the use of agent technology in intelligent, multisensory and distributed surveillance. The Knowledge Engineering Review 26(2), 191–208 (2011)
11. Gascueña, J.M., Fernández-Caballero, A., López, M.T., Delgado, A.E.: Knowledge modeling through computational agents: application to surveillance systems. Expert Systems 28(4), 306–323 (2011)
12. Gascueña, J.M., Navarro, E., Fernández-Caballero, A.: Model-driven engineering techniques for the development of multi-agent systems. Engineering Applications of Artificial Intelligence 25(1), 159–173 (2012)
13. KaewTraKulPong, P., Bowden, R.: An improved adaptive background mixture model for real-time tracking with shadow detection. In: Video Based Surveillance Systems: Computer Vision and Distributed Processing, pp. 1–5 (2001)
14. Kieran, D., Yan, W.: A framework for an event driven video surveillance system. Journal of Multimedia 6(1), 3–13 (2011)
15. Pavón, J., Gómez-Sanz, J., Fernández-Caballero, A., Valencia-Jiménez, J.J.: Development of intelligent multi-sensor surveillance systems with agents. Robotics and Autonomous Systems 55(12), 892–903 (2007)
16. Rao, R., Taylor, C., Kumar, V.: Experiments in robot control from uncalibrated overhead imagery. In: Experimental Robotics IX, pp. 491–500 (2006)
17. Reenskaug, T.: Thing-model-view-editor an example from a planning system. XEROX PARC Technical Note (May 1979)
18. Rivas-Casado, A., Martinez-Tomás, R., Fernández-Caballero, A.: Multiagent system for knowledge-based event recognition and composition. Expert Systems 28(5), 488–501 (2012)
19. Serrano-Cuerda, J., Castillo, J.C., Sokolova, M.V., Fernández-Caballero, A.: Efficient people counting from indoor overhead video camera. In: Pérez, J.B., et al. (eds.) Trends in Pract. Appl. of Agents & Multiagent Syst. AISC, vol. 221, pp. 129–137. Springer, Heidelberg (2013)
20. Sokolova, M.V., Castillo, J.C., Fernández-Caballero, A., Serrano-Cuerda, J.: Intelligent monitoring and activity interpretation framework - INT3-Horus general description. In: Advances in Knowledge-Based and Intelligent Information and Engineering Systems, pp. 970–979 (2012)

Comparison of Finite Difference and B-Spline Deformable Models in Characterization of 3D Data

Rafael Berenguer-Vidal[1], Rafael Verdú-Monedero[2,*],
Rosa-María Menchón-Lara[2], and Álvar Legaz-Aparicio[2]

[1] Departamento de Ciencias Politécnicas,
Universidad Católica San Antonio, 30107, Murcia, Spain
rberenguer@ucam.edu
[2] Departamento de Tecnologías de la Información y las Comunicaciones,
Universidad Politécnica de Cartagena, 30202, Cartagena, Spain
rafael.verdu@upct.es, rmml@alu.upct.es, alvarlegaz@gmail.com

Abstract. This paper shows a new matrix formulation for three-dimensional deformable models. The matrix formulation is developed in the spatial domain for the cases of B-*spline* and finite difference shape function. Then, the spatial equations are translated into the frequency domain by means of the discrete Fourier transform. Once the iterative algorithm of the multidimensional deformable models is arranged, a comparison between B-*spline* and finite differences deformable models is performed. As expected, results confirm the robustness of B-*splines* in noisy environments whereas finite differences shows a sharper and faster behaviour.

1 Introduction

Deformable models are mathematical methods for describing boundaries of objects by means of curves or surfaces that deform under the influence of internal and external forces. The internal forces provide physical characteristics to the model, that is, elasticity, stiffness and mass. In contrast, the external forces drive the model to the boundaries of the dataset. The shape of the model is usually determined from the minimization of a functional through an iterative process. Thus, the final state is achieved when equilibrium is reached between internal and external forces.

Introduced with active contours or *snakes* by Kass et al. [1] and generalized to three dimensions by Terzopoulos et al. [2], deformable models are well-know tools in the literature [3]. Most approaches are raised in the space domain. This produces a tangled extension to multi-dimensional systems, due to the resulting high computational load for models with a large number of degrees of freedom [4]. This paper presents a formulation of multidimensional parametric models from a generalization of [5], using finite differences and B-*spline* as polynomial

---

* This work is partially supported by the Spanish *Ministerio de Ciencia y Tecnología*, under grant TEC2006-13338/TCM.

J.M. Ferrández Vicente et al. (Eds.): IWINAC 2013, Part II, LNCS 7931, pp. 230–240, 2013.

finite support. The paper is organized as follows. Section 2 reviews the continuous formulation of the generalized model. Section 3 describes the spatial and temporal discretization applied to the model, resulting in an iterative process in the spatial domain with an implementation in the spatial domain. Section 4 details the formulation for both shape functions. Results with real data is shown in Section 5, and finally Section 6 presents the conclusions.

2   Deformable Models

A e-dimensional parametric deformable model is a time-varying curve, surface or hypersurface defined as a parametric function in the space $\mathbb{R}^d$,

$$\mathbf{v} \equiv \mathbf{v}(\mathbf{s}, t) = [v_1(\mathbf{s}, t), \dots, v_d(\mathbf{s}, t)]^\top, \tag{1}$$

where $\mathbf{s} \equiv [s_1, \dots, s_e]$ with $s_j \in [0, L_j]$, $e \leqslant d$ and $e, d \in \mathbb{N}$ is the vector of parametric variables in the spatial domain, t is time, $v_i(\mathbf{s}, t)$ represents the coordinate function for dimension i and the symbol $^\top$ denotes transposition. The shape of the model is governed by the following energy functional,

$$\mathcal{E}(\mathbf{v}) = \mathcal{S}(\mathbf{v}) + \mathcal{P}(\mathbf{v}), \tag{2}$$

where $\mathcal{S}(\mathbf{v})$ is the internal energy of deformation which characterizes the deformation of an elastic and flexible model,

$$\mathcal{S}(\mathbf{v}) = \frac{1}{2} \sum_{i=1}^{d} \int_\Omega \left(\alpha(\mathbf{s}) \left\| \nabla v_i(\mathbf{s}) \right\|^2 + \beta(\mathbf{s}) \left(\Delta v_i(\mathbf{s}) \right)^2 \right) d\mathbf{s}, \tag{3}$$

where $\Omega := [0, L_1] \times \dots \times [0, L_e]$ is the integration domain of the multiple integral, v_i are the coordinate functions, and $\alpha(\mathbf{s})$ and $\beta(\mathbf{s})$ control, respectively, the elasticity and rigidity of the model. The term of external energies $\mathcal{P}(\mathbf{v})$ commonly generates forces to attract the model to the edges of the dataset. The energy functional (2) reaches a minimum when $\mathcal{S}(\mathbf{v})$ is balanced by $\mathcal{P}(\mathbf{v})$.

According to the variational calculus, the model $\mathbf{v}(\mathbf{s})$ that minimizes (2) must satisfy the Euler-Lagrange equation, which generalized for the multidimensional case produces the following system of d partial differential equations (PDE) [6]:

$$\mu(\mathbf{s})\partial_{tt}\mathbf{v}(\mathbf{s}, t) + \gamma(\mathbf{s})\partial_t\mathbf{v}(\mathbf{s}, t) - \nabla \cdot \left(\alpha(\mathbf{s})\nabla\mathbf{v}(\mathbf{s}, t) \right) + \Delta\big(\beta(\mathbf{s})\Delta\mathbf{v}(\mathbf{s}, t)\big) = \mathbf{q}(\mathbf{v}(\mathbf{s}, t)), \tag{4}$$

where ∂_t and ∂_{tt} denote respectively first and second partial derivative with respect to time, $\mu(\mathbf{s})$ is the mass density and $\gamma(\mathbf{s})$ is the damping density, $\nabla\cdot$ stands for the divergence operator, and the right-hand side

$$\mathbf{q}(\mathbf{v}(\mathbf{s}, t)) = -\nabla_{v_i}P(\mathbf{v}(\mathbf{s}, t)) + \mathbf{f}(\mathbf{v}(\mathbf{s}, t)) \in \mathbb{R}^d, \tag{5}$$

stands for the external forces that couple the model to the external data. $\nabla_{v_i}P$ symbolizes the gradient of the potential function P with respect to coordinate functions v_i. Finally, $\mathbf{f} \in \mathbb{R}^d$ indicates the vector that provides the hard restrictions for each coordinate function.

By assuming that each coordinate function of $\mathbf{f}$ can be determined independently of the other coordinates, Eq. (4) becomes a system of d decoupled partial differential equations (PDEs),

$$
\mu(\mathbf{s})\partial_{tt}v_i(\mathbf{s},t)+\gamma(\mathbf{s})\partial_t v_i(\mathbf{s},t)-\partial_{s_1}\big(\alpha(\mathbf{s})\partial_{s_1}v_i(\mathbf{s},t)\big)-\cdots-\partial_{s_e}\big(\alpha(\mathbf{s})\partial_{s_e}v_i(\mathbf{s},t)\big)+
$$
$$
\big(\partial_{s_1 s_1}+\cdots+\partial_{s_e s_e}\big)\big(\beta(\mathbf{s})\partial_{s_1 s_1}v_i(\mathbf{s},t)+\cdots+\beta(\mathbf{s})\partial_{s_e s_e}v_1(\mathbf{s},t)\big)=
$$
$$
\partial_{v_i}P\big(\mathbf{v}(\mathbf{s},t)\big)+f_i\big(\mathbf{v}(\mathbf{s},t)\big),\quad 1\leqslant i\leqslant d,\quad i\in\mathbb{N},\quad (6)
$$

where ∂_{s_j} and $\partial_{s_j s_j}$ are the first and second partial derivatives with respect to the parametric variables s_j respectively. f_i is the coordinate function i of the vector of constraints $\mathbf{f}$.

3   Spatial and Temporal Discretizations

3.1   Spatial Discretization

The discretization in the spatial domain is done by means of finite elements. The domain $[0, L_j]$ of each parametric variable s_j is partitioned into N_j finite subdomains. Hence, the model $\mathbf{v}(\mathbf{s},t)$ can be expressed as the union of $N = N_1 N_2 \cdots N_e$ elements. Given that $\mathbf{v}(\mathbf{s},t)$ is defined by means of d coordinate functions, which are mutually independent, this procedure may be applied separately to each coordinate function:

$$
v_i(\mathbf{s},t) = \sum_{n_1=0}^{N_1-1} \cdots \sum_{n_e=0}^{N_e-1} v_i^{\overline{n}}(\mathbf{s},t), \quad (7)
$$

where $\overline{n} = [n_1, \ldots, n_e]$ are the indexes of the element in each coordinate function. Each element $v_i^{\overline{n}}$ is represented geometrically using shape functions $\mathbf{N}$ and nodal variables $\mathbf{u}$. Hence, $v_i^{\overline{n}}(\mathbf{s},t) = \mathbf{N}_i^{\overline{n}}(\mathbf{s})\mathbf{u}_i^{\overline{n}}(\mathbf{s},t)$. If we assume closed model and use the same shape function for all elements [6], this expression can be simplified to $v_i^{\overline{n}}(\mathbf{s},t) = \mathbf{N}(\mathbf{s})\mathbf{u}_i^{\overline{n}}(t)$.

The derivation of the matrix equation for one-dimensional models or *snakes* is described in [5]. Similarly, we apply the Galerkin's method [7] to the equation for the multidimensional case (6). This equation stands for the necessary condition for the model at equilibrium. The average weighted residual of this expression can be calculated as,

$$
I = \int_{\Omega} \Psi(\mathbf{s},t)w(\mathbf{s})ds = 0, \quad (8)
$$

where $\Psi(\mathbf{s},t)$ represents the left-hand side of (6) and $w(\mathbf{s})$ is a arbitrary test function. Assuming for this mathematical reasoning a two-dimensional model or active mesh, i.e. $e = 2$. By applying two-dimensional integration by parts once to the third term and two times for the fourth term, we arrive at the weak form of the deformable model,

$$
\int_{\Omega} w(\mathbf{s})\mu(\mathbf{s})\partial_{tt}v_i(\mathbf{s},t)ds + \int_{\Omega} w(\mathbf{s})\gamma(\mathbf{s})\partial_t v_i(\mathbf{s},t)ds
$$

$$
+ \int_{\Omega} \partial_{s_1}(w(\mathbf{s}))\alpha(\mathbf{s})\partial_{s_1}v_i(\mathbf{s},t)ds + \int_{\Omega} \partial_{s_2}(w(\mathbf{s}))\alpha(\mathbf{s})\partial_{s_2}v_i(\mathbf{s},t)ds
$$

$$
+ \int_{\Omega} \partial_{s_1 s_1}(w(\mathbf{s}))\beta(\mathbf{s})\partial_{s_1 s_1}v_i(\mathbf{s},t)ds + \int_{\Omega} \partial_{s_2 s_2}(w(\mathbf{s}))\beta(\mathbf{s})\partial_{s_2 s_2}v_i(\mathbf{s},t)ds \qquad (9)
$$

$$
+ \int_{\Omega} \partial_{s_1 s_1}(w(\mathbf{s}))\beta(\mathbf{s})\partial_{s_2 s_2}v_i(\mathbf{s},t)ds + \int_{\Omega} \partial_{s_2 s_2}(w(\mathbf{s}))\beta(\mathbf{s})\partial_{s_1 s_1}v_i(\mathbf{s},t)ds
$$

$$
+ \int_{\Omega} w(\mathbf{s})q(\mathbf{s},t)ds + b = 0,
$$

where

$$
b = -\int_0^{L_2} \left[w(\mathbf{s})\alpha(\mathbf{s})\partial_{s_1}v_i(\mathbf{s},t) \right]_0^{L_1} ds_2 - \int_0^{L_1} \left[w(\mathbf{s})\alpha(\mathbf{s})\partial_{s_2}v_i(\mathbf{s},t) \right]_0^{L_2} ds_1
$$

$$
+ \int_0^{L_2} \left[w(\mathbf{s})\partial_{s_1}(\beta(\mathbf{s})\partial_{s_1 s_1}v_i(\mathbf{s},t)) \right]_0^{L_1} ds_2 + \int_0^{L_1} \left[w(\mathbf{s})\partial_{s_2}(\beta(\mathbf{s})\partial_{s_2 s_2}v_i(\mathbf{s},t)) \right]_0^{L_2} ds_1
$$

$$
+ \int_0^{L_2} \left[w(\mathbf{s})\partial_{s_1}(\beta(\mathbf{s})\partial_{s_2 s_2}v_i(\mathbf{s},t)) \right]_0^{L_1} ds_2 + \int_0^{L_1} \left[w(\mathbf{s})\partial_{s_2}(\beta(\mathbf{s})\partial_{s_1 s_1}v_i(\mathbf{s},t)) \right]_0^{L_2} ds_1
$$

$$
- \int_0^{L_2} \left[\partial_{s_1}(w(\mathbf{s}))\beta(\mathbf{s})\partial_{s_1 s_1}v_i(\mathbf{s},t) \right]_0^{L_1} ds_2 - \int_0^{L_1} \left[\partial_{s_2}(w(\mathbf{s}))\beta(\mathbf{s})\partial_{s_2 s_2}v_i(\mathbf{s},t) \right]_0^{L_2} ds_1
$$

$$
- \int_0^{L_2} \left[\partial_{s_1}(w(\mathbf{s}))\beta(\mathbf{s})\partial_{s_2 s_2}v_i(\mathbf{s},t) \right]_0^{L_1} ds_2 - \int_0^{L_1} \left[\partial_{s_2}(w(\mathbf{s}))\beta(\mathbf{s})\partial_{s_1 s_1}v_i(\mathbf{s},t) \right]_0^{L_2} ds_1,
$$

represents the boundary conditions in the integration domain $[0, L_1] \times [0, L_2]$.

In this section, we assume that both the array of shape functions $\mathbf{N}(\mathbf{s})$ and the array of parameters $\mathbf{u}_i(t)$, can be written in the form of row and column vectors respectively. B-*splines* and finite differences satisfy this property.

First, a model with a single element $N_1 = N_2 = 1$ is examined. Then, $v_i(\mathbf{s},t)$ is a scalar function of several independent variables obtained as the scalar product of $\mathbf{N}(\mathbf{s})$ and $\mathbf{u}_i(t)$. The size of both vectors are $m = m_1 m_2$, where m_j is the number of nodal variables for the parameter s_j of the model. Garlekin's method uses an arbitrary test function $w(\mathbf{s})$ with the form $w(\mathbf{s}) = \mathbf{N}(\mathbf{s})\mathbf{c}$, where $\mathbf{N}(\mathbf{s})$ is the same shape function vector and $\mathbf{c}$ is an arbitrary vector. Note that $w(\mathbf{s})$ is a scalar, so $w(\mathbf{s}) = w^\top(\mathbf{s}) = \mathbf{c}^\top \mathbf{N}^\top(\mathbf{s})$. Substituting these expressions in (9) and simplifying, we obtain the equation of motion of the elemental model:

$$
\mathbf{M}^e d_{tt}\mathbf{u}_i(t) + \mathbf{C}^e d_t\mathbf{u}_i(t) + \mathbf{K}^e \mathbf{u}_i(t) - \mathbf{Q}_i^e(t) + \mathbf{P}_i^e(t) = \mathbf{0}, \qquad (10)
$$

where $\mathbf{M}^e$ is the mass matrix, $\mathbf{C}^e$ is the damping matrix, $\mathbf{K}^e$ is the stiffness matrix, $\mathbf{Q}_i^e$ is the force vector and $\mathbf{P}_i^e$ are the boundary forces. These matrices can be calculated analytically by,

$$
\mathbf{M}^e = \int_{\Omega} \mathbf{N}^\top(\mathbf{s})\mu(\mathbf{s})\mathbf{N}(\mathbf{s})ds, \quad \mathbf{C}^e = \int_{\Omega} \mathbf{N}^\top(\mathbf{s})\gamma(\mathbf{s})\mathbf{N}(\mathbf{s})ds, \quad \mathbf{K}^e = \mathbf{K}_\alpha^e + \mathbf{K}_\beta^e, \qquad (11)
$$

$$
\mathbf{K}_\alpha^e = \int_{\Omega} \partial_{s_1}(\mathbf{N}^\top(\mathbf{s}))\alpha(\mathbf{s})\partial_{s_1}\mathbf{N}(\mathbf{s})ds + \int_{\Omega} \partial_{s_2}(\mathbf{N}^\top(\mathbf{s}))\alpha(\mathbf{s})\partial_{s_2}\mathbf{N}(\mathbf{s})ds, \qquad (12)
$$

$$\mathbf{K}_\beta^e = \int_\Omega \partial_{s_1 s_1}\left(\mathbf{N}^\top(\mathbf{s})\right)\beta(\mathbf{s})\partial_{s_1 s_1}\mathbf{N}(\mathbf{s})ds + \int_\Omega \partial_{s_2 s_2}\left(\mathbf{N}^\top(\mathbf{s})\right)\beta(\mathbf{s})\partial_{s_2 s_2}\mathbf{N}(\mathbf{s})ds$$
$$+ \int_\Omega \partial_{s_1 s_1}\left(\mathbf{N}^\top(\mathbf{s})\right)\beta(\mathbf{s})\partial_{s_2 s_2}\mathbf{N}(\mathbf{s})ds + \int_\Omega \partial_{s_2 s_2}\left(\mathbf{N}^\top(\mathbf{s})\right)\beta(\mathbf{s})\partial_{s_1 s_1}\mathbf{N}(\mathbf{s})ds, \tag{13}$$

$$\mathbf{Q}_i^e = \int_\Omega \mathbf{N}^\top(\mathbf{s})q_i(\mathbf{s},t)ds, \tag{14}$$

$$\mathbf{P}_i^e = \left(-\int_0^{L_2}\left[\mathbf{N}^\top(\mathbf{s})\alpha(\mathbf{s})\partial_{s_1}\mathbf{N}(\mathbf{s})\right]_0^{L_1}ds_2 - \int_0^{L_1}\left[\mathbf{N}^\top(\mathbf{s})\alpha(\mathbf{s})\partial_{s_2}\mathbf{N}(\mathbf{s})\right]_0^{L_2}ds_1\right.$$
$$+ \int_0^{L_2}\left[\mathbf{N}^\top(\mathbf{s})\partial_{s_1}\left(\beta(\mathbf{s})\partial_{s_1 s_1}\mathbf{N}(\mathbf{s})\right)\right]_0^{L_1}ds_2 + \int_0^{L_1}\left[\mathbf{N}^\top(\mathbf{s})\partial_{s_2}\left(\beta(\mathbf{s})\partial_{s_2 s_2}\mathbf{N}(\mathbf{s})\right)\right]_0^{L_2}ds_1$$
$$+ \int_0^{L_2}\left[\mathbf{N}^\top(\mathbf{s})\partial_{s_1}\left(\beta(\mathbf{s})\partial_{s_2 s_2}\mathbf{N}(\mathbf{s})\right)\right]_0^{L_1}ds_2 + \int_0^{L_1}\left[\mathbf{N}^\top(\mathbf{s})\partial_{s_2}\left(\beta(\mathbf{s})\partial_{s_1 s_1}\mathbf{N}(\mathbf{s})\right)\right]_0^{L_2}ds_1$$
$$- \int_0^{L_2}\left[\partial_{s_1}\left(\mathbf{N}^\top(\mathbf{s})\right)\beta(\mathbf{s})\partial_{s_1 s_1}\mathbf{N}(\mathbf{s})\right]_0^{L_1}ds_2 - \int_0^{L_1}\left[\partial_{s_2}\left(\mathbf{N}^\top(\mathbf{s})\right)\beta(\mathbf{s})\partial_{s_2 s_2}\mathbf{N}(\mathbf{s})\right]_0^{L_2}ds_1$$
$$\left.- \int_0^{L_2}\left[\partial_{s_1}\left(\mathbf{N}^\top(\mathbf{s})\right)\beta(\mathbf{s})\partial_{s_2 s_2}\mathbf{N}(\mathbf{s})\right]_0^{L_1}ds_2 - \int_0^{L_1}\left[\partial_{s_2}\left(\mathbf{N}^\top(\mathbf{s})\right)\beta(\mathbf{s})\partial_{s_1 s_1}\mathbf{N}(\mathbf{s})\right]_0^{L_2}ds_1\right)\mathbf{u}_i(t).$$

Equation (10) provides the formulation for a one-segment model. In practical implementations, deformable models are defined with multiple segments. By assuming a closed model, the boundary forces of each segment $\mathbf{P}_i^e$ cancel each other. By assembling both matrices and nodal variables of all elements, we get the motion equation for the whole system,

$$\mathbf{M}d_{tt}\mathbf{u}_i(t) + \mathbf{C}d_t\mathbf{u}_i(t) + \mathbf{K}\mathbf{u}_i(t) = \mathbf{q}_i(t), \tag{15}$$

where $\mathbf{M}$, $\mathbf{C}$ and $\mathbf{K}$ represent the global matrices of mass, damping and stiffness, respectively, and $\mathbf{q}_i(t)$ is the external forces vector. Note that the dimensions of the original e-dimensional array $\mathbf{u}_i(t)$ are $N_1 \times \cdots \times N_e$, but this array must be reshaped into a column vector to be used in this equation. The global matrices of mass $\mathbf{M}$, damping $\mathbf{C}$ and stiffness $\mathbf{K}$ are usually constructed by means of the elementary matrices $\mathbf{M}^e$, $\mathbf{C}^e$ and $\mathbf{K}^e$ and assembling matrices $\mathbf{G}$ as follows,

$$\mathbf{R} = \sum_{n_1=0}^{N_1-1} \cdots \sum_{n_e=0}^{N_e-1} (\mathbf{G}^{\bar{n}})^\top \mathbf{R}^e (\mathbf{G}^{\bar{n}}), \tag{16}$$

where $\mathbf{R}$ and $\mathbf{R}^e$ stands for the matrices $\mathbf{M}$, $\mathbf{C}$ and $\mathbf{K}$ and the elementary matrices $\mathbf{M}^e$, $\mathbf{C}^e$ and $\mathbf{K}^e$. The assembling matrix $\mathbf{G}^{\bar{n}} = \mathbf{G}^{n_1} \otimes \cdots \otimes \mathbf{G}^{n_e}$, depends on the shape function and is constructed by means of the Kronecker product of one-dimension assembling matrices $\mathbf{G}^{n_l}$.

The matrices $\mathbf{M}$ and $\mathbf{C}$ satisfy $\mathbf{M} = m\mathbf{F}$ and $\mathbf{C} = c\mathbf{F}$, where m and c represent the mass and damping of the model and $\mathbf{F}$ is the shape matrix. So, Eq. (15) can be written as $m\mathbf{F}d_{tt}\mathbf{u}_i(t) + c\mathbf{F}d_t\mathbf{u}_i(t) + \mathbf{K}\mathbf{u}_i(t) = \mathbf{q}_i(t)$. Both $\mathbf{F}$ and $\mathbf{K}$ are nested circulant. In consequence, $\mathbf{F}$ and $\mathbf{K}$ are completely defined by their first rows. Section 4 shows the first rows of the global matrices for one- and two-dimensional models.

3.2 Temporal Discretizacion and Implementation in the Frequency Domain

Time is discretized $t = \xi \Delta t$, where Δt is the time step and $\xi \in \mathbb{N}$ the iteration index[1]. The time derivatives of $\mathbf{u}(t)$ are replaced by their regressive discrete approximations [6]. Thus, Eq. (15) can be written as,

$$\left(\left(\frac{m}{\Delta t^2} + \frac{c}{\Delta t} \right) \mathbf{F} + \mathbf{K} \right) \mathbf{u}_\xi = \left(\frac{2m}{\Delta t^2} + \frac{c}{\Delta t} \right) \mathbf{F} \, \mathbf{u}_{\xi-1} + \left(\frac{-m}{\Delta t^2} \right) \mathbf{F} \, \mathbf{u}_{\xi-2} + \mathbf{q}_{\xi-1}.$$
(17)

A circulant matrix is the algebraic representation of a circular discrete convolution [8]. Hence, the result of multiplying a circulant matrix and a vector is equivalent to a circulant convolution of two vectors. Similarly, it can be proved that the product between a nested circulant matrix by a vector reshaped as $\mathbf{u}_\xi$ is equivalent to an e-dimensional circular convolution of two e-dimensional arrays. Since $\mathbf{F}$ and $\mathbf{K}$ are nested circulant, $\mathbf{F} \mathbf{u}_\xi = \mathbf{f} \circledast u_\xi$ and $\mathbf{K} \mathbf{u}_\xi = \mathbf{k} \circledast u_\xi$, where u_ξ must be arranged in the original layout, i.e. an e-dimensional array of size $N_1 \times \cdots \times N_e$. The arrays $\mathbf{f}$ and $\mathbf{k}$ can be assembled rearranging the first rows of the respective global matrices $\mathbf{F}$ and $\mathbf{K}$ as detailed in Sec. 4.

Once defined the equation of motion (17) by means of e-dimensional circulant convolutions, these convolutions are transformed into the product of their e-dimensional discrete Fourier transforms ($e\mathcal{DFT}$). Thus, the equation of motion can be formulated completely in the frequency domain as described in [9].

4 Development of the System Matrices

4.1 B-Spline Shape Function

Multidimensional B-*spline* allow the representation of each coordinate function $v_i(\mathbf{s})$ as a weighted sum of basis functions $B_{\overline{n}}(\mathbf{s})$. Additionally, these shape functions have the property of separability [10], then

$$v_i(\mathbf{s}, t) = \sum_{n_1=0}^{N_1-1} \cdots \sum_{n_e=0}^{N_e-1} B_{n_1,b}(s_1) \cdots B_{n_e,b}(s_e) \mathbf{u}_i^{\overline{n}}(t) = \mathbf{N}(\mathbf{s}) \mathbf{u}_i^{\overline{n}}(t),$$
(18)

where $B_{n,b}(s)$ is the uni-dimensional basis function of order b [10]. When model domain is divided into a number of small subdomains, each subdomain can be considered as an elementary deformable model. Hence, using cubic B-*spline* ($b = 4$) for all spans and dimensions, $\mathbf{N}(\mathbf{s})$ can be computed as:

$$\mathbf{N}^{1D}(s_1) = [B_0(s_1) \; B_1(s_1) \; B_2(s_1) \; B_3(s_1)]$$
$$\mathbf{N}^{2D}(s_1, s_2) = [B_0(s_1)B_0(s_2) \; B_0(s_1)B_1(s_2) \; B_0(s_1)B_2(s_2) \; B_1(s_1)B_0(s_2) \; \cdots \; B_3(s_1)B_3(s_2)],$$

where $\cdots$ represents elements $B_{n_1}(s_1)B_{n_2}(s_2)$ in the order expressed in these equations. Vectors $\mathbf{N}^{1D}$ and $\mathbf{N}^{2D}$ have a length 4 and 4^2 respectively.

---

[1] In the following, the notation $\mathbf{u}_i(\xi \Delta t) = \mathbf{u}_\xi$ is used. For the sake of simplicity, we omit the subscript i.

The elementary matrices $\mathbf{M}^e$, $\mathbf{C}^e$ and $\mathbf{K}^e$ can be calculated for any span using these shape vectors $\mathbf{N(s)}$ and Eqs. (11-13). The global matrices $\mathbf{M}$, $\mathbf{C}$ and $\mathbf{K}$ and the vector of forces $\mathbf{Q}$ can be calculated through the elementary matrices by means of (16) and the assembling matrices $\mathbf{G}^{\overline{n}}$. For B-*splines*, the one-dimension assembling matrices $\mathbf{G}^{n_l}$ can be calculated as

$$(\mathbf{G}^{n_l})_{rs} = \begin{cases} 1 & \text{if } (r + b_{\sigma_l}) \bmod N_l = s, \\ 0 & \text{otherwise.} \end{cases} \tag{19}$$

The e-dimensional arrays $\mathbf{f}$ and $\mathbf{k}$ obtained in Sec. 3.2 must be derived from global matrices $\mathbf{F}$ and $\mathbf{K}$. For 1-D models these arrays are defined as follows

$$\mathbf{f}_B^{1D} = \frac{1}{5040} \begin{bmatrix} 2416 & 1191 & 120 & 1 & 0 & \cdots & 0 & 1 & 120 & 1191 \end{bmatrix},$$

$$\mathbf{k}_{B\alpha}^{1D} = \frac{1}{120} \begin{bmatrix} 80 & -15 & -24 & -1 & 0 & \cdots & 0 & -1 & -24 & -15 \end{bmatrix},$$

$$\mathbf{k}_{B\beta}^{1D} = \frac{1}{6} \begin{bmatrix} 16 & -9 & 0 & 1 & 0 & \cdots & 0 & 1 & 0 & -9 \end{bmatrix},$$

where $\cdots$ represents null values to provide vectors of size N_1. For 2-D models:

$$\mathbf{f}_B^{2D} = \frac{1}{5040^2} \begin{bmatrix}
5837056 & 2877456 & 289920 & 2416 & 0 & \cdots & 0 & 2416 & 289920 & 2877456 \\
2877456 & 1418481 & 142920 & 1191 & 0 & \cdots & 0 & 1191 & 142920 & 1418481 \\
289920 & 142920 & 14400 & 120 & 0 & \cdots & 0 & 120 & 14400 & 142920 \\
2416 & 1191 & 120 & 1 & 0 & \cdots & 0 & 1 & 120 & 1191 \\
0 & 0 & 0 & 0 & 0 & \cdots & 0 & 0 & 0 & 0 \\
\vdots & \vdots & \vdots & \vdots & \vdots & & \vdots & \vdots & \vdots & \vdots \\
0 & 0 & 0 & 0 & 0 & \cdots & 0 & 0 & 0 & 0 \\
2416 & 1191 & 120 & 1 & 0 & \cdots & 0 & 1 & 120 & 1191 \\
289920 & 142920 & 14400 & 120 & 0 & \cdots & 0 & 120 & 14400 & 142920 \\
2877456 & 1418481 & 142920 & 1191 & 0 & \cdots & 0 & 1191 & 142920 & 1418481
\end{bmatrix},$$

$$\mathbf{k}_{B\alpha}^{2D} = \frac{1}{120^2 \cdot 21} \begin{bmatrix}
193280 & 29520 & -24192 & -1168 & 0 & \cdots & 0 & -1168 & -24192 & 29520 \\
29520 & -17865 & -15192 & -603 & 0 & \cdots & 0 & -603 & -15192 & -17865 \\
-24192 & -15192 & -2880 & -72 & 0 & \cdots & 0 & -72 & -2880 & -15192 \\
-1168 & -603 & -72 & -1 & 0 & \cdots & 0 & -1 & -72 & -603 \\
0 & 0 & 0 & 0 & 0 & \cdots & 0 & 0 & 0 & 0 \\
\vdots & \vdots & \vdots & \vdots & \vdots & & \vdots & \vdots & \vdots & \vdots \\
0 & 0 & 0 & 0 & 0 & \cdots & 0 & 0 & 0 & 0 \\
-1168 & -603 & -72 & -1 & 0 & \cdots & 0 & -1 & -72 & -603 \\
-24192 & -15192 & -2880 & -72 & 0 & \cdots & 0 & -72 & -2880 & -15192 \\
29520 & -17865 & -15192 & -603 & 0 & \cdots & 0 & -603 & -15192 & -17865
\end{bmatrix}$$

$$\mathbf{k}_{B_\beta}^{2D} = \frac{1}{6^2 \cdot 4200} \begin{bmatrix} 520960 & -38640 & -30720 & 10480 & 0 & \cdots & 0 & 10480 & -30720 & -38640 \\ -38640 & -102465 & 2160 & 6225 & 0 & \cdots & 0 & 6225 & 2160 & -102465 \\ -30720 & 2160 & 12096 & 1104 & 0 & \cdots & 0 & 1104 & 12096 & 2160 \\ 10480 & 6225 & 1104 & 31 & 0 & \cdots & 0 & 31 & 1104 & 6225 \\ 0 & 0 & 0 & 0 & 0 & \cdots & 0 & 0 & 0 & 0 \\ \vdots & \vdots & \vdots & \vdots & \vdots & \vdots & \vdots & \vdots & \vdots & \vdots \\ 0 & 0 & 0 & 0 & 0 & \cdots & 0 & 0 & 0 & 0 \\ 10480 & 6225 & 1104 & 31 & 0 & \cdots & 0 & 31 & 1104 & 6225 \\ -30720 & 2160 & 12096 & 1104 & 0 & \cdots & 0 & 1104 & 12096 & 2160 \\ -38640 & -102465 & 2160 & 6225 & 0 & \cdots & 0 & 6225 & 2160 & -102465 \end{bmatrix},$$

where $\cdots$ and $\vdots$ represent null values to provide vectors of length $N = N_1 N_2$. In both dimensions, $\mathbf{k}_B = \alpha \mathbf{k}_{B_\alpha} + \beta \mathbf{k}_{B_\beta}$.

4.2 Finite Difference Shape Function

Despite the substantial differences between the models defined by B-*splines* and finite differences, finite difference models can also be constructed by the described procedure by defining the Dirac delta as shape function,

$$v_i(\mathbf{s}, t) = \sum_{n_1=0}^{N_1-1} \cdots \sum_{n_e=0}^{N_e-1} \delta\left(s_1 - n_1 L_1\right) \cdots \delta\left(s_e - n_e L_e\right) \mathbf{u}_i^{\overline{\mathbf{n}}}(t) = \mathbf{N}(\mathbf{s}) \mathbf{u}_i^{\overline{\mathbf{n}}}(t), \quad (21)$$

where L_j is the length of segments for dimension j. Nevertheless, the calculation of elementary matrices with the continuos delta $\delta(\overline{\mathbf{s}})$ has no analytical solution. Therefore, the shape function must be approximated by a discrete delta $\delta[\overline{\mathbf{n}}]$ and its derivatives by their discrete approximations. By applying the discrete approximations to Eqs. (11-13) elementary matrices $\mathbf{M}^e$, $\mathbf{C}^e$ and $\mathbf{K}^e$, and global matrices $\mathbf{M}$, $\mathbf{C}$ and $\mathbf{K}$ and the vector of forces $\mathbf{Q}$ can be calculated. For finite differences, the one-dimension assembling matrices are defined in [5]. Finally, the e-dimensional arrays $\mathbf{f}$ and $\mathbf{k}$ derived from global matrices $\mathbf{F}$ and $\mathbf{K}$ are for one-dimensional models,

$$\mathbf{f}_{FD}^{1D} = \begin{bmatrix} 1 & 0 & 0 & \ldots & 0 & 0 \end{bmatrix},$$
$$\mathbf{k}_{FD_\alpha}^{1D} = \begin{bmatrix} 2 & -1 & 0 & 0 & \ldots & 0 & 0 & -1 \end{bmatrix},$$
$$\mathbf{k}_{FD_\beta}^{1D} = \begin{bmatrix} 6 & -4 & 1 & 0 & \ldots & 0 & 1 & -4 \end{bmatrix},$$

where $\cdots$ represents null values to provide vectors of size N_1. For 2-D models:

$$\mathbf{f}_{FD}^{2D} = \begin{bmatrix} 1 & 0 & \cdots & 0 \\ 0 & 0 & \cdots & 0 \\ \vdots & \vdots & & \vdots \\ 0 & 0 & \cdots & 0 \end{bmatrix}, \quad \mathbf{k}_{FD_\alpha}^{2D} = \begin{bmatrix} 4 & -1 & 0 & \cdots & 0 & -1 \\ -1 & 0 & 0 & \cdots & 0 & 0 \\ 0 & 0 & 0 & \cdots & 0 & 0 \\ \vdots & \vdots & \vdots & & \vdots & \vdots \\ 0 & 0 & 0 & \cdots & 0 & 0 \\ -1 & 0 & 0 & \cdots & 0 & 0 \end{bmatrix}, \quad \mathbf{k}_{FD_\beta}^{2D} = \begin{bmatrix} 20 & -8 & 1 & 0 & \cdots & 0 & 1 & -8 \\ -8 & 2 & 0 & 0 & \cdots & 0 & 0 & 2 \\ 1 & 0 & 0 & 0 & \cdots & 0 & 0 & 0 \\ 0 & 0 & 0 & 0 & \cdots & 0 & 0 & 0 \\ \vdots & \vdots & \vdots & \vdots & & \vdots & \vdots & \vdots \\ 0 & 0 & 0 & 0 & \cdots & 0 & 0 & 0 \\ 1 & 0 & 0 & 0 & \cdots & 0 & 0 & 0 \\ -8 & 2 & 0 & 0 & \cdots & 0 & 0 & 2 \end{bmatrix},$$

where $\cdots$ and $\vdots$ represent null values to provide vectors of length $N = N_1N_2$. In both dimensions, $\mathbf{k}_{FD} = \alpha\mathbf{k}_{FD_\alpha} + \beta\mathbf{k}_{FD_\beta}$.

5   Results

In order to compare the effectiveness of each shape function, 2-D models are used for the estimation of the surface of 3-D data. That is, $e = 2$ and $d = 3$. Four datasets $\mathcal{DS}_1$-$\mathcal{DS}_4$ are used for this experiment. $\mathcal{DS}_1$ is the dataset with lowest bandwidth and $\mathcal{DS}_4$ is the model with highest bandwidth. In all cases a model with $N_1 = 64$ and $N_2 = 64$ nodes is used to segment the surface of these data. Fig. 1 depicts the temporal evolution of the model for each dataset. The first column shows the initialization of the model, the second and third columns portray an intermediate state and the model in equilibrium, respectively.

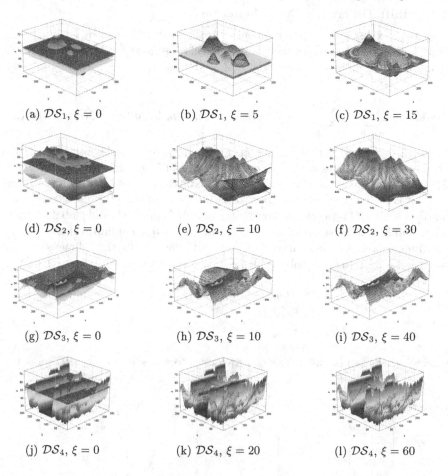

 (a) $\mathcal{DS}_1$, $\xi = 0$ (b) $\mathcal{DS}_1$, $\xi = 5$ (c) $\mathcal{DS}_1$, $\xi = 15$

 (d) $\mathcal{DS}_2$, $\xi = 0$ (e) $\mathcal{DS}_2$, $\xi = 10$ (f) $\mathcal{DS}_2$, $\xi = 30$

 (g) $\mathcal{DS}_3$, $\xi = 0$ (h) $\mathcal{DS}_3$, $\xi = 10$ (i) $\mathcal{DS}_3$, $\xi = 40$

 (j) $\mathcal{DS}_4$, $\xi = 0$ (k) $\mathcal{DS}_4$, $\xi = 20$ (l) $\mathcal{DS}_4$, $\xi = 60$

Fig. 1. Temporal evolution of the deformable model for datasets $\mathcal{DS}_1 - \mathcal{DS}_4$.

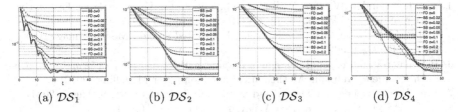

(a) $\mathcal{DS}_1$ (b) $\mathcal{DS}_2$ (c) $\mathcal{DS}_3$ (d) $\mathcal{DS}_4$

Fig. 2. Mean error between the model and datasets $\mathcal{DS}_1 - \mathcal{DS}_4$

To analyze the robustness against noise, the adaptive process is simulated by adding Gaussian white noise with standard deviation between $\sigma = 0.02 D_{max}$ and $\sigma = 0.2 D_{max}$, where D_{max} is the largest value of the z coordinate of the data. Fig. 2 depicts the mean error for all nodes in each dataset, shape function and level of added noise along the iterative process. As can be drawn from the picture, in all cases the use of the B-*spline* shape function leads to a small average error, and a better adjustment between the model and the data surface. When no noise is applied, the results are fairly similar between the two shape functions. However, as more noise is added to the signal, the more significant is the difference between both shape functions. Regarding to the bandwidth of the signal, although the differences are not significant, the advantage of the B-*splines* over finite differences is greater for signals with lower bandwidth. When no noise is applied, the results are fairly similar between the two shape functions. However, as more noise is added to the signal, the more significant is the difference between both shape functions. Regarding to the bandwidth of the signal, although the differences are not significant, the advantage of the B-*splines* over finite differences is greater for signals with lower bandwidth.

6   Conclusions

This paper describes a formulation of two-dimensional deformable models by means of a system matrix based on finite elements. B-*splines* and finite differences are used as shape functions. These models have been applied to the characterization of 3D noisy datasets. Results show that these models can be used effectively to outline the shape of 3D objects, noting that B-*splines* are able to minimize the effect of noise on a higher level than finite differences.

References

1. Kass, M., Witkin, A., Terzopoulos, D.: Snakes: Active contour models. Int. J. of Computer Vision 1, 321–331 (1988)
2. Terzopoulos, D., Witkin, A., Kass, M.: Constraints on deformable models: recovering 3D shape and nonrigid motion. Artificial Intelligence 36, 91–123 (1988)
3. Montagnat, J., Delingette, H., Ayache, N.: A review of deformable surfaces: Topology, geometry and deformation. Image and Vision Computing 19, 1023–1040 (2001)

4. Neuenschwander, W., Fua, P., Székely, G., Kübler, O.: Velcro surfaces: Fast initialization of deformable models. In: CVGIU, pp. 237–245 (1997)
5. Liang, J., McInerney, T., Terzopoulos, D.: United snakes. Medical Image Analysis 10, 215–233 (2006)
6. Berenguer-Vidal, R., et al.: Design of B-spline multidimensional deformable models in the frequency domain. Mathematical and Computer Modelling (2012)
7. Ern, A., Guermond, J.: Theory and practice of finite elements. Springer (2004)
8. Davis, P.J.: Circulant Matrices. Wiley-Interscience, NY (1994)
9. Berenguer-Vidal, R., et al.: Characterization of 3D data with multidimensional deformable models based on B-splines in the fourier domain. In: ICASSP (2011)
10. Unser, M.: Splines: a perfect fit for signal and image processing. IEEE Signal Processing Magazine 16, 22–38 (1999)

Automatic Evaluation
of Carotid Intima-Media Thickness
in Ultrasounds Using Machine Learning

Rosa-María Menchón-Lara[1], María-Consuelo Bastida-Jumilla[1],
Antonio González-López[2], and José Luis Sancho-Gómez[1]

[1] Dpto. Tecnologías de la Información y las Comunicaciones.
Universidad Politécnica de Cartagena.
Plaza del Hospital, 1, 30202, Cartagena (Murcia), Spain
[2] Hospital Universitario Virgen de la Arrixaca.
Ctra. Madrid-Cartagena s/n, 30120, El Palmar (Murcia), Spain
rmml@alu.upct.es

Abstract. Cardiovascular diseases (CVD) are the main cause of death
and disability in the world. Atherosclerosis is responsible for a large proportion of cardiovascular diseases. The atherosclerotic process is a degenerative condition, mainly affecting the medium- and large-size arteries,
that develops over many years. It causes thickening and the reduction
of elasticity in the blood vessels. An early diagnosis of this condition is
crucial to prevent patients from suffering more serious pathologies. The
evaluation of the Intima-Media Thickness (IMT) of the Common Carotid
Artery (CCA) in B-mode ultrasound images is considered the most useful
tool for the investigation of preclinical atherosclerosis. This paper proposes an effective image segmentation procedure for the measurement
of the IMT in an automatic way, avoiding the user dependence and the
inter-rater variability. Segmentation is raised as a pattern recognition
problem and a neural network ensemble has been trained to classify the
image pixels. The suggested approach is tested on a set of 25 ultrasound
images and its validation is performed by comparing the automatic segmentations with manual tracings.

1 Introduction

Cardiovascular diseases (CVD) are a major cause of death worldwide. The leading underlying pathological process that results in a large proportion of CVD
is atherosclerosis [16]. It is characterized by the accumulation of fatty material
and cholesterol at the arterial walls, which causes thickening and the reduction of elasticity of these. Since it may remain unnoticed for decades, the study
of preclinical atherosclerosis is crucial for preventive purposes. In this sense,
the Intima-Media Thickness (IMT) of the Common Carotid Artery (CCA) has
emerged as an early and reliable indicator of atherosclerosis [13].

J.M. Ferrández Vicente et al. (Eds.): IWINAC 2013, Part II, LNCS 7931, pp. 241–249, 2013.
© Springer-Verlag Berlin Heidelberg 2013

The IMT is measured by means of a B-mode ultrasound scan, which is a noninvasive , relatively inexpensive, and widely available technique. The use of different protocols and the variability between observers are recurrent problems in the IMT measurement procedure. Repeatability and reproducibility of the process are of great significance to study the IMT [7]. For these reasons, IMT should be measured preferably on the far wall of the CCA within a region free of plaque [13]. The optimal measurement section (1-cm-long) is located at least 5 mm below the carotid bifurcation, where a double-line pattern corresponding to the intima-media-adventitia layers can be clearly observed. As can be seen in Fig. 1, the IMT is the distance between the lumen-intima (LI) interface and the media-adventitia (MA) interface.

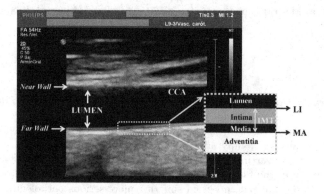

Fig. 1. B-mode ultrasonography of the CCA (longitudinal view)

In the last two decades, with the purpose of making more efficient and faster the measurement process of IMT, several image processing techniques have been proposed. Most of the proposed methods are not completely automatic and they require user interaction to start the algorithm, such as [2, 4, 11, 12]. However, some fully automatic approaches were recently published [1, 3, 9]. It is possible to realize a classification of techniques according to the used methodology. In this sense, we can find algorithms based on gradient-based techniques [4, 12], and other proposals based on dynamic programming [15], active contours [1–3, 9] or in a combination of both [11]. We can also find highlight techniques in which Hough transform is employed [6].

In this work, a fully automatic segmentation technique based on the use of neural networks is proposed to measure IMT from carotid ultrasounds. Firstly, a given image is pre-processed to detect the region of interest (ROI). Then, a network ensemble perform a classification of the pixels belonging to the ROI in either 'IMT-boundary' pixels or 'non-IMT-boundary' pixels, resulting to a binary output image. Finally, the obtained classification results are post-processed to extract the final contours corresponding to the LI and MA interfaces. The automatic measurements of the IMT have been compared with the values obtained from a manual segmentation and the statistical analysis of this comparison shows the accuracy of the proposed method.

The remainder of this paper is structured as follows: *Section* 2.1 describes the dataset of ultrasound images. The proposed segmentation method is explained in detail in *Section* 2.2. The obtained results are shown in *Section* 3. Finally, the main conclusions of our work close the paper.

2 Materials and Methods

2.1 Carotid Ultrasounds

A set of 25 longitudinal B-mode ultrasound images of the CCA have been used in the validation of our segmentation technique. All of them were provided by the Radiology Department of Hospital Universitario Virgen de la Arrixaca (Murcia, Spain). Fig. 1 shows an example of the tested ultrasound images. Ultrasound scans were acquired using a Philips iU22 Ultrasound System according to the measurement protocol proposed in [14] and recorded digitally with 256 gray levels. The spatial resolution of the images ranges from 0.029 to 0.081 mm/pixel.

To assess the accuracy of the obtained measurements, it is necessary to compare the automatic results with some indication of ground-truth. Although it is not possible to define the perfect segmentation, we use a manual segmentation from a experienced radiologist to perform the comparison with our method. Manual segmentations include the manually tracings of the LI and MA interfaces on the far carotid wall.

2.2 Proposed Method

Fig. 2 shows an overview of the proposed methodology. Firstly, a given ultrasound image of the CCA is pre-processed to automatically detect the region of interest (ROI). Then, a binary classification of the intensity patterns corresponding to the image pixels is performed using a network ensemble. Finally, classification results are post-processed to extract the final contours for the LI interface and the MA interface.

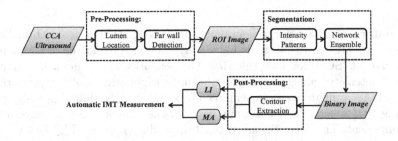

Fig. 2. Overview of the proposed methodology

ROI Detection: With the aim of avoiding the mis-segmented pixels outside the far wall of the CCA, we propose a pre-processing stage in which the referred wall is identified. Following this idea, detection of the lumen vessel is carried out in a completely automatic way. In particular, a binary mask is built using morphological operations [8] to locate the carotid lumen (see Fig. 3).

Within the ultrasonography (see Fig. 1), the lumen corresponds to a dark region (low echogenicity) delimited by the arterial walls. Over the lumen in the picture, at less depth, it is observed the echo corresponding to the near wall. The far wall, where the IMT is measured, is located below the lumen and it constitutes our ROI.

Once the lumen has been located, we focus on its lower limit corresponding to the far wall of the CCA and the boundaries of the ROI are established. The superior boundary is fixed to 0.6 mm above the upper point of the far wall detected in the binary mask, whereas the bottom boundary is fixed to 1.5 mm below the lower point. In Fig. 3 we can observe the selected ROI for the ultrasound image in Fig. 1.

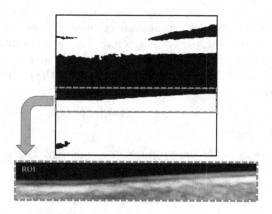

Fig. 3. Binary mask and ROI selection for the image in Fig.1

Segmentation: The artificial neural networks used in this work are Multi-Layer Perceptrons (MLP), with a single hidden layer, trained under the Scaled Conjugate Gradient (SCG) learning rule [10]. The networks take as input information only the intensity values of the pixels from a neighborhood of the pixel to be classified. Accordingly, a square window ($W \times W$) must be shifted pixel-by-pixel over the ROI image. For each input pattern, the network outputs have a single component. The networks are trained to produce a value of '1' for an input pattern with a target '*IMT-boundary*' at its central position, and '0' otherwise ('*non-IMT-boundary*').

To perform the supervised neural network training, a labelled dataset is needed. In our case, this dataset was assembled by taking samples from different manually segmented images. Finally, it consists of 6,000 patterns: 2,500 of them are from class 'IMT-boundary', and the remaining 3,500 are from class 'non-IMT-boundary'. During the learning process, the dataset was randomly divided into three subsets: 60% of samples for training, 20% for validation and 20% for testing.

In this paper, we propose a classification strategy in two stages to solve the posed segmentation task (see Fig. 4). In the first stage, each MLP is trained using a different window size to construct the input patterns (W = 3, 5, 7, and 9). However, the number of inputs is equal to 9 for all these networks (center pixel and the eight pixels which define the limits of the corresponding window). The reconstruction of the whole image is needed at the output of each MLP in stage 1. Then, a new windowing process (3 × 3) is applied to these images. Thus, the network in the second stage (MLP 5 in Fig. 4) consist of 36 inputs (9 from each output image of the first stage).

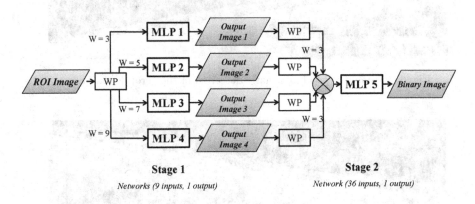

Fig. 4. Classification strategy adopted to solve the segmentation task

Contour Extraction: The output binary image in the classification stage should be debugged to measure the IMT. To begin with this post-processing stage, it is necessary to identify and separate the LI and MA interfaces. For this purpose, the binary image is processed column by column. Only those columns in which two boundaries have been found and in which the separation between them is not atypical will be considered.

Furthermore, due to the poor contrast of the ultrasound images, we obtain thick boundaries instead of one-pixel contours. In order to define the final contours on which the IMT evaluation is performed, we formulate a nonlinear least squares problem. Thus, the best model, which minimizes the Root Mean Squared Error (RMSE) between the white pixels in the binary image and the approximated contour, is found for each interface (LI and MA).

3   Results

Some examples of successfully segmented images are shown in Figs. 5-7. As can be seen, our automatic segmentation approach is robust against the orientation and appearance of the CCA in the image.

Fig. 5. Example of good automatic detection of the carotid wall layers

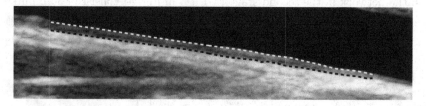

Fig. 6. Example of good automatic detection of the carotid wall layers

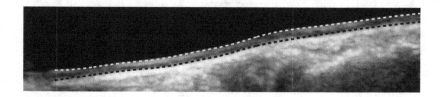

Fig. 7. Example of good automatic detection of the carotid wall layers

The mean value of the IMT, i.e. the average of the vertical distance between the segmented LI and MA interfaces along the longitudinal direction of the image, is assessed for each one of the ($n = 25$) processed images. The automatic evaluations of IMT have been compared with the considered 'ground-truth' (IMT values from manual tracings) by applying the Mean Absolute Distance (MAD) in the following form:

$$error^{IMT} = \frac{1}{n} \sum_{i=1}^{n} |IMT_i^{automatic} - IMT_i^{manual}| \qquad (1)$$

The obtained IMT measurement error (mean and standard deviation) is 0.041 ± 0.025 mm. Fig. 8 shows the regression analysis for the automatic and manual evaluations of IMT. As can be seen, a high degree of agreement (R = 0.98, correlation coefficient) is obtained when comparing the automatic measures with the manual ones. The Bland-Altman plot of the differences between the IMT of the corresponding two segmentations against their average can be seen in Fig. 9. Note that it shows the following limits of agreement (mean ± 2 standard deviation): - 0.034 ± 0.068 mm between the manual segmentation procedure and the proposed method.

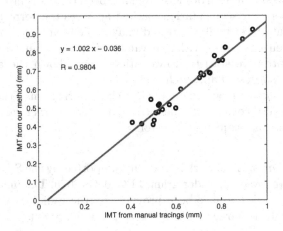

Fig. 8. IMT from our automatic method versus IMT from manual tracings

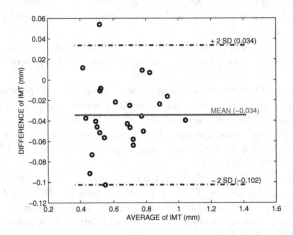

Fig. 9. Bland-Altman plot of IMT between the automatic and manual procedures

4 Conclusions

This paper proposes a fully automatic segmentation method of the CCA far wall using neural networks in order to measure the Intima-Media Thickness (IMT). The main step of the proposed method is a classification stage. In this stage, a combination of Multi-Layer Perceptrons (MLP) must perform a binary classification of the pixels to find the IMT contours. Our system is completed with a pre-processing stage in which the ROI is automatically selected and with a post-processing stage for the extraction of the final LI and MA contours.

The proposed configuration of the system has been tested using a dataset of 25 ultrasound images. Our segmentation method achieves the correct detection of the LI and MA interfaces in all the tested images. The automatic measurements of IMT have been compared with the values obtained from manual tracings and several quantitative statistical evaluations have shown the accuracy and robustness of the suggested approach.

Future works must be focussed on a better characterization of the results on a larger dataset by also considering the intra and inter-observer variability of the measurements and the segmentation errors.

Acknowledgements. This work is partially supported by the Spanish Ministerio de Ciencia e Innovación, under grant TEC2009-12675. The images used and the anatomical knowledge have been provided by the radiology department of 'Hospital Universitario Virgen de la Arrixaca', Murcia, Spain.

References

1. Bastida-Jumilla, M.C., et al.: Segmentation of the common carotid artery walls based on a frequency implementation of active contours. Journal of Digital Imaging 26(1), 129–139 (2013)
2. Cheng, D.C., et al.: Using snakes to detect the intimal and adventitial layers of the common carotid artery wall in sonographic images. Computers Methods and Programs in Biomedicine 67(1), 27–37 (2002)
3. Delsanto, S., et al.: Characterization of a completely user-independent algorithm for carotid artery segmentation in 2-d ultrasound images. IEEE Transactions on Instrumentation and Measurement 56(4), 1265–1274 (2007)
4. Faita, F., et al.: Real-time measurement system for evaluation of the carotid intima-media thickness with a robust edge operator. Journal of Ultrasound in Medicine 27(9), 1353–1361 (2008)
5. Giacinto, G., Roli, F.: Desing of effective neural network ensembles for image classification purposes. Image and Vision Computing 19(9-10), 699–707 (2001)
6. Golemati, S., et al.: Using of the hough transform to segment ultrasound images of longitudinal and transverse sections of the carotid artery. Ultrasound in Medicine & Biology 33(12), 1918–1932 (2007)
7. Gonzalez, J., Wood, J., Dorey, F.J., Wren, T.A.L., Gilsanz, V.: Reproducibility of carotid intima-media thickness measurements in young adults. Radiology 247(2), 465–471 (2008)

8. González, R.C., Woods, R.E., Eddins, S.L.: Digital Image Processing using Matlab. Pentice Hall (2004)
9. Molinari, F., Zeng, G., Suri, J.S.: An integrated approach to computerbased automated tracing and its validation for 200 common carotid arterial wall ultrasound images: A new technique. Journal of Ultrasound in Medicine 29(3), 399–418 (2010)
10. Moller, M.: A scaled conjugate gradient algorithm for fast supervised learning. Neural Networks 6, 525–533 (1993)
11. Rocha, R., et al.: Segmentation of the carotid intima-media region in b-mode ultrasound images. Image and Vision Computing 28(4), 614–625 (2010)
12. Stein, J.H., et al.: A semiautomated ultrasound border detection program that facilitates clinical measurement of ultrasound carotid intima-media thickness. Journal of the American Society of Echocardiography 18(3), 244–251 (2005)
13. Touboul, P.J., et al.: Mannheim carotid intima-media thickness and plaque consensus (2004-2006-2011). Cerebrovascular Diseases 34, 290–296 (2012)
14. Velázquez, F., et al.: Reproducibility of sonographic measurements of carotid intima-media thickness. Acta Radiologica 49(10), 1162–1166 (2008)
15. Wendelhag, I., et al.: A new automated computerized analysis system simplifies readings and reduces the variability in ultrasound measurement of intima-media thickness. Stroke 28, 2195–2200 (1997)
16. WHO: Global atlas on cardiovascular disease prevention and control, http://www.who.int/cardiovascular_diseases/en/

Active Contours Tool for the Common Carotid Artery Layers Segmentation in Ultrasound Images

María-Consuelo Bastida-Jumilla[1], Rosa-María Menchón-Lara[1],
Juan Morales-Sánchez[1], and Rafael Berenguer-Vidal[2]

[1] Dpto. Tecnologías de la Información y las Comunicaciones.
Universidad Politécnica de Cartagena.
Plaza del Hospital, 1, 30202, Cartagena (Murcia), Spain
[2] Dpto. de Ciencias Politécnicas. Universidad Católica de Murcia.
Campus de los Jerónimos, s/n, 30107, Guadalupe (Murcia), Spain
consuelo.bastida@upct.es

Abstract. Among all cardiovascular diseases, atherosclerosis is considered as an alarm bell before other pathologies appear. Intima-Media Thickness (IMT) of the Common Carotid Artery (CCA) is widely used as an early and reliable indicator of atherosclerosis. Another element that affects cardiovascular risk estimation is the CCA diameter (CCAD). In clinical practice, these parameters are manually measured in ultrasound images to assess the cardiovascular risk in the corresponding patient. Ultrasound images present poor contrast and are quite affected by speckle noise, which introduces uncertainty in the manual measurement.

This paper presents a fully-automatic approach based on active contours to segment the artery layers involved in the IMT and the CCA diameter in ultrasound images. The proposed method reduces the subjectivity of the measurement and gives as results mean IMT and CCAD along a certain artery length.

1 Introduction

Atherosclerosis consists of a progressive fat accumulation on blood vessels. As a result, the vessel tissue thickens, which reduces its elasticity and hinders blood circulation. Atherosclerosis remains asymptomatic for decades before triggering more severe affections, such as stroke, ischemia or embolism.

Being CVD's the main cause of death in developed countries [1] and atherosclerosis a trigger of CVD's, the early detection of atherosclerosis is of paramount importance in the prevention and treatment of more serious CVD's. With that purpose, the diameter and, especially, the intima-media thickness (IMT) of the common carotid artery (CCA) have emerged as early and reliable indicators of cardiovascular risk [2–4]. Medical efforts are focused on finding a repeatable and reproducible method to measure IMT [5, 6].

The method here presented is based on the manual protocol proposed by the Radiology Department from Hospital Universitario Virgen de la Arrixaca

J.M. Ferrández Vicente et al. (Eds.): IWINAC 2013, Part II, LNCS 7931, pp. 250–257, 2013.
© Springer-Verlag Berlin Heidelberg 2013

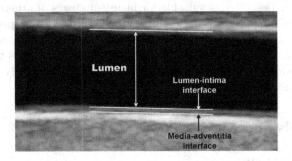

Fig. 1. Longitudinal cut of the CCA

[6], which has provided the images used. In the manual procedure, an expert manually marks from one to three pair of points to assess IMT around 1 cm after the carotid bifurcation, which is always placed at the left of the image. Finally, the maximum value is considered as the IMT of the patient.

From innermost to outtermost, blood vessels present three different layers, intima, media and adventitia. In Fig. 1 the interfaces to detect are superimposed over a longitudinal cut of the CCA. The lumen, the channel where the blood flows, appears in the middle of the image as a dark region. On top of it, the near wall interface (NWI) can be seen. At the bottom lies the far wall, where the IMT is measured as the distance between lumen-intima interface (LII) and media-adventitia interface (MAI).

This paper presents a fully-automated method, in which no user interaction is needed, to segment the artery layers in all its length. The proposed segmentation provides more accurate results because it takes into acccount more pairs of points together with the near wall interface segmentation. Since the results are segments instead of pair of points, more information can be extracted from the images, such as mean or maximum values, or the ecogenicity between two interfaces. Besides, the automatization characteristic of the method provides reproducibility, reduces subjectivity and is suitable for studies over a large population.

Different approaches appear in the literature to provide IMT measurement of certain automation. The first approaches [7, 8] were based on the analysis of vertical cuts of the image. To assure the continuity of the contours, these methods used a cost function minimized by means of dynamic programming. Other methods used active contours to detect the LII and MAI [9–11]. We can also find methods based on the Hough transform [12] or in a statistic model [13]. All these methods require some user interaction. It is not until 2007 that a fully automatic method can be found [14]. Other completely automatic methods are developed by the same group of authors [15, 16]. These methods usually work always with the same spatial resolution.

We propose a fully automatic method based on a frecuency implementation of active contours to segment LII, MAI and, additionally, near wall interface.

The method here described is suitable for incipient stages of atherosclerosis and works in a range of spatial resolutions from 0.029 to 0.081 mm/pixel.

The document is structured as follows; after the current introduction section, a Materials and Methodology section presents and describes the developed method. In section 3 results are presented and discussed. Finally, a conclusion section closes the paper.

2   Materials and Methodology

2.1   Image Database

The results over 25 images were compared to a manual segmentation made by an expert observer considered as ground truth. All images were acquired with a Philips iU22 Ultrasound System. The resolution of the images is determined by the radiologist. In our database, minimum resolution is 0.081 mm/pixel and the maximum is 0.029 mm/pixel. Most of the patients were healthy, being only four of them diagnosed with atherosclerosis after the IMT assessment.

2.2   CCA Layers Segmentation

The method implements mainly two steps. The first one consists of a coarse segmentation to locate the near and far walls of the lumen. After that, a fine segmentation of the CCA layers is performed thanks to a frequency implementation of active contours algorithm [17]. Additionally, we provide an automatic validation stage to avoid the inclusion of IMT measurements in sections where the IMT value is not reliable.

Initial Contours Calculation. The steps to automatically initialize the contours are depicted in Fig. 2.

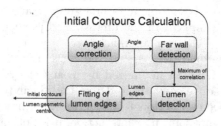

Fig. 2. Initial contours calculation diagram

Thanks to a Hough transform [18], the artery inclination is corrected to improve the far wall detection step, which is performed by means of correlation with a model of the typical bright-dark-bright pattern of the far wall [19].

After that, a median filtering is applied to the image with its orientation corrected to eliminate speckle noise while homogenizing regions. With a simple thresholding process, we achieve a binary image containing different regions. Among all the regions, lumen must include the maximum of correlation obtained during the far wall detection step. Finally, upper and lower lumen edges are softened with an order 3 interpolation.

As initial contours, we use the upper boundary of the lumen for the near wall interface detection, while the lower boundary is split into two, corresponding to the LII and MAI, respectively.

Fourier Domain Active Contours Implementation. Our active contour implementation starts with a B-spline interpolation of the initial nodes (see Fig. 3). B-splines have been chosen as function form because they produce soft final contours, avoiding the characteristic rough texture in ultrasound images, and presents a good performance-running time ratio [20].

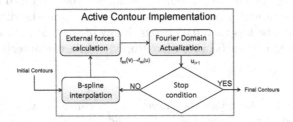

Fig. 3. Flow chart of the frequency-implementation of active contour algorithm

After that, external forces affecting each node are calculated. As external force image, a combination of the positive and negative vertical gradient is used. In our case, since we are interested in increasing gradient transitions (from dark to white) for the far wall interfaces and in decreasing gradient (from bright to dark) for the near wall, both gradients are combined in a single image considering the center line of the lumen as a border between near and far wall (see Fig. 4). In order to eliminate more undesired structures, a morphological reconstruction is performed where the combined gradient image is the marker image. To generate the mask, a Hough transform is performed over the softened gradient image. Once the main three directions are extracted, an accumulative opening by the Hough lines as structuring elements is performed. The result of the accumulative openings is used as mask in the reconstruction. After the reconstruction, the resulting image is clearer and less noisy. Finally, a gradient operation over the reconstructed image is computed to get the external forces which will drive the active contour algorithm.

After the node interpolation, the Laplacian affecting each control point v is evaluated and the position of the nodes u in the next iteration is calculated in the Fourier domain [17] (see Fig.3). The process continues until the stop condition

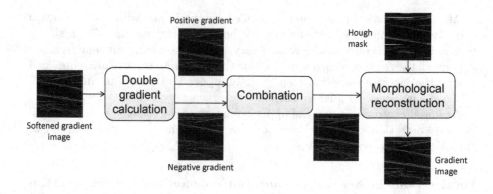

Fig. 4. External forces image calculation process

is fulfilled or the algorithm reaches 1000 iterations, enough in all cases for the contours to converge. The stop condition is double, the combined displacement of LI and MA contours must be less than 0.1 pixels and the displacement of NW contour must be less than 0.05 pixels. If any of these conditions is fulfilled, the corresponding contour(s) evolution stop(s), while the other(s) continue(s) evolving.

Validation Stage. Although the curves always converge, they may converge to a non desired edge. Hence, before the validation of the results by medical experts, an automatic validation of the final contours is performed. Thanks to this validation stage, the regions where the contours have reached the correct interfaces with certainty are stablished. Therefore, final measurements do not include the information corresponding to the sections considered as non-valid.

In the vertical direction, the intensity of far wall presents a bimodal profile. It is over the peaks of this bimodal distribution that the contours should lie. Since IMT is the difference of the peaks in the bimodal profile, IMT distribution presents a normal distribution. To assure that the curves adjust to this normal distribution, we adjust the IMT histogram to a Gauss window. With this, IMT outliers are discarded (see Fig. 5). In cases of incipient plaques, a thickened section can be considered as outlier. To avoid it, curves must lie over intensity peaks too. To evaluate if the curves lie over intensity peaks, a two classes K-means classification of the gradient image is performed. Around the nodes two separated intensity peaks are searched in the vertical direction. Notice that only this intensity-based strategy is not valid by itself. It could give wrong validation results in cases where the curves reach a nearby edge (presenting thus an intensity peak) that is not the desired one.

Both strategies, statistical and intensity-based ones are combined with and AND operator. After that, the position of the nodes in the wrong sections is corrected. The corrected node positions are again intensity-based validated, but with a gamma correction over the gradient image. The result of this validation

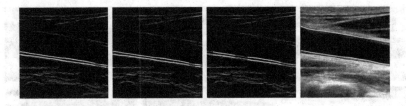

Fig. 5. Validation process (from left to right): statistic, intensity-based, combined validation and result after validating

stage is depicted in Fig. 5, where the different validation results are placed over the gradient image (three first images in Fig. 5 (from left to right)) or over the original image (Fig. 5 (right)). Continuous line represents validated sections and dotted line non-validated sections.

3   Results and Discussion

Figure 6 shows the results for some of the database images. Visually, both, the segmentation and the automatic validation perform well. A comparison with a manual segmentation of the far wall made by doctors has been done. The error between manual and automatic contours is measured as the Mean Absolute Distance (MAD). The IMT is also evaluated as the MAD between LII and MAI for both, manual and automatic contours (see Table 1). The mean absolute distance or error between two contours C_1 and C_2 with N points each is defined as

$$e_{C_1,C_2} = \frac{1}{N} \sum_{i=1}^{N} |C_1(i) - C_2(i)| \qquad (1)$$

For 25 tested images, the mean error between manual and automatic IMT is 25 μm, comparable to other fully-automatic methods, which present errors from 10 to 60 μm. The error for the MAI is greater than for the LII mainly because the MAI is more difuse than the LII. The Pearson's correlation coefficient shows a great correspondence (of 99.19%) between manual and automatic IMT measurement. Finally, the lumen diameter for the tested images presents a mean $\pm$ standard deviation of $6.45 \pm 1.025 mm$, which is a normal value for our mostly healthy population.

Table 1. Mean Absolute Error of the automatic measurements in μm

	e_{IMT}	e_{MAI}	e_{LII}
mean	25	39	29
standard deviation	19	18	14

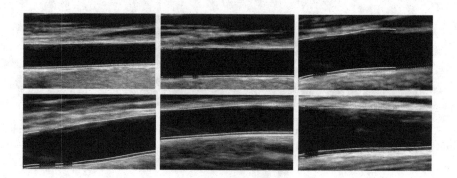

Fig. 6. Results over the original image

4 Conclusions

Thanks to an automatic rough segmentation of the lumen, active contours are initialized, which makes the method fully automatic. The frecuency implementation of active contours together with B-splines as function form achieves a considerable computational reduction and soft final contours, respectively. Besides the segmentation itself, a validation stage determines where the results are more reliable, avoiding the inclusion of wrong measurements in the statistics.

Finally, the method here presented achieves the automatic segmentation of the LII and MAI of the far wall as well as the NWI, which allows the extraction of the IMT and the lumen diameter of the CCA. Quantitatively, the results present high correlation with ground-truth segmentation and produce errors comparable to those of other fully-automated methods.

Future work includes extending our database image and present a more exhaustive result characterization with linear regressions, Bland-Altman's plots, etc.

Acknowledgements. This work is supported by the Spanish Ministerio de Ciencia e Innovación, under grant TEC2009-12675, and by the Séneca Foundation (09505/FPI/08). The authors would like to thank the Radiology Department of Hospital Universitario Virgen de la Arrixaca for providing all the ultrasound images used and the medical knowledge.

References

1. World Health Organization: Global atlas on cardiovascular disease prevention and control (2011)
2. Bots, M.L., et al.: Isolated systolic hypertension and vessel wall thickness of the carotid artery. The rotterdam elderly study. Arterioscl. Throm. Vas. 13(1), 64–69 (1993)
3. de Groot, E., et al.: Measurement of arterial wall thickness as a surrogate marker for atherosclerosis. Circulation III, 33–38 (2004)

4. Lorenz, M., et al.: Prediction of clinical cardiovascular events with carotid intima-media thickness: A systematic review and meta-analysis. Circulation 115, 459–467 (2007)
5. Toubol, P.-J., et al.: Mannheim carotid intima-media thickness consensus (2004-2006). Cerebrovasc. Dis. 23, 75–80 (2007)
6. Velázquez, F., et al.: Reproducibility of sonographic measurements of carotid intima-media thickness. Acta Radiol. 49(10), 1162–1166 (2008)
7. Gustavsson, T., Liang, Q., Wendelhag, I., Wikstrand, J.: A dynamic programming procedure for automated ultrasonic measurement of the carotid artery. In: Comput. Cardiol. pp. 297–300 (September 1994)
8. Liang, Q., Wendelhag, I., Wikstrand, J., Gustavsson, T.: A multiscale dynamic programming procedure for boundary detection in ultrasonic artery images. IEEE Transactions on Medical Imaging 19(2), 127–142 (2000)
9. Cheng, D.C., Schmidt-Trucksass, A., Cheng, K.S., Sandrock, M., Pu, Q., Burkhardt, H.: Automatic detection of the intimal and the adventitial layers of the common carotid artery wall in ultrasound b-mode images using snakes. In: Proceedings of the International Conference on Image Analysis and Processing, pp. 452–457 (1999)
10. Schmidt-Trucksäss, A., et al.: Computerized analysing system using the active contour in ultrasound measurement of carotid artery intima-media thickness. Clin. Physiol. 21(5), 561–569 (2001)
11. Ceccarelli, M., De Luca, N., Morganella, A.: Automatic measurement of the intima-media thickness with active contour based image segmentation. In: IEEE International Workshop on Medical Measurement and Applications, MEMEA 2007, pp. 1–5 (May 2007)
12. Loizou, C.P., et al.: Snakes based segmentation of the common carotid artery intima media. Med. Biol. Eng. Comput. 45(1), 35–49 (2007)
13. Destrempes, F., Meunier, J., Giroux, M.F., Soulez, G., Cloutier, G.: Segmentation in ultrasonic B-mode images of healthy carotid arteries using mixtures of Nakagami distributions and stochastic optimization. IEEE Trans. Medical Imaging 28(2), 215–229 (2009)
14. Delsanto, S., et al.: Characterization of a completely user-independent algorithm for carotid artery segmentation in 2-d ultrasound images. IEEE Trans. Instrum. Meas. 56(4), 1265–1274 (2007)
15. Molinari, F., et al.: Automatic computer-based tracings (act) in longitudinal 2-d ultrasound images using different scanners. J. Mech. Med. Biol. 9(04), 481–505 (2009)
16. Meiburger, K.M., et al.: Automated carotid artery intima layer regional segmentation. Phys. Med. Biol. 56(13), 4073–4090 (2011)
17. Weruaga, L., Verdu, R., Morales, J.: Frequency domain formulation of active parametric deformable models. IEEE Transactions on Pattern Analysis and Machine Intelligence 26(12), 1568–1578 (2004)
18. Duda, R.O., Hart, P.E.: Use of the hough transformation to detect lines and curves in pictures. Commun. ACM 15(1), 11–15 (1972)
19. Bastida-Jumilla, M.C., et al.: Segmentation of the common carotid artery walls based on a frequency implementation of active contours. J. Digit. Imaging 26, 129–139 (2013)
20. Unser, M.: Splines: a perfect fit for medical imaging. Pro. Biomed. Opt. Imag., 225–236 (2002)

Early Computer Aided Diagnosis of Parkinson's Disease Based on Nearest Neighbor Strategy and *striatum* Activation Threshold

Pablo Padilla, Juan Manuel Górriz, Javier Ramírez, Diego Salas-González, and Ignacio Álvarez

Granada University-CITIC, Granada 18071, Spain
pablopadilla@ugr.es

Abstract. In this document, a complete computer aided diagnosis procedure is presented, for the early diagnosis of Parkinson's disease. The method is applied to single-photon emission computed tomography (SPECT) brain images in order to identify parkinsonian patterns on them. Two strategies are proposed for the identification: a nearest neighbors classification based on similitude, and an activation threshold identification procedure. Both of them present a good performance, although some drawbacks must be taken into account. Validation results over a SPECT dataset are presented in this work.

Keywords: Computer Aided Diagnosis, Parkinson's Disease, Nearest Neighbors.

1 Introduction

The Parkinson's disease (PD) is the second most frequent neurodegenerative disorder, only being surpassed by the Alzheimer's disease [1,2]. The Parkinson's disease consists of a severe chronic neurodegenerative disorder that affects gradually the movement of the individual, generating involuntary spasmodic muscular movements, but also an affection in the normal motor function in the body [3,4]. The disease usually also includes alterations in the cognitive function, in the demonstration of emotions and in the autonomous function. The Parkinsons disease is directly related to a progressive loss of dopaminergic neurons of the nigrostriatal pathway: this cell reduction involves a substantial decrease in the dopamine content of the *striatum* region of the brain and a loss of dopamine transporters [5].

The diagnosis of the disease is based on the deep study of brain functional or structural images (such as positron emission tomography (PET), single-photon emission computed tomography (SPECT), magnetic resonance imaging (MRI), etc.) by expert clinicians. Their study is commonly based on the visual analysis, their experience, and in some cases, their intuition [6]. The recent developed techniques of diagnosis based on computer-aided analysis have the aim not only of improving the assertiveness of this diagnosis process, but also of anticipating

J.M. Ferrández Vicente et al. (Eds.): IWINAC 2013, Part II, LNCS 7931, pp. 258–265, 2013.
© Springer-Verlag Berlin Heidelberg 2013

the diagnosis to an early stage of the disease [7]. These machine-learning techniques let obtain, when applied over medical images, three-dimensional brain activity maps to be analyzed and compared.

2   3D Brain SPECT Images for the Experiments

The analysis provided in this work is carried out over a set of SPECT brain images. This set of SPECT brain images has been collected in *Virgen de las Nieves* Hospital, in Granada. A total of 80 subjects (41 PD patients and 39 controls) are considered in this dataset. The SPECT study with $_{123}$I-FP-CIT was carried out using a Siemens Gamma camera, Symbia model, dual head, low energy and high resolution collimator. Again, a 360-degree circular acquisition is made around the skull, at intervals of 3 degrees, having 60 images with a 35 second duration per interval, and a 128x128 matrix.

The images have been acquired according to the transaxial sections and the orbito-meatal line. They are oriented from posterior (top) to anterior (bottom) in each one, and from ventral (top left) to dorsal (bottom right), in each set of slices. Thus, a 45x73x73 three-dimensional functional activity map for each patient is registered. An example of a PD subject and a control one is provided in Fig. 1.

3   Computer Aided Diagnosis Based on Nearest Neighbor Strategy and *Striatum* Activation Threshold

The Computer Aided Diagnosis (CAD) tools are becoming a powerful complementary tool for the physicians in their diagnosis activity [1]. In this way, there are some steps to be performed in order to proceed with the proper analysis. These steps are described below.

3.1   Spatial Normalization

Once the set of brain images is collected, it is necessary to process these images, in order to have a correct brain shape and voxel distribution in the 3D image. All the images in the study must be spatially normalized to ensure that a given voxel in one patient refers to the same brain position than the same voxel in another patient. The spatial normalization is typically carried out by means of the Statistical Parametric Mapping (SPM) software [8], which is the case in this work.

3.2   Intensity Normalization

Once the set of images has been properly normalized in space, it must be also normalized in terms of intensity. This intensity normalization guarantees that the differences between images of different subjects are due to physiological reasons and brain functioning, and not due to the baseline calibration of the Gamma

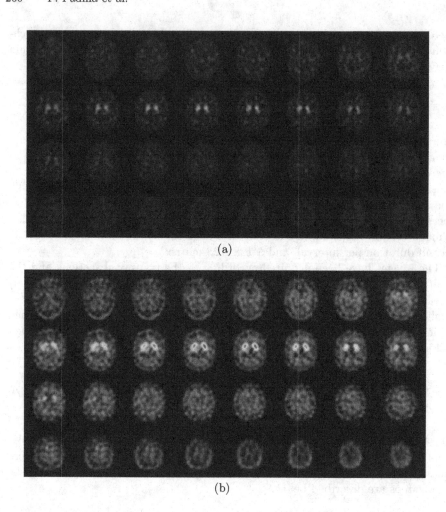

Fig. 1. Example of SPECT transaxial brain slices of the *Virgen de las Nieves* dataset, oriented from posterior (top) to anterior (bottom) in each slice, and from ventral (top left) to dorsal (bottom right) in the complete set, a) for a PD patient, b) for a control subject

camera applied for the acquisition. The conventional way of carrying out the intensity normalization is to consider as a reference for all the images the brain region that is not significant to differentiate between ill subjects and healthy ones. In the case of Parkinson's disease, the discriminant region is the *striatum* and, thus, for the intensity normalization, the rest of the brain is considered. Among the different options available [9,10], in this work the integral-based approach is considered: the integral of the intensity values of the brain except for the *striatum* is computed for all the images, the mean value of this integral is obtained and each image is renormalized so that their new integral value for each one of them is the same as the mean value previously computed [9].

3.3   Selection of the Region of Interest (ROI)

Once the images are ready for their proper processing, the relevant information has to be extracted: only the voxels that contain relevant information in terms of discrimination ability must be chosen. In the case of Parkinson's disease, this region is, as previously mentioned, the *striatum*. Thus, after the selection of the region of interest (ROI), a set of intensity values is obtained for each subject, arranged in a 1D array. This array is the key data to be processed in the final steps of the CAD tool.

3.4   Classification Method Based on Nearest Neighbors

The nearest neighbors approach (NN) has been widely employed in classification, identification and performance evaluation. It is based on the calculation of the lowest distance from one particular sample to the rest of them, considering them as a reference set. The classification of the sample under test (S_{test}) is done by selecting the label of the majority of the N nearest neighbors among all the I samples (S_i) in the reference set. In this work, each sample is formed by a set of M voxels located in the *striatum* region. In this case, euclidean metric is selected, and the distance to the neighbors is computed as in (1), for each sample under test.

$$d_i = \sum_{m=1}^{M} abs(S_{test}(m) - S_i(m)) \tag{1}$$

3.5   Classification Method Based on *Striatum* Activation Threshold

An additional procedure is proposed for the proper identification of the subjects. In this case, the summation of the intensity values of all the voxels in the *striatum* region is computed. If this cumulative value is over an activation threshold, the subject is classified as normal, otherwise, as PD. This reference threshold is set according to the mean values of the cumulative value of healthy subjects and PD ones. In the experimentation, the threshold is defined in terms of % regarding the complete range of variation between means.

4   Validation and Results

After the proper spatial and intensity normalization, the set of SPECT images are ready to be analyzed. As mentioned, the relevant data for the diagnosis of Parkinson's disease is located in the *striatum* area. For this reason, the voxels located in this area are extracted according to the adequate brain mask. Fig. 2 provides the mask applied for every brain slice. Thus, 3D image data are translated into an array of intensity values, formed by the significant voxels selected by the mask, relocated in a row.

Fig. 2. Mask for the extraction of the data corresponding to the *striatum* region

Once all the dataset information is condensed in one array of intensity values for each one of the M subjects (M equal to 80), the proper analysis can be performed. For the proper study of the CAD tool described, a *train and test* strategy based on *leave-one-out* is selected. Under these circumstances, M iterations are carried out: subject m (from 1 to M) is extracted and the resulting *M-1* subject data set is used as the reference dataset to be compared with.

In order to evaluate the performance of the CAD tool, some statistics are computed (2): the success rate (Acc), the sensitivity (Sens) and the specificity (Spec), obtained for each iteration.

$$Acc = \frac{TP + TN}{TP + TN + TN + FP} \quad Sens = \frac{TP}{TP + FN} \quad Spec = \frac{TN}{TN + FP} \quad (2)$$

where TP is the number of true positives (PD patients correctly classified); TN is the number of true negatives (healthy patients correctly classified); FP is the number of false positives (healthy patients classified as PD); FN is the number of false negatives (PD classified as healthy). The mean value of these statistics for all the iterations is computed and the final validation results are derived.

4.1   Results with Nearest Neighbors

In Fig. 3 and Table 1, the values of Acc, Sens and Spec are provided, for a different number of neighbors considered in the NN approach. As it can be seen, the strategy of comparing the sample under test with a set of samples taken as a reference provides very good results, not only for the accuracy, but also for the sensitivity an the specificity. However, the main drawback of the method is that the reference set is essential and must be properly labelled for its proper application, and this is not the typical situation in most of the cases.

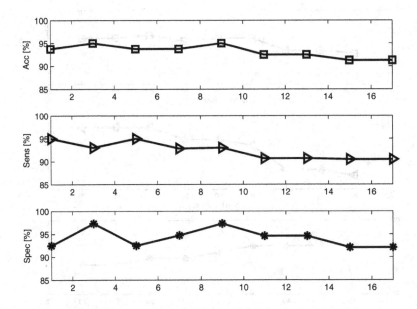

Fig. 3. Acc, Sens and Spec results in the NN approach, for a different number of neighbors (1 to 17)

Table 1. Performance [%] of the NN approach, for a different number of neighbors (1 to 17)

	Number of Neighbors								
	1	3	5	7	9	11	13	15	17
Accuracy [%]	93,75	95,00	93,75	93,75	95,00	92,50	92,50	91,25	91,25
Sensitivity [%]	95,00	93,02	95,00	92,86	93,02	90,70	90,70	90,48	90,48
Specificity [%]	92,50	97,30	92,50	94,74	97,30	94,59	94,59	92,11	92,11

4.2   Results with *striatum* Activation Threshold

In Fig. 4 and Table 2, the results in terms of Acc, Sens and Spec are provided, for a variable threshold (% in total possible range), in the *striatum* activation threshold approach. For the sake of completeness, Fig. 5 provides the reference to the activation threshold regarding the activation levels of the subjects in the dataset. As it can be identified, the success rate is not as relevant as in the case of the NN strategy. Also relevant is the complementary tendency of the sensitivity and specificity: it makes reference to the possibility of having more false positives or false negatives depending on the threshold selection, as it is directly identified in Fig. 5.

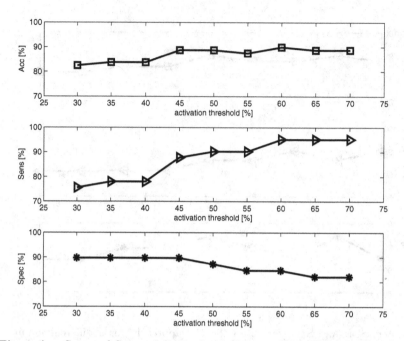

Fig. 4. Acc, Sens and Spec results in the *striatum* activation threshold approach

Table 2. Performance [%] in the *striatum* activation threshold approach, for different activation thresholds [%]

	activation threshold [%]								
	30%	35%	40%	45%	50%	55%	60%	65%	70%
Accuracy [%]	82,50	83,75	83,75	88,75	88,75	87,50	90,00	88,75	88,75
Sensitivity [%]	75,61	78,05	78,05	87,80	90,24	90,24	95,12	95,12	95,12
Specificity [%]	89,74	89,74	89,74	89,74	87,18	84,62	84,62	82,05	82,05

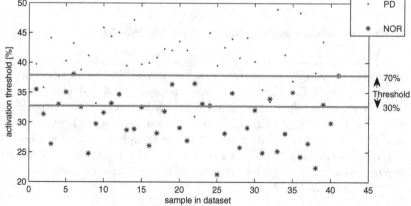

Fig. 5. Activation threshold regarding activation levels of the subjects in the dataset

5   Conclusions

In this work, it is provided a CAD method applied to the analysis of SPECT brain images for the identification of parkinsonian symptoms and the early aided diagnosis of Parkinson's disease. Once the data is properly normalized (in space and intensity) and the discriminant features are extracted, two strategies are proposed for the identification: a nearest neighbors classification and an activation threshold identification. Both of them provide good results, but have different drawbacks to be considered in advance. The results presented in this work are promising and provide a good feedback for further work.

References

1. Padilla, P., López, M., Górriz, J.M., Ramírez, J., Salas-Gonzalez, D., Illán, I.A.: NMF-SVM Based CAD Tool Applied to Functional Brain Images for the Diagnosis of Alzheimer's Disease. IEEE Trans. Medical Imaging 31(2), 207–216 (2012)
2. López, M., Ramírez, J., Górriz, J.M., Illán, I.A., Salas-Gonzalez, D., Segovia, F., Chaves, R., Padilla, P., Gómez-Río, M.: Principal component analysis-based techniques and supervised classification schemes for the early detection of Alzheimer's disease. Neurocomputing 74(8), 1260–1271 (2011)
3. Colosimo, C., et al.: Task force report on scales to assess dyskinesia in Parkinson's disease: critique and recommendations. Mov. Disord. 25(9), 1131–1142 (2010)
4. Goetz, C.G., Fahn, S., Martñez-Martin, P., et al.: Movement Disorder Society-Sponsored Revision of the Unified Parkinson's Disease Rating Scale (MDS-UPDRS): Process, Format, and Clinimetric Testing Plan. Movement Disorders 22(1), 41–47 (2007)
5. Booij, J., Tissingh, G., Boer, G.J., Speelman, J.D., Stoof, J.C., Janssen, A.G., Wolters, E.C., van Royen, E.A.: [123i]FP-CIT SPECT shows a pronounced decline of striatal dopamine transporter labelling in early and advanced Parkinson's disease. Journal of Neurology, Neurosurgery and Psychiatry 62, 133–140 (1997)
6. Hughes, A.J., Daniel, S.E., Kilford, L., Lees, A.J.: Accuracy of clinical diagnosis of idiopathic Parkinson's disease: a clinico-pathological study of 100 cases. Journal of Neurology, Neurosurgery and Psychiatry 55, 181–184 (1992)
7. Rojas, A., Górriz, J.M., Ramírez, J., Illán, I.A., Martinez-Murcia, F.J., Ortiz, A., Gómez-Río, M., Moreno-Caballero, M.: Application of Empirical Mode Decomposition (EMD) on DaTSCAN SPECT images to explore Parkinson Disease. Expert Systems with Applications 40(7), 2756–2766 (2013)
8. Friston, K., et al.: Statistical Parametric Mapping: The Analysis of Functional Brain Images. Academic, New York (2007)
9. Illán, I.A., Górriz, J.M., Ramírez, J., Segovia, F., Jimenez-Hoyuela, J.M., Ortega-Lozano, S.J.: Automatic assistance to Parkinson's disease diagnosis in DaTSCAN SPECT imaging. Med. Phys. 39(10) (2012)
10. Salas-Gonzalez, D., Juan Górriz, J.M., Ramírez, J., Illán, I.A., Lang, E.W.: Linear intensity normalization of FP-CIT SPECT brain images using the alpha-stable distribution. NeuroImage 65, 449–455 (2013)

Texture Features Based Detection of Parkinson's Disease on DaTSCAN Images*

Francisco Jesús Martínez-Murcia[1], Juan Manuel Górriz[1],
Javier Ramírez[1], I. Alvarez Illán[1], and C.G. Puntonet[2]

[1] Department of Signal Theory, Networking and Communications,
Universidad of Granada, Spain
[2] Department of Computer Architecture and Technology,
Universidad de Granada, Spain

Abstract. In this work, a novel approach to Computer Aided Diagnosis (CAD) system for the Parkinson's Disease (PD) is proposed. This tool is intended for physicians, and is based on fully automated methods that lead to the classification of Ioflupane/FP-CIT-I-123 (DaTSCAN) SPECT images. DaTSCAN images from the Parkinson Progression Markers Initiative (PPMI) are used to have in vivo information of the dopamine transporter density. These images are normalized, reduced (using a mask), and then a GLC matrix is computed over the whole image, extracting several Haralick texture features which will be used as a feature vector in the classification task. Using the leave-one-out cross-validation technique over the whole PPMI database, the system achieves results up to a 95.9% of accuracy, and 97.3% of sensitivity, with positive likelihood ratios over 19, demonstrating our system's ability on the detection of the Parkinson's Disease by providing robust and accurate results for clinical practical use, as well as being fast and automatic.

Keywords: Parkinson's Disease, DaTSCAN images, Computer Aided Diagnosis, Haralick Texture Features, Support Vector Machines, Supervised Learning.

1 Introduction

Parkinsonian Syndrome (PS), also known as Parkinsonism, is a neurological syndrome characterized by tremor, hypokinesia, rigidity and postural instability [3]. It is considered as the second most common neurodegenerative disease, with a prevalence of 1-3% in the population over 65 years of age [11]. A wide range of etiologies may lead to the PS, while the most common cause is the neurodegenerative condition called Parkinson's Disease (PD). This disease originates due to

---

* Data used in the preparation of this article were obtained from the Parkinson's Progression Markers Initiative (PPMI) database (www.ppmi-info.org/data). As such, the investigators within PPMI contributed to the design and implementation of PPMI and/or provided data but did not participate in the analysis or writing of this report. PPMI investigators include (complete listing at PPMI site).

J.M. Ferrández Vicente et al. (Eds.): IWINAC 2013, Part II, LNCS 7931, pp. 266–277, 2013.

the progressive loss of dopaminergic neurons of the nigrostriatal pathway, which connects the substantia nigra to the striatum. As a result, the dopamine content of the striatum decreases, and consequently, dopamine transporters (DAT) are lost. Other possible causes include some toxins, a few metabolic diseases, and a handful of non-PD neurological conditions [2].

As the PD is related to a loss of dopamine transporters in the nigrostriatal pathway, the study of its status by means of brain imaging techniques has been suggested to increase the diagnostic accuracy in the case of parkinsonian syndromes [3]. Ioflupane/FP-CIT-I-123 (better known as DaTSCAN) is a tracer that binds to the dopamine transporters in the striatum, allowing the obtention of Single Photon Emission Computed Tomography (SPECT) images that show a reduced uptake of the tracer in the striatum in patients with PS [1].

A wide range of supervised learning techniques have been combined to generate Computer Aided Diagnosis (CAD) systems that allow to detect neurodegenerative diseases, such as Alzheimer's Disease [9] or Parkinson [16]. Techniques range from the use of selection of Regions of Interest (ROIs) [5], or Single Value Decomposition strategies (SVD) [14] to more complex approaches such as Empirical Mode Decomposition (EMD) combined with Principal Component Analysis (PCA) combined method in [13].

In this work we use the Haralick Texture analysis, proposed in [4], that provides several texture features, which we will use to characterize some patterns of the Parkinson's Disease. To do so, we calculate a 3D Gray Level Co-occurrence (GLC) matrix [12], required for the computation of these features. Finally we make use of a Support Vector Machine (SVM) [18] binary classifier to test the ability of these features in the PD pattern detection.

2   Methodology

2.1   Test Data and Preprocessing

Data used in the preparation of this article were obtained from the Parkinson's Progression Markers Initiative (**PPMI**) database (www.ppmi-info.org/data). For up-to-date information on the study, visit www.ppmi-info.org.

The images in this database were imaged 4 ± 0.5 hours after the injection of between 111 and 185 MBq of DaTSCAN. Subjects were also pretreated with saturated iodine solution (10 drops in water) or perchlorate (1000 mg) prior to the injection. All subjects had a supplied [57]Co line marker affixed along the canthomeatal line, which will facilitate subsequent image processing and allow the core lab to accurately distinguish left and right in the face of multiple image file transfers. These markers are only evident in the [57]Co window and hence do not contaminate the ^{123}I-DaTSCAN brain data [8,6].

Raw projection data are acquired into a 128×128 matrix stepping each 3 degrees for a total of 120 projection into two 20% symmetric photopeak windows centered on 159 KeV and 122 KeV with a total scan duration of approximately 30 - 45 minutes. Other scan parameters (collimation, acquisition mode, etc) are selected for each site. The images of both the subject's data and the cobalt striatal phantom are reconstructed and attenuation corrected, implementing either filtered back-projection or an iterative reconstruction algorithm using standardized approaches [6]. After the processing, images used are spatially and intensity normalized, and of a $91 \times 109 \times 91$ size.

All images in the databases have been spatially normalized (using the **SPM8** software and a custom DaTSCAN template -see Fig. 1, which depicts some cuts of both normal and affected subjects-) and intensity normalized using the **Integral Normalization** algorithm. This method is based on the obtainment of an intrinsic parameter from the image, I_p, and the estimation of the binding activity as:

$$t' = t/I_p \qquad (1)$$

where t denotes the spatially normalized image, and t' the image normalized spatially and in intensity. In this case, all intensity values on the image are summed as an aproximation of the expression $I_p = \int t$, which results in an integral value of the intensity.

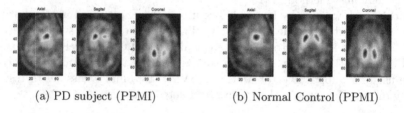

(a) PD subject (PPMI)   (b) Normal Control (PPMI)

Fig. 1. Sample image from (a) a PD patient from PPMI database and (b) a healthy subject from PPMI database

2.2   Mask

At this point, all the images are spatially and intensity normalized. These images feature two major changes in their intensity levels: the change between the noisy background and the whole brain, and the increase difference between brain intensity levels and those of the striatum.

We assume that the change between the background and the brain should not be significative enough for the diagnosis of the PD, though its influence on the texture of the image is clear. Given this assumption, it would be desirable to remove all background pixels from the images. To do so, we use a **masking process**.

As the computation of the GLC matrix (which will be explained more extensively in the following section) needs an cuboid image, our purpose is to extract the biggest box which contains only brain pixels. Thus, we establish an intensity threshold, I_{th}, and extract a mask from the mean image of all images in the database using:

$$Mask = I > I_{th} \tag{2}$$

Then, the coordinates of the largest box that fits into that mask is obtained, and this area is selected in all images. We have used intensity thresholds ranging from 0% (all image is selected) to 50% of the highest intensity of the mean image.

2.3   Haralick Texture Features

A co-occurrence matrix is a matrix that is defined over an image to be the distribution of co-occurring values at a given offset. Mathematically, a co-occurrence matrix C is defined over an $n \times m$ image I, parameterized by an offset $(\Delta x, \Delta y)$, as:

$$\mathbf{C}_{\Delta x, \Delta y}(i, j) = \sum_{p=1}^{n} \sum_{q=1}^{m} \begin{cases} 1, & \text{if } \mathbf{I}(p, q) = i \text{ and } \mathbf{I}(p + \Delta x, q + \Delta y) = j \\ 0, & \text{otherwise} \end{cases} \tag{3}$$

Note that the (x,y) parameterization makes the co-occurrence matrix sensitive to rotation. We choose one offset vector, so a rotation of the image not equal to 180 degrees will result in a different co-occurrence distribution for the same (rotated) image. This is rarely desirable in the applications co-occurrence matrices are used in, so the co-occurrence matrix is often formed using a set of offsets sweeping through 180 degrees (i.e. 0, 45, 90, and 135 degrees) at the same distance to achieve a degree of rotational invariance.

The method used here to expand the co-occurrence matrix to a tridimensional space is defined in [12], introducing another variable in Eq. 3. A 3D co-occurrence matrix C is defined over an $n \times m \times k$ three-dimensional image I, parameterized by an offset $(\Delta x, \Delta y, \Delta z)$, as:

$$\mathbf{C}_{\Delta}(i, j) = \sum_{p=(1,1,1)}^{(n,m,k)} \begin{cases} 1, & \text{if } \mathbf{I}(\mathbf{p}) = i \text{ and } \mathbf{I}(\mathbf{p} + \Delta) = j \\ 0, & \text{otherwise} \end{cases} \tag{4}$$

where i and j are different gray levels, $\mathbf{p} = (x, y, z)$ is the spatial position and Δ is the offset vector.

$$\mathbf{p} = (p, q, r), \text{ and } \Delta = (\Delta x, \Delta y, \Delta z) \tag{5}$$

Twelve Haralick texture features [4] are then extracted from the GLC matrix computed in Eq. 4, in this order: Energy, Entropy, Correlation, Contrast, Variance, Sum Average, Inertia, Cluster shade, Cluster prominence, Homogeneity, Maximum probability and Inverse variance. As 13 spatial directions and 10 distances are considered, a total number of 130 co-ocurrence matrices are computed, and therefore, 130 values for each of the Haralick texture features have been computed. These values have been used as an input vector to the following classifier.

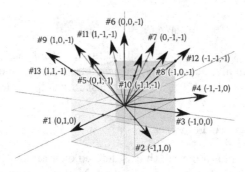

Fig. 2. Spatial representation of the thirteen direction vectors used to compute the thirteen different GLC matrices

2.4   Classifier

The classification step is performed as follows. A predictive model is derived from a set of training data from two different classes -in our case, patients without dopaminergic deficit (NOR) and Parkinson's Disease (PD)-, and the test image is then classified. To build our predictive model, we make use of the Support Vector Machines paradigm. Support Vector Machine (SVM) [17] is a recent class of statistical classification and regression techniques playing an increasing role in applications to detection problems in various engineering problems, notably in statistical signal processing, pattern recognition, image analysis [9], and communication systems. SVM with linear discriminant functions define decision hypersurfaces or hyperplanes in a multidimensional feature space, that is:

$$g(\mathbf{x}) = \mathbf{w}^T \mathbf{x} + \omega_0 = 0, \tag{6}$$

where $\mathbf{w}$ is known as the weight vector and ω_0 as the threshold. The weight vector $\mathbf{w}$ is orthogonal to the decision hyperplane and the optimization task consists of finding the unknown parameters $\omega_i, i = 1, ..., n$ defining the decision hyperplane.

Let $\mathbf{x_i}, i = 1, 2, ..., n$ be the feature vectors of the training set, X. These belong to either ω_1 or ω_2, the two classes. If the classes were linearly separable, the objective would be to design a hyperplane that classifies correctly all the training vectors. Among the different design criteria, the maximal margin hyperplane is usually selected since it leaves the maximum margin of separation between the two classes. Since the distance from a point $\mathbf{x}$ to the hyperplane is given by $z = |g(\mathbf{x})|/\|\mathbf{w}\|$, scaling w and $\mathbf{w_0}$ so that the value of $g(\mathbf{x})$ is +1 for the nearest point in ω_1 and -1 for the nearest points in ω_2, the optimization problem is reduced to minimizing a cost function $J(\boldsymbol{\omega}) = 1/2\|\boldsymbol{\omega}\|^2$ subject to:

$$f_{svm}(\mathbf{x}) = \sum_{i=1}^{N_S} \alpha_i \omega_i \Phi(\mathbf{s}_i) \cdot \Phi(\mathbf{x}) + \omega_0 \tag{7}$$

where α_i are the solution of a quadratic optimization problem that is usually determined by quadratic programming or the well-known sequential minimal optimization algorithm, and $\Phi(\mathbf{s})$ or $\Phi(\mathbf{x})$ denote the transformation of the feature vectors into the effective feature space. This basic SVM classifier produces a linear separation hyperplane.

SVM classifiers with linear kernels are based on a solid theoretical background thus leading to reproducible performance and finally, showing a good robustness to noisy or mislabeled data. Therefore, they are usually applied to evaluate the separability of different features, providing low generalization error even with small learning sample datasets.

3   Results and Discussion

To perform an evaluation of our proposed CAD system, we have independently evaluated the effect of using each of the 12 Haralick Texture Features. Thus, regarding their ability to correctly interpret the texture of the normalized DaT-SCAN image. All directions are considered but the influence of the distance d at whith the GLC matrix is calculated has been also considered, using all the feature values extracted in a range of $1 < d$ distance from the central voxel.

In each of these experiments, the images are previously reduced using masks, with an specific intensity threshold I_{th} starting at 0% (whole image) up to 50% of the highest intensity value, as commented in Section 2.2.

3.1   Evaluation

The proposed methodology has been tested on the PPMI databse (see Sec. 2.1). We have used a cross-validation method called leave-one-out to extract some evaluation parameters which will allow us to compare and evaluate the performance of these proposed systems. Parameters like accuracy, sensitivity, specificity, Positive Likelihood (PL) and Negative Likelihood (NL) ratios have been estimated using this method.

Leave-one-out method provides us with a mean of using almost all images for the training of the classifier and still get an unbiased error estimate [7]. However this estimate might be affected by the database topology and the classifier used. Anyway, this is one of the most used methods for system validation, and so will be used in this work.

The accuracy, sensitivity and specificity parameters are calculated as:

$$\text{Acc} = \frac{TP+TN}{TP+TN+FP+FN}, \text{Sens} = \frac{TP}{TP+FN}, \text{Spec} = \frac{TN}{TN+FP} \quad (8)$$

where TP, TN, FP and FN are the number of true positives, true negatives, false positives and false negatives, respectively. Accuracy measures the proportion of correctly classified samples. Sensitivity and specificity are used to measure the proportion of actual positives or negatives which are identified correctly (e.g. the percentage of PS patients, or normal controls who are identified as such).

Although sensitivity and specificity are very important to reveal the ability of a system on detecting PD/NOR patterns, they are prevalence-dependent. This means that in an positive prevalent database (those where there is a higher number of positives), sensitivity will be higher, and similarly occurs with specificity. So even if they are a good estimate of the goodness of the classifier, other parameters such as Positive and Negative Likelihood ratios (PL and NL), which are prevalence independent, are computed.

$$PL = \text{Sensitivity}/(1 - \text{Specificity}) \tag{9}$$

$$NL = (1 - \text{Sensitivity})/\text{Specificity} \tag{10}$$

These parameters are also widely used in clinical medicine, where values of PL greater than 5 or NL values less than 0.2 can be applied to the pre-test probability of a patient having the disease tested to estimate a post-test probability of the disease state existing [10]. A positive result for a test with PL of 8 adds approximately 40% to the pre-test probability that a patient has a specific diagnosis. These parameters are computed with Eq. 9 and 10.

3.2   Results and Discussion

In this work, we have tested the performance of a CAD system which makes use of the Haralick Texture Features to extract relevant textural information from images, and then using these features to classify each image. We have used a mask to reduce dimensionality and improve the GLC matrix computation, and the whole system has been evaluated using the widely available PPMI database, to better facilitate the reproducibility of results.

As commented in Section 2.2, a mask has been added to the normalized images, to extract smaller box-shaped images that allows us to both perform the following tests more efficiently and optimize the calculation of the GLC matrix by removing all non-brain pixels from the source image. Therefore, this masking step might seem profitable.

However, it shows one major drawback: its dependence on the manual operation of setting the mask threshold. Although large number of border detection algorithms exist, and most of them could be applied here to select the brain area, and thus, the intensity threshold, in this work we have established some manual thresholds -see Sec. 2.2- for better evaluating the performance of our system.

To better illustrate the influence of using masks, Table 1 shows the accuracy results obtained by our system using either no mask (mask = 0%) or different levels for the intensity threshold I_{th}.

Most of the best values are obtained using the Homogeneity as an input vector, except for when $I_{th} > 0.35 * I_{max}$. This can be easily explained due to the image characteristics. As we commented before, one of the reasons of using masks was to select the biggest image box that contains only brain pixels. This is achieved in these integral-normalized images when the intensity threshold values

Table 1. Accuracy results for each of the 12 Haralick texture features (see Sec. 2.3), using different intensity thresholds (in %)

Feature	0	5%	15%	25%	35%	45%
Energy	0.822	0.822	0.784	0.862	0.929	0.952
Entropy	0.807	0.807	0.792	0.881	0.933	0.914
Correlation	0.851	0.851	0.870	0.929	0.933	0.955
Contrast	0.818	0.818	0.833	0.937	0.948	0.937
Variance	0.762	0.762	0.784	0.825	0.888	0.914
Sum Average	0.810	0.810	0.844	0.900	0.937	0.933
Inertia	0.773	0.773	0.810	0.833	0.937	0.941
ClusterShade	0.818	0.818	0.833	0.937	0.948	0.937
ClusterTendency	0.926	0.926	0.911	0.941	0.933	0.918
Homogeneity	0.926	0.926	0.929	0.944	0.952	0.907
MaxProbability	0.766	0.766	0.747	0.770	0.818	0.814
InverseVariance	0.743	0.743	0.796	0.870	0.874	0.914

are around this percentage of the maximum intensity of the image. Therefore, more automatic methods to detect this threshold in all types of images should be desirable. In Figure 3, the resulting area of using an $I_{th} = 0.30 I_{max}$ is depicted, applied to the same subject on Fig. 1b. It is interesting to note that our goal of selecting only internal brain voxels has been achieved with this threshold.

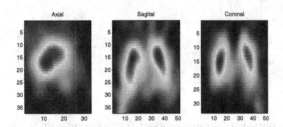

Fig. 3. Resulting image box when using a mask with a $I_{th} = 0.30 I_{max}$

As commented before, all features improve their accuracy results as the intensity threshold increases, but there are some of them that perform particularly well in this task, e.g. Homogeneity, Cluster Shade or Energy. Five of these best features are depicted on Fig. 4. It is noticeable that, while some features increase their performance significantly once I_{th} exceeds some values (which we can consider as an indication that only brain voxels have been selected), there are others that offer good values almost independently of the chosen value for I_{th}.

We have already focused on the performance of the system depending on the mask applied and the feature used for classifying. Nevertheless, there is one more

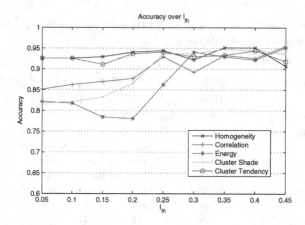

Fig. 4. Accuracy obtained by our proposed system using each of the texture features listed in the legend, using different values for I_{th}

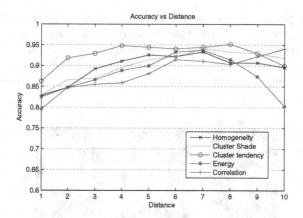

Fig. 5. Accuracy obtained by our proposed system using only the features extracted from the 13 GLC matrices (one in each direction) computed at a distance d of the central voxel

parameter on the system that can affect the results: the distance d at which the GLC matrix is calculated. To perform a deeper analysis of this parameter, we have evaluated the behavior of the system using only the 13 values of each Haralick texture feature extracted from the 13 GLC matrices (one in each direction) computed at a distance d of the central voxel, with $1 < d < 10$. When using each of the five aforementioned texture features, the performance results on our system are displayed on Fig. 5. Notice that, once a maximum is achieved generally at $d = 7, 8$, the accuracy decreases, what might point out the softness degree of the texture pattern of these images.

Table 2. Comparison of our proposed system (using different texture features) and some other methods in the bibliography: VAF system using the intensity-normalized images, a combination of intensity normalization strategies and classifiers (VAF-IN) [5], a SVD-based approach [14] and EMD using the third independent mode function (IMF3) [13].

System	Acc	Sens	Spec	PL	NL
Homogeneity	0.959	0.973	0.949	19.22	0.028
Cluster Shade	0.955	0.964	0.949	19.01	0.038
Cluster Tendency	0.955	0.973	0.943	17.10	0.029
Correlation	0.941	0.946	0.937	14.92	0.058
Energy	0.937	0.964	0.918	11.73	0.039
VAF	0.840	0.807	0.862	5.88	0.224
VAF-IN	0,913	0.890	0.932	13.08	0.118
SVD	0.940	0.962	0.918	11.73	0.041
EMD-IMF3	0.950	0.951	0.948	18.28	0.051

In Table 2 we compare this method with others from the bibliography and the Voxels-as-Features approach, commonly used as an estimation of the performance of visual analysis performed by experts [15]. Some of the methods evaluated include a combination of intensity normalization strategies and classifiers (VAF-IN) [5], a SVD approach that independently decomposes each side of the brain (as PD often show asymmetrical dopamine deficit) [14] and a EMD of the images, then modeled using PCA or ICA to classify the images [13].

Our method based on the Haralick Texture Features and a mask obtains better results in almost every case (specially when using one of the three best features: Homogeneity, Cluster Shade or Cluster Tendency), which also show a high independence of the value used for the intensity threshold of the mask (I_{th}). All methods clearly outperforms the VAF approach, and only the EMD based method obtains similar values to our system. In the latter case, our system shows a clear advantage: its simplicity in terms of computation. While we make use of the directly-computed texture features from the GLC matrix, after a reduction of the image using a mask, the EMD-based method combines gaussian filtering, a Multidimensional Ensemble EMD extraction of some slices, feature reduction using PCA and classification using SVM.

4   Conclussion

The development of new computer-based diagnosis software is a promising area of research. The proposed system aims to provide a fully automated method for physicians to help them in the diagnosis task of Parkinson's Disease (PD), eliminating expensive manual operations, in the sense of requiring an expertise degree of the operator, as well as reducing time costs. Moreover, as the process is automatic, it might not suffer from the pitfalls of investigator-dependent methods.

The presented work makes use of several widely known techniques that, combined, demonstrate their ability in the detection of some PD patterns in DaTSCAN imaging. Particularly, the use of a mask to select subimages that contains only brain voxels has a great impact on the computation of a GLC matrix, facilitating a subsequent texture analysis. Therefore, our system that combines a voxel selection based on mask and the analysis of the textural features of the image demonstrates its ability on the detection of PD patterns in DaTSCAN imaging, providing robust and accurate results up to a 95.9% of accuracy, and 97.3% of sensitivity, with a Positive likelihood ratio over 19, a very good indicator of its robustness on PD detection.

The results hereby presented are very promising, so new approaches to the usage of textural information for the pattern characterization of neurodegenerative disorders can be made.

Acknowledgements. PPMI –a public-private partnership– is funded by The Michael J. Fox Foundation for Parkinson's Research and funding partners, including Abbott, Biogen Idec, F. Hoffman-La Roche Ltd., GE Healthcare, Genentech and Pfizer Inc.

This work was partly supported by the MICINN under the TEC2008-02113 and TEC2012-34306 projects and the Consejería de Innovación, Ciencia y Empresa (Junta de Andalucía, Spain) under the Excellence Projects P07-TIC-02566, P09-TIC-4530 and P11-TIC-7103.

References

1. Bhidayasiri, R.: How useful is (123I) beta-CIT SPECT in the diagnosis of parkinson's disease? Reviews in Neurological Diseases 3(1), 19–22 (2006) PMID: 16596082, http://www.ncbi.nlm.nih.gov/pubmed/16596082
2. Christine, C.W., Aminoff, M.J.: Clinical differentiation of parkinsonian syndromes: Prognostic and therapeutic relevance. The American Journal of Medicine 117(6), 412–419 (2004),
http://www.sciencedirect.com/science/article/pii/S0002934304003626
3. Eckert, T., Edwards, C.: The application of network mapping in differential diagnosis of parkinsonian disorders. Clinical Neuroscience Research 6(6), 359–366 (2007), neural Networks in the Imaging of Neuropsychiatric Diseases,
http://www.sciencedirect.com/science/article/pii/S1566277207000023
4. Haralick, R., Shanmugam, K., Dinstein, I.: Textural features for image classification. IEEE Transactions on Systems, Man and Cybernetics 3(6), 610–621 (1973)
5. Illán, I., Górriz, J., Ramírez, J., Segovia, F., Jiménez-Hoyuela, J., Ortega Lozano, S.: Automatic assistance to parkinsons disease diagnosis in datscan spect imaging. Medical Physics 39(10), 5971–5980 (2012)
6. The Parkinson Progression Markers Initiative: PPMI. Imaging Technical Operations Manual, 2 edn. (June 2010)
7. Kohavi, R.: A study of cross-validation and bootstrap for accuracy estimation and model selection. In: Proceedings of International Joint Conference on AI, pp. 1137–1145 (1995), http://citeseer.ist.psu.edu/kohavi95study.html

8. Marek, K., Jennings, D., Lasch, S., Siderowf, A., Tanner, C., Simuni, T., Coffey, C., Kieburtz, K., Flagg, E., Chowdhury, S., Poewe, W., Mollenhauer, B., Klinik, P., Sherer, T., Frasier, M., Meunier, C., Rudolph, A., Casaceli, C., Seibyl, J., Mendick, S., Schuff, N., Zhang, Y., Toga, A., Crawford, K., Ansbach, A., De Blasio, P., Piovella, M., Trojanowski, J., Shaw, L., Singleton, A., Hawkins, K., Eberling, J., Brooks, D., Russell, D., Leary, L., Factor, S., Sommerfeld, B., Hogarth, P., Pighetti, E., Williams, K., Standaert, D., Guthrie, S., Hauser, R., Delgado, H., Jankovic, J., Hunter, C., Stern, M., Tran, B., Leverenz, J., Baca, M., Frank, S., Thomas, C., Richard, I., Deeley, C., Rees, L., Sprenger, F., Lang, E., Shill, H., Obradov, S., Fernandez, H., Winters, A., Berg, D., Gauss, K., Galasko, D., Fontaine, D., Mari, Z., Gerstenhaber, M., Brooks, D., Malloy, S., Barone, P., Longo, K., Comery, T., Ravina, B., Grachev, I., Gallagher, K., Collins, M., Widnell, K.L., Ostrowizki, S., Fontoura, P., Ho, T., Luthman, J., van der Brug, M., Reith, A.D., Taylor, P.: The parkinson progression marker initiative (PPMI). Progress in Neurobiology 95(4), 629–635 (2011),
http://www.sciencedirect.com/science/article/pii/S0301008211001651
9. Martínez-Murcia, F., Górriz, J., Ramírez, J., Puntonet, C., Salas-González, D.: Computer aided diagnosis tool for Alzheimer's disease based on Mann-Whitney-Wilcoxon U-test. Expert Systems with Applications 39(10), 9676–9685 (2012)
10. McGee, S.: Simplifying likelihood ratios. Journal of General Internal Medicine 17(8), 646–649 (2002)
11. Moghal, S., Rajput, A.H., D'Arcy, C., Rajput, R.: Prevalence of movement disorders in elderly community residents. Neuroepidemiology 13(4), 175–178 (1994) PMID: 8090259, http://www.ncbi.nlm.nih.gov/pubmed/8090259
12. Philips, C., Li, D., Raicu, D., Furst, J.: Directional invariance of co-occurrence matrices within the liver. In: International Conference on Biocomputation, Bioinformatics, and Biomedical Technologies, pp. 29–34 (2008)
13. Rojas, A., Górriz, J., Ramírez, J., Illán, I., Martínez-Murcia, F., Ortiz, A., Río, M.G., Moreno-Caballero, M.: Application of empirical mode decomposition (emd) on datscan spect images to explore parkinson disease. Expert Systems with Applications 40(7), 2756–2766 (2013),
http://www.sciencedirect.com/science/article/pii/S0957417412012274
14. Segovia, F., Górriz, J.M., Ramírez, J., Álvarez, I., Jiménez-Hoyuela, J.M., Ortega, S.J.: Improved parkinsonism diagnosis using a partial least squares based approach. Medical Physics 39(7), 4395–4403 (2012)
15. Stoeckel, J., Ayache, N., Malandain, G., Malick Koulibaly, P., Ebmeier, K.P., Darcourt, J.: Automatic classification of SPECT images of alzheimer's disease patients and control subjects. In: Barillot, C., Haynor, D.R., Hellier, P. (eds.) MICCAI 2004. LNCS, vol. 3217, pp. 654–662. Springer, Heidelberg (2004)
16. Towey, D.J., Bain, P.G., Nijran, K.S.: Automatic classification of 123I-FP-CIT (DaTSCAN) SPECT images. Nuclear Medicine Communications 32(8), 699–707 (2011) PMID: 21659911, http://www.ncbi.nlm.nih.gov/pubmed/21659911
17. Vapnik, V.N.: Estimation of Dependences Based on Empirical Data. Springer, New York (1982)
18. Vapnik, V.N.: Statistical Learning Theory. John Wiley and Sons, Inc., New York (1998)

Automatic ROI Selection Using SOM Modelling in Structural Brain MRI

Andrés Ortiz García[1], Juan Manuel Górriz[2],
Javier Ramírez[2], and Diego Salas-González[2]

[1] Communications Engineering Department
University of Malaga. 29004 Malaga, Spain
[2] Department of Signal Theory, Communications and Networking
University of Granada. 18060 Granada, Spain

Abstract. This paper presents an automatic method for selecting Regions of Interest (ROI) related to the Alzheimer's Disease (AD) using the information contained in 3D structural MRIs. Normal and AD images are modelled by a number of prototypes provided by Vector Quantization (VQ) algorithms (specifically, Self-Organizing Maps, SOM) which model the volumetric probability histogram and describe the intensity profile of the image. The receptive field of each SOM unit represent a different region of interest on the brain associated to the peaks on the probability histogram. Thus, the space is quantized and the activation level of each SOM unit is associated to the probability of occurrence of the modelled gray level. Additionally, this method can be used to extract a reduced and discriminative features for AD classification, as it compress the information contained in the brain in a reduced number of models. The proposed method has been assessed using the computed ROIs to classify a set of images from the Alzheimer's disease Neuroimaging Initiative (ADNI).

1 Introduction

Nowadays, Alzheimer's disease (AD) is the most common cause of dementia which affects more than 30 million people worldwide. Due to the increasing life expectancy and the aging of the population on developed nations, it is expected AD to affect 60 million people worldwide over the next 50 years. AD is a slow degenerative disease, with different evolution on every individual, but usually starting with mild memory problems and turning to severe brain damage in several years. Since currently there is no a known cure for the AD, the early diagnosis may help to slow down the rapid advance of the disease. Although the development of the disease depends on the individual, aging, etc. there are many common symptoms in addition to structural changes in the brain. In order to deal with objective diagnosis of the AD, Brain Magnetic Resonance Images (MRI).

[4,17,3] as well as functional images [6,10,15,16] can be used to reveal common patterns in AD and healthy patients in order to diagnose the AD even before the manifestation of any cognitive symptoms. Most of these techniques use functional

J.M. Ferrández Vicente et al. (Eds.): IWINAC 2013, Part II, LNCS 7931, pp. 278–285, 2013.

images such as Single Emission Computerized Tomography (SPECT) [14,6,10] or Positron Emission Tomography (PET) [1,15]. Thus, Computer Aided Diagnosis systems aim to exploit the information contained in the images to learn patterns associated to cerebral neurodegeneration.

In this work structural MRIs from normal and AD patients are used to reveal 3D brain regions associated to AD and to generate a reduced but discriminative set of features based on the intensity profile of regions. This is accomplished by computing a number of prototype vectors using the SOM algorithm to quantize the space and the intensity profile. The receptive field of each SOM unit represents a ROI on the image and its relative importance is proportional to the probability of occurrence of the modelled intensity level.

The rest of the paper is organized as follows. Section 2 shows the necessary background on SOM to support the next sections. Section 3 describes preprocessing and registration processes applied to each image in the dataset, as well as the modelling using SOM. This section also shows the method we used to compute ROIs associated to AD. Moreover, the feature extraction process based on the computed ROIs and the model selection are shown in Section 4. Next, the proposed method is assessed by classiying a subset of images from the ADNI database in Section 5 and the conclusions are drawn in Section 6.

2 Background in SOM

The Self-Organizing Map (SOM) [8] is a well-known, peculiar clustering algorithm, inspired in the animal brain which is always seeking for the most representative and most economic representation of data and its relationships [8,7]. During the training stage, the prototypes retain the most representative part of the input data, while the units on the output space (i.e. 2D or 3D lattice) holding similar prototypes (in terms of euclidean distance) are moved closed together. Thus, some important features of the input space can be inferred from the output space [7], regarding the input space modelling, density distribution of the data space and feature selection.

SOM training is performed in a competitive way, where only one neuron wins (i.e. its prototype vector is the most similar to the input data instance) with each input data instance, such that

$$\|v_k - \omega_i\| \le \|v_k - \omega_j\| \quad \forall j \in S \tag{1}$$

where S is the output SOM space.

Moreover, prototypes of neurons belonging to the neighbourhood of the wining unit (also called Best Matching Unit (BMU)) are also updated according to

$$\omega_j(t+1) = \omega_j(t) + \alpha(t)h_{i,j}(t)(v_k - \omega_j(t)) \tag{2}$$

where $\alpha(t)$ is the learning factor and $h_{i,j}(t)$ is the neighbourhood function defining the unit surrounding the BMU ω_i. Both $\alpha(t)$ and $h_{ij}(t)$ decrease exponentially with t. Thus, the prototype vectors ω_i quantize the data manifold and represent the cluster center of the data mapped on each BMU.

3   Image Preprocessing and Modelling

This section describes the use of SOM algorithm to compute a number of prototypes corresponding to ROIs in the brain as representative regions of AD.

3.1   Image Preprocessing and Registration

First, it is necessary to normalize the all the images in the database in order to warp the brains to match a standard size, orientation and shape. This allows to compare neuroanatomical positions of different brains and it is performed through a co-registration process using SPM software [9] and the built-in templates. After image registration, all the images from the database were resized to 91x109x91 voxels with voxel-sizes of 1.5 mm (Sagittal) x 1.5 mm (coronal) x 1.5 mm (axial). Moreover, as we are not interested in non-brain structures, they are removed using BET [5]. Additionally, the images are intensity normalized using the method presented in [13] for PET images, where the 0.1% of the voxels in each image with the maximum intensity values are used to normalize all the image set.

3.2   3D Image Modelling

Each image is modelled by quantizing the space and taking into account the intensity value of each voxel and its probability of occurrence. Thus, an image is characterized by its volumetric histogram that contains information regarding the probability distribution of voxel intensities, and the volumetric histogram can be computed using the expression

$$p(l) = \frac{N(l)}{N} \tag{3}$$

where $N(l)$ is the number of voxels with an intensity level l and N is the total number of voxels present in the image. Moreover, each voxel is associated to a bin in the histogram as in [12], and the feature vectors comprising the input space of the SOM are computed as

$$v_n = (x_i, x_j, x_k, b_n, p_n) \tag{4}$$

where (x_i, x_j, x_k) are the three coordinates of the voxel n, b_n its intensity level and p_n is the probability of occurrence of that intensity level. Thus, $I(x_i^n, x_j^n, x_k^n) = b_n$ and each histogram bin has an associated set of voxels.

3.3   ROIs Computation

The goal of using SOM prototyping in this work is to parcel the ROIs on a structural MRI. Thus, the image used to extract the features described in the previous section is computed as the difference between control and AD images, creating the difference image defined as

$$I_D = I_N^\mu - I_{AD}^\mu \tag{5}$$

where I_N^μ and I_{AD}^μ are the mean image of normal and AD images, respectively.

Then, the SOM is trained, generating a number of model vectors representing the cluster centers. In other words, vectors v_n are quantized and clustered by the SOM algorithm. However, only $1^{st}, 2^{nd}, 3^{rd}$ and 5th coordinates are taken into account for BMU calculation, so that intensity values of the difference image are not taken into account for clustering.

Once the SOM is trained, the receptive field of each unit is defined as

$$RF_i = \{v_k \in M : \| v_k - \omega_i \| \leq \| v_k - \omega_j \| \ \forall j \in G\} \tag{6}$$

where M is the data manifold and G is the output space. Each receptive field consist of a set of voxels defining a ROI, and its relative importance is defined by the modelled probability value, p_n.

3.4   SOM Topology Selection

The number of units and the topology of the output space plays an important role in the generalization properties of the SOM. In order to determine the optimum number of SOM units, we measure the quantization error by means of the reconstruction Mean Square Error (MSE) of a MRI using the computed prototypes. In addition, a 3D output space (infinite plane) have been used in order to provide a better adaptation to the input data.

Figure 1 shown that the reconstruction error tends to be constant when the number of SOM units. In this work we used a 9x9 map, i.e. 81 SOM units.

4   Feature Extraction

Once the ROIs have been defined as described in Section 3.3, the receptive fields act as a mask to select voxels in images to be used. Moreover, the receptive field of the unit i is defined as the data subset closer to the unit i in terms of the Euclidean distance taking into account the $1^{st}, 2^{nd}, 3^{rd}$ and 4th coordinates of v_n. Thus, for each image, a feature vector c_n is extracted, containing as many features as units in the SOM map. These feature vectors are defined as

$$c_n = h_n * p_n * \sum_{(\vec{x_n}) \in RF_n} I(\vec{x_n}) \tag{7}$$

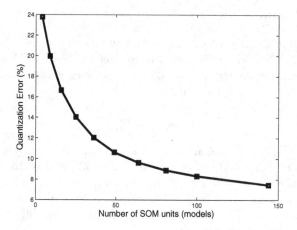

Fig. 1. Reconstruction error of a MRI as a function of the number of SOM units used in the model

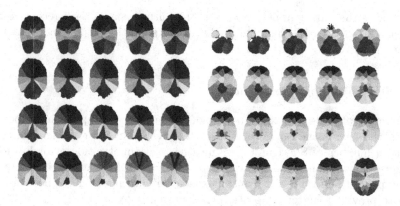

Fig. 2. ROIs computed. Some relevant slices from (a) axial plane and (b) coronal plane are shown. Note that dark red regions corresponds to representative regions of AD [11]

where RF_n is the receptive field of the unit n, p_n is the probability associated to the unit n, and h_n is the height of the corresponding bar of the SOM hit histogram, also defined as the cardinality of RF_n set, $\#\{RF_n\}$.

Then, the probability value computed by means of the histogram and associated to each voxel is used to model the relative importance of each ROI. Figure 2 shows some representative axial and coronal slices in terms of their discriminative capabilities according to the FDR selection criteria [13,18].

Furthermore, Figure 3 shows the three-dimensional representation of the receptive fields of each SOM unit. in this figure, the color of each region is related

(a)   (b)   (c)

Fig. 3. Three different views of the 3D representation of the ROIs obtained with the SOM-based method. Each color represents the receptive field of a SOM unit and it is related to the relative importance of the ROI. Note that dark red regions corresponds to representative regions of AD [11]

to the height of the associated histogram bar. Thus, dark red regions are the most representative, according to the colorbar in Figure 3.

5   Experiments

The computed ROIs have been used along a subset of images from the ADNI database [2] to assess the method presented in this paper. Thus, a set of 50 MRI images have been used for training and testing through cross-validation using k-fold ($k = 5$). The feature vectors computed as described in Section 4 are used to train a Support Vector Classifier [19] with Radial Basis Function (RBF) kernel and $\sigma = 5$. Table 1 shows the results obtained by averaging the accuracy, specificity and sensitivity values when 10 images were used for testing in each k-fold iteration.

Table 1. Classification results

Sensitivity	0.88
Specificity	0.84
Accuracy	0.86

6   Conclusions

In this paper a method based on unsupervised vector quantization techniques for automatic ROI calculation is presented. Specifically, the Self-Organizing Map is used to model MRI images, computing a number of prototype vectors from features extracted taking into account the spatial relationship among voxels. This deals to a quantization of the space, where the relative importance of each cluster is related to the probability of occurrence of the modelled gray level of the voxels belonging to that cluster (i.e. modelled by the corresponding SOM unit). The spatial relationship among clusters is also preserved in the SOM output space due to the topology preserving properties of the SOM. The method have been assessed by classifying a set of images from the ADNI [2] database, using the ROIs autoimatically computed by the presented approach. The classification results provide average accuracy, sensitivity and specificity values of up to 88%, 86% and 84%, respectively, for cross-validation using k-fold.

Acknowledments. This work has been funded by the Consejería Innovación Ciencia y Empresa, Junta de Andalucía, under Project No. P11-TIC-7103.

References

1. Álvarez, I., Górriz, J.M., Ramírez, J., Salas-González, D., López, M.M., Segovia, F., Chaves, R., Gomez-Río, M., García-Puntonet, C.: 18f-fdg pet imaging analysis for computer aided Alzheimer's diagnosis. Information Sciences 184(4), 903–916 (2011)
2. Alzheimer's Disease Neuroimaging Initiative (ADNI) (2012)
3. Chyzhyk, D., Graña, M., Savio, A., Maiora, J.: Hybrid dendritic computing with kernel-lica applied to Alzheimer's disease detection in mri. Neurocomputing 75(1), 72–77 (2012)
4. Cuingnet, R., Gerardin, E., Tessieras, J., Auzias, G., Lehéricy, S., Habert, M.-O., Chupin, M., Benali, H., Colliot, O., Alzheimer's Disease Neuroimaging Initiative: Automatic classification of patients with Alzheimer's disease from structural MRI: a comparison of ten methods using the adni database. Neuroimage 56(2), 766–781 (2010)
5. FMRIB Centre. Nuffield Department of Clinical Neurosciences. University of Oxford. Fmrib software library (2012)
6. Górriz, J.M., Segovia, F., Ramírez, J., Lassl, A., Salas-González, D.: GMM based SPECT image classification for the diagnosis of Alzheimer's disease. Applied Soft Computing 11, 2313–2325 (2011)
7. Haykin, S.: Neural Networks, 2nd edn. Prentice-Hall (1999)
8. Kohonen, T.: Self-Organizing Maps. Springer (2001)
9. London Institute of Neurology (UCL). Statatistical parametric mapping (SPM), ver 8.0 (2012)
10. López, M., Ramírez, J., Górriz, J., Álvarez, I., Salas-González, D., Segovia, F., Chaves, R., Padilla, P., Gómez-Río, M.: Principal component analysis-based techniques and supervised classification schemes for the early detection of Alzheimer's disease. Neurocomputing 74(8), 1260–1271 (2011)
11. Minoshima, S., Giordani, B., Berent, S., Frey, K.A., Foster, N.L., Kuhl, D.E.: Metabolic reduction in the posterior cingulate cortex in very early Alzheimer's disease. Annals of Neurology 42, 85–94 (1997)
12. Ortiz, A., Górriz, J.M., Ramírez, J., Salas-González, D.: Two fully-unsupervised methods for MR brain image segmentation using SOM-based strategies. Applied Soft Computing (2012)
13. Padilla, P., López, M., Górriz, J.M., Ramírez, J., Salas-González, D., Álvarez, I., The Alzheimer's Disease Neuroimaging Initiative: NMF-SVM based CAD tool applied to functional brain images for the diagnosis of Alzheimer's disease. IEEE Transactions on medical imaging 2, 207–216 (2012)

14. Ramirez, J., Chaves, R., Gorriz, J.M., Lopez, M., Alvarez, I.A., Salas-Gonzalez, D., Segovia, F., Padilla, P.: Computer aided diagnosis of the Alzheimer's disease combining spect-based feature selection and random forest classifiers. In: Proc. IEEE Nuclear Science Symp. Conf. Record (NSS/MIC), pp. 2738–2742 (2009)
15. Segovia, F., Górriz, J.M., Ramírez, J., Salas-González, D., Álvarez, I., López, M., Chaves, R., The Alzheimer's Disease Neuroimaging Initiative: A comparative study of the feature extraction methods for the diagnosis of Alzheimer's disease using the adni database. Neurocomputing 75, 64–71 (2012)
16. Stoeckel, J., Fung, G.: Svm feature selection for classification of spect images of Alzheimer's disease using spatial information. In: Proc. Fifth IEEE Int. Data Mining Conf. (2005)
17. Termenon, M., Graña, M.: A two stage sequential ensemble applied to the classification of Alzheimer's disease based on mri features. Neural Processing Letters 35(1), 1–12 (2012)
18. Theodoridis, S., Koutroumbas, K.: Pattern Recognition. Academic Press (2009)
19. Vapnik, V.N.: Statistical Learning Theory. Wiley-Interscience (1998)

Onboard Vision System for Bus Lane Monitoring

David Fernández-López, Antonio S. Montemayor, Juan José Pantrigo,
Mará Luisa Delgado, and R. Cabido

Universidad Rey Juan Carlos, C/Tulipán s/n, 28933 Móstoles, Spain

Abstract. Improving the mobility is one of the most important challenges the cities face. The coexistence of public and private vehicles sometimes force the city governments to designate reserved lanes for bus use only. However, not all the private drivers respect these reserved spaces and they use them. Therefore, it is necessary to provide a surveillance mechanism. This work presents a visual system devoted to perform automatic surveillance of a bus lane. This system proposal consists of a heuristic combination of filtered images of the road for the bus lane change detection. We show how to refine the strategy for reducing false positives as well as improving its computational performance. The resulting system is able to run in real time on an Intel Atom platform without the use of any programming optimization technique.

Keywords: Mobility, Bus Lane, Automatic Surveillance System, On-board Vision System.

1 Introduction

The Smart City concept has been growing during the last few years. City governments has perceived the importance of this philosophy and many of them has elaborated a plan to become a smart city. This concept emerges from the search of a sustainable city where the citizen is the most important part [5]. The six main axes which define a smart city (i.e., people, economy, living, mobility, environment and governance) are represented in Fig. 1. These areas are used as indicators of the city progress. There also exist different rankings based on the score obtained by the cities in each one of those indicators (see for example [5]).

In this work we mainly focus on the mobility axe. It depends on different indicators like the sustainable transport or the intelligent trafic control. These indicators are directly associated to traffic problems like traffic jams, vehicles crashes, atmospheric and noise pollution and the loss of time associated to an inefficient transport. In the opinion of various experts in mobility, the solution for all these problems relies on the correct management of the public transport [9][11].

Urban buses is the public transport that most affects the city traffic. This is due to the fact that urban buses share the road with private vehicles. With the aim of improving the public transport service exclusive bus lanes have been introduced. However, in many cities these lanes are indiscriminately occupied by private vehicles. It does not only make the public transport inefficient but also it affects the overall mobility, as the buses are forced to use the common lanes. In the last years, several attempts to avoid the illegal

J.M. Ferrández Vicente et al. (Eds.): IWINAC 2013, Part II, LNCS 7931, pp. 286–295, 2013.

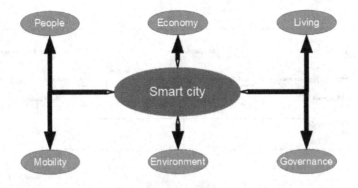

Fig. 1. Main axes of the smart city concept (adapted from [5])

use of the bus lane have been carried out. Some of the measures are only dissuasive while others physically separates the bus lane from the common road. Despite the fact that these separators improve the bus lane traffic flow, they are dangerous and cause vehicle crashes.

In this work, we present an on-board automatic system which is able to detect vehicles parked in the bus lane. This system has a great dissuasive effect because it can be installed on all buses serving the bus lane network. The use of this technology would contribute to improve the urban transport traffic. Furthermore, the use of separators will not longer be required, improving the general mobility and road safety of the cities. Our system hadrware consists of a camera used to detect the lane separator, a GPS receptor and an on-board computer.

The remainder of the paper is organized as follows. Section 2 presents a system overview. In Section 3, the lane detection change algorithms are detailed. In Section 4, the obtained results in a testing under real conditions are shown. Finally, Section 5 depicts the most important conclusions extracted from this work.

2 System Overview

This section presents an overview of the system hardware and software. The hardware system view is depicted in Fig. 2 and consists of two main subsystems: the on-board system and the control center. The former is the object of this paper because the control center is different in each city. The on-board systems consists of a camera, a GPS receptor and an on-board computer installed in each bus. As shown in Fig. 2, each bus sends information concerning the offending vehicles through a wireless network to the central server. This information is sent to the Control Center, where a human operator performs a manual validation. For this task, the operator uses a front-end, where the sequence of the offense can be visualized to help him to make the decision of filing a complaint or not.

Fig. 3 presents the software view of the on-board System. It describes the relationship among the four main modules: setup, GPS data analyzer, lane change detection and

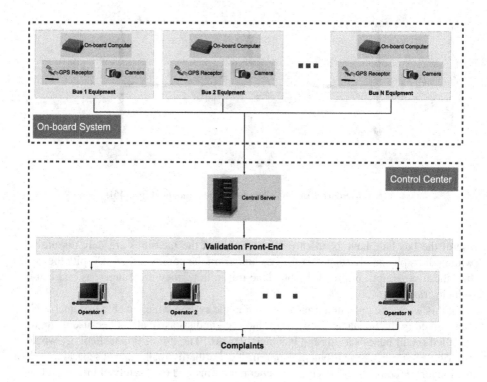

Fig. 2. Hardware System Overview

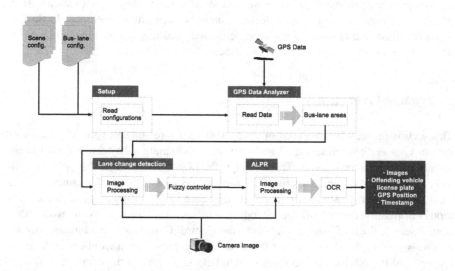

Fig. 3. Software System Overview

ALPR modules. Setup module reads the scene configuration and the bus-lane configuration. GPS data analyzer is devoted to parse the GPS data, allowing us to categorize the different lanes in the city. Through the camera images, the lane change detection module is able to distinguish the lane in which the bus is located. Additionally, an ALPR module provides the license plate information about the offending vehicles.

3   Bus Lane Change Detector

A main requisite of our system is that we need a real time execution on a commodity PC. It is necessary to avoid the processing of the whole image. To this aim, we define a region of interest (ROI) covering the width of the lane and enough distance for evaluating the lane delimiters as in [1]. The lane segmentation and the lane change detection will be only applied into this ROI. An example of this ROI can be seen in the Figure 4.a.

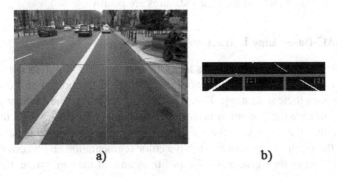

a)   b)

Fig. 4. a)Example of a ROI. b) Image division to compute the horizontal projection.

3.1   Image Filtering Framework and Lane Change Heuristic

We apply the following heuristic to detect a lane change: *"If the onboard camera is centered in the vehicle, the lane delimiter will cross the image from one side to the other when the vehicle performs a lane change"*. Based on this assumption the segmented image is split into three regions ([0], [1] and [2] in Fig. 4.b) corresponding to the lower half of the left, central and right zones, respectively.

In order to decide where the lane delimiters are, we count the labelled pixels in each region. Once these values are obtained, we determine that a lane change occurs when most of the significant pixels are located in the central part of the image (i.e., when the line marker is in front of the vehicle). Additionally, to avoid false positives produced by other markings, the method compares the number of labelled pixels in the left and right zones to those labelled into central zone.

3.2   Decision Fusion Approach

In order to refine the decision process we perform a fusion approach. It involves different visual features and combines them to decide if a lane change is produced. Specifically, we consider four different features to extract the lane delimiters:

- Adaptive color segmentation: this filter is based on the road properties presented in [8]. These properties define the valid road colors based on a gray color with very similar RGB components. During the execution we update the maximum and minimum gray values using a 3D color histogram like in [6].
- Fixed thresholding: The lane delimiters in the city of Madrid are white. We experimentaly determine the threshold to segment the white road markers as in [15], [13]. Note that this approach does not only detects lane delimiters but all the road markers.
- Horizontal motion detection using image difference: When a lane change happens the lane delimiter moves horizontally so that this measure can segment it.
- Background subtraction: This measure is the result of a subtraction operation between the current frame and an adaptive background computed as described in [7].

Once these measures have been evaluated, the system decides that a lane change is taking place if at least three of the four measures are positive.

3.3   RANSAC-Based Line Extraction Approach

The decision fusion approach produces many false positives and it does not run in real time on a low cost PC when not using optimized SIMD instructions. For these reasons, we need a more efficient strategy. This new approach reduces the number necessary measures to improve the execution performance and includes a RANSAC method [4] to decide whether there is a lane change or not. The proposed method is based on two measures of the previous approach: adaptive color segmentation and thresholding. We select these filters as they give less false positives and, at the same time, they offer a good performance.

The adaptive color segmentation is the same one that was used in the previous approach. This measure has the best rate between true positives and false positives (see Section 4). The thresholding method is improved using the Otsu method [10], that computes the threshold that minimizes the intraclass variance between two classes in a histogram. Otsu thresholding has been used in the literature to separate the gray part of the road from the white lane delimiters [3][14]. It is important that the ROI image contains lane delimiters before applying the Otsu method. Otherwise, the method separates two gray levels, resulting in an incorrect segmentation. To avoid this, we previously perform an image thresholding with a high threshold value. Then, we compute the percentage of the pixels that survives to this threshold. If this percentage is sufficient, the Otsu method is applied. Otherwise, we apply a thresholding with a fixed predefined value.

Figure 5 shows the difference among a fixed thresholding, the original Otsu method and our improved Otsu method. Note that the original Otsu method fails in the third and forth considered examples resulting in a wrong segmentation (Fig.5.c) while our proposed improved method obtains very accurate results (Fig.5.d).

We fit a line in the segmented image using RANSAC method, based on the selection of points belonging to the line at a predefined set of heights. RANdom SAmple Consensus (RANSAC) [4] is a method for fitting a model to a point cloud with outliers. We use a RANSAC approach instead of others methods (for example, linear regression or hough transform) for two main reasons. The first one is that we expect, at most,

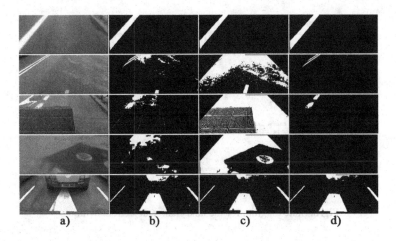

Fig. 5. a) Original image, b) fixed threshold, c) original Otsu method, and d) improved Otsu method

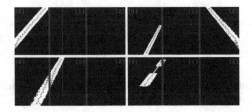

Fig. 6. Example of fitted lines using the RANSAC method

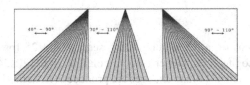

Fig. 7. Supported line angles for each zone

one lane delimiter in each part of the ROI and we need to fit a line for each one. The second one relies in the performance of RANSAC method especially when compared to the Hough transform [12]. Figure 6 shows some examples of line fitting using the RANSAC method after the point candidates selection. Once the line which best fits the centroids is obtained we compute the angle between the line and the horizontal. Then, we consider the fitted line is a lane separator if its angle belongs to the range of predefined valid angles. Figure 7 depicts the considered valid angles for each region.

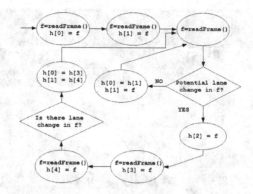

Fig. 8. State diagram for the lane change detection method

To decide if a lane change is taking place, we use the state diagram depicted in Figure 8, where $h[x]$ ($x = 1..5$), stores a history of the last five consecutive video frames. The following two conditional states are shown in the mentioned figure:

- "Potential lane change in f?": in this state, the algorithm evaluates if a potential lane change is taking place. This event is characterized by the presence of a line in the central section and, at the same time, no lines in the left and right regions of the ROI.
- "Is there a lane change?": in this state the algorithm confirms the previous potential lane change. To this aim, the algorithm considers several previous frames stored in the process recent history buffer, and determines the position of the line in the center zone of each frame. Then, if these positions show a continuous lateral shift, we consider that a lane change is taking place.

4   Results

We test our algorithms on a 57 minutes video sequence at 25 fps taken from a real urban bus using a Logitech consumer camera configured at 320x240 pixels of video resolution. In this sequence 44 lane changes are produced (from bus lane to normal lane or vice versa), 16 of them from right to left and 28 from left to right. Additionally, to test the performance of each individual filter, we have used a subsequence of 5130 frames (about 4 minutes at 25 fps). In this sequence, seven bus lane changes are produced.

4.1   Lane Change Detection Results

In our experimental design we include a preliminary experiment and a final experiment. The objective of the preliminar experiment is to identify the best filter method to be coupled with the RANSAC method. To this aim, we test the performance of each filter presented in Section 3 on the image subsequence described above. Figure 9 illustrates the results obtained. As it can be seen from the figure, the background subtraction and our proposed motion detection are the worst methods. The improved Otsu method performs

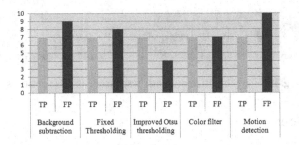

Fig. 9. Quality estimation of individual filters applied to the test sequence with a ground truth of 7 lane changes. True positives (TP) are correct lane change detections and false positives (FP or ϵ_{fp} in Eq.1) are incorrect ones.

reasonably well, as it is able to significantly reduce the number of false positives. We finally conclude that the RANSAC-based approach should use the Otsu thresholding and the color adaptive filters.

Table 1 resumes the results of our three approaches for the considered video sequence. It shows the number of real bus lane changes (from right to left and from left to right), the number of false positives (ϵ_{fp}) and an error measure that combines the difference between real and detected lane changes as well as false positives given by Eq.1:

$$\epsilon = \epsilon_{tn} + \epsilon_{fp} \tag{1}$$

where ϵ_{tn} denotes the true negatives.

This error measure combines two error sources, however, for practical situations, false positives are more inconvenient that loosing real lane changes, as we need to minimize false alarms when possible.

Table 1 shows that the RANSAC-based approach obtains the better results (the lower the value of ϵ, the better the method), followed by the decision fusion strategy. Finally, the single improved Otsu thresholding produces the highest number of false detections among the considered methods. Figure 10 depicts typical cases for miss detection or false positives. The most challenging situations are given by roads in bad conditions with patches or cracks.

Table 1. Results of our three approaches

	Ground truth	Improved Otsu	Fusion approach	RANSAC approach
Left changes	16	14	11	8
Right changes	28	24	16	15
ϵ_{fp}	–	55	34	2
ϵ	–	61	51	23

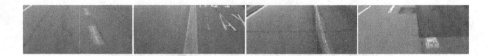

Fig. 10. Views of different frames responsible of false positives

5   Conclusions

In this work, we present a visual system devoted to perform automatic surveillance of a bus lane. This system proposal consists of a heuristic combination of filtered images of the road for the bus lane change detection. We show how to refine the strategy for reducing false positives as well as improving its computational performance. Experimental results show that the system performance is reasonably adequate for solving this task, as it can detect the lane changes of the buses with a low rate of false negatives. The resulting system is computationally efficient as it is able to run in real time on an Intel Atom platform without the use of any programming optimization technique.

Acknowledgments. This work has been supported by the Cátedra de Ecotransporte, Tecnología y Movilidad between University Rey Juan Carlos and the Empresa Municipal de Transportes de Madrid (EMT) through the BusVigia project, by the Spanish Ministry of Economy and Competitiveness grant TIN2011-28151 and by the Government of the Community of Madrid grant ref S2009/TIC-1542.

References

1. Aly, M.: Real time detection of lane markers in urban streets. In: Intelligent Vehicles Symposium, pp. 7–12 (2008)
2. Caragliu, A., Del Bo, C., Nijkamp, P.: Smart cities in Europe. VU University Amsterdam, Faculty of Economics, Business Administration and Econometrics (2009)
3. D'Cruz, C., Zou, J.J.: Lane detection for driver assistance and intelligent vehicle applications. In: International Symposium on Communications and Information Technologies, pp. 1291–1296 (2007)
4. Fischler, M.A., Bolles, R.C.: Random sample consensus: a paradigm for model fitting with applications to image analysis and automated cartography. ACM Commun. 24(6), 381–395 (1981)
5. Giffinger, R.: Smart cities, Ranking of European medium-sized cities (2007),
   http://www.smart-cities.eu
6. Cheng, H.-Y., Jeng, B.-S., Tseng, P.-T., Fan, K.-C.: Lane Detection With Moving Vehicles in the Traffic Scenes. Intelligent Transportation Systems 4(7), 571–582 (2006)
7. Wang, J.-M., Chung, Y.-C., Chang, S.-L., Chen, S.-W.: Lane Marks Detection Using Steerable Filters. In: IPPR Conference on Computer Vision (2003)
8. Kuo-Yu, C., Sheng-Fuu, L.: Lane detection using color-based segmentation. In: Intelligent Vehicles Symposium, pp. 706–711 (2005)
9. Odeck, J.: Congestion, ownership, region of operation, and scale: Their impact on bus operator performance in Norway. Socio-Economic Planning Sciences 40(1), 52–69 (2006)

10. Otsu, N.: A Threshold Selection Method from Gray-Level Histograms. Systems, Man and Cybernetics 1(9), 62–66 (1979)
11. Santos, G., Behrendt, H., Teytelboym, A.: Part II: Policy instruments for sustainable road transport. Research in Transportation Economics 28, 46–91 (2010)
12. Se, S., Lowe, D., Little, J.: Global localization using distinctive visual features. In: International Conference on Intelligent Robots and Systems, vol. 1, pp. 226–231 (2002)
13. Su, C.-Y., Fan, G.-H.: An Effective and Fast Lane Detection Algorithm. In: Bebis, G., et al. (eds.) ISVC 2008, Part II. LNCS, vol. 5359, pp. 942–948. Springer, Heidelberg (2008)
14. Yanqing, W., Deyun, C., Chaoxia, S., Peidong, W.: Vision-based road detection by monte carlo method. Information Technology Journal 9, 481–487 (2010)
15. Zhou, X., Huang, X.-Y.: Multi lane line reconstruction for highway application with a signal view. In: Third International Conference on Image and Graphics, pp. 35–38 (2004)

Urban Traffic Surveillance in Smart Cities Using Radar Images

J. Sánchez-Oro, David Fernández-López, R. Cabido,
Antonio S. Montemayor, and Juan José Pantrigo

Dept. Ciencias de la Computación
Universidad Rey Juan Carlos
Spain
jesus.sanchezoro@urjc.es

Abstract. The Smart City concept arises from the need to provide more intelligent and optimized applications for the development of future urban centers. Traffic monitoring including surveillance is becoming a problem as cities are getting larger and crowded with vehicles. Intelligent video applications for outdoor scenarios need for good quality, stable and robust signal in every moment or climate condition. In this paper we present a radar signal surveillance application that works in real-time, in 360 degrees, with long range up to 400 meters away from the detector, with daylight or night, or even with adverse climatology like fog presence, detecting and tracking high speed vehicles in urban areas.

Keywords: smart city, particle filter, computer vision, radar processing, visual tracking.

1 Introduction

The detection and tracking of moving objects are two of the most important tasks in the context of video surveillance systems. With the appearance of smart cities, the desirable features for video surveillance systems have been increased to satisfy the user needs. Traffic monitoring including surveillance is becoming a problem as cities are getting larger and crowded with vehicles. For this reason, new video surveillance systems should be able to analyze the scene and extract more information than the classical systems, mostly based on the scene context. Furthermore, lighting and weather conditions usually limit the functionality of the common techniques. Radar devices are robust against adverse climatological conditions, and they present long and wide range detection (up to 400 meters far away from the device and 360°).

This work describes a complete system for detecting and tracking vehicles in urban environments using marine radar images, with the objective of generating alerts when an unusual situation occurs. The visual detection and tracking problem has been tackled using different algorithm techniques, mostly based on the particle filter framework [4], Kalman filters [9] and combinations of probabilistic and evolutionary strategies [7]. Object detection and tracking are two areas of interest in the radar-assisted remote surveillance context [6].

J.M. Ferrández Vicente et al. (Eds.): IWINAC 2013, Part II, LNCS 7931, pp. 296–305, 2013.

Most of the existing radar-based detection and tracking methods in the literature work in aerial or marine environments, in which there are very low level of noise caused by structural elements. The main aim of this work is to present a new radar-based target detection and tracking system to control unusual vehicle behaviors in urban areas, which contain obstacles that can produce interference and noise in the radar signal. Furthermore, the system must be able to tackle with adverse climatological and lighting conditions, as it should be working all the day. The main proposals of this work can be summarized as a representation of vehicles in radar images; a detection system based on the definition of relevant areas in the image, a dual background model and an adaptive sliding window algorithm to save computing time in the detection; and a radar tracking algorithm based on a particle filter scheme, with the improvements of an intelligent diffusion stage and an optimized moving average model devoted to be robust against occlusions produced by the signal.

2   System Overview

The system proposed in this work is divided in two views (hardware and software). The hardware view consists of the deployment of fourteen radars placed in strategical locations of the city of Torrevieja (Spain), and two central servers placed in the control center. The radar locations have been selected by the local police department in order to cover the most problematic places of the city. Each radar is connected through a WiMAX dedicated network to one of the central servers.

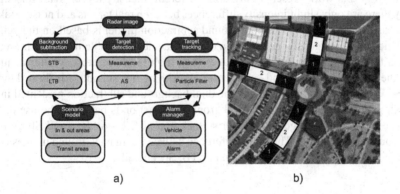

a)   b)

Fig. 1. a) Software architecture overview and b) expert knowledge modeling of a roundabout with in & out (1) and transit (2) areas

Software architecture of the system is depicted in Figure 1.a, describing the relations among the five main modules: scenario modeling, target detection, target tracking, background subtraction and alert manager.

3 Scenario Modeling

Radar devices are placed in strategical locations, fixed by the police department, usually in roundabouts where we can observe different ways with the same detector (see Figure 1.b). Therefore the system should be able to make the most of the location knowledge and use the structural information of the area to define regions of interest. In the context of detection and tracking, there are two important areas: the in and out areas and the transit areas. The former represents the zones where the vehicles can enter and exit of the scenario, while the latter defines the areas where vehicles can move along (generally roads). The radar device used also presents the function to stop receiving data for some angle aperture. This feature is used by the system to eliminate non-interesting parts of the image, reducing its noise and the size of the data that must be analyzed. Figure 1.b shows an example of an scenario modeled by an expert. Scenario represents a round-about with six in-out areas (represented with a "1" label, highlighted in black) and three transit areas (represented with a "2" label, highlighted in white). The scenario models are configured using a user-friendly interface provided with the system.

4 Target Detection

Acquisition of the radar signal results in images with high level of noise, which is difficult to delete. Therefore, it is necessary to pre-process the image to deal with the noise without deleting real targets. The signal obtained from the KODEN MDC-2000 radar used in this work presents only eight different intensity levels. With this signal quality there are only slight intensity differences between real targets and noise, making the filtering more difficult. The background subtraction model is based on the vehicle kinematics for the target segmentation in order to tackle this problem. Furthermore, the artifacts that can be found in radar images (i.e., high levels of noise, clutter or jitter) make the classical background subtraction algorithms not suitable for these kind of images. We propose an adaptation of the dual background subtraction presented in [8] to tackle these problems. The dual background subtraction is based on the use of two different background images: Long Term Background (LTB or B_L) and Short Term Background (STB or B_S). The main difference between both background images is the updating time, being in LTB higher than in STB. Specifically,

$$B_X(x,y,t) = \begin{cases} B_X(x,y,t-\delta t(B_X)) + 1 \text{ if } I(x,y,t) > B_X(x,y,t-\delta t(B_X)) \\ B_X(x,y,t-\delta t(B_X)) - 1 \text{ if } I(x,y,t) < B_X(x,y,t-\delta t(B_X)) \end{cases}$$

where I_t is the radar image at time t, and $\delta t(B_X)$ is the updating time period for B_X (which represents both images, B_L and B_S).

The background subtraction results in two foreground images for each radar image: Long Term Foreground (LTF or I_L) and Short Term Foreground (STF or I_S), which are computed as the difference between their corresponding background image and the current frame (I). The combination of the active pixels in each foreground generates the whole set of possible events that can be detected, depicted in Figure 2.

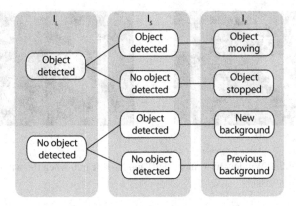

Fig. 2. Event detection using dual background subtraction

The comparison of both foreground images generates the final foreground segmentation, which is a binary image where a pixel is activated only if the value of the corresponding pixel in both outputs, STF and LTF, has a higher value than an experimental threshold. As the radar presents only eight intensity levels, the experimental threshold can only vary in a range of eight values, which makes it easy to configure.

The combination of these two foreground images can also be used for modeling traffic jams or vehicle crashes. This kind of events are detected when an object has been marked as a static object, and it remains static enough time to be added to the background image.

Vehicles are detected in the binary foreground image using the sliding window algorithm presented in [10], varying the size of the window according to the expected dimensions of the targets. The algorithm detects an object if the density of pixels of the region enclosed by the window exceeds a predefined threshold. The weight (π) of the region enclosed by a window centered in (x_p, y_p) is calculated as the area (i.e., zero-order moment) of the window as follows:

$$\pi = \sum_{x=x_p-\frac{1}{2}w}^{x_p+\frac{1}{2}w} \sum_{y=y_p-\frac{1}{2}h}^{y_p+\frac{1}{2}h} I_F(x,y) \tag{1}$$

We propose two different strategies to slide the window through the in-out areas: a standard method and an Adaptive Sliding Window (ASW) method. The former traverses the areas sequentially, resulting in an exhaustive search, which is inefficient in terms of computing time. The ASW method uses the features of the binary image to efficiently explore the in-out areas. Specifically, if the weight associated to a region is very low, then the method discards the exploration of all the consecutive windows which overlap with the former window. This strategy results in a considerable reduction of the computing time needed to explore the image, especially when the foreground image is correctly filtered and low noise is present. Furthermore, when the weight associated with a region exceeds the threshold, the method performs a local search to find the best

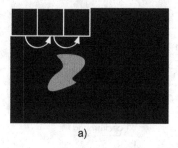 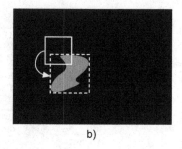

a)   b)

Fig. 3. Adaptive Sliding Window movement (left) and local search when the window detects part of a foreground object (right)

fitting window. When the local search has finished, the method continues sliding the window in the next non-overlapping region.

Figure 3 shows the movement of the sliding window when no objects are detected (left side) and the local search procedure (right side). The dotted rectangle in the right image represents the final output of the local search (larger than the original window and centered for covering the entire object).

5 Target Tracking

This work proposes a tracking method based on the Sampling Importance Resampling (SIR) Particle Filter (PF) [3], very popular in visual tracking applications, coupled with a local search to adapt the size of the estimated region of interest. The local search method is an adaptation of the LSPF presented in [2] for radar images.

Tracked vehicles are represented by a state vector, which consists of the position (x_i^t, y_i^t) and velocity (vx_i^t, vy_i^t) of the vehicle i at time t. The state vector has an associated weight π^t, related to its likelihood [3,11], which is computed considering the foreground image I_F obtained in the detection stage. The computation is carried out by calculating the number of active pixels inside the bounding box centered in the position of the target.

The selection stage starts when all the weights have been evaluated. The particle filter replaces the particles with the lowest weights with better estimators, preserving the best particles, thus improving the quality of the particle set. The selection uses a roulette wheel selection procedure, which will probably eliminate the particles with the lowest weights, but maintaining a low probability to select them in order to escape from local minima. Once the selection stage has finished, the filter starts the diffusion stage with the aim of boosting diversity. This stage consists of altering the position of the particles to simulate the movement. This work proposes an exploitation of the scenario knowledge to improve the performance of the stage. Specifically, it uses the information stored in the scenario modeling (Section 3) to adapt the diffusion and fit it inside the transit areas. Figure 4 shows an example of this intelligent diffusion.

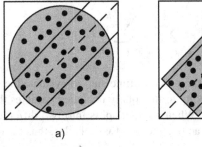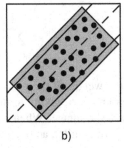

a)   b)

Fig. 4. Particle Filter diffusion without roads information (a) and with roads information (b)

The information stored in the scenario modeling is also used to create a heat map during the execution of the algorithm. This map contains the areas where vehicles are detected more often, based on the position of the vehicles detected and tracked. This information is used to extend the functionality of the system, like reporting statistics about the most problematic areas, or where most traffic jams are located, for instance, which can be transmitted to the authorities, and eventually resulting in improvements of the traffic, creating a kind of "smart roads".

The main problems that can be found in radar images are: jitter, occlusions and disappearances, correspondence between targets and the intensity difference. The first problem (jitter), caused by the reception of the signal, produces a drastic change on the measure taken with respect to the previous measure. This change is represented in the image as displacements of the objects in the scene, although they have not really move, which may lead to errors in the measure. The occlusions and disappearances are caused by an object that cannot be reached by the radar pulse because there is another object in the trajectory of the pulse. In radar images, two vehicles moving with similar velocity can be seen as only one object, until their velocity varies with respect to each other, which makes the identification of the targets quite difficult. Finally, the shape, materials and distance to the antenna of an object determines its intensity in the signal, which can cause that the radar signal of an object affects the signal of another object.

These problems need to be tackled in order to provide an useful system for vehicle tracking in real conditions. The jitter problem is solved with the dual background subtraction proposed, since the method is able to difference between real moving objects or jitter noise. The occlusions and changes in the intensity of the signal are tackled with an adapted computation of the particle weight. It is called Exponentially Weighted Moving Average (EWMA), where the particle weight is computed taking into account the previous weight. Specifically,

$$\pi_i^t = \alpha M_{00} + (1 - \alpha)\pi_i^{t-1}$$

where M_{00} is the area in a binary region given by the position of the bounding box associated to the particle state coordinates (zero-order moment), calculated as follows:

$$M_{00} = \sum_{x=x_i^t - \frac{1}{2}w}^{x_i^t + \frac{1}{2}w} \sum_{y=y_i^t - \frac{1}{2}h}^{y_i^t + \frac{1}{2}h} I_F(x, y)$$

The relevance of the weight of the previous time step to evaluate the new one is represented by the parameter α, giving to the totally occluded particles the possibility to appear afterwards in the image. When a target splits into two different vehicles, it is removed from the particle filter, adding the new two vehicles, thus overcoming the correspondence problem.

The generation of alerts uses the information of the vehicles tracked. The reduction of false positives given by the tracking system must be reduced, in order to avoid sending non relevant alerts. This improvement is carried out by assuming continuity in the movement of the vehicles. Therefore, abrupt changes in the movement indicate that the tracked object is not a vehicle, and it will not generate an alarm.

6   Experimental Results

This section describes the results obtained by the system in real locations. The experiments were performed on an Intel Core 2 Duo E8400 3GHz with 3 GB of RAM and Windows 7 32 bits OS. The radar images were obtained from low-cost radar devices (KODEN Electronics model MDC-2000) oriented to marine applications. The system was entirely developed in C++, using the OpenCV 2.1 (Open Source Computer Vision) library. The parameters have been experimentally obtained, and they are summarized in Table 1. See [10] for more details on the parameter selection.

Table 1. Values of the system parameters

Parameter	Value
Short term background update period - $\delta_t(B_S)$	2 frames
Long term background update period - $\delta_t(B_L)$	5 frames
Bounding box size - $w \times h$	100×100 pixels
Learning factor - α	0.75
Frames needed to generate an alarm - F	3 frames
Number of particles for each particle filter - N	100 particles

The experiments have been carried out using six different video sequences obtained directly from a radar device. The first experiment is devoted to test the quality of the target detection method, compared to a human expert. The results of the background subtraction for the two main sequences can be seen in Figure 5. Sequence 1 presents an ideal scenario with no structural elements that interfere in the radar signal. As a result, most of the foreground pixels correspond to the two targets labeled by the expert. Although some noisy pixels remain active in the foreground image, they will be discarded in the tracking stage.

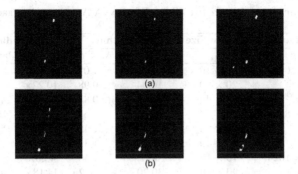

Fig. 5. Background subtraction stage over two different sequences: Sequence 1 (a) and 2 (b)

Sequence 2 shows a more complex scenario where some structural elements like buildings are present introducing higher levels of noise in the foreground. The noisy areas in this sequence are large enough to avoid being discarded in the tracking stage, resulting in some false positives. This kind of false positive is filtered by the continuity premise, as presented in Section 5. Table 2 shows the detection performance of the algorithm, including the ground truth, and the false positives rate. The system is able to detect the 92.9% of real objects (labeled by experts), while the number of false positives is about 0-2 per sequence.

Table 2. Objects detected in the analyzed sequences

Sequence	#Frames	Real Objects	True Detections	False Detections
1	37	2	2	0
2	24	5	5	0
3	43	5	5	2
4	57	2	2	1
5	33	9	7	1
6	72	5	5	1

The aim of the next experiment is the analysis of the performance produced by the Adaptive Sliding Windows (ASW) algorithm described in Section 4. The experiment analyzes it for different radar image resolutions and window sizes. Table 3 shows the results obtained by executing the detection stage for each sequence, reporting the average time per frame obtained with the adaptive proposal versus the standard sliding window method.

The ASW method obtains the highest speedup (43.36×) in the most computing time consuming configurations (the highest image resolution and the largest window size), with an average of 32.05×. The increase of the window size makes the standard method slower than the ASW, as it needs to evaluate larger areas, specially if there is no noise in the image.

Table 3. Performance of the Adaptive Sliding Window (ASW) versus the standard method

Image Resolution	Window Size	Algorithm		Speedup	
		Standard (ms)	ASW (ms)	Average	Maximum
640 × 640	10 × 10	3.75	1.08	3.60×	5.32×
	20 × 20	10.04	0.98	10.22×	14.70×
	40 × 40	15.82	0.85	17.88×	24.72×
800 × 800	10 × 10	6.09	1.64	3.77×	4.89×
	20 × 20	17.04	1.45	11.63×	15.55×
	40 × 40	31.76	1.31	23.83×	31.15×
1024 × 1024	10 × 10	9.97	2.46	4.20×	5.95×
	20 × 20	30.20	2.12	14.18×	18.36×
	40 × 40	63.91	1.97	32.05×	43.36×

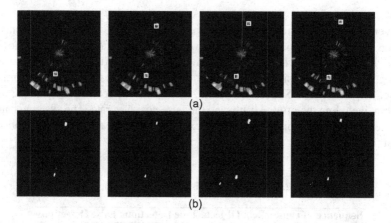

Fig. 6. Visual tracking results for Particle Filter (PF) in the first sequence: target tracking (a) and foreground image (b)

Figure 6.a shows the visual tracking of the particle filter over the first sequence with the targets tracked highlighted with a white rectangle. Figure 6.b shows the foreground image obtained in the background subtraction stage. This sequence shows how an occluded vehicle (upper area in the third frame) is recovered by the particle filter.

7   Conclusions

This work presents a complete intelligent traffic surveillance system based on low-cost marine radar devices, which is able to extract information of the controlled area in real time. This solution supports the concept of smart cities as it can be connected to other systems, to control and improve the traffic automatically. The main aim of this system is traffic monitoring and the generation of alarms in case of non-common situations (i.e., speed exceeds, traffic jams, etc.). The system is robust against adverse lighting and weather conditions thanks to the nature of the signal received, which makes it better than camera-based systems in those kind of conditions, extending its functionality. The system is divided into five main modules: signal filtering, detection, tracking, alarm manager and scenario modeling. The target detection module uses a dual background subtraction to segment vehicles and an adaptive sliding window method to detect them in the image. The target tracking module uses a particle filter algorithm to track the vehicles previously detected. Finally, the alarm manager filters false positives from the tracking system and generates alarms for unusual situations. The computational results show that the system is able to detect and track vehicles in real time even in difficult scenarios. Furthermore, the number of false positive alerts generated is very low in average, which is a really important feature of systems that are part of bigger ones, as would occur in smart cities. Currently the proposed system is working in a real urban environment, presenting a high performance based on the end-user analysis.

Acknowledgments. This research has been partially supported by the Spanish companies Inteomedia Mobile SLNE and Vigidar SL and the Spanish Government research funding ref. TIN 2011-28151 and TIN2012-35632, and by the Government of the Community of Madrid, grant ref S2009/TIC-1542.

References

1. Bernardin, K., Stiefelhagen, R.: Evaluating multiple object tracking performance: the clear mot metrics. EURASIP Journal on Image and Video Processing 2008(1), 1–10 (2008)
2. Cabido, R., Montemayor, A.S., Pantrigo, J.J., Payne, B.R.: Multiscale and local search methods for real time region tracking with particle filters: local search driven by adaptive scale estimation on GPUs. Machine Vision & Applications 21(1), 43–58 (2009)
3. Gordon, N., Salmond, D., Smith, A.: Novel approach to nonlinear/non-Gaussian Bayesian state estimation. IEEE Proceedings F. Radar and Singal Processing 140(2), 107–113 (1993)
4. Jing, L., Vadakkepat, P.: Interacting MCMC particle filter for tracking maneuvering target. Digital Signal Processing 22(1), 54–65 (2010)
5. Kasturi, R., Goldgof, D., Soundararajan, P., Manohar, V., Garofolo, J., Boonstra, M., Korzhova, V., Zhang, J.: Framework for performance evaluation of face, text, and vehicle detection and tracking in video: data, metrics and protocol. IEEE Transactions on Pattern Analysis and Machine Intelligence 31(2), 319–336 (2009)
6. Lanterman, A.D.: Passive radar imaging and target recognition using illuminators of opportunity. In: RTO SET Symposium on "Target Identification and Recognition Using RF Systems" (2004)
7. Pantrigo, J.J., Sánchez, A., Mira, J.: Representation spaces in a visual based human action recognition system. Neurocomputing 72(4-6), 901–915 (2009)
8. Porikli, F., Ivanov, Y., Haga, T.: Robust abandoned object detection using dual foregrounds. EURASIP J. Adv. Signal Process 2008 (2008)
9. Rossi, C., Abderrahim, M., Dìaz, J.C.: Tracking moving optima using Kalman-based predictions. Evolutionary Computation 16(1), 1–30 (2008)
10. Sánchez-Oro, J., Fernández-López, D., Cabido, R., Montemayor, A.S.: Radar-based road-traffic monitoring in urban environments. Digital Signal Processing 23(1), 364–374 (2013)
11. Zotkin, D.N., Duraiswami, R., Davis, L.S.: Joint audio-visual tracking using particle filters. EURASIP Journal on Applied Signal Processing 2002(1), 1154–1164 (2002)

Vehicle Tracking by Simultaneous Detection and Viewpoint Estimation

Ricardo Guerrero-Gómez-Olmedo[1], Roberto J. López-Sastre[1],
Saturnino Maldonado-Bascón[1], and Antonio Fernández-Caballero[2]

[1] GRAM, Department of Signal Theory and Communications, UAH, Alcalá de Henares, Spain
[2] Department of Computing Systems, UCLM, Albacete, Spain

Abstract. We address the problem of vehicle detection and tracking for traffic monitoring in Smart City applications. We introduce a novel approach for vehicle tracking by simultaneous detection and viewpoint estimation. An Extended Kalman Filter (EKF) is adapted to describe the vehicle's motion when not only the pose of the object is measured, but also its viewpoint with respect to the camera. Specifically, we enhance the motion model with observations of the vehicle viewpoint jointly extracted by the detection step. The approach is evaluated on a novel and challenging dataset with different video sequences recorded at urban environments, which is released with the paper. Our experimental validation confirms that the integration of an EKF with both detections and viewpoint estimations results beneficial.

Keywords: vehicle tracking, vehicle detection, tracking by detection, viewpoint estimation, Smart City.

1 Introduction

Within the context of Smart Cities, there are several relevant applications which need a robust system for detecting and tracking the vehicles in the scene. Some examples are vehicle speed estimation [1] or illegal parking detection [2].

In a vehicle tracking application, a fundamental part of the pipeline is the object detection step. However, we want to argue that it is also beneficial to incorporate to the tracking model the observation of the object's viewpoint. Can we recover this information jointly during the detection step? How can we efficiently integrate these pose observations into the tracking approach? These are just two of the questions we want to answer with this work.

Object class detection and recognition in images and videos has been a very popular research theme over the last years (*e.g.* [3,4,5,6,7]). That is, the objective of all these works has been to estimate the bounding boxes in the images in order to localize object of interest within them. Although it is a much less researched area, some recent works propose to deal with the problem of estimating the viewpoint of the objects (*e.g.* [8,9,10,11,12]). We do believe that this viewpoint observation can be beneficial for a tracking model. For instance, if we humans look at the car shown in Figure 1, we are able to infer its pose, and consequently to predict a *logical* direction for its movement.

J.M. Ferrández Vicente et al. (Eds.): IWINAC 2013, Part II, LNCS 7931, pp. 306–316, 2013.

Fig. 1. We humans are able not only to detect an object, but also to estimate its viewpoint or orientation. Furthermore, we are able to use this *semantic* information to estimate a logical direction for the movement of the object of interest. For instance, if a car is observed under a frontal orientation, we will predict that it will move towards the camera position.

In this paper we propose a novel approach for vehicle tracking, using the Extended Kalman Filter (EKF), which is able to simultaneously integrate into the motion model both the position and the viewpoint of the object observed. The approach is evaluated on a novel and challenging dataset with three video sequences recorded at urban environments. We publicly release the dataset, with the ground truth annotations, to provide a common framework for evaluating the performance of vehicle detection and tracking systems within the context of smart city applications.

The paper is structured as follows. In Section 2 we review related work. Section 3 introduces a detailed description of the tracking system. Section 4 presents experimental results and Section 5 concludes the paper.

2 Related Work

The tracking-by-detection approach has become very popular recently [13,14,15]. A common problem of most of this type of works is that the bounding boxes are not adequate to constrain the object motion sufficiently. This really complicates the estimation of a robust trajectory. On the other hand, following the tracking-by-detection philosophy one is able to work in complex scenes and to provide automatic reinitialization by continuous application of an object detector.

For the tracker, we use an EKF [16]. Other approaches have been proposed, like the color based Mean-Shift[17] and Cam-Shift[18]. They are lightweight and robust, but in a traffic urban scene, typically crowded with vehicles, they are not the best choice.

Some model-based tracking approaches use a 3D model, of a particular target object, in order to estimate its precise pose [19,20]. However, it is hard to run them in complex outdoor settings where many different objects are present.

This paper builds on state-of-the-art object detection and viewpoint estimation approaches and leverages recent work in this area [11,7]. Specifically, we propose to learn a system for simultaneous detection and viewpoint estimation, following a similar learning strategy to the one introduced in [11], but using the ground-HOG detector [7]. This system is further integrated into the dynamic motion model of an EKF, which is used to track the vehicles in the scene.

3   Tracking by Detection and Viewpoint Estimation

We present a new approach to address the problem of vehicle tracking via simultaneous detection and viewpoint estimation of the target objects. Essentially, ours is a
tracking-by-detection approach which incorporates the observations of the viewpoint
of the objects into the EKF motion model. This way, by adequately parameterizing the
scene, an object detection can be accompanied by a viewpoint, which is subsequently
associated to an orientation of the movement (see Figure 1).

For the tracking system, we decided to use the EKF [16]. A simple motion model
considering just the position and the speed of the vehicles using a Kalman Filter (KF)
is enough in some cases. To enhance its performance, we use the discrete and non-
linear version of the KF, *i.e.* the EKF, with the Ackermann steering model [21] for
vehicle's non-holonomic motion. One of its main disadvantage is that, as the EKF is a
Taylor's linearized version of the KF, it quickly diverges if the process is not perfectly
modeled, or if we are not able to get measures in a certain interval. We try to avoid
these limitations using the pose recovered from the detector to estimate the orientation
of the object movement. In order to track vehicles in crowded scenes, where occlusions
are one of the main problems to deal with, the EKF, with an adequate motion model,
results very convenient, specially if we compare it with other tracking approaches based
in color features (*e.g.* [18,17]).

We start briefly describing the object detection and viewpoint estimation step in Section 3.1. Then, we offer a detailed description of how we integrate these observations
into the object tracking pipeline (Section 3.2).

3.1   Vehicle Detection and Pose Observation

The basis for a tracking-by-detection and viewpoint estimation approach are the object
detections, which are defined as follows,

$$\bar{d}_t^{(i)} = [x_t^{(i)}, y_t^{(i)}, \theta_t^{(i)}]^T, \tag{1}$$

where, $(x_t^{(i)}, y_t^{(i)})$ encodes the 2D position of the object, and $\theta_t^{(i)}$ corresponds to the
estimation of the viewpoint, for an object i at time stamp t. For the sake of clarity, we
shall mostly omit the superscript i in the following.

Inspired by [11], we propose to to learn a set of viewpoint vehicle detectors for four
particular viewpoints: frontal, rear, left and right. Instead of using the Deformable Part
Model [6], we use the HOG based model described in [7]. Specifically, for learning the
set of detectors we follow the next approach. We learn a HOG template for each viewpoint independently. Each template is refined using the hard negatives, as described in
[3]. Because the objective is to provide a precise viewpoint estimation, when training
for a particular pose (*e.g.* frontal), the negative examples may be extracted from images
with the same object class but from the opposite viewpoint (*e.g.* rear). During detection, and in order to combine all the outputs for the different viewpoints into a single
detection response, we follow a bounding box based non-maximum-suppression step
on the individual outputs. This way, the object detection step is able to feed the tracking
motion model with object detections like $\bar{d}_t$.

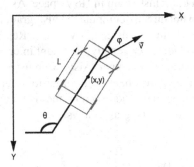

Fig. 2. Ackermann Steering model [21]

3.2 Vehicle Tracking

We define a measurement vector $\bar{z}_t \in \mathbb{R}^3$ as $\bar{z}_t = [x_t, y_t, \theta_t]^T$, and we assume that the tracking process has a state vector $\bar{x}_t \in \mathbb{R}^6$ as $\bar{x}_t = [x_t, y_t, \theta_t, v_t, \phi_t, a_t]^T$, where: x_t and y_t encode the position of the object in the image (*i.e.* the center of the bounding box), θ_t defines the orientation of the movement, ϕ_t is the steering angle, and v_t and a_t are the linear speed and the tangential acceleration, respectively. This formulation corresponds to the Ackermann steering model for cars [21]. See Figure 2 for a graphical representation of the dynamic model.

We use an EKF in combination with this dynamic model to describe the motion of the vehicles. The EKF is a recursive Bayesian filter which iteratively repeats two steps at each frame: first, it estimates the object state $\bar{x}_t^-$ by applying the dynamic model to the previous state $\bar{x}_{t-1}$; second, it updates the resulting state to the corrected state $\bar{x}_t$ for the current frame by fusing it with the new observation $\bar{d}_t$.

According to the dynamic model proposed, we define the following state transition function $f : \mathbb{R}^6 \to \mathbb{R}^6$ as

$$\bar{x}_t = f(\bar{x}_{t-1}) = \begin{bmatrix} x_{t-1} + v_{t-1}cos(\theta_{t-1})\Delta t + \frac{1}{2}a_{t-1}cos(\theta_{t-1})\Delta t^2 \\ y_{t-1} + v_{t-1}sin(\theta_{t-1})\Delta t + \frac{1}{2}a_{t-1}sin(\theta_{t-1})\Delta t^2 \\ \theta_{t-1} + \frac{1}{L}v_{t-1}tan(\phi_{t-1})\Delta t \\ v_{t-1} + a_{t-1}\Delta t \\ \phi_{k-1} \\ a_{t-1} \end{bmatrix} . \quad (2)$$

Note that L is the distance between the axles of the car, we fixed it to the value of 3.2 m.

We follow a hypothesize-and-verify framework for the proposed tracker. Each vehicle trajectory hypothesis is defined as $H^{(i)} = [D^{(i)}, A^{(i)}]$, where $D^{(i)}$ denotes its supporting detections, and $A^{(i)}$ is the appearance model for the vehicle. For $A^{(i)}$, we choose an $(8 \times 8 \times 8)$-bin color histogram in HSV space. As in [15], given a bounding box via the detector, we do not directly compute the histogram for all its pixels. Instead, we preprocess the image within this detection window. Rejecting the portion of the bounding box that is not vehicle is extremely important in order to be able to perform a good matching from one frame to another. Also, we must be resistant to small color variations produced by illumination changes. In order to accentuate the pixels located at the center, we process the image inside each bounding box using a Gaussian kernel to weight each pixel at position x and y as follows,

$$\alpha_{x,y} = e^{\frac{(x-x_c)^2}{(\frac{w}{\delta})^2} + \frac{(y-y_c)^2}{(\frac{h}{\delta})^2}}, \tag{3}$$

where w and h are respectively the width and the height of the bounding box where the histogram is computed, x_c and y_c encode the center of the bounding box, and δ is an empirical constant with a value of 2. Finally, we compute the histogram in HSV color space only for those weighted pixels inside an ellipse fitted to the bounding box. This final step allows us to ignore portions of the image located at the corners that are typically asphalt. Figure 3 graphically shows how the preprocessing pipeline works.

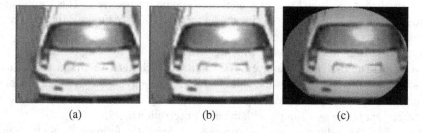

(a)   (b)   (c)

Fig. 3. Preprocessing of the detection window before computing its histogram. a) Original image. b) Image weighted with a Gaussian kernel. c) Image masked with an ellipse.

Every time a new observation $\overline{d}_t^{(i)}$ is added to a trajectory, we update the appearance model as follows, $A^{(i)} = a_t^{(i)}$, where $a_t^{(i)}$ encodes the appearance of the new measure. If there is no measure available, the appearance model is updated using the last estimation. For the data association step, we measure the similarity of an object and its hypothesis using the Bhattacharyya distance [22] between the histograms,

$$d(a_t^{(i)}, A^{(i)}) = \sqrt{1 - \sum_j \frac{a_t^{(i)}(j) \cdot A^{(i)}(j)}{\sqrt{\sum_j a_t^{(i)}(j) \cdot \sum_j A^{(i)}(j)}}}. \tag{4}$$

4 Results

4.1 Experimental Setup

We have tested our approach in the novel *GRAM Road-Traffic Monitoring* (GRAM-RTM) dataset. It consists of 3 challenging video sequences, recorded under different conditions and with different platforms. The first video, called M-30 (7520 frames), has been recorded in a sunny day with a Nikon Coolpix L20 camera, with a resolution of 640×480 @30 fps. The video was scaled to 2000×1200. The second sequence, called M-30-HD (9390 frames), has been recorded in the same place but during a cloudy day, and with a high resolution camera: a Nikon DX3100 at 1280×720 @30 fps. In this case, the video was scaled at 2980×1788. The third video sequence, called Urban1 (23435 frames), has been recorded in a busy intersection with a low-quality traffic camera with a resolution of 480×320 @25fps. This video was also scaled to 2980×1788. Figure 4 shows some examples of the images provided in the dataset. All the vehicles in the GRAM- RTM dataset have been manually annotated using the tool described in [23]. The following categories are provided: car, truck, van, and big-truck. The total number of different objects in each sequence is: 256 for M-30, 235 for M-30-HD and 237 for Urban1. Note that we provide a unique identifier for each vehicle. All the annotations included in the GRAM-RTM were created in an XML format PASCAL VOC compatible [24]. The vehicles that appear within the red areas shown in the second row of Figure 4 have not been annotated, hence any detection in these areas must be discarded before the experimental evaluation. We publicly release the GRAM-RTM dataset[1], including the images, the annotations, and a set of tools for accessing and managing the database. Our aim is to establish a new benchmark for evaluating vehicle tracking and road- traffic monitoring algorithms within the context of Smart City applications.

For establishing further fair comparisons, we encourage to use for training the systems any data except the provided sequences. The test data provided with these sequences must be used strictly for reporting of results alone - it must not be used in any way to train or tune systems, for example by running multiple parameter choices and reporting the best results obtained.

We define two evaluation metrics. First, for detection, we propose to use the Average Precision (AP), which is the standard metric used in the object detection competition of the last PASCAL VOC challenges [24]. For the evaluation of the tracking, inspired by [25], we propose to use the AP and precision/recall curves too. That is, for each estimated bounding box given by the tracking ROI_T, we measure its overlap with the ground truth bounding box provided ROI_G. An estimation is considered valid if it is over a threshold τ,

$$\frac{area(ROI_T \cap ROI_G)}{area(ROI_T \cup ROI_G)} > \tau. \tag{5}$$

Normally, for object detection performance evaluations in the PASCAL VOC challenge, $\tau = 0.5$. In our tracking scenario, we propose to use a less restrictive threshold of

---

[1] http://agamenon.tsc.uah.es/Personales/rlopez/data/rtm/

(a)   (b)   (c)

(d)   (e)   (f)

Fig. 4. GRAM Road-Traffic Monitoring dataset images. Row 1: Examples of images for the sequences: a) M-30, b) M-30-HD, c) Urban1. Row 2: Exclusion areas (in red). Vehicles within these areas have not been annotated.

0.2 (for both the detection and tracking AP computations). This strategy has been also considered within the context of object detection [26].

Note that with this new criterion, when a tracking systems is going to be evaluated, we can consider the four cases represented in Figure 5. Basically, the AP is adequate for measuring the four situations proposed. However, for penalizing the situation drawn in Figure 5(d), we propose to also compare the different methods by providing the number of vehicles counted during the sequence. We do believe that with these three evaluations metrics, AP for detection, AP for tracking, and number of vehicles tracked, we provide a rigorous benchmark.

Technical Details. As it was described in Section 3.1, we train a multi-view detector using the groundHOG [7]. We use the PASCAL VOC 2007 dataset [24] as training set to learn four viewpoints for the category car: frontal, left, right and rear. The groundHOG was parameterized with different values of HOG window and HOG descriptor for each viewpoint. See Table 1 for all the details. Depending on the sequence, detections with a score below a threshold are discarded: M-30 (0.3), M-30-HD (0.21) and Urban1 (0.17).

4.2   Results

Detection Results. First, we analyze the results of our detector in order to provide a baseline to establish further comparisons, specially with the tracking results. As it can

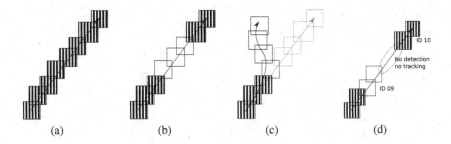

Fig. 5. Tracking Evaluation cases: a) Detector and Tracker are correct. b) Detector fails and Tracker is correct. c) Detector and Tracker fail (only one vehicle is counted). d) Detector and Tracker fail (more than one vehicle is counted).

Table 1. groundHOG's settings

	Frontal	Left	Rear	Right
HOG window (width, height)	(145,107)	(145,57)	(145,107)	(145,57)
HOG descriptor (width, height)	(16,12)	(16,6)	(16,12)	(16,6)

be seen in Figure 6, with the thresholds considered in our detectors, we have a very conservative approach where the number of false positives is under control. The recall is low, and the precision is high, *i.e.* our system only provides very confident detections. Note that within the context of smart cities and surveillance applications, it is important to control the number of false positives. The Urban1 sequence is more challenging due to the variation in viewpoint and scale of the objects within it. Furthermore, it seems that our detector is able to deal with very different image resolutions and qualities. Surprisingly, the best average precision was obtained in the most difficult sequence: Urban1, with a value of 0.4872.

Tracking Results. With the introduced benchmark, we can evaluate the tracking precision of the proposed EKF with pose information. As a baseline, we also report the results of a simple EKF, with the same dynamic model, but where the pose of the object is not observed through the detector. We call this second approach EKF-NP (no pose). Figure 7 shows the results obtained. In general, the EKF obtains higher APs than the EKF-NP for all the sequences, which confirms our hypothesis: to observe the orientation of the object during the detection step improves the tracking. This increment is very relevant for both M-30 sequences. It is also relevant the increment of the recall with respect to the detection curves (see Table 2), which confirms the benefits of using the proposed tracking approaches. The robustness of the tracker relies on the video quality due to the data association step. As resolution and quality decay, histograms of close regions will be more similar and in that way, more incorrect matchings will be performed, increasing the false positive ratio. This explains the false positive obtained for the Urban 1 sequence. Finally, in Table 3 we show that our approach is the best one counting the cars in the scene.

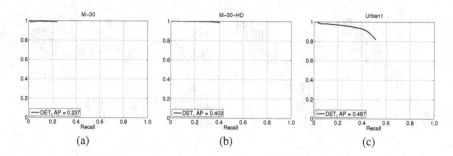

Fig. 6. Precision/Recall curves for object detection: a) M-30, b) M-30-HD c) Urban1

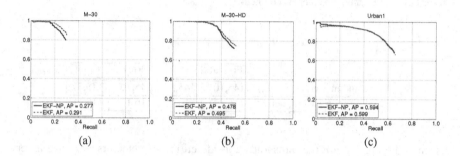

Fig. 7. Precision/Recall curves for vehicle tracking: a) M-30, b) M-30-HD c) Urban1

Table 2. Maximum recall

	M-30	M-30-HD	Urban1
Detection	0.2384	0.4044	0.5132
EKF-NP	0.2916	0.5074	0.6518
EKF with Pose	**0.3009**	**0.5241**	**0.6616**

Table 3. Counted cars

	M-30	M-30-HD	Urban1
Annotated	256	235	237
EKF-NP	300	353	877
EKF with Pose	290	339	789

5 Conclusion

We have proposed a novel approach for tracking vehicles. Using the EKF, our architecture is able to simultaneously integrate into the motion model both the detections and the viewpoint estimations of the objects observed. Our experimental evaluation in a novel and challenging dataset confirms that this semantic information is beneficial for the tracking.

Acknowledgements. This work was partially supported by projects TIN2010-20845-C03-01, TIN2010-20845-C03-03, IPT-2011-1366-390000 and IPT-2012-0808-370000. We wish to thank Fernando García and Laura Martín for their help with the annotation of the GRAM-RTM dataset.

References

1. Zhu, J., Yuan, L., Zheng, Y., Ewing, R.: Stereo visual tracking within structured environments for measuring vehicle speed. IEEE TCSVT 22, 1471–1484 (2012)
2. Lee, J., Ryoo, M., Riley, M., Aggarwal, J.: Real-time illegal parking detection in outdoor environments using 1-d transformation. IEEE TCSVT 19, 1014–1024 (2009)
3. Dalal, N., Triggs, B.: Histograms of oriented gradients for human detection. In: CVPR (2005)
4. Leibe, B., Leonardis, A., Schiele, B.: Robust object detection with interleaved categorization and segmentation. IJCV 77(1-3), 259–289 (2008)
5. Chang, W.C., Cho, C.W.: Online boosting for vehicle detection. IEEE Transactions on Systems, Man, and Cybernetics, Part B: Cybernetics 40, 892–902 (2010)
6. Felzenszwalb, P.F., Girshick, R.B., McAllester, D., Ramanan, D.: Object detection with discriminatively trained part-based models. PAMI 32, 1627–1645 (2010)
7. Sudowe, P., Leibe, B.: Efficient use of geometric constraints for sliding-window object detection in video. In: Crowley, J.L., Draper, B.A., Thonnat, M. (eds.) ICVS 2011. LNCS, vol. 6962, pp. 11–20. Springer, Heidelberg (2011)
8. Thomas, A., Ferrari, V., Leibe, B., Tuytelaars, T., Schiele, B., Van Gool, L.: Towards multi-view object class detection. In: CVPR, vol. 2, pp. 1589–1596 (2006)
9. Savarese, S., Fei-Fei, L.: 3D generic object categorization, localization and pose estimation. In: ICCV, pp. 1–8 (2007)
10. Sun, M., Su, H., Savarese, S., Fei-Fei, L.: A multi-view probabilistic model for 3D object classes. In: CVPR (2009)
11. Lopez-Sastre, R.J., Tuytelaars, T., Savarese, S.: Deformable part models revisited: A performance evaluation for object category pose estimation. In: 1st IEEE Workshop on Challenges and Opportunities in Robot Perception, ICCV 2011 (2011)
12. Pepik, B., Gehler, P., Stark, M., Schiele, B.: 3D ^ 2PM - 3D deformable part models. In: Fitzgibbon, A., Lazebnik, S., Perona, P., Sato, Y., Schmid, C. (eds.) ECCV 2012, Part VI. LNCS, vol. 7577, pp. 356–370. Springer, Heidelberg (2012)
13. Gavrila, D.M., Munder, S.: Multi-cue pedestrian detection and tracking from a moving vehicle. IJCV 73(1), 41–59 (2007)
14. Leibe, B., Schindler, K., Cornelis, N., Van Gool, L.: Coupled object detection and tracking from static cameras and moving vehicles. PAMI 30(10), 1683–1698 (2008)
15. Ess, A., Schindler, K., Leibe, B., Van Gool, L.: Object detection and tracking for autonomous navigation in dynamic environments. Int. J. Rob. Res. 29, 1707–1725 (2010)
16. Welch, G., Bishop, G.: An introduction to the kalman filter. Technical Report TR 95-041, University of North Carolina at Chapel Hill (2006)
17. Comaniciu, D., Meer, P.: Mean shift analysis and applications. In: ICCV (1999)
18. Bradsky, G.: Computer vision face tracking for use in a perceptual user interface. Intel Technology Journal, Q2 (1998)
19. Koller, D., Danilidis, K., Nagel, H.H.: Model-based object tracking in monocular image sequences of road traffic scenes. IJCV 10(3), 257–281 (1993)
20. Dellaert, F., Thorpe, C.: Robust car tracking using kalman filtering and bayesian templates. In: Intelligent Transportation Systems (1997)
21. Cameron, S., Proberdt, P.: Advanced guided vehicles, aspects of the oxford agv project. World Scientific, Singapore (1994)
22. Bradsky, G., Kaehler, A.: Learning OpenCV. Computer Vision with the OpenCV Library. O'Reilly (2008)
23. Vondrick, C., Patterson, D., Ramanan, D.: Efficiently scaling up crowdsourced video annotation - a set of best practices for high quality, economical video labeling. IJCV 101(1), 184–204 (2013)

24. Everingham, M., Van Gool, L., Williams, C.K.I., Winn, J., Zisserman, A.: The PASCAL Visual Object Classes Challenge 2007 (VOC2007) Results (2007)
25. Wang, Q., Chen, F., Xu, W., Yang, M.H.: An experimental comparison of online object tracking algorithms. In: SPIE (2011)
26. Hoiem, D., Chodpathumwan, Y., Dai, Q.: Diagnosing error in object detectors. In: Fitzgibbon, A., Lazebnik, S., Perona, P., Sato, Y., Schmid, C. (eds.) ECCV 2012, Part III. LNCS, vol. 7574, pp. 340–353. Springer, Heidelberg (2012)

A Neural Network Approximation of L-MCRS Dynamics for Reinforcement Learning Experiments

Jose Manuel Lopez-Guede, Manuel Graña, Jose Antonio Ramos-Hernanz, and Fernando Oterino

Grupo de Inteligencia Computacional (GIC), Universidad del País Vasco (UPV/EHU), San Sebastian, Spain

Abstract. The autonomous learning of the control of Linked Multi-component Robotic Systems (L-MCRS) is an open research issue. We are pursuing the application of Reinforcement Learning algorithms to achieve such control. However, accurate simulations needed for RL trials are time consuming, so that the process of training and validation becomes excesively long. In order to obtain results in affordable time, we perform the approximation of the detailed dynamic model of the L-MCRS by Artificial Neural Networks (ANN).

1 Introduction

In this paper we address a practical obstacle to perform the autonomous learning of the control of Linked Multi-Component Robotic Systems (L-MCRS) [1] to complete specific tasks. A L-MCRS can be described as a collection of autonomous robots linked by a non-rigid unidimensional link which introduces additional non-linearities and uncertainty in the control of the robots aimed to accomplish a given task. The tasks are often related to some function the non-rigid link itself. The paradigm is illustrated by the transportation of a hose-like object [2, 3], more specifically the transportation of of the hose tip to an given location in the working space while the other extreme is attached to a source. Reinforcement Learning (RL) requires the repetition of the experiments a large number of times, which is time consuming even for simulated systems. We use an accurate simulation of the hose dynamics based on a Geometrically Exact Dynamic Splines (GEDS) model [2] to compute the next environment state after a robot action. The GEDS model is computationally expensive, so that the realization of the reported experimental results [3–7] has required long times even gathering all available computational resources. To solve this issue, we have proposed [8] training approximations to the GEDS model by Artificial Neural Networks (ANN). The trained ANN model provides very fast responses allowing to perform exhaustive simulations for RL. In previous publications we provided the definition of the GEDS model, the Q-learning algorithm and its application to the single-robot hose transportation problem [3–7, 9]. In this paper

J.M. Ferrández Vicente et al. (Eds.): IWINAC 2013, Part II, LNCS 7931, pp. 317–325, 2013.
© Springer-Verlag Berlin Heidelberg 2013

we focus on the development of the ANN approximation to the GEDS model and its impact on RL simulation.

The content of the paper is as follows. Section 2 introduces the computational cost problem that we aim to solve. Section 2 describes the proposed approach, oriented to speed up the computation of the system's next state. Finally, section 4 gives our conclusions.

2   The Problem

2.1   Hose-Robot System

The system under consideration consists of an unidimensional object (the hose) that has one end attached to a fixed point (which is set as the middle point of the ground working space), and the other end (the tip) is transported by a mobile robot. The task for the robot is to bring the tip of the unidimensional object to a designated destination point through discrete actions of a predefined duration. The working space is a square of 2×2 m^2. We have applied a spatial discretization of $0, 5\,m$, so each discrete action of the robot is the translation of the robot to a neighboring box. We have designed a procedure to generate a number of different initial configurations, and figure 1(a) illustrates a possible initial configuration of the entire system, where P_{ri} is the initial position of the robot and P_d is the desired position of the robot. The main goal is that the robot learns autonomously how to reach any arbitrary point carrying the hose attached. To achieve this goal, the RL algorithm has to perform multiple simulations of the system based on a GEDS model. Figure 1(b) shows one of these simulations in which the robot carrying the tip of the unidimensional object goes from an initial position $P_{r\,initial}$ to a final position $P_{r\,final}$, that corresponds with P_d, through five intermediate points.

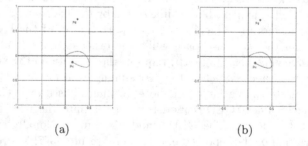

(a)   (b)

Fig. 1. (a) A possible initial configuration of the entire system. P_{ri} is the initial position of the robot driving the tip of the hose. P_d is the goal position. (b) Evolution of the hose starting from a different initial position. The tip of the unidimensional object robot reaches the goal. Arrows are used to indicate the motion of the hose.

Algorithm 1. Hose model learning procedure

Given algorithm finding all feasible paths between any two arbitrarily designed points, without crossings, whose inputs are: exact path length, a set of actions, and the distance that the robot moves with each movement.

1. Use the path finding algorithm to produce all the paths from the fixed point of the hose at $(0, 0)$ to any point reachable by the tip of a flexible hose of $1\,m$.
2. for each path,
    (a) for each available robot action,
        i. Simulate the effect of the action on the system robot-hose by the GEDS initialized with selected path.
        ii. Save the initial position of the hose and the movement as inputs, and the final position of the hose as output.
3. Train an ANN with the input/output patterns obtained in the previous step, partitioning all examples in three data sets (train, validation and test).
4. Test the model learned by each artificial neural network on the test data set, to validate that that model can be used instead the analytical model.

2.2 On the Computational Cost of Experimental Simulation

Due to the complexity of the hose GEDS model, which is implemented in Matlab, running it is acceptable if only a few repetitions are neede, but not when much more simulations (several millions) must be done. Specifically, on a Dell Optiplex 760 personal computer, equipped with a processor Intel(R) Core(TM) 2 Duo CPU E8400 @ 3,00Ghz with 3,00 GB of RAM memory and a Microsoft Windows XP Professional v. 2002 SP 3 operating system, a simulation of a movement of $0,5\,m$ lasts 4 seconds. This response time is unacceptable for the convergence conditions of the RL algorithm, because it requires to repeat all possible movements in each reachable situation of the system many times. This implies that a simulation of one million of movements may last over 45 days running 24 hours a day. The GEDS model and the RL algorithm are not suitable for parallization.

3 Hose Model Learning

3.1 Experimental Design

Algorithm 1 provides a detailed specification of the process followed for the training of the accurate approximation of the GEDS model by the ANN. In essence, all feasible paths between any two points are given to obtain the set of input/ouput train and test samples, where inputs consist of the hose configuration and the action performed, and the output is the final hose configuration. These datasets are used to validate the ANN approximation of the GEDS model.

In the practical application, we choose the smallest spatial discretization step of $0,2\,m$. The trained ANN are feedforward single hidden layer models, trained with the classical backpropagation. The size of the hidden layer was chosen by experimentation. Two activation functions of the neural units has been tested: linear and tan-sigmoid. Five independent training/test processes have been performed for each network, to assess its generalization. Training is performed applying Levenberg-Marquardt algorithm for speed reasons.

Each hose configuration is specified by 11 points in the 2D space where the hose is deployed, therefore, input patterns are 23 dimensional vectors (one dimension corresponding to the executed action), and output patterns are 22 dimensional real vectors. The approximation of the GEDS model is a multivariate regression problem, which strong no-linearities. Finally, we have to determine which data and how they are used for the training process. All the ANN input vectors are presented once per iteration in a batch. We have used the raw data, i.e., without normalization. The input vectors and target vectors have been divided into three sets using random indices as follows: 60% are used for training, 20% are used for validation, and finally, the last 20% are used for testing.

3.2   Results

We present two kinds of results, on one hand the accuracy of the ANN approximation, on the other hand the time gain achieved by this approximation.

ANN Learning Performance Versant. Figure 2(a) shows the best case mean squared error (MSE) achieved by ANNs of different hidden layer sizes when the activation function is linear. Small networks (i.e. 30 neurons in the hidden layer) reach a MSE of $1,5.10^{-5}$ on the training data, $5,13.10^{-4}$ on validation data, and $2,06.10^{-5}$ on test data. Taking into account that the discretization step is $0,2\,m$, these results can be assumed as perfect approximation, test error is below 0.01% of the discretization step. Figura 2(b) shows similar results for ANN trained with tan-sigmoid activation function.

Taking as reference the best initialization of the ANN with linear activation function and 30 neurons in the hidden layer, figure 3 shows the linear regression of target values relative to output values for the training, validation and test datasets, as well as the whole datasets. The fit of the ANN is very good as shown in the figures, with excellent generalization properties.

As a practical illustration, using the 30 hidden neurons ANN, we show in figure 4 the following results for a test pattern: (a) the original path corresponding to the initial configuration of the hose, (b) the hose configuration after discretizing the continuous GEDS to 11 points in the plane, (c) the result of the exact GEDS simulation of the action selected (90° turn of the lead robot), (d) the output hose configurations after discretization to 11 points in the plane,

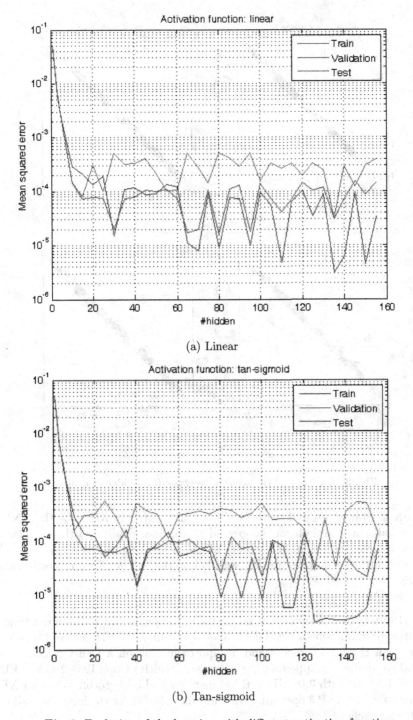

(a) Linear

(b) Tan-sigmoid

Fig. 2. Evolution of the learning with different activation functions

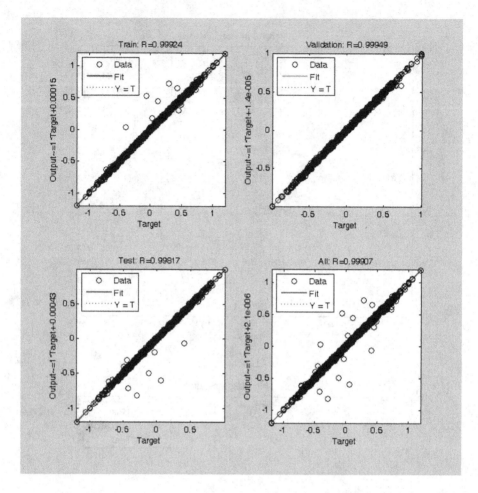

Fig. 3. Accuracy of the selected network (30 hidden nodes)

which are the target ground truth values, (e) the output provided by the ANN, which is indistinguisable from the ground truth, and, finally, (f) the euclidean distance between corresponding points of the ground truth discretization and the ANN output. These errors fall well below the $0,2\,m$ space discretization resolution.

Time Reduction Versant. The main goal of the paper was to provide a time-efficient approximation of the GEDS model by an ANN. To assess the time gain, we run the GEDS model and its approximation on a Dell Optiplex 760 personal computer, equipped with a processor Intel(R) Core(TM) 2 Duo CPU E8400 @ 3,00Ghz with 3,00 GB of RAM memory and a Microsoft Windows XP Professional v. 2002 SP 3 operating system, using Matlab for the implementation of both approaches.

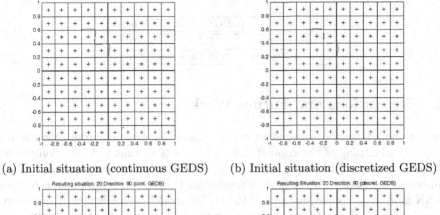

(a) Initial situation (continuous GEDS)   (b) Initial situation (discretized GEDS)

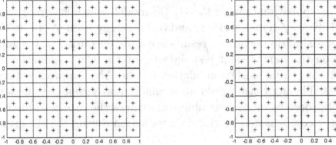

(c) Resulting situation (continuous GEDS) (d) Resulting situation (discretized GEDS)

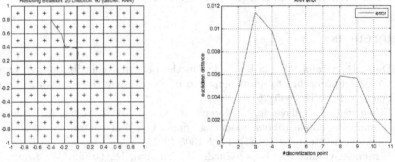

(e) Resulting situation (ANN)   (f) Error of the 11 discretization points

Fig. 4. Experimento con la situación del LMCRS 20 (test)

Table 1 shows the times spent by each model to provide a response on the simulation of an isolated action and a sequence of 10^6 actions, which is the order of the number of actions in a conventional RL simulation. It can be said that the time gain is at least a 99,6% speedup of the ANN process against the GEDS model.

Table 1. Tiempos de simulación de ambos modelos

	GEDS model	ANN model
Time 1 action	> 2,5 s	< 0,01 s
Time 10^6 actions	> 28 days	< 2,8 hours

4   Conclusions and Future Work

In this paper we have discussed one of the main practical problems for autonomous learning of the control of mobile robots in Linked Multi-Component Robotic Systems (L-MCRS) [1], i.e., the need of a cheap computational model allowing extensive experimentation needed for learning. Our approach is to train ANN models as approximations to the GEDS model, which is a multivariate regression problem. We devise a training and test strategy train to obtain the most exhaustive approximation. The test results are quite encouraging, reaching MSE below 2.10^{-5} on the independent test dataset, and a 99,6% reduction in computation time comparing GEDS simulation to the ANN approximation inside the RL process. The approach would allow much more extensive simulations to obtain control processes in a reasonable time where the overhead of the initial ANN training is compensated by the time reduction during the RL execution.

Acknowledgements. The research was supported by Grant UFI11-07 of the Research Vicerectorship, Basque Country University (UPV/EHU).

References

1. Duro, R., Graña, M., de Lope, J.: On the potential contributions of hybrid intelligent approaches to multicomponen robotic system development. Information Sciences 180(14), 2635–2648 (2010)
2. Echegoyen, Z., Villaverde, I., Moreno, R., Graña, M., d'Anjou, A.: Linked multi-component mobile robots: modeling, simulation and control. Robotics and Autonomous Systems 58(12, SI), 1292–1305 (2010)
3. Fernandez-Gauna, B., Lopez-Guede, J., Zulueta, E., Graña, M.: Learning hose transport control with q-learning. Neural Network World 20(7), 913–923 (2010)
4. Fernandez-Gauna, B., Lopez-Guede, J.M., Graña, M.: Towards concurrent Q-learning on linked multi-component robotic systems. In: Corchado, E., Kurzyński, M., Woźniak, M. (eds.) HAIS 2011, Part II. LNCS, vol. 6679, pp. 463–470. Springer, Heidelberg (2011)
5. Lopez-Guede, J.M., Fernandez-Gauna, B., Graña, M., Zulueta, E.: Empirical study of Q-learning based elemental hose transport control. In: Corchado, E., Kurzyński, M., Woźniak, M. (eds.) HAIS 2011, Part II. LNCS, vol. 6679, pp. 455–462. Springer, Heidelberg (2011)
6. Fernandez-Gauna, B., Lopez-Guede, J., Zulueta, E., Graña, M.: Further results learning hose transport control with q-learning. Journal of Physical Agents (2012) (in press)

7. Fernandez-Gauna, B., Lopez-Guede, J., Zulueta, E., Graña, M.: Improving the control of single robot hose transport. Cybernetics and Systems 43(4), 261–275 (2012)
8. Lopez-Guede, J.M., Fernandez-Gauna, B., Zulueta, E.: Towards a real time simulation of linked multi-component robotic systems. In: KES, pp. 2019–2027 (2012)
9. Lopez-Guede, J.M., Fernandez-Gauna, B., Graña, M.: State-action value modeled by elm in reinforcement learning for hose control problems. International Journal of Uncertainty Fuzziness and Knowledge-Based Systems (submitted, 2013)

Addressing Remitting Behavior
Using an Ordinal Classification Approach

Pilar Campoy-Muñoz[1], P.A. Gutiérrez[2], and C. Hervás-Martínez[2]

[1] Universidad Loyola Andalucía, Córdoba, Spain
mpcampoy@uloyola.es
[2] Department of Computer Science and Numerical Analysis, University of Cordóba,
Campus de Rabanales, C2 building, 1004
{pagutierrez,chervas}@uco.es

Abstract. Remittance flows have drawn the attention of international development community interested in enhancing their potential benefits in the recipient communities. This papers deals with the migrants' remitting patterns, addressing this economic behavior by a classification approach rather than the traditional regression one. Five nominal and two ordinal classifiers were compared in order to verify the nature of the problem and to obtain a model which predicts the remittance levels sent by migrants according to their individual characteristics. The best performance was achieved by the support vector machine with ordered partitions, an ordinal classifier based on binary decomposition, and thus three remitting profiles for immigrants were drawn from the support vectors obtained. As result, the proposed model can be used as a tool for better factoring remittances flows into the design of policies and programs in the migrants' home country.

1 Introduction

According to the estimates of the World Bank, the more than 215 million migrants around the world sent back home US\$ 440,000 MM, of which more than 70% was received in developing countries [1]. These flows remained resilient during the economic downturn compared to other private capital flows, therefore ameliorating the macroeconomic stability [2] and contributing to the long-term economic growth of the many developing countries [3]. Besides this, remittance flows have an even more important role in the reduction of the depth and severity of poverty for the recipients [3].

Given its inherent potential for development, remitting behavior has been largely analyzed in the light of the theoretical models proposed to explain the motivations underlying this behavior [4]. More recent studies analyze how variables highlighted by those theories influence the remittance levels by using ordinal regression models [5]. In line with this latter strand of the literature, this papers aims at answering why some migrants send much more (less) money to their home country than others do but tackling the remitting behavior study from a classification perspective rather than the widely used regression approach.

J.M. Ferrández Vicente et al. (Eds.): IWINAC 2013, Part II, LNCS 7931, pp. 326–335, 2013.

Remitting behavior would be, in principle, an ordinal outcome, in the same way as other economic behaviors such saving [6], investment [7] or credit consume [8]. However, to test its ordering nature both nominal and ordinal classifiers are compared in this study by using performance measures.

Specifically, five nominal classifiers were employed in this our study, encompassing the most widely used approaches in machine learning. Namely, MLogistic and SLogistc, from the logistic regression perspective; Multilayer Perceptron Network and Radial Basis Function Network, two well-known artificial neural networks; and, finally, the Support Vector Machine [9] with "1 versus 1" configuration [10], as a representative of kernel methods. Several extensions have been recently developed to apply Support Vector Machine to ordinal outcomes, among them two classifiers were selected, the Support Vector Machine with Ordered Partitions (SVMOP) model [11], based on binary decomposition of the target variable, and the Support Vector Machine for Ordinal Regression with Explicit Constraints [12], belonging to the threshold model category.

The results show that ordinal classifiers yield better results than nominal ones, confirming the ordinal nature of the remitting behavior. The performance achieved by these classifiers allows policymakers and other stakeholders to forecast remittance levels and assess their developmental impacts on receiving communities.

After this introductory Section, the methodological approach is set out in Section 2. The experimental study is carried out with the description of the database employed and the classifiers in Section 3. Results are presented in Section 4. Finally, main conclusions and policy implications are drawn in Section 5.

2 Methodological Approach

2.1 Nominal and Ordinal Classification

Nominal classification is a problem of building a classification rule that correctly assigns the class label for each individual or pattern in the sample and then, by using this rule, to accurately predict the class label for each new pattern. Therefore, the problem in this paper can be stated as follows: let $\mathbf{x}$ be the input vector of an immigrant with K characteristics, where $\mathbf{x} \in X \subseteq \Re^K$, and $\mathbf{y}$ the label class vector of remittance levels, where $\mathbf{y} \in Y = \{C_1, C_2, \ldots, C_Q\}$. The objective is to find a classification rule or function $f \colon X \to Y$ by using a training set $D = \{(\mathbf{x}_i, y_i)\}, 1 \leq i \leq N_1$ of N_1 randomly selected patterns, so that for a given value of $\mathbf{x}_i$ an objective function f estimates the corresponding value of y_i, i.e. $f(\mathbf{x}_i) \to y_i$, in such manner that the function f minimizes an error measure in an independent generalization set $G = \{(\mathbf{x}_i, y_i), 1 \leq i \leq N_2\}$ within the remaining N_2 patterns, to ensure the good generalization performance of the function.

When an inherent or natural order exists among classes ($C_1 \prec C_2 \prec C_Q$), the nominal classification problem turns into an ordinal classification one, which can be tackled by means of regression techniques. Under the regression approach, the different class labels, $C_1, C_2, \ldots, C_Q$, are casted into real values $r_1, r_2, \ldots, r_Q$, where $r_i \in \Re$, and then standard regression techniques can be applied. However, mapping this ordinal scale by assigning numerical values hamper the performance of the regression models because the distance between classes are unknown. Ordinal classifiers are specifically built to exploit the existing order among classes of the target variable, as seems to be the case here where remittance levels could be established among immigrants. To do this, ordinal classifiers take into account the rank of the label, that is, the position of the label into an ordinal scale, which usually is expressed by the following function $O(C_q) = q$, $1 \leq q \leq Q$. But when the ordinal nature of the target variable is not obvious or has been defined *a posteriori*, nominal classifiers can be also applied to ordinal problems. These algorithms completely ignore the ordering of the labels by considering them as independent classes and hence they usually require larger training sets with respect to the ordinal approaches [13].

2.2 Binary Decomposition Method

Among ordinal approaches, special attention is given to the binary decomposition method since it yields the best performance for the problem under study, as will be shown below in Section 4. This method is based on decomposing the ordinal target variable into several binary ones, which are then estimated by a single or multiple models. Ordinal information gives us the possibility of comparing the different labels. For a given rank k, a direct question can be the following, "Is the label of pattern $\mathbf{x}$ greater than k?" [14]. This question is clearly a binary classification problem, so ordinal classification can be solved by considering each binary classification problem independently and combining the binary outputs into a label, which is the approach followed by Frank and Hall in [15] (we name this decomposition Ordered Partitions). In their work, Frank and Hall considered C4.5 as the binary classifier and the decision of the different binary classifiers were combined by using associated probabilities, $p_q = P(y \succ C_q | \mathbf{x}), 1 \leq q \leq Q - 1$. In the work of Waegeman et al. [11], this framework is used but explicit weights over the patterns of each binary system are imposed, in such a way that errors on training objects are penalized proportionally to the absolute difference between their rank and k. Additionally, labels for the test set are obtained by combining the estimated outcomes y_q of all the $Q-1$ binary classifiers. The interpretation of these binary outcomes $y_{jq} \in \{-1, +1\}$, for $1 \leq q \leq Q - 1, 1 \leq i \leq N$, intuitively leads to the fact that $y_i \succ C_j$ if $y_{qi} = 1$. In this way, the rank k is assigned to pattern $\mathbf{x}_i$ so that $y_{iq} = -1$, $\forall j < q$, and $y_{iq} = 1$, $\forall j \geq k$. As stated by the authors, this strategy can result in ambiguities for some test patterns, and they should be solved by using techniques similar to those considered for multiclass classification. In this paper, the Ordered Partitions decomposition was applied to the C-Support Vector Classifier classification algorithm (SVMOP), but including different weights as proposed by Waegeman et al. [11]. Given the problems of possible ambiguities recognized by the authors, probability estimates are obtained following the method presented in [16], considering the fusion of probabilities presented by Frank and Hall [15].

2.3 Performance Evaluation

The performance of both ordinal and nominal classifiers (f) can be evaluated by several metrics, and three of them have been selected in order to quantify the accuracy of the predicted labels with respect to the true targets. Each one covers a different feature that can be considered when a classification problem is analyzed. The correct classification rate (CCR) evaluates if patterns (immigrants) are globally well classified. The sensitivity for the q-th class (S_q) assess if the classifier is independently suitable for each class, and finally the Mean Absolute Error (MAE) examines if the classifier tends to predict a class as close as possible to the real one (in the ordinal scale), in a way that the order relation on the classes is taken into account. The CCR is the proportion of patterns whose class label has been correctly predicted by the classifier:

$$CCR = \frac{1}{N_2} \sum_{i=1}^{N_2} I(y_i^* = y_i), \tag{1}$$

where $I(\cdot)$ is the zero-one loss function, y_i^* the predicted class and y_i is the real one.

S_q estimates the probability of correct classification within each class q:

$$S_q = \frac{n_{qq}}{n_{q\cdot}}, \tag{2}$$

where $n_{q\cdot} = \sum_{j=1}^{Q} n_{qj}$, $1 \leq q \leq Q$, denotes the total number of patterns of class q, n_{qj} is the number of patterns of class q classified as class j and $N_2 = \sum_{q=1}^{Q} n_{q\cdot}$.

The MAE is the average deviation in absolute value of the predicted class y_i^* with respect to the real one y_i:

$$MAE = \frac{1}{N_2} \sum_{i=1}^{N_2} |O(y_i) - O(y_i^*)|, \tag{3}$$

$|O(y_i) - O(y_i^*)|$ being the distance between the true and the predicted class. MAE values range from 0 to $Q-1$, which is the maximum deviation in number of classes, while the former two metrics range from 0 to 1. MAE and S_q have to be maximized, while MAE has to be minimized.

3   Experimental Framework

3.1   Database

The data for this analysis come from the National Immigrant Survey (NIS) carried out by Spanish Statistical Institute between November 2006 and February

2007. The NIS is a cross-sectional survey administered to 15,465 foreign-born individuals, aged 16 or over and residing in Spain for at least one year. We use sample evidence from recent Ecuadorian immigrants (599 individuals), who make up one of the largest immigrant collective in Spain along with Moroc-cans and Romanians. The NIS includes twenty one variables about a wide range of subjects including the socio-demographic and family characteristics of immi-grants, conditions of departure from and arrival in Spain, housing conditions and residential mobility across the host country, labor activity at origin and destination country, and relationship with their origin country and with Spanish civil society. This latter section informs on the remitting behavior of immigrants throughout questions regarding the frequency, the recipients and the amount of money (euros) sent overseas. The selection of which individual characteristics to control for was based on [4]. The variables were grouped in five sets: socio-economic variables, labor market situation, linkages with home country (HC), migratory strategy and remittances information.

For the purpose of this study, to estimate the effects of various migrant char-acteristics on remittance levels, the response variable of interest is the amount of remittances sent overseas during the last year. This amount is separated in three categories. The first category or class (C_1) encompasses migrants who re-mit little or no money, up to 250 euros per year; the second class (C_2) concerns to migrants who remit a little bit more, although they sent below the average amount of $1,718.3$ euros; and the third one (C_3) those who sent above than the annual average. From our sample of Ecuadorian immigrants, 189 (31.55%) were in class C_1, while 194 (32.39%) were in class C_2 and 216 (36.06%) were in C_3, thus the dataset exhibits a quite balanced class distribution.

3.2   Machine Learning Methods

For comparison purpose, five nominal and two ordinal classifiers have been in-cluded in the experimentation. The first four algorithms were implement in the Weka machine-learning software [17] while the remaining three were run in Mat-Lab. Their characteristics are the following:

Nominal Classifiers

1. MLogistic algorithm builds a multinomial logistic regression model by using a ridge estimator to avoid over-fitting the estimates for the parameter matrix.
2. SLogistic algorithm also builds a multinomial logistic regression model but by using the LogitBoost algorithm, which performs an attribute selection
3. Multilayer Perceptron Network is a feed-forward Artificial Neural Network (ANN) with sigmoidal transfer functions in nodes of the hidden layers.
4. Radial Basis Function Network is also a feed-forward ANN with non-linear Gaussian radial basis functions (RBF) as transfer functions in the nodes of the hidden layer.
5. Support Vector Machine (SVM) [9] is a kernel learning method for classi-fication problems in which linear separation is not possible into the input

space. Taking into account the recommendations of Hsu and Li [10], the "1 Vs 1" approach is considered to tackle the problem under study. Under this approach (SVC1V1), a binary SVM classifier is built for each pair of classes, resulting in $Q(Q-1)/2$ SVMs trained to differentiate the patterns of one class from the ones of another.

Ordinal Classifiers

1. Support Vector Machine with Ordered Partitions (SVMOP) [11] decomposes the target variable into several binary ones, but considering its ordinal nature by comparison between the label of a pattern and a given rank k.
2. Support Vector Machine for Ordinal Regression with Explicit Constraints (SVOREX) [12] handles classification task based on threshold models which assume that the ordinal response variable is the indicator of an unobserved continuous variable. In our case, the latent variable is the migrant's utility derived from sending remittances whatever their purpose.

Model Selection. All the algorithms run in Weka were configured with their default values. For the rest of methods (SVM methods), it was necessary to adjust hyperparameter values in the best possible way. This was done by nested cross-validation (i.e. performing a 5-fold cross-validation process using only training data, and retraining the model after deciding the lowest MAE parameter configuration) using the following ranges for the two existing parameters: cost parameter $C \in \{10^{-3}, 10^{-2}, \ldots, 10^3\}$, and width of the Gaussian kernel $\sigma \in \{10^{-3}, 10^{-2}, \ldots, 10^3\}$.

4 Results

The five groups of input variables previously mentioned were studied in relation to the classification performance obtained. Each group included the variables of the previous group, in such a way that socio-economic variables (which are supposed to be the less important ones) were used alone, labor market situation set included also socio-economic variables, and so on. The results showed in Table 1 highlight that ordinal classifiers are better than nominal ones for determining the migrant belonging to the remittances classes. Namely, the SVMOP classifier performs the best in terms of $CCR_G = 71.13\%$ and $MAE_G = 0.3087$, and the SVOREX yields the second best results for these metrics, once all the sets of variables are considered. Therefore, the cost of misclassification is lower for ordinal classifiers compared to nominal ones.

Besides, ordinal methods also exhibit the most balanced results in terms of S_q (see Eq. (2)). Thus, although the CCR difference between the ordinal classifiers is tiny, SVMOP with a $CCR_G = 71.13\%$ and SVOREX with $CCR_G = 69.96\%$, and the best nominal one, $SVC1V1$ with a $CCR_G = 69.95\%$, the formers yields more balanced outcomes by class than the latter.

Table 1. Generalization results of the different methods evaluated

Variable set	Classifier	CCR_G	S_{1G}	S_{2G}	S_{3G}	MAE_G
Socio-economic variables	SVMOP	45.74%	11.64%	64.39%	58.77%	61.94%
	SVC1V1	44.74%	33.95%	41.16%	57.27%	68.29%
	SVOREX	44.90%	17.46%	76.87%	40.13%	59.44%
	Logistic	47.41%	43.92%	62.03%	*9.08%*	65.95%
	Slogistic	40.05%	46.52%	42.92%	14.07%	76.97%
	MLP	40.72%	43.92%	42.58%	15.08%	74.66%
	RBF	46.41%	39.09%	61.10%	8.62%	67.29%
Labor market situation	SVMOP	46.25%	16.43%	57.79%	61.97%	62.44%
	SVC1V1	48.56%	46.61%	37.13%	60.48%	64.97%
	SVOREX	47.41%	34.53%	63.39%	44.35%	60.27%
	Logistic	50.75%	48.77%	61.47%	14.83%	62.44%
	Slogistic	35.89%	35.44%	47.60%	11.11%	82.32%
	MLP	37.38%	64.06%	30.45%	19.74%	86.68%
	RBF	51.42%	47.69%	65.63%	13.38%	60.94%
Linkages with HC	SVMOP	48.92%	31.75%	48.45%	64.29%	61.77%
	SVC1V1	51.75%	44.50%	40.21%	68.42%	60.61%
	SVOREX	47.42%	33.89%	63.34%	44.91%	59.93%
	Logistic	52.75%	50.91%	65.65%	11.09%	59.77%
	Slogistic	43.24%	42.34%	53.55%	8.84%	71.12%
	MLP	38.72%	66.17%	30.97%	18.15%	84.34%
	RBF	52.60%	48.77%	68.44%	11.42%	59.93%
Migratory strategy	SVMOP	52.91%	50.76%	38.66%	67.49%	57.77%
	SVC1V1	54.24%	58.10%	37.53%	65.63%	56.95%
	SVOREX	50.74%	51.75%	43.37%	56.45%	58.11%
	Logistic	55.42%	53.98%	66.56%	9.75%	55.78%
	Slogistic	42.07%	42.84%	53.61%	8.50%	72.29%
	MLP	41.55%	80.91%	30.50%	7.04%	81.50%
	RBF	54.42%	51.81%	68.83%	10.66%	57.44%
Remittaces information	SVMOP	**71.13%**	77.78%	65.97%	*69.91%*	**30.87%**
	SVC1V1	69.95%	75.67%	59.24%	**74.50%**	32.05%
	SVOREX	*69.96%*	78.30%	*66.84%*	65.26%	*31.37%*
	Logistic	66.77%	79.39%	64.65%	8.50%	36.74%
	Slogistic	62.61%	*80.96%*	62.03%	14.62%	40.39%
	MLP	56.58%	**87.87%**	52.25%	6.59%	51.94%
	RBF	69.29%	76.73%	**70.26%**	8.36%	32.54%

Note: The best result is in bold face and the second best result in italics

Moreover, evaluation metrics also provide relevant information in order to examine to what extent the difference among remittance levels is caused by various immigrant's characteristics. Taking the results obtained with the socio-economic set as baseline, the increase of accuracy provided for each set of variables, as percentage of the previous one, is calculated in Table 2. Variations on S_q for the corresponding class indicate that variables regarding to market labor situation at the host country are crucial to identify migrants belonging to C_3. For class C_2, the remittances information leads to the higher increase in the

Table 2. Average generalization improvement considering all methods and depending on the set of variables included in the model (percentage of performance difference with respect to the previous set)

Variable set	CCR_G	S_{1G}	S_{2G}	S_{3G}	MAE_G
Labor market situation	2.07%	33.06%	-6.95%	21.43%	0.80%
Linkages with HC	6.09%	16.61%	2.48%	-7.17%	-4.37%
Migratory strategy	4.73%	25.55%	-8.27%	-8.19%	-4.02%
Remittaces information	33.45%	46.79%	38.41%	9.15%	-42.03%

classification (38%), whereas all the sets matter for class C_1. Therefore, it seems that the migrants who do not remit or remit little amount are the hardest to identify.

All the above leads to the conclusion that tackling the problem as an ordinal classification task can provide accurate information of the remitting pattern among immigrants. In addition, based on the support vectors identified by the SVMOP classifier, the specific profile of the migrants for each remittance class has been drawn up. According to this information, migrants who remit little or no money (C_1) can be defined as a young individual, with secondary education or below, living with his/her spouse, at least one of his/her parents and sometimes any one of her children, although the migrants in this group have in average more than one child. He/she has stayed in Spain for the longest period and has legal status in a higher percentage compared to the other two groups. Wealth in the host country mainly consists on his/her own house. He/she is employed into construction or service sectors with a monthly income about 900 euros, quite close to the average monthly income of the sample, but mostly in elementary occupations. He/she keeps in touch with their family at home country, where he/she sometimes owns a house or rarely lands, and has visited them slightly above the average. Although his/her migratory plan is mostly to remain in Spain, the option to not regroup family is preferred by a higher percent of migrants compared to other groups. Finally, he/she mainly remit to his/her parents and sometimes his/her children by using agency services with a monthly or quarterly frequency and to a larger extent only occasionally.

The profile of migrants belonging to C_2 is easier to draw. A young woman with secondary studies that lives with her spouse, one of her parents and sometimes any one of her children in her own house at the host country, so two out of five migrants in this group own a house in the host country, the largest proportion compared to the other two groups. She is mainly employed in services sector and household activities developing elementary occupations in most cases. She mainly sent money back to her parents and to a lesser extent to her children monthly, quarterly or occasionally, especially agency services but also post office service.

Finally, the highest level remitters C_3 can be characterized as an older man with more than two children and living with his spouse and to a lesser extent with other family members at the host country. He has spent shorter time in Spain compared to the other two groups, but his monthly earnings are the highest. He is mainly employed in construction sector as well as services sector but performing more qualified works than in the other two groups. He monthly remits to his parents, as in previous cases, but mostly to his children by using official channels.

5   Conclusions

Knowing the remitting behavior of migrants is of the major interest for policy makers due to it facilitates correctly factoring them into the design of policies and programs aimed at leveraging their potential benefits for receiving communities. In this vein, five nominal and two ordinal classification methods from the machine-learning field have been compared resulting in a better performance of ordinal ones, namely the SVMOP classifier. Thus a model to predict the remittance level sent by migrants according to a set of characteristics has been obtained.

In addition, the observed profiles, based on the support vectors, reveal the key role of gender labor segmentation in the economic outcomes of migrants as the differences highlighted between remitters in class C_2 and C_3. Since all the remitters have stronger ties with their origin countries, legal status and longer length of stay make the difference between those migrants who remit a little or no money and the remaining groups.

The above mentioned points out possible effects of host country policy measures on migrants, thus remaining illegal migrants could lead to larger amount of remittances, given the higher uncertainty faced by those migrants, although the repatriation would lead to a larger fall in remittances flows. Therefore the migrant receiving countries should implement amnesty programs as long as with measures designed to foster financial and commercial relationship as more efficient way of development co-operation. For their part, sending countries should facilitate migrant's investment at home country as remittances sent for this purpose become more likely once the needs of family at home country are covered and migrants have settled on the host country.

Acknowledgments. This work was partially subsidized by the Spanish Inter-Ministerial Commission of Science and Technology under Project TIN2011-22794, the European Regional Development fund, and the "Junta de Andalucía", Spain, under Project P2011-TIC-7508.

References

1. World Bank: Migrations and remittances: Factbook 2011. World Bank, Washington (2010)
2. Ratha, D., Mohapatra, S., Scheja, E.: Impact of Migration on Economic and Social Development: A Review of Evidence and Emerging Issues. Policy research working paper No. 5558, World Bank (2011)
3. Adams, R., Page, J.: Do international migration and remittances reduce poverty in developing countries? World Development 33, 1645–1669 (2005)

4. Rapoport, H., Docquier, F.: The Economics of Migrants' Remittances. In: Kolm, S., Ythier, J.M. (eds.) Handbook of the Economics of Giving, Altruism and Reciprocity, pp. 1135–1198. Elsevier, North-Holland (2006)
5. Arun, T., Ulka, H.: Determinants of Remittances: The Case of the South Asian Community in Manchester. The Journal of Development Studies 47, 894–912 (2011)
6. Eckel, C., Johnson, C., Montmarquette, C.: Saving decisions of the working poor: short- and long-term horizons. In: Glenn, W., Carpenter, J., List, J.A. (eds.) Field Experiments in Economics. Research in Experimental Economics, vol. 10, pp. 219–260. Emerald Group Publishing Limited, Bingley (2005)
7. Vella, F.: Gender Roles and Human Capital Investment: The Relationship between Traditional Attitudes and Female Labour Market Performance. Economica 61, 191–211 (1994)
8. Huang, C., Chen, M., Wang, C.: Credit scoring with a data mining approach based on support vector machines. Expert Systems with Applications 33, 847–856 (2007)
9. Cortes, C., Vapnik, V.: Support-vector networks. Machine Learning 20, 273–297 (1995)
10. Hsu, C., Lin, C.: A comparison of methods for multi-class support vector machines. IEEE Transaction on Neural Networks 13, 415–425 (2002)
11. Waegeman, W., Boullart, L.: An ensemble of Weighted Support Vector Machines for Ordinal Regression. International Journal of Computer Systems Science and Engineering 3, 7–11 (2009)
12. Chu, W., Keerthi, S.S.: Support vector ordinal regression. Neural Computation 19, 792–815 (2007)
13. Kramer, S., Widmer, G., Pfahringer, B., DeGroeve, M.: Prediction of ordinal classes using regression trees. Fundamenta Informaticae 47, 1–13 (2001)
14. Lin, H.-T., Li, L.: Reduction from cost-sensitive ordinal ranking to weighted binary classification. Neural Computation 24, 1329–1367 (2012)
15. Frank, E., Hall, M.: A simple approach to ordinal classification. In: Flach, P.A., De Raedt, L. (eds.) ECML 2001. LNCS (LNAI), vol. 2167, pp. 145–156. Springer, Heidelberg (2001)
16. Wu, T.-F., Lin, C.-J., Weng, R.C.: Probability estimates for multi-class classification by pairwise coupling. J. of Machine Learning Research 5, 975–1005 (2004)
17. Hall, M., Frank, E., Holmes, G., Pfahringer, B., Reutemann, P., Witten, I.: The WEKA Data Mining Software: An update. ACM SIGKDD Explorations News 11, 10–18 (2009)

Greedy Sparsification WM Algorithm for Endmember Induction in Hyperspectral Images

Ion Marques* and Manuel Graña

Grupo de Inteligencia Computacional (GIC), Universidad del País Vasco
(UPV/EHU), San Sebastian, Spain

Abstract. The Linear Mixing Model (LMM) of hyperspectral images asumes that pixel spectra are affine combinations of basic spectral signatures, called endmembers, which are the vertices of a convex polytope covering the image data. Endmember induction algorithms (EIA) extract the endmembers from the image data, obtaining a precise spectral characterization of the image. The WM algorithm assumes that a set of Affine Independent vectors can be extracted from the rows and columns of dual Lattice Autoassociative Memories (LAAM) built on the image spectra. Indeed, the set of endmembers induced by this algorithm defines a convex polytope covering the hyperspectral image data. However, the number of induced endmembers obtained by this procedure is too high for practical purposes, besides they are highly correlated. In this paper, we apply a greedy sparsification algorithm aiming to select the minimal set of endmembers that explains the data in the image. We report results on a well known benchmark image.

1 Introduction

The Linear Mixing Model (LMM) [1] assumes that the spectral signature of one pixel of the hyperspectral image is a linear combination of the endmember spectra corresponding to the aggregation of materials in the scene due to reduced sensor spatial resolution. Given the endmembers, unmixing algorithms compute the abundance coefficients providing sub-pixel resolution segmentations of the image. The endmembers can be extracted from some spectral library by an expert user, or can be extracted from the image by an Endmember induction algorithms (EIA) [2]. The latter approach has the advantage of not requiring user expertise, and of being automatically fitted to the image content. For instance, endmembers can be used as features for content based image retrieval [3].

The WM algorithm [4] is a Lattice Computing based EIA finding a collection of affine independent vectors that define a convex polytope covering the data of the image in high dimensional spectral space. The algorithm is very fast, using only lattice operators and the resulting endmember set has a direct relation with the image data. However, it has the inconvenient of producing too many endmembers, which are strongly correlated. Therefore, some endmember selection

---

* Ion Marques has a predoctoral grant from the Basque Goverment.

J.M. Ferrández Vicente et al. (Eds.): IWINAC 2013, Part II, LNCS 7931, pp. 336–344, 2013.
© Springer-Verlag Berlin Heidelberg 2013

method is needed to find the relevant endmembers which produce the most parsimonious explanation of the data. Previous works have applied Multi-Objective Genetic Algorithms for this task [5]. In this paper we propose the application of greedy sparse methods, based on gradient pursuit [6]. The aim of the sparse methods is to find the minimal set of contributions from a dictionary that make up the data with minimal loss. In this regard, the WM provides the data dictionary, and sparse method performs the selection of the endmembers for the optimal unmixing of the image.

The contents of the paper are the following: Section 2 recalls the definition of the WM algorithm. Section 3 recalls the sparse estimation process based on gradient pursuit. Section 4 gives some experimental results on a well known hyperspectral image. Finally, Section 5 gives some conclusions of our work.

2 WM Algorithm

Algorithm 1 shows a pseudo-code specification for the WM algorithm[4,7,5]. Given an hyperspectral image H, it is reshaped to form a matrix X of dimension $N \times L$, where N is the number of image pixels, and L is the number of spectral bands. The algorithm starts by computing the minimal hyperbox covering the data, $\mathcal{B}(\mathbf{v}, \mathbf{u})$, where $\mathbf{v}$ and $\mathbf{u}$ are the *minimal* and *maximal corners*, respectively, whose components are computed as follows:

$$v_k = \min_{\xi} x_k^{\xi} \text{ and } u_k = \max_{\xi} x_k^{\xi}; \ k = 1, \ldots, L; \ \xi = 1, \ldots, N. \tag{1}$$

Next, the WM algorithm computes the dual erosive and dilative Lattice Auto-Associative Memories (LAAMs), $\mathbf{W}_{XX}$ and $\mathbf{M}_{XX}$ [8]. The columns of $\mathbf{W}_{XX}$ and $\mathbf{M}_{XX}$ are scaled by $\mathbf{v}$ and $\mathbf{u}$, forming the additive scaled sets $W = \{\mathbf{w}^k\}_{k=1}^{L}$ and $M = \{\mathbf{m}^k\}_{k=1}^{L}$:

$$\mathbf{w}^k = u_k + \mathbf{W}^k; \ \mathbf{m}^k = v_k + \mathbf{M}^k, \ \forall k = 1, \ldots, L, \tag{2}$$

where $\mathbf{W}^k$ and $\mathbf{M}^k$ denote the k-th column of $\mathbf{W}_{XX}$ and $\mathbf{M}_{XX}$, respectively. Finally, the set $V = W \cup M \cup \{\mathbf{v}, \mathbf{u}\}$ contains the vertices of the convex polytope covering all the image pixel spectra represented as points in the high dimensional space. The algorithm is simple and fast but the number of induced endmembers, the amount of column vectors in V, can be too large for practical purposes. Furthermore, some of the endmembers induced that way can show high correlation even if they are affine independent. To obtain a meaningful set of endmembers, we search for an optimal subset of V in the sense of minimizing the unmixing residual error and the number of endmembers.

3 Directional Pursuit for Sparse Approximation

The sparse signal approximation problem can be summarized as follows: Let have a data matrix X (i.e. as defined as in section 2). We define a matrix $\mathbf{\Phi} \in \mathbb{R}^{\tilde{N} \times L}$

Algorithm 1. Pseudo-code specification of the WM algorithm.

1. L is the number of the spectral bands and N is the number of data samples.
2. Compute $\mathbf{v} = [v_1, \ldots, v_L]$ and $\mathbf{u} = [u_1, \ldots, u_L]$,

$$v_k = \min_\xi x_k^\xi; u_k = \max_\xi x_k^\xi$$

   for all $k = 1, \ldots, L$ and $\xi = 1, \ldots, N$,
3. Compute the LAAMs

$$\mathbf{W}_{XX} = \bigwedge_{\xi=1}^N \left[\mathbf{x}^\xi \times \left(-\mathbf{x}^\xi\right)' \right]; \mathbf{M}_{XX} = \bigvee_{\xi=1}^N \left[\mathbf{x}^\xi \times \left(-\mathbf{x}^\xi\right)' \right]$$

   where $\times$ is any of the ⊠ or ⊠ operators.
4. Build $W = \{\mathbf{w}^1, \ldots, \mathbf{w}^L\}$ and $M = \{\mathbf{m}^1, \ldots, \mathbf{m}^L\}$ such that

$$\mathbf{w}^k = u_k + \mathbf{W}^k; \mathbf{m}^k = v_k + \mathbf{M}^k; k = 1, \ldots, L.$$

5. Return the set $V = W \cup M \cup \{\mathbf{v}, \mathbf{u}\}$.

---

called the dictionary. The $\tilde{N}$ columns of $\mathbf{\Phi}$ are referred as atoms. In this paper, Ritter's WM algorithm provides the dictionary $\mathbf{\Phi}$. Therefore, each of the $\tilde{N}$ induced endmembers corresponds to one atom of the dictionary. The problem is to find a mixing matrix Y so that

$$X = \mathbf{\Phi}Y + \varepsilon, \tag{3}$$

where matrix Y optimizes certain sparsity measure. This matrix Y is in fact the collection of abundance images obtained by the unmixing process. One of many methods to achieve this sparsification is to use Conjugate Gradient Pursuit [6].

Conjugate Gradient Pursuit. The conjugate gradient method is a popular directional optimization method. This method calculates a similar decomposition as the QR factorization; and it's guaranteed to solve quadratic problems in as many steps as the dimension of the problem. The conjugate gradient method, used as a directional pursuit method, is explained in algorithm 2. For clarity, we denote $\mathbf{y}$ the set of elements that compose matrix Y.

4 Results

The Salinas hyperspectral image was collected by the 224-band AVIRIS sensor over Salinas Valley, California, and is characterized by high spatial resolution (3.7-meter pixels). The area covered comprises 512 lines by 217 samples. As with Indian Pines scene, we discarded the 20 water absorption bands, in this case bands $[108 - 112]$, $[154 - 167]$ and 224. It includes vegetables, bare soils, and vineyard fields. Salinas ground truth contains 16 classes. Figure 1 shows a pseudo-color visualization and the ground truth of Salinas dataset.

Algorithm 2. Pseudo-code specification of the Conjugate Gradient Pursuit algorithm.

1. $\mathbf{r}^0 = X$ is the initial residual error. $\boldsymbol{\Gamma}^0 = \emptyset$ is an index set. $\mathbf{y}_{\Gamma^0}^0 = 0$ is the initial set of output sparse vectors. $b_0 = 1$ is a term needed to calculate new conjugate gradients.

2. For $i = 1, 2, 3, \ldots$ util stopping criterion is met:

   (a) Calculate gradient $\mathbf{g}$ for $\mathbf{y}$ restricted to $\boldsymbol{\Gamma}^i$:

   $$\mathbf{g}_{\Gamma^i} = \boldsymbol{\Phi}_{\Gamma^i}^T \left(X - \boldsymbol{\Phi}_{\Gamma^i} \mathbf{y}_{\Gamma^i}^{i-1} \right).$$

   (b) Select the best element index:

   $$\gamma^i = \arg_j \max |\mathbf{g}_{\Gamma^i}|.$$

   (c) Update the index set:

   $$\boldsymbol{\Gamma}^i = \boldsymbol{\Gamma}^{i-1} \cup \gamma^i.$$

   (d) Calculate the gram matrix $\mathbf{G}_{\Gamma^i}$:

   $$\mathbf{G}_{\Gamma^i} = \boldsymbol{\Phi}_{\Gamma^i}^T \boldsymbol{\Phi}_{\Gamma^i}.$$

   (e) We denote $\mathbf{D}_{\Gamma^i}$ the matrix containing all conjugate update directions from iteration $i - 1$, with an additional row all zeros. We calculate the update direction $\mathbf{d}_{\Gamma^i}$:

   $$\mathbf{b} = \left(\mathbf{D}_{\Gamma^i}^T \mathbf{G}_{\Gamma^i} \mathbf{D}_{\Gamma^i} \right)^{-1} \left(\mathbf{G}_{\Gamma^i}^T \mathbf{D}_{\Gamma^i} \mathbf{g}_{\Gamma^i} \right),$$

   $$\mathbf{d}_{\Gamma^i} = b_0 \mathbf{g}_{\Gamma^i} + \mathbf{D}_{\Gamma^i} \mathbf{b}.$$

   (f) Calculate new set of vectors $\mathbf{y}_{\Gamma^i}^i$:

   $$\mathbf{c}^i = \boldsymbol{\Phi}_{\Gamma^i} \mathbf{d}_{\Gamma^i},$$

   $$a^i = \frac{\left\langle \mathbf{r}^i, \mathbf{c}^i \right\rangle}{\|\mathbf{c}^i\|_2^2},$$

   $$\mathbf{y}_{\Gamma^i}^i = \mathbf{y}_{\Gamma^i}^{i-1} + a^i \mathbf{d}_{\Gamma^i}.$$

   (g) Calculate new residual error $\mathbf{r}^i$:

   $$\mathbf{r}^i = \mathbf{r}^{i-1} - a^i \mathbf{c}^i.$$

3. Output $\mathbf{r}$ and $\mathbf{y}$.

We compute the gradient pursuit on the collection of endmembers obtained from the image, obtaining a sparse mixing matrix which corresponds to the abundance images for each selected endmember. To determine the most salient endmembers we compute the abundance magnitude of each endmember as the norm of the abundance image. Table 1 shows the endmember number in the WM resulting endmember set, and its magnitude, ordered by decreasing magnitude. It can be appreciated that after the 4th endmember, the remaining ones

(a)

(b)

Fig. 1. (a) Salinas pseudo-color image visualization. (b) Salinas spatial distribution of ground truth classes.

Table 1. 10 most significant endmembers and their sparse magnitude of Salinas image.

#Endmember	Sparse abundance magnitude
449	29963.62
46	14632.85
45	160.182
67	21.14
303	8.92
258	7.82
6	7.30
257	5.21
5	4.76
39	4.39

are residual. This sparsification process performs a kind of principal component selection from the WM results, finding the ones that contribute most to the image.

It is important to note that the endmember induction processes are unsupervised, therefore the meaning of the endmembers found and their relation to the ground-truth classes is unknown. However, we want to support our work on the knowledge of a given ground-truth for the benchmark images. The evaluation process looks for the best match between the abundance images produced by the

Table 2. Correlation coefficients of the abundance images of the 10 most significant endmembers with the ground truth spatial region of each class of Salinas image. Bold entries are positive correlations above 0.1.

	1	2	3	4	5	6	7	8	9	10
Background	**0.3338**	-0.3981	0.0958	-0.0150	-0.0481	-0.0516	-0.0351	-0.0505	0.0315	0.0169
Broccoli_green_weeds_1	-0.1924	**0.4342**	0.0853	-0.0083	0.0067	0.0072	0.0049	0.0070	-0.0044	0.0083
Brocoli_green_weeds_2	-0.2276	**0.4131**	-0.0235	-0.0114	0.0092	0.0099	0.0067	0.0097	-0.0060	0.0114
Fallow	0.0566	-0.0902	-0.0170	-0.0082	0.0066	0.0071	0.0048	0.0070	-0.0043	0.0083
Fallow_rough_plow	-0.0798	-0.0756	-0.0142	-0.0069	0.0056	0.0060	0.0041	0.0058	-0.0036	0.0069
Fallow_smooth	0.0185	-0.1054	-0.0199	-0.0096	0.0078	0.0083	0.0057	0.0081	-0.0051	0.0097
Stubble	**0.3495**	-0.0023	-0.0243	-0.0118	0.0095	0.0102	0.0069	0.0100	-0.0062	0.0118
Celery	-0.2974	**0.3841**	-0.0231	-0.0112	0.0090	0.0097	0.0066	0.0095	-0.0059	0.0112
Grapes_untrained	-0.3506	**0.1744**	-0.0425	-0.0058	0.0166	0.0178	0.0121	0.0174	-0.0108	-0.0007
Soil_vineyard_develop	**0.2175**	-0.1626	-0.0307	-0.0149	0.0120	0.0129	0.0088	0.0126	-0.0078	0.0149
Corn_senesced_green_weeds	0.0917	-0.0731	-0.0220	**0.1601**	0.0086	0.0092	0.0063	0.0090	-0.0056	-0.1518
Lettuce_romaine_4wk	0.1036	-0.0661	-0.0124	-0.0060	0.0049	0.0052	0.0035	0.0051	-0.0032	0.0061
Lettuce_romaine_5wk	0.0560	-0.0853	-0.0168	-0.0081	0.0066	0.0070	0.0048	0.0069	-0.0043	0.0082
Lettuce_romaine_6wk	0.0005	-0.0048	-0.0115	-0.0056	0.0045	0.0048	0.0033	0.0047	-0.0029	0.0056
Lettuce_romaine_7wk	-0.0459	0.0903	-0.0125	-0.0060	0.0049	0.0052	0.0036	0.0051	-0.0032	0.0061
Vineyard_untrained	-0.2640	**0.1121**	-0.0334	0	0.0131	0.0140	0.0095	0.0137	-0.0085	-0.0013
Vineyard_vertical_trellis	-0.1681	**0.1136**	-0.0162	-0.0079	0.0063	0.0068	0.0046	0.0067	-0.0041	0.0070

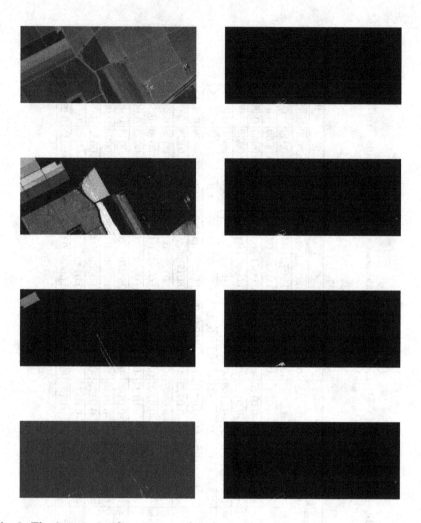

Fig. 2. The 8 most significant sparse abundance images for Salinas image. Significance decreases by rows and columns.

unmixing and the image regions identified with each class in the ground truth[1]. We compute all the possible spatial correlation coefficients between them, obtaining a matrix of correlation indices. The examination of this matrix gives information about the discovery of the ground-truth classes, which endmembers are associated to them and the uncertainty or ambiguity of this association. The Table 2 shows the spatial correlations between the endmember abundance image (columns) and the ground truth spatial region corresponding to each class (rows). It can be appreciated that the endmembers #1, #2, and #4 account for most of the positive correlations, so that they can be interpreted as some kind of

---

[1] We do not have knowledge of the ground truth endmembers.

representative endmembers of these classes, or even the abundance coefficients can be used as discriminant features for these classes. Endmember #1 appears to be related to soil and mineral features, covering much of the background class. The other endmembers correspond to vegetal covers.

The figure 2 depicts the abundance images of the first eight endmembers. They are sort by decreasing significance from top to bottom, left to right, so the most significant image is the top leftmost image. It can be appreciated that the first two endmembers have corresponding abundance images that contain most of the information in the image. The remaining abundances have small regions of interest with non-zero coefficients, or are merely marginal noise-like images. Therefore, our approach in this paper performs some kind of principal component selection on the endmembers.

5   Conclusions

The WM Algorithm proposed by Ritter at al. [4,7] is a fast procedure to obtain a set of affine independent vectors which are the vertices of a convex polytope covering the sample data. Applied to hyperspectral images, WM Algorithm produces a large set of candidate endmembers. We propose the application of an specific greedy sparsification algorithm following the gradient pursuit to compute sparse abundance images, producing a selection of the relevant endmembers in the image. Experiments on a well known hyperspectral image show that the approach works as some kind of non-linear principal component analysis, where the non-linearity is introduced by the WM algorithm, which a Lattice Computing base algorithm. Further work must look to other sparsification algorithms, trying to obtain locally meaningful sparse representations, following the idea that not all endmembers are relevant to the entire image, that is, endmembers will only be representing scene materials locally present in some regions.

Acknowledgments. This work has been supported by the MINECO grant TIN2011-23823, and UFI 11/07 of the UPV/EHU.

References

1. Keshava, N., Mustard, J.F.: Spectral unmixing. IEEE Signal Processing Magazine 19(1), 44–57 (2002)
2. Graña, M., Villaverde, I., Maldonado, J.O., Hernandez, C.: Two lattice computing approaches for the unsupervised segmentation of hyperspectral images. Neurocomput. 72(10-12), 2111–2120 (2009)
3. Graña, M., Veganzones, M.A.: An endmember-based distance for content based hyperspectral image retrieval. Pattern Recognition 49(9), 3472–3489 (2012)
4. Ritter, G.X., Urcid, G.: A lattice matrix method for hyperspectral image unmixing. Information Sciences 181(10), 1787–1803 (2010)
5. Graña, M., Veganzones, M.A.: Endmember induction by lattice associative memories and multi-objective genetic algorithms. EURASIP Journal on Advances in Signal Processing 2012, 64 (2012)

6. Blumensath, T., Davies, M.: Gradient pursuits. IEEE Transactions on Signal Processing 56(6), 2370–2382 (2008)
7. Ritter, G., Urcid, G.: Lattice algebra approach to endmember determination in hyperspectral imagery. In: Hawkes, P.W. (ed.) Advances in Imaging and Electron Physics, vol. 160, pp. 113–169. Academic Press, Burlington (2010)
8. Ritter, G.X., Sussner, P., Diaz-de-Leon, J.L.: Morphological associative memories. IEEE Transactions on Neural Networks 9(2), 281–293 (1998)

Dynamic Saliency from Adaptative Whitening

Víctor Leborán Alvarez[1], Antón García-Díaz[1,2], Xosé R. Fdez-Vidal[1],
and Xosé M. Pardo[1]

[1] Centro de Investigación en Tecnoloxías da Información (CITIUS),
University of Santiago de Compostela, Spain
[2] AIMEN Technology Center - O Porriño, Spain

Abstract. This paper describes a unified framework for the static and dynamic saliency detection by whitening both the chromatic characteristics and the spatio-temporal frequencies. This approach is grounded in the statistical adaptation to the input data, resembling the human visual system's early codification. Our approach, AWS-D, outperforms state-of-the-art models in the ability to predict human eye fixations while freely viewing a set of videos from three open access datasets (task free). We used as assessment measure an adaptation of the shuffling-AUC metric to spatio-temporal stimulus, together with a permutation test. Under this criterion, AWS-D not only reaches the highest AUC values, but also holds significant AUC figures for longer periods of time (more frames), over all the videos used in the test. The model also reproduces psychophysical results obtained in pop-out experiments in agreement with human behavior.

Keywords: Spatio-temporal saliency, visual attention, adaptive whitening, short-term adaptation, eye fixations.

1 Introduction

Selective visual attention allows primates to select an affordable amount of the incoming visual data that is most relevant to their ongoing behavior. As provided by this mechanism, visual attention is both attracted by visual stimuli (bottom-up) and biased by the task (top-down). Over the last decade a great deal of research effort has been devoted to the development of computational models of bottom-up saliency, mainly aimed both at trying to describe and/or predict the human visual system behaviour, and also at easing the computational burden in computer vision applications.

The literature on visual attention models is rapidly growing. Thorough reviews of computational models and computer vision applications can be found in [1,3,11]. Models can be classified depending on the basic hypothesis assumed [14]. Here we focus on dynamic saliency map models, a subgroup that has its roots in Feature Integration Theory [12] and the common link of these models is the presence of feature maps associated to some stimulus characteristics. Some of the first samples of these models are the Itti/Itti-CIOFM/Itti-Surprise [6,7], models based on previous biological descriptive models, that adds the movement

J.M. Ferrández Vicente et al. (Eds.): IWINAC 2013, Part II, LNCS 7931, pp. 345–354, 2013.

to the list of processed characteristics through the addition of movement conspicuity maps. Other models concentrate on the idea of interesting points or regions, like SUNDAY [17] which extracts spatio-temporal statistical information from natural scenes that is compared with the incoming information or the SEOD model [9] that uses the self-resemblance of each pixel to its surroundings to predict saliency locations. In addition there are models like GBVS [5] that employ dissimilarity and graph based algorithms to find an equilibrium distribution that raises to the saliency map. There are many others and both the count and the possible applications are growing every day.

Recently, we have proved the hypothesis that the short-term statistical adaptation to incoming visual data provides a good measurement of spatial saliency [4]. That approach has been demonstrated to outperform previous models in predicting the eye fixations of a group of humans, while freely viewing a set of images, under ROC (receiver operating characteristic) measure [1]. In this paper, we claim that the adaptive whitening of visual data makes up a unifying mechanism to saliency computation that combines color and other spatial and temporal features. The contributions of this paper are threefold: (i) we present a whole (static and dynamic) computational saliency model under a unifying theoretical framework, (ii) we improve existing methodology to measure the capability of dynamic saliency methods for predicting human fixations, by adding a permutation test that determines the time intervals with reliable measures. And (iii) we have built a new public available video database that contains 72 videos and the data of the 40.558 fixations of 22 humans, to overcome some deficiencies exhibited by other public databases. These deficiencies mainly come in three types: to few fixations per frame to provide reliable statistical measures, narrow diversity in video types (synthetic, natural, pop-out effects or different motion types), or the presence of competing stimuli, like emotional content or audio, that bias the observers attention through the interplay between bottom-up and top-down processes.

The rest of the paper is organized as follows: Section 2 outlines the main components of the AWS-D model and its bioinspired ground. Section 3 describes the characteristics of the novel open access eye-tracking dataset and the methodology employed to make the comparison with other state-of-the-art models, as well as visual and quantitative results. Finally, section 4 summarizes the main conclusions.

2   The AWS-D Model

Figure 1 depicts the whole framework of how adaptive whitening saliency unifies the processing of chromatic, spatial and temporal features. The common whitening procedure used on different stages of the AWS-D model mainly consists of two steps; first a decorrelation is done through PCA (principal component analysis) followed by a normalization by their variance, getting a whitened and still ordered representation of the incoming data [4]. Formally this procedure can be outlined as:

Let $\mathbf{X} = (x_{i1}, ..., x_{iM})$ be the original centered data representation, with i being the pixel index varying from 1 to N pixels (either $N = width * height$ both for chromatic and static scale whitening or $N = Width * Height * Block\_Size$ for dynamic scale whitening, see fig. 1) and M the number of input components (either three for the chromatic step or the number of scales both in the static and dynamic paths) and let $\Sigma = E\{\mathbf{X}^T\mathbf{X}\}$ be the covariance matrix.

Let $\mathbf{Z} = (z_{i1}, ..., z_{iM})$ be the corresponding representation in the whitened coordinates, then the whitening transformation can be expressed as:

$$\mathbf{Z} = \mathbf{X}\Phi\Lambda^{-1/2} \quad \text{with} \quad E\{\mathbf{Z}^T\mathbf{Z}\} = \mathbf{I} \tag{1}$$

being the columns of Φ the eigenvectors that diagonalize the covariance matrix and being Λ the autovalue diagonal matrix with $\Sigma\Phi = \Phi\Lambda$. Based on this definition of whitened components, the model computes data driven visual saliency for each frame in a video stream going through five main steps.

Input stage. To allow temporal processing, the video sequence is first divided in groups of frames (shots in figure 1) that are sequentially processed. Previous and posterior frames are added during the processing, and eliminated at the end, to minimize transition effects between blocks and Fourier wrap-around artifacts[1].

Chromatic stage. In this stage the AWS-D algorithm applies a chromatic decomposition and whitening of RGB (red, green and blue) color components for each frame (eq 1). The three whitenned color components $\mathbf{Z}^c, c \in \{1, 2, 3\}$ are the inputs to the next stages. This represents a short-term plausible contextual adaptation mechanism that provides an efficient representation of the image. It allows an efficient use of the available dynamic range, like the observed behavior of lateral geniculate nucleus (LGN) and visual area V1. From then on, like in the human visual system, the data flow divides in two parts, the magnocellular flow, responsible of the movement detection, and the parvocellular flow, responsible of the color and spatial coding processing.

Temporal path. In the dynamic part of the AWS-D model, the luminance related frames (the first component after the chromatic whitening, $\mathbf{Z}^1$) are processed in blocks of seven frames. After eliminating the zero-frequency plane, corresponding with static information, a 3D spatio-temporal filtering (with 41 uniform 3D distributed orientations and 6 scales) is performed over each block in the frequency domain. To that end a 3D FFT (three dimensional fast Fourier transform) is applied to each block (see fig 1) and the inverse Fourier transform (IFFT) is applied at the end of the filtering step.

The filter's transfer function $T = R(\rho; \rho_s) \cdot A(\alpha)$ detailed in eq 2 is designed in spherical coordinates, being R the radial and A the angular components. These filters only have analytical expression in the frecuency domain. The radial term, given by the log Gabor function, and the angular term are designed to achieve

---

[1] The optimal block size was observed to be seven frames similar to the previously value of five frames used by Tsotsos [13].

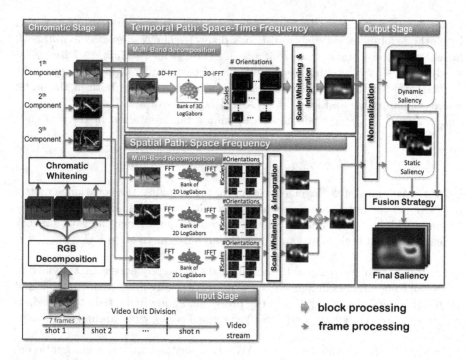

Fig. 1. A general diagram showing the data flow through the AWS-D model

high orientation selectivity [2]. That means selectivity in spatial orientation, speed and direction of motion.

$$T = \exp\left(-\frac{(log(\rho/\rho_s))^2}{2(log(\sigma_{\rho s}/\rho_s))^2}\right) \cdot \exp\left(-\frac{\alpha^2}{2\sigma_\alpha^2}\right) \qquad (2)$$

being (ρ, α) the frequency coordinates, α depends on the azimuthal, ϕ, and the elevation, θ, components. Where α represents the angular distance between the position vector $\mathbf{f}$ of a given point in the spectral domain and the filter direction $\mathbf{v}$ defined as $\alpha(\phi, \theta) = acos(\mathbf{f} \cdot \mathbf{v}/\|\mathbf{f}\|)$, being ρ_s the central frequency of the filter and s the scale index. The width parameters $\sigma_{\rho s}$ and σ_α are the radial and angular standard deviations. The convolution of the log-Gabor functions gives a complex response where the norm represents the local energy of the image analyzed by the log-Gabor filter.

A multi-scale whitening operation is applied to the resulting blocks (eq. 3) decorrelation is achieved by applying PCA over a set of multi-scale low level features and then calculating the scale contribution to the saliency as the squared distance of each feature vector associated to each image point to the center of the distribution (average feature vector of the scene). Finally the dynamic component of the saliency S^D is obtained through a σ width Gaussian smoothing, $G(Z, \sigma)$, and addition of the orientation maps.

$$S^D(x,y,t) = \sum_{O=1}^{N_{or}} \frac{G\left(\sum_{s=1}^{N_{scales}} \left(Z_{o,s}(x,y,t) - \bar{Z}_o(x,y,t)\right)^2, \sigma\right)}{\max\limits_{x,y\in\{1\cdots N_{pix}\}}\left\{G\left(\sum_{s=1}^{N_{scales}} \left(Z_{o,s}(x,y,t) - \bar{Z}_o(x,y,t)\right)^2, \sigma\right)\right\}}$$
(3)

being $\bar{Z}_o(x,y,t) = \frac{1}{N_{scales}} \sum_{s=1}^{N_{scales}} Z_{o,s}(x,y,t)$. The temporal path of the AWS-D model is based on the scale whitening of the spatio-temporal filter responses which resembles the behavior observed in V1, MT (middle temporal), MST (medial superior temporal) and 7a [13].

Spatial path. Across the spatial path, each of the three whitened chromatic components is decomposed into a multi-oriented, multi-resolution representation by means of a 2D Gabor-like bank of filters (with 7 scales and 4 orientations for the first component and, 5 scales and 4 orientations for the other two). The first component is filtered with a higher number of scales because of its richer spatial detail. Resembling the dynamic part of the model, an efficient measure of saliency is obtained by decorrelating the multi-scale responses and then calculating the scale contribution to the saliency as the distance to the center of the distribution [4]. The filter transfer function described in equation 4 is used with (ρ, β) being the polar frequency coordinates instead of the spherical coordinates of eq. 2 and being (ρ_s, β_o) the central frequency of the filter, s is the scale index and o the orientation index. These filters only have analytical expression in the frecuency domain, thereafter a FFT is aplied to the input images and the inverse (IFFT) is applied at the end of the filtering step.

$$T = \exp\left(-\frac{(log(\rho/\rho_s))^2}{2(log(\sigma_{\rho s}/\rho_s))^2}\right) \cdot \exp\left(-\frac{(\beta - \beta_o)^2}{2\sigma_{\beta_o}^2}\right)$$
(4)

Finally the static component of the saliency S^S is obtained through a σ width Gaussian smoothing of each component, $G(Z, \sigma)$, and addition of the orientations is performed analog to the dynamic path, being $Z_{o,s}$ on eq. 5 the whittened and normalized scale components for each orientation.

$$S^S(x,y,t) = \sum_{c=1}^{3} \sum_{O=1}^{N_{or}} \frac{G\left(\sum_{s=1}^{N_{scales}} \left(Z_{o,s}^c(x,y,t) - \bar{Z}_o^c(x,y,t)\right)^2, \sigma\right)}{\max\limits_{x,y\in\{1\cdots N_{pix}\}}\left\{G\left(\sum_{s=1}^{N_{scales}} \left(Z_{o,s}^c(x,y,t) - \bar{Z}_o^c(x,y,t)\right)^2, \sigma\right)\right\}}$$
(5)

The adaptive whitening of local scales with orientation specificity is based on contextual adaptation to spatial frequencies and orientations observed in visual areas V1 and V2 [4,13] and it uses the variability in local energy as a measure of saliency estimation [4].

Output stage. The goal of this stage is to normalize and fusion static, $S^S(x,y,t)$, and dynamic, $S^D(x,y,t)$, saliency maps of each video frame $t \in [1, N_{frames}]$. Normalization step (Eq. 6) scales both saliency maps to $[0,1]$ and thereafter, each pixel value represents the probability to receive attention, being w and h the frame witdth and height in pixels.

$$\bar{S}^k(x,y,t) = \frac{S^k(x,y,t)}{\sum\limits_{x=1}^{w}\sum\limits_{y=1}^{h} S^k(x,y,t)} \qquad \text{with} \quad k \in \{S, D\}. \qquad (6)$$

Fusion step (Eq. 7) is implemented by the following priority function:

$$S^P(x,y,t) = \begin{cases} \bar{S}^S(x,y,t) & \text{if } \frac{n_d}{n_s} > \varepsilon \\ \bar{S}^D(x,y,t) & \text{if } \frac{n_s}{n_d} > \varepsilon \\ \frac{1}{n_s}\bar{S}^S(x,y,t) + \frac{1}{n_d}\bar{S}^D(x,y,t) & \text{otherwise} \end{cases} \qquad (7)$$

where n_s and n_d are the number of maxima in static and dynamic maps, respectively, for each frame t. This step builds a final saliency map $S^P(x,y,t)$ through the competition between $\bar{S}^S(x,y,t)$ and $\bar{S}^D(x,y,t)$, depending on their numbers of local maxima (the lower the number of local maxima, the more conspicuous the maxima are). When both numbers are similar, the combined map is computed as the weighted sum of the normalized maps.

3   Capability of Predicting Human Fixations

Model performance can be assessed measuring its capability to predict spatio-temporal fixation patterns of humans. Human fixation patterns can be obtained using an eye tracker, while a set of humans freely view a collection of videos. ROC analysis is widely considered to yield the most reliable measure for benchmarking [1,15]. In that context, a saliency map is considered as the output of a binary classifier that distinguishes between fixations (TP) and non-fixations (FP) image points. Human fixations are the ground-truth, and the AUC (area under curve) of the ROC curve plotted for each frame is computed by thresholding the saliency map values rescaled to 0-255 range. As points close to image center are more often fixated by humans than others, Zhang [16,17] proposed the shuffling-AUC to cancel this center-bias in order to avoid a simple Gaussian centered model obtaining high AUC scores. In this work, we extended this methodology to cope with videos. The shuffling to obtain the FP's is done by a random selection of fixations on a frame from an also randomly selected video from the database. AUC scores are global (average of the AUC values for every frame of each video on the whole database), so the temporal behavior of the saliency model is not observable. We propose to compute a local AUC measure to assess the reliability of the global measure. Besides, a permutation test can deliver the periods of time (frames) with significant AUC scores. Together, we have a global AUC score plus a confidence interval for each saliency model ($p < 0.01$). Furthermore, we have compared the AWS-D model to some classic reference models like the Itti et all. [6,7], and some state of the art models, like GBVS[5], SEOD [9] and SUNDAY [17]. All the code of those models can be downloaded from their respective web pages. Figure 2 depicts some frames for one sample video compared with all the models. In addition, we have included in the comparison two base-line reference models. Firstly, the Gauss model represents the fit of all the fixations for the whole database to a Gaussian function.

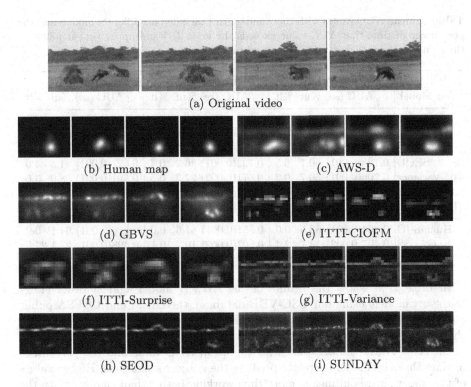

(a) Original video

(b) Human map

(c) AWS-D

(d) GBVS

(e) ITTI-CIOFM

(f) ITTI-Surprise

(g) ITTI-Variance

(h) SEOD

(i) SUNDAY

Fig. 2. Frames for video RCD-Lion4 part of the USC-VDB video database

It allows to check that the ROC has correctly compensated the center bias, giving an AUC value similar to 0.5. Secondly, the inter-observer human model (Human-IO), it measures the capability of a subset of humans to predict the fixations of other humans. It was created using the 50% of the subject's fixations at each instant to predict the fixations of all the subjects. It represents a reference point to compare the computational models with the results obtained from human-responses.

The comparison between the subject data and the models has been done with the public database DIEM[2], the UCFSA database (UCF Sports action dataset) [8,10] and the newly created video database USC-VDB. It contains the eye fixations of 22 subjects aged 11-43 years while freely viewing 72 videos. The SMI EyeTracker and BeGaze[TM] software were used to record eye fixations for both eyes while subjects free viewed some videos. Those videos were classified in four categories obtained from both synthetic and natural scenes, with static or movement camera. Fixation information and videos are freely available to download[3].

---

[2] This database is part of the a DIEM project (Dynamic Images and Eye Movements).

[3] All the videos and fixation data for the USC-VDB can be downloaded at http://www-gva.dec.usc.es/persoal/xose.vidal/code.php?person=xose

Table 1. Mean AUC values with the standar error included for different models and the percentage of time that AUC value exceeds the lower (lt) and upper (ut) significance threshold level calculated using the permutation test

Test Model	USC-VDB AUC (se) %ut %lt			UCFSA-DB AUC (se) %ut %lt			DIEM-DB AUC (se) %ut %lt		
AWS-D	**0.801**(0.001)	76.0	0.1	**0.745**(0.001)	47.2	0.1	**0.659**(0.001)	58.9	4.6
GBVSm	0.758(0.001)	63.7	0.6	0.715(0.002)	42.6	0.7	0.628(0.001)	51.3	6.4
SUNDAY	0.757(0.001)	63.8	0.1	0.726(0.001)	41.0	0.2	0.651(0.001)	55.6	2.1
SEOD	0.728(0.001)	60.1	0.5	0.733(0.001)	46.7	0.4	0.621(0.001)	41.5	4.0
Itti-Variance	0.706(0.001)	50.7	0.2	0.671(0.002)	27.3	0.5	0.591(0.001)	38.0	3.6
Itti-Surprise	0.702(0.001)	52.2	1.0	0.674(0.002)	33.7	0.9	0.606(0.001)	44.8	6.4
Itti-CIOFM	0.677(0.001)	47.0	1.8	0.656(0.002)	33.6	1.7	0.605(0.001)	46.0	5.7
Human-IO	0.854(0.001)	88.5	0.0	0.851(0.001)	81.6	0.0	0.816(0.001)	94.1	0.0
GAUSS	0.497(0.002)	13.7	12.4	0.502(0.002)	10.7	10.1	0.498(0.001)	27.1	24.6

In table 1 we show the results of the AWS-D model against other state-of-the-art models using our USC-VDB database, the DIEM and UCFSA public databases. The Gauss model gives a mean AUC value similar to 0.5 meaning that the center bias compensation works fine with all the tested databases. Results of the AWS-D are the best, both for the mean AUC value and the percentage of time that the model correctly predicts the subject's fixations. Higher values of the upper threshold means more time working better than chance. With the USC-VDB and also with the UCFSA-DB the values obtained by all the models are much similar to the Human-IO model than those obtained with the DIEM database. This fact confirms that the subject's visual patterns are more correlated. A possible explanation is that the stimulus (video contents) selection and all the experimentation carried on the building of our database managed to avoid the influence of top-down effects among humans, which is not the objective of other public-access databases. Also, the DIEM database includes audio during presentation, and that characteristic is not included in the saliency processing. All the parameters for all the tested models has been kept the same for all the videos of the tested databases.

3.1   Reproduction of Psychophysical Results

Psychological patterns are widely used in attention experiments not only to explore the mechanism of visual search, but also to test effectiveness of saliency maps [1,9,13]. The AWS-D model solves classical examples of search asymmetry, linearity against corner angle and other previously described effects with static scenes [4]. In addition, the AWS-D model is able to capture successfully perceptual differences where the movement competes with other features.

In figure 3 we can see four samples of this experiments: 3a) one object changes the movement direction, 3b) one object with different color competes for the saliency with one moving object, 3c) some objects form a group generating

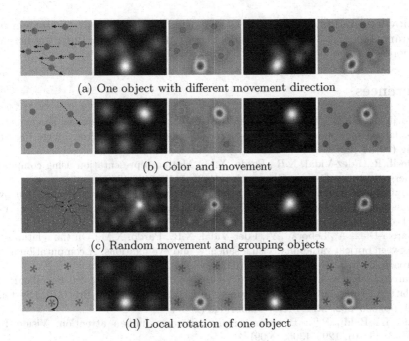

(a) One object with different movement direction

(b) Color and movement

(c) Random movement and grouping objects

(d) Local rotation of one object

Fig. 3. Different characteristics changing. From left to right on each row, the original frame, the map for the AWSD model alone and superimposed, the human fixation map and superimposed to the original frame.

saliency as soon they are grouped and 3d) one object rotating around itself generates saliency immediately.

4 Conclusions

We have proposed a unified framework for both static and dynamic saliency detection based on the short-term statistical adaptation to the input data. The AWS-D model can predict human fixations for longer periods of time with more accuracy than the other state-of-the-art algorithms tested in terms of ROC metric, reaching a percentage of significant time similar to which humans obtain predicting other human fixations. The AWS-D model can detect salient movements on video sequences and the static component of the model can identify salient points and regions in static images. It successfully captures perceptual differences in cases where the movement competes with other features. The proposed methodology to evaluate the human fixations over video sequences offers a complete tool that allows us to measure a center-bias free global ROC. Additionally, it allows us to evaluate the temporal statistical reliability, determined by confidence intervals.

The open source USC-VDB database provides a benchmark with increased scene variability (i.e. statistical variability) intended to challenge the models by minimizing biases. Moreover, the higher number of observers (i.e. of fixations) strengths the reliability of comparisons between models.

Acknowledgments. This work was supported by the Spanish government under grant TIN2012-32262. We would like to thank Paul L. Bach for helpfull comments.

References

1. Borji, A., Sihite, D., Itti, L.: Quantitative analysis of human-model agreement in visual saliency modeling: A comparative study. IEEE Trans. on Image Processing (99), 1 (2012)
2. Dosil, R., Fdez-Vidal, X.R., Pardo, X.M.: Motion representation using composite energy features. Pattern Recognition 41(3), 1110–1123 (2008)
3. Frintrop, S., Rome, E., Christensen, H.I.: Computational visual attention systems and their cognitive foundations: A survey. ACM Trans. Appl. Percept. 7(1), 6:1–6:39 (2010)
4. García-Díaz, A., Leborán, V., Fdez-Vidal, X.R., Pardo, X.M.: On the relationship between optical variability, visual saliency, and eye fixations: A computational approach. Journal of Vision 12(6) (2012)
5. Harel, J., Koch, C., Perona, P.: Graph-based visual saliency. In: Schölkopf, B., Platt, J., Hoffman, T. (eds.) Advances in Neural Information Processing Systems 19, pp. 545–552. MIT Press, Cambridge (2007)
6. Itti, L., Baldi, P.F.: Bayesian surprise attracts human attention. Vision Research 49(10), 1295–1306 (2009)
7. Itti, L., Koch, C., Niebur, E.: A model of saliency-based visual attention for rapid scene analysis. IEEE Trans. on PAMI 20(11), 1254–1259 (1998)
8. Mathe, S., Sminchisescu, C.: Actions in the eye: Dynamic gaze datasets and learnt saliency models for visual recognition. Tech. rep., Institute of Mathematics of the Romanian Academy and University of Bonn (2012)
9. Seo, H.J., Milanfar, P.: Static and space-time visual saliency detection by self-resemblance. Journal of Vision 9(12) (2009)
10. Mathe, S., Sminchisescu, C.: Dynamic eye movement datasets and learnt saliency models for visual action recognition. In: Fitzgibbon, A., Lazebnik, S., Perona, P., Sato, Y., Schmid, C. (eds.) ECCV 2012, Part II. LNCS, vol. 7573, pp. 842–856. Springer, Heidelberg (2012)
11. Toet, A.: Computational versus psychophysical bottom-up image saliency: A comparative evaluation study. IEEE Trans. on PAMI 33(11), 2131–2146 (2011)
12. Treisman, A.M., Gelade, G.: A feature-integration theory of attention. Cognitive Psychology 12(1), 97–136 (1980)
13. Tsotsos, J.K., Pomplun, M., Liu, Y., Martinez-Trujillo, J.C., Simine, E.: Attending to motion: Localizing and classifying motion patterns in image sequences. In: Bülthoff, H.H., Lee, S.-W., Poggio, T.A., Wallraven, C. (eds.) BMCV 2002. LNCS, vol. 2525, pp. 439–452. Springer, Heidelberg (2002)
14. Tsotsos, J.K.: A Computational Perspective on Visual Attention. MIT Press (2011)
15. Wilming, N., Betz, T., Kietzmann, T.C., König, P.: Measures and limits of models of fixation selection. PLoS ONE 6(9), e24038 (2011)
16. Zhang, L., Tong, M.H., Marks, T.K., Shan, H., Cottrell, G.W.: Sun: A bayesian framework for saliency using natural statistics. Journal of Vision 8(7) (2008)
17. Zhang, L., Tong, M.H.: W, G.: Sunday: Saliency using natural statistics for dynamic analysis of scenes. In: Proc. of the Thirty-First Annual Cognitive Science Society Conference (2009)

PIR-Based Motion Patterns Classification for AmI Systems

Francisco Fernandez-Luque, Juan Zapata, and Ramón Ruiz

Depto. Electrónica, Tecnología de Computadoras y Proyectos
ETSIT- Escuela Técnica Superior de Ingeniería de Telecomunicación
Universidad Politécnica de Cartagena
Antiguo Cuartel de Antigones. Plaza del Hospital 1, 30202 Cartagena, Spain
{ff.luque,juan.zapata,ramon.ruiz}@upct.es
http://www.detcp.upct.es

Abstract. The analysis of human motion has become a major application area in computer vision. The vast majority of applications require using video cameras as main sensor for acquisition, analysis and detection of human motion. Some applications are very sensitive to the acquisition of images of people that may feel violated his right to privacy. Especially in ambient intelligence applications, non intrusive nature, whose scope is the user's home. The scope of this work was constrained to the analysis of human motion obtained from signals acquired for PIR sensors. We propose a method to classify PIR signals regarding two parameters: speed and distance from sensor to the subject. Our method allows to get compact devices that perform local computation and are able to transmit middle abstraction level context information in an efficient way. Signals from a PIR sensor array generated by several subjects have been registered to get patterns. Subjects moved at a certain speed at a given distance from sensors, so patterns are indexed by these two parameters. The patterns are then used to classify other tests signals regarding the mentioned parameters. The hit rate, for the method which combines information from all the sensors in the array, results in 79% for distance classification, 96% for speed classification and 77% for classification regarding both parameters simultaneously. We consider that this is a satisfying result for a non intrusive method.

Keywords: pattern recognition, PIR, signal transforms, AmI, ubiquitous monitoring, WSN.

1 Introduction

Recent computational and electronic advances have increased the level of autonomous semi-intelligent behaviour exhibited by systems so much that new terms like ambient intelligence (AmI) [1], started to emerge. Step by step the personal computer is becoming practically obsolete because computing access can be everywhere: in the environment, on human body, and in things lying about to be used as needed [2]. This paradigm is also described as pervasive

J.M. Ferrández Vicente et al. (Eds.): IWINAC 2013, Part II, LNCS 7931, pp. 355–364, 2013.
© Springer-Verlag Berlin Heidelberg 2013

computing [3] or, more recently, everyware [4], where each term emphasises slightly different aspects. When primarily concerning the objects involved, it is also physical computing, the Internet of Things, haptic computing, and things that think.

As computing becomes more ubiquitous and distributed, healthcare systems, in particular for the elderly, have attracted enormous attention worldwide [5]. Caring for the elderly people cause that much of the social resources in the national health insurance system are aimed at improving the quality of life of these people. A data only as an example, the estimated expenditure on medical care for the falls suffered by elderly residents and their related injuries will reach 32 billion euros in 2020 [6]. There is no doubt that elderly healthcare is one of the main items of expenditure for social security systems therefore it is imperative for social security and healthcare systems to take advantage of the prevailing assistive technologies [7], such as embedded devices (EDs), wireless sensor networks (WSNs), human-computer interaction (HCI), ambient intelligence (AmI) and ubiquitous computing (UC).

By embedding sensors, ubiquitous computation, and wireless communication into ordinary things, future computing applications can adapt to human users rather than the other way around. However, it is currently difficult to develop this type of ubiquitous computing in these ordinary things because of the lack of devices integrating both the required hardware and software. In this sense, we are creating a class of small computers equipped with wireless communication and sensors-actuators to make it possible to create low power smart objects.

A passive infrared sensor (PIR sensor) is an electronic sensor that measures infrared (IR) light radiating from objects in its field of view without generating any energy for this purpose. The basic principle of operation is as follows: all objects emit energy (if they are situated above 0 K, or absolute zero) in the form of infrared radiation. Infrared radiation enters through the front of the electronic device which conforms the sensor. The core of a PIR sensor is a solid state transducer in the form of a thin film (or set of transducers), made from pyroelectric materials. Pyroelectric material generates a temporary voltage when is exposed to heat. The change in temperature modifies the positions of the atoms slightly within the crystal structure of pyroelectric material causing the polarization of the material changes. This polarization change gives rise to a voltage across the crystal. If the temperature remains constant at its new value, the pyroelectric voltage gradually disappears due to leakage current. Materials commonly used in PIR sensors include gallium nitride (GaN), caesium nitrate (CsNO3), polyvinyl fluorides, derivatives of phenylpyridine, and cobalt phthalocyanine. The sensor is manufactured as an integrated circuit based on pyroelectric transducers.

In this way, PIR sensors are used as a motion sensor. A PIR sensor can detect abrupt changes in temperature at a given point. When an object, such as a human, passes in front of the field background of view then the temperature at that point will change abruptly. When the object goes out from the field background of view then the temperature change back again. Any quick change can trigger the detection, although moving objects of identical temperature will not trigger a detection.

The motion sensor can be enhanced by means of additional electronics. Pairs of sensor elements may be employed as opposite inputs to a differential amplifier. In such a configuration, the PIR measurements cancel each other so that the average temperature of the field of view is removed from the electrical signal. In this way, an increase of infrared energy across the entire sensor is self-cancelled and will not trigger the device. This allows the device to resist false indications of change in the event of being exposed to brief flashes of light or field-wide illumination. At the same time, this differential arrangement minimizes common-mode interference, allowing the device to resist triggering due to nearby electric fields.

Different mechanisms can be used to focus the distant infrared energy onto the sensor surface. The most common mechanism have numerous Fresnel lenses although some large PIR sensor are made with a single Fresnel lens.

The analysis of human motion has become a major application area in computer vision. This development has been driven by the many interesting applications that lie ahead in this area and the recent technological advances involving the real-time acquisition, transfer, and processing of images on widely available low-cost hardware platforms. A number of promising application scenarios have appeared for this field: virtual reality, surveillance systems, advanced user interfaces and so on. However, the vast majority of applications require using video cameras as main sensor for acquisition, analysis and detection of human motion. But some applications are very sensitive to the acquisition of images of people that may feel violated his right to privacy. Especially in ambient intelligence applications, non intrusive, nature whose scope is the user's home. A clear example of this type of system to be particularly intrusive are systems based Ambient Assisted Living (AAL) applications. The scope of this work was limited to the analysis of human motion obtained from signals acquired for PIR sensors, and its main objective is to determine the velocity and distance of a person to go through the field of view of a PIR sensor.

Related work is presented in Section 2. It includes work related to classification of motion signals generated by PIR sensors. The proposed classification method is presented in Section 3. Results in terms of hit rate are shown in Section 4. Conclusions and possible future work lines are referenced in Section 5.

2   Related Work

PIR sensors are widely used for basic motion detection. Some advances can be found with the aim of getting richer information from these sensors, but this research field has not been yet fully exploited and no much literature has been found.

Byunghun Song et al. analyse the performance and the applicability of the PIR sensors for security systems [8]. They propose a region-based human tracking algorithm with actual implementation and experiment in real environment, and show that their human tracking algorithm based on the PIR sensors performs very well with proper sensor deployment. Ren C. Luo et al. propose an indoor monitoring system based on wireless and pyroelectric infrared (WPIR) sensory fusion system which can be embedded and integrated with the traditional fire/smoke detector where they are usually installed on the ceiling by the use of pyroelectric motion sensor and low power wireless communication device [9]. Basically, each of radio frequency (RF) localization system and pyroelectric infrared (PIR) monitoring system provides the coarse information of individuals' location respectively. They developed a sensor network based localization method called WPIR intersection algorithm to determine the fused position from PIR and radio frequency signal localization system which utilizes received signal strength (RSS) model. Jian-Shuen Fang et al. propose a real-time human identification system using a pyroelectric infrared (PIR) detector array and hidden Markov models (HMMs) [10]. A PIR detector array with masked Fresnel lens arrays is used to generate digital sequential data that can represent a human motion feature. HMMs are trained to statistically model the motion features of individuals through an expectation-maximization (EM) learning process. Human subjects are recognized by evaluating a set of new feature data against the trained HMMs using the maximum-likelihood (ML) criterion. Tong Liu et al. present a distributed direction-sensitive infrared sensing approach for fall detection in elderly healthcare applications [11]. Pyroelectric infrared (PIR) sensors are employed in sensing human activities. For capturing the characteristics of human normal and abnormal activities, three modules of a direction-sensitive PIR sensor are organized using a distributed sensing structure. The advantage of using the distributed sensing paradigm is that the synergistic motion patterns of head, upper-limb and lower-limb can be efficiently encoded and thus the more discriminative features can be captured. This is the new consideration of using PIR sensors in building a full detection system. In addition, a two-layer hidden Markov model is developed for recognizing a fall event based on the multidimensional signals of the distributed infrared sensing system. Experimental studies are conducted to validate the proposed method.

We propose a method to classify PIR signals regarding two parameters: speed and distance from sensor to the subject. These are generalist parameters with potential application in most of AmI systems. The proposed method uses information from an array of sensors which are not needed to be disposed using an specific distribution, simply as close as possible. Other proposed methods need from PIR sensors to be positioned in large vertical arrays or distributed in the room. Vertical arrays are not easy to fit in a real deployment, while sensors distributed in the room often need from high communication rates to gather information from several sensors before computation. This is a problem for applications where low-power wireless sensors are recommended. Our method allows to get compact devices that perform local computation and are able to transmit middle abstraction level context information in an efficient way.

3   Methodology

We propose a method to classify PIR signals regarding two parameters: speed and distance from sensor to the subject. Signals from PIR sensors generated by several subjects have been registered to get patterns. Subjects moved at a certain speed at some distance from sensors, so patterns are indexed by these two parameters. The patterns are then used to classify other tests signals regarding the mentioned parameters.

3.1   Signal Capture

A sensor array has been designed using 5 different PIR sensors (S1 to S5). Main components in a PIR sensor are the Fresnel lens and the piroelectric device. Sensor S1 is assembled using the Fresnel lens GLOLAB FL-65 and the piroelectric device taken from the Panasonic AMN23111 sensor. Sensor S2 is assembled using the same Fresnel lens (FL-65) and an own built piroelectric device. This device has been designed using the piroelectric transducer MURATA IRA-E712ST3 and an own implementation of a band pass filter-amplifier, typically used for these sensor devices. The filter is centred at $10\,\mathrm{Rad/s}$ ($1.59\,\mathrm{Hz}$), with gain $G = 250$ and quality factor $Q = 0.5$. The gain of the amplifier has been tuned to avoid signal saturation during the tests. Sensors S3 is assembled using the piroelectric device from AMN23111 and Fresnel lens MURATA IML0636. Sensor S4 is the AMN23111 with its original Fresnel lens. Finally, S5 is assembled using the piroelectric device from AMN23111 and Fresnel lens MURATA PPGI0626A+B. All the 5 sensors are fed at $5\,\mathrm{V}$ and their output signals are bounded between $0\,\mathrm{V}$ and $5\,\mathrm{V}$, centred about $2.5\,\mathrm{V}$. Sensors are not needed to be disposed using an specific distribution, simply as close as possible. A picture of the sensor array is shown in Figure 1, where they are identified by labels from S1 to S5.

Sensors data have been sampled by means of a data acquisition board model NI PCI-6035E. The National Instruments PCI-6035E data acquisition board gets up to $200\,\mathrm{kS/s}$ sampling and 16-bit resolution with 16 single-ended or eight differential analog inputs. Sensor data was sampled at 100 samples per second, which is enough because PIR devices incorporates a pass band filter centred at $10\,\mathrm{Rad/s}$ ($1.59\,\mathrm{Hz}$) and quality factor $Q = 0.5$.

To perform a test, the subject moves following a straight line in front of the sensorsThe route is perpendicular to the central axis of the sensor vision range, at a certain distance and speed. Sensors are placed $1\,\mathrm{m}$ above the ground.

A mobile electromechanical device has been built in order to guide the subject and mark his speed. This mobile device moves on rails and its engine is fed with a pulse-width modulation signal (PWM). PWM is a commonly used technique for controlling power to inertial electrical devices. The average value of voltage fed to the load is controlled by turning the switch between supply and load on and off at a fast pace. The longer the switch is on compared to the off periods, the higher the power supplied to the load is. The mobile device incorporates a PID (proportional-integral-derivative) controller for speed and position. Feedback for the controller is solved by a rotatory electromechanical encoder coupled to the axle of a wheel. Therefore, the mobile device responds to consigns with the desired position to reach and the desired speed. To perform a test, the subject just have to follow the mobile device. Figure 1 shows a picture of the testing bench; composed by the PIR sensor array, marked by a circle, and the controlled mobile device, marked by a square.

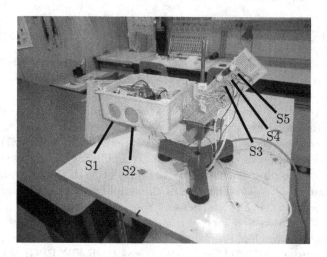

Fig. 1. PIR sensor array. Sensors are identified by labels from S1 to S5

A Matlab application manages the test procedure. It generates the consigns for speed and position of the mobile device and performs the sampling of the data from the sensor array. Data are automatically structured, indexed and stored by the application.

Tests have been performed at different distances between the sensor array and the middle of the subject route. Specifically tests have been carried out at distances from 0.5 m to 3.5 m with 0.5 m steps. These distance values have been considered of interest because they are typical in an indoor situation. Test have been carried out at speed values from 200 mm/s to 800 mm/s, with 200 mm/s steps, for each distance value. These speeds have been considered representative of the motion in people of all ages in indoor locations. For each combination of both parameters, the test was repeated 4 times.

Thus, a data base has been generated from motion traces generated by the sensor array for a set of subjects.

Data set generated by every subject is composed by signals registered by every one of the 5 sensors, from 4 repetitions for every combination of distance and speed. In total: 5 sensors ×4 repetitions ×7 distances ×4 speeds = 560 signals.

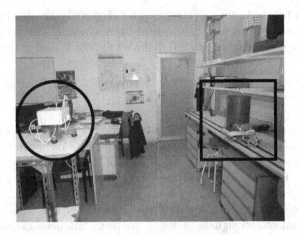

Fig. 2. Testing bench for PIR signal capture. Inside the circle, the PIR sensor array. Inside the square, the controlled mobile device.

3.2 Pattern Generation and Signal Classification

Signals from the 4 repetition of every test are fused to get a pattern set composed by 28 signals for every subject and every sensor. Patterns are obtained using the mean of the signals calculated sample by sample. Signals are previously aligned by means of cross-correlation to avoid shifts.

Once pattern sets are obtained, it is possible to classify signals generated by one subject using the pattern set generated by other subject. Classification is performed regarding the likeness between the test signal and every pattern in the set. The likeness between the test signal and one pattern signal is quantified by means of correlation as follows: (1) DC component is removed from both signals. (2) Signals are aligned using the shift which gets highest correlation value in the cross-correlation operation. (3) This correlation value is multiplied by the result of correlation between the FFT (fast Fourier transform) of both aligned signals. (4) Resulting value is multiplied by the result of correlation between the derivative of the aligned signals. (5) Resulting value is multiplied by the result of 2-dimensional (2-D) correlation between the Wigner-Ville distribution of the aligned signals. (7) Resulting value is stored as the likeness between the test signal and the pattern signal. This procedure is performed for the signals from the 5 sensors. Test signals from one sensor are compared only with patterns generated by the same sensor.

An likeness array is generated for the test signal, for every sensor, containing the likeness value with respect to 28 pattern signals from the same sensor. The signal is classified with the distance and speed parameters of the pattern with the highest likeness value. This classification is performed for every sensor separately. Finally, the likeness array of the different sensors are multiplied element by element, obtaining a single likeness array where information from all the 5 sensors is fused. Again, the signal is classified with the distance and speed parameters of the pattern with the highest likeness value. Results are obtained for every sensor separately (S1 to S5) and also for the fusion of them (SFus).

4   Results

The aim of this study is to obtain a method for classifying a motion trace, or signal, with respect to the subject's speed and distance from the sensor. This is achieved by comparing the motion trace with of motion patterns generated by other subjects. Test sets were generated from 3 subjects, in order to carry out the validation of the method. These test sets include 4 replicates for each of the 28 considered combinations of speed and distance. 112 signals for each of the 5 sensors, a total of 560 signals. A set of patterns has been obtained for each individual and each sensor, consisting of 28 signals, one for each combination of speed and distance. Figure 3 illustrates graphics for the set of patterns obtained for an individual and a given sensor. Speed is indexed by columns (A to D), from 200 mm/s to 800 mm/s in steps of 200 mm/s. Distance is indexed by rows (1 to 7), from 0.5 m to 3.5 m in steps of 0.5 m.

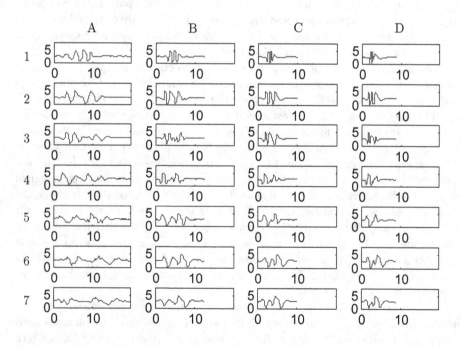

Fig. 3. Set of patterns obtained for an individual and a given sensor, consisting of 28 signals, one for each combination of speed and distance. Speed is indexed by columns (A to D), from 200 mm/s to 800 mm/s in steps of 200 mm/s. Distance is indexed by rows (1 to 7), from 0.5 m to 3.5 m in steps of 0.5 m.

We proceeded to classify each of the signals from the test set for each individual on the pattern sets of the other subjects. The signals generated by a given sensor is compared only with the patterns generated by the same sensor. Classification is performed for every sensor separately (S1 to S5) and using fused likeness information from the 5 sensors (SFus), in the way exposed in Setcion 3.2.

The performance of the method was quantified by the hit rate in the classification of the signals of the study database. Hit rate is calculated for the classification with respect to the distance parameter, for classification with respect to the speed parameter, and finally for classification with respect to both parameters. Table 1 shows the hit rate of the classification method for different sensors individually (S1 to S5) and by the method which uses fused information from all sensors (SFus).

Table 1. Hit rate of the method for classification regarding the parameters separately and simultaneously. Rows show the hit rate for given sensors (S1 to S5) and for the method which uses fused information from all of them (SFus). Columns show hit rate regarding the distance from the subject to the sensor, the hit rate regarding the speed and the hit rate regarding both parameters.

Sensor	Classif. Parameter		
	Distance	Speed	Both
S1	34%	97%	33%
S2	70%	98%	70%
S3	63%	83%	58%
S4	75%	90%	71%
S5	60%	84%	56%
SFus	79%	96%	77%

5   Conclussions and Future Work

A method has been developed for classification of PIR sensor signals regarding speed and distance between the subject and a sensor array. This method allows to get compact devices that perform local computation and are able to transmit middle abstraction level context information in an efficient way. The hit rate, for the method which combines information from all the sensors in the cluster, results in 79% for distance classification, 96% for speed classification and 77% for classification regarding both parameters simultaneously. We consider that this is a satisfying result for a non intrusive method.

A possible immediate application of this method could be to identify age groups at home. To this end, the array of sensors would be placed on the ceiling of a room, next to the door. Information about distance from the sensor to the subject would allow to estimate the individual's height. This information can be a first approach to the age group, while speed information would allow to refine the classification. For instance, an adult is usually high and moves fast, while an elderly is usually medium height and moves slowly. Finally, a child is usually low height and moves fast.

Current and future research includes: (1) Expansion of the population of test subjects. (2) Fusion of signals from different subjects in order to obtain richer patterns. (3) Reduction of the dimensionality of the classification problem by means of principal components-based techniques to decrease redundancy. (4) Adaptation of the method for its implementation on a microcontroller or DSP (digital signal processor).

Acknowledgment. This work was supported by the Spanish Ministry of Ciencia e Innovación (MICINN) under grant TIN2009-14372-C0302 and for the Fundación Séneca under grant 15303/PI/10.

References

1. Cook, D.J., Augusto, J.C., Jakkula, V.R.: Ambient intelligence: Technologies, applications, and opportunities. Pervasive and Mobile Computing 5(4), 277–298 (2009)
2. Weiser, M.: Ubiquitous computing. Computer 26, 71–72 (1993)
3. Hansmann, U., Merk, L., Nicklous, M.S., Stober, T.: Pervasive Computing: The Mobile World. Springer (August 2003)
4. Greenfield, A.: Everyware: The Dawning Age of Ubiquitous Computing, 1st edn. New Riders Publishing (March 2006)
5. Cortés, U., Urdiales, C., Annicchiarico, R.: Intelligent healthcare managing: An assistive technology approach. In: Sandoval, F., Prieto, A.G., Cabestany, J., Graña, M. (eds.) IWANN 2007. LNCS, vol. 4507, pp. 1045–1051. Springer, Heidelberg (2007)
6. Brewer, K., Ciolek, C., Delaune, M.F.: Falls in community dwelling older adults: Introduction to the problem. APTA Continuing Education Series, pp. 38–46 (July 2007)
7. Fernández-Luque, F., Zapata, J., Ruiz, R., Iborra, E.: A wireless sensor network for assisted living at home of elderly people. In: Mira, J., Ferrández, J.M., Álvarez, J.R., de la Paz, F., Toledo, F.J. (eds.) IWINAC 2009, Part II. LNCS, vol. 5602, pp. 65–74. Springer, Heidelberg (2009)
8. Song, B., Choi, H., Lee, H.S.: Surveillance tracking system using passive infrared motion sensors in wireless sensor network. In: International Conference on Information Networking, ICOIN 2008, pp. 1–5 (January 2008)
9. Luo, R., Chen, O., Lin, C.W.: Indoor human monitoring system using wireless and pyroelectric sensory fusion system. In: 2010 IEEE/RSJ International Conference on Intelligent Robots and Systems (IROS), pp. 1507–1512 (October 2010)
10. Fang, J.S., Hao, Q., Brady, D.J., Guenther, B.D., Hsu, K.Y.: Real-time human identification using a pyroelectric infrared detector array and hidden markov models. Optics Express 14(15), 6643–6658 (2006)
11. Liu, T., Guo, X., Wang, G.: Elderly-falling detection using distributed direction-sensitive pyroelectric infrared sensor arrays. Multidimensional Systems and Signal Processing 23, 451–467 (2012)

Motor Imagery Classification for BCI Using Common Spatial Patterns and Feature Relevance Analysis

Luisa F. Velásquez-Martínez, A.M. Álvarez-Meza, and C.G. Castellanos-Domínguez*

Universidad Nacional de Colombia sede Manizales,
Signal Processing and Recognition Group, Universidad Nacional de Colombia, Campus La
Nubia, km 7 via al Magdalena, Manizales-Colombia.
{lfvelasquezma,amalvarezme,cgcastellanosd}@unal.edu.co

Abstract. Recently, there have been many efforts to develop Brain Computer
Interface (BCI) systems, allowing to identify and discriminate brain activity. In
this work, a Motor Imagery (MI) discrimination framework is proposed, which
employs Common Spatial Patterns (CSP) as preprocessing stage, and a feature
relevance analysis approach based on an eigendecomposition method to identify
the main features that allow to discriminate the studied EEG signals. The CSP
is employed to reveal the dynamics of interest from EEG signals, and then we
select a set of features representing the best as possible the studied process. EEG
signals modeling is done by feature estimation of three frequency-based and one
time-based. Besides, a relevance analysis over the EEG channels is performed,
which gives to the user an idea about the channels that mainly contribute for the
MI discrimination. Our approach is tested over a well known MI dataset. Attained
results (95.21 ± 4.21 [%] mean accuracy) show that presented framework can be
used as a tool to support the discrimination of MI brain activity.

Keywords: Motor Imagery, Common Spatial Patterns, Feature Relevance
Analysis.

1 Introduction

The electroencephalography (EEG) is the most commonly employed method for monitoring brain activity and it has been used for several applications, such as: epilepsy
detection, analysis of cognitive behaviors, game controlling, among others. Brain Computer Interfaces (BCI) take advantage of the extracted information from EEG signals
to establish a direct communication channel between the human brain and the machine [1]. BCI is used to help people with disability by means of the analysis of the
human sensorimotor functions, which are based on the paradigm in cognitive neuroscience named as Motor Imagery (MI). However, the analysis of the EEG signals requires to develop suitable preprocessing, feature representation, feature selection, and
classification methodologies to improve the performance of real-world BCI applications. Regarding to the preprocessing stage, Common Spatial Pattern (CSP) is a popular algorithm for MI-based BCI systems [1]. CSP method constructs spatial filters that

---

* Under grants provided by a Msc. scholarship funded by Universidad Nacional de Colombia,
and by a PhD. scholarship and the project 111045426008 funded by Colciencias.

J.M. Ferrández Vicente et al. (Eds.): IWINAC 2013, Part II, LNCS 7931, pp. 365–374, 2013.
© Springer-Verlag Berlin Heidelberg 2013

maximize the variance of one kind of task and simultaneously minimize the variance of another. In order to achieve high classification accuracy, a pre-filtered broad band or subject-specific frequency bands are fixed to highlight the dynamics of interest. To find such optimal bands, several algorithms have been proposed, such as: Common Spatio-Spectral Pattern, Sub-band CSP and Filter Bank Common Spatial [1, 2]. In [1] the CSP preprocessing stage is matched with an Empirical Mode Decomposition (EMD) based method to select informative frequency bands from EEG. However, a direct and an automatic framework that allows to find such bands of interest is still an open issue.

Now, with respect to feature representation methodologies for BCI systems, the attributes are estimated by different methods such as Adaptive Autoregressive (AAR) coefficients, Hjorth parameters, Power Spectral Density (PSD), and continuous and discrete wavelet transforms (CWT and DWT) [3]. Although, many features may be extracted from different methods, several features may not contain relevant information introducing redundancy. Therefore, it is necessary to find a subset of attributes that preserving, as well as possible, the input data variability, allows to identify the most important information that helps to recognize different classes from EEG data. Several approaches have been used to identify the relevance of the computed features in BCI systems [3, 4]. Nevertheless, most of these feature selection methods are computationally expensive and they are not able to find directly a measure that relates each feature with its discriminative contribution.

In this work, an MI discrimination framework is proposed, which employs CSP as preprocessing stage, and a feature relevance analysis approach based on an eigendecomposition method to identify discriminative features. The CSP is matched with EMD to reveal the dynamics of interest. Then, we select a set of features representing the best as possible the studied process. For such purpose, a variability analysis is presented to identify relevant features. EEG signals modeling is done by feature estimation of three frequency-based and one time-based. Moreover, a relevance analysis over the EEG channels is performed, which gives to the user an idea about the channels that mainly contribute for the MI discrimination.

2 Materials and Methods

2.1 Preprocessing

Let $Y_r \in \mathbb{R}^{C \times T_Y}$ represents the raw EEG signal of the r-th single trial; being $r = 1, \ldots, n$; C the number of channels, and T_Y the length of the samples. The CSP method is employed to analyze multi-channel EEG data based on recordings from two classes [5], producing spatial filters $W \in \mathbb{R}^{C \times C}$, which project the original signal to a space where the differences in variances of two kinds of tasks can be maximized [1]. The projected signal $Z_r \in \mathbb{R}^{C \times T_Y}$ is given by $Z_r = W Y_r$. Given the projected signal Z_r by CSP, an EMD is performed to find out the main components of each Z_r. Thus, EMD decomposes Z_r into a residual and intrinsic modes as $Y_r = \sum_{i=1}^{N} c_i + \epsilon_n$, where $c_i \in \mathbb{R}^{C \times T_Y} : i = 1, \ldots, N$ stands for Intrinsic Mode Functions (IMFs) and $\epsilon_n \in \mathbb{R}^{C \times T_Y}$ indicates a residual. The zero-mean amplitude IMFs are obtained by a sifting process according to the characterizing conditions of the IMFs. The process can be finished when residual becomes a monotonic component or a constant [1]. In this regard, the main bands $\hat{\boldsymbol{Y}}_r \in \mathbb{R}^{C \times T_Y}$ of $\boldsymbol{Z}_r$ can be highlighted by considering N_c IMFs as $\hat{\boldsymbol{Y}}_r = \sum_{i=1}^{N_c} \boldsymbol{c}_i$.

2.2 Feature Representation

From the preprocessed EEG signal matrix $\hat{\boldsymbol{Y}}_r$, the Power Spectral Density (PSD), the Continuous and Discrete Wavelet Transforms (CWT)-(DWT), and the Hjorth parameters are computed for each row vector $\hat{\boldsymbol{y}}_r \in \mathbb{R}^{1 \times T_Y}$ as follows. Let $\boldsymbol{p} = \{p_f : f = 0, \ldots, F_s/2\}$ the PSD of input signal $\hat{\boldsymbol{y}}_r$ that, in the concrete case, is computed by means of the nonparametric Welch's method, being F_s the sample frequency [4]. Particularly, the fast Fourier transform algorithm is employed to estimate the PSD, by dividing the time-series into M overlapped segments of length L, and applying a smooth time weighting window $\boldsymbol{w} = \{w_i : i = 1, \ldots, L\}$, obtaining the windowed segments $\boldsymbol{v}^{(m)} = \{v_i^{(m)} : i = 1, \ldots, L\}$, with $m = 1, \ldots, M$. The main goal is to deal with the non-stationary nature of the EEG, assuming a piece-wise stationarity into each overlapped segment. So, inspired by singular spectrum analysis-based approaches for analyzing one-dimensional time-series, the length of the segments is fixed as $L > F_s/F_r$, with F_r the minimum frequency to be considered within the analysis [6]. Thus, the modified periodogram vector $\boldsymbol{u} = \{u_f : f = 0, \ldots, F_s/2\}$ is calculated by the Discrete Fourier Transform as $u_f = \sum_{m=1}^{M} |\sum_{i=1}^{L} v_i^{(m)} \exp(-j2\pi i f)|^2$. Afterwards, each element of PSD vector $\boldsymbol{p}$ can be computed as $p_f = u_f/(MLU)$, with $U = \mathbf{E}\{|w_i|^2 : \forall i \in L\}$, where notation $\mathbf{E}\{\cdot\}$ stands for expectation operator. The motor imagery discrimination analysis is mostly provided for μ ($8 - 13\,Hz$) and β ($13 - 30\,Hz$) bands. Therefore, their PSD bands (noted as S_μ and S_β, respectively) are calculated from $\boldsymbol{p}$, for which the PSD magnitude is parameterized based on the first and second statistical moments.

Now, regarding to CWT, noted that this inner-product-based transformation quantifies the similarity between a given signal ($\hat{\boldsymbol{y}}_r$) and the considered base function (termed mothers wavelets). Therefore, the wavelet transform of a EEG signal, at time t and frequency f, is provided by their convolution with the scaled and shifted wavelet [4]. The short-time instantaneous amplitude of the CWT of EEG data is accomplished, where two Morlet wavelets centered at the bands of interest ($10\,Hz$ and $22\,Hz$) to highlight the μ and β bands, respectively. After that, the first and second statistical moments, as well as the maximum value of the coefficients magnitude are estimated; those values are considered as the CWT based features. With respect to DWT, this transformation is assumed to provide a multi-resolution decomposition and non-redundant representation of the input signal $\hat{\boldsymbol{y}}_r$. DWT has a wide application in biomedical signal processing, especially, for non-stationary signals such as EEG [7]. A seventh order Symlet mother wavelet is used, for which the detail coefficients of the third and fourth level are obtained (DWT4 and DWT3) to compute the required frequency bands α and β. Namely, the estimated frequency bands for each wavelet level are $62.5 - 125\,Hz$; $31.3 - 62.5\,Hz$; $15.7 - 31.3\,Hz$ (including the β rhythm); and $7.9 - 15.7\,Hz$; $0.5 - 7.9\,Hz$ (including the α rhythm) [4]. From the detail coefficient sets, DWT4 and DWT3, the first and second statistical moments, and the maximum value are estimated.

Lastly, a time-domain based characterization is also employed to describe the EEG data. Particularly, from the input signal $\hat{\boldsymbol{y}}_r$, the following short-time Hjorth parameters are estimated: activity, mobility, and complexity [3]. The activity is directly described by the variance that is related to the signal power, $\sigma^2(\hat{\boldsymbol{y}}_r)$. The mobility is a measure of the signal mean frequency, defined as $\phi(\hat{\boldsymbol{y}}_r) = \sqrt{\sigma^2(\hat{\boldsymbol{y}}_r')/\sigma^2(\hat{\boldsymbol{y}}_r)}$, being $\hat{\boldsymbol{y}}_r'$ the derivative of $\hat{\boldsymbol{y}}_r$. Finally, the complexity measures the deviation of the signal from the sine shape, that is, the change in frequency and it can be computed as $\vartheta(\hat{\boldsymbol{y}}_r) = \phi(\hat{\boldsymbol{y}}_r')/\phi(\hat{\boldsymbol{y}}_r)$. From the estimated short-time Hjorth parameter sets, the first and second statistical moments, and the maximum value are obtained as features.

2.3   Feature Relevance Analysis

From the above mentioned EEG representations, a feature space matrix $\boldsymbol{X} \in \mathbb{R}^{n \times D}$ is obtained, assuming that a set of preprocessed EEG signals $\{\hat{\boldsymbol{Y}}_r : r = 1, \ldots, n\}$ is provided, being n the number of training trails of a given subject in a BCI system, and D the number of estimated features. Particularly, each row, $\hat{\boldsymbol{y}}_r$ of $\hat{\boldsymbol{Y}}_r$ holds the c-th studied EEG channel, with $c = 1, \ldots, n_c$ and being n_c the number of analyzed channels. To carry out a low-dimensional representation of the original feature representation space, this work uses Principal Component Analysis (PCA) as a statistical eigendecomposition, searching for directions with greater variance to project the data. Although, PCA is commonly used as a feature extraction method, it can be useful to properly select a relevant subset of original features that better represent the studied process [8]. In this sense, given a set of features $\Xi = \{\boldsymbol{\xi}_d : d = 1, \ldots, D\}$, where $\boldsymbol{\xi}_d$ corresponds to each column of the input data matrix $\boldsymbol{X}$, the relevance of each feature can be analyzed by the PCA mapping. More precisely, the relevance of $\boldsymbol{\xi}_d$ can be identified by computing the corresponding weighting term $\rho = \{\rho_d : d = 1, \ldots, D\}$, where ρ is defined as $\rho = \mathbf{E}\{|\lambda_d \boldsymbol{\alpha}_d| : \forall d \in D'\}$, being λ_d and $\boldsymbol{\alpha}_d$ the eigenvalues and eigenvectors of the covariance matrix $\boldsymbol{\Sigma} \in \mathbb{R}^{p \times p}$, which is estimated as $\boldsymbol{\Sigma} = \boldsymbol{X}^{\top} \boldsymbol{X}$. The main assumption is that the largest values of ρ_d point out to the best input attributes, since they exhibit higher overall correlations with principal components. The D' value is fixed as the number of dimensions needed to conserve a percentage of the input data variability.

3   Experiments and Results

In order to test the proposed framework, experimental tests were done over the a well known Motor Imagery (MI) dataset. The EEG data collection is provided by the Berlin Brain-Computer Interface (BCI competition IV 2008 - Data sets 1)[1]. This database is based on the paradigm in cognitive neuroscience of MI, e.g. imagination of hand movements, whole body activities, relaxation, etc. For each subject the first two classes of motor imagery were selected from the three classes left hand, right hand, and foot (side chosen by the subject). The EEG signals were obtained from seven subjects. For each subject, the signals from 59 EEG positions were measured, being the sensorimotor

---

[1] http://bbci.de/competition/iv/desc_1.html

area the most densely covered area by the electrodes. Signals were band-pass filtered between 0.05 and 200 Hz and then digitized at 1000 Hz. Moreover, the database was downsampled at $F_s = 100$ Hz, but previously a low.-pass Chevyshev II filter (order 10) was employed with stopband ripple 50dB down and stopband edge frequency 49 Hz. The whole motor imagery session was performed without feedback. The data base contains 100 repetitions of each MI class per person. Particulary, the EEG segments were extracted while a cue (indicating a side) is presented, i.e. an arrow pointing left or right were presented on a screen, the duration of each extracted segment is 4 s during which the subject was instructed to perform the cued motor imagery task. These periods were interleaved with 2 s of blank screen and 2 s with a fixation cross shown in the center of the screen. All EEG channels per subject of the above mentioned MI dataset were used. We test four frameworks mainly changing the preprocessing stage, in order to validate the performance of the proposed approach (see Fig. 1). The first one does not uses any kind of preprocessing method of the data before the characterization stage, the next two frameworks are conceived to use either CSP or EMD techniques (Framework 2 (FW2) and Framework 3 (FW3) respectively) as part of the preprocessing stage of the EEG recordings. The last framework (FW4 - proposed approach) uses together EMD and CSP techniques as a preprocess of the data.

For a given subject, a set of signals $\{Y_r : r = 1, \ldots, 200\}$ was obtained, with $Y_r \in \mathbb{R}^{C \times T_Y}$, $C = 59$ and $T_Y = 400$. For FW3 and FW4 the number of IMFs in EMD is fixed as $N_c = 3$ [1]. Thereby, for each framework the following analysis is performed. According to the described features in section 2.2, three frequency-based (PSD, CWT, and DWT) and one time-based (Hjorth parameters) kind of features are estimated for each channel of a given trial $\hat{y}_r$. Hence, a feature space representation matrix $X \in \mathbb{R}^{400 \times 1593}$ is calculated. It is important to note that for the segment length value L in PSD and Hjorth parameters based features, the minimum frequency to be analyzed is fixed as $F_r = 8$ Hz, taking into account that the band of interest for the analyzed BCI application is $8 - 30$Hz (containing the α and β rhythms).

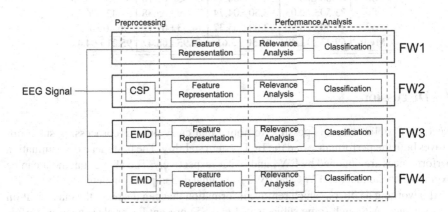

Fig. 1. Tested Frameworks. FW4-proposal.

Regarding to the eigendecomposition-based feature relevance analysis presented in section 2.3, the number of dimensions D' in PCA is calculated looking for a 95% of the input data variability. Therefore, the inferred relevance vector $\rho \in \mathbb{R}^{1593 \times 1}$ is employed to sort the original features. In addition, a soft-margin Support Vector Machine (SVM) classifier is trained using a regularization parameter $C \in \mathbb{R}^+$, and a Gaussian kernel $k(\mathbf{x}_a, \mathbf{x}_b) = \exp(-||\mathbf{x}_a - \mathbf{x}_b||/2\delta^2)$, with band-width $\delta \in \mathbb{R}^+$; and being $\mathbf{x}_a, \mathbf{x}_b \in \mathbb{R}^{1 \times D}$ two given samples of the feature representation space. We generate a curve of performance adding one by one the characteristics obtained in each subspace representation based on the order given by the relevance vector ρ. For a given subset, the optimum working point has been searched using a 10-fold cross validation scheme to fix the C and δ values. The C value is selected from the set $\{1, 10, 100, 1000\}$; and the δ value from $\{\delta_s, 10\delta_s, 100\delta_s, 1000\delta_s\}$; being $\delta_s = 0.9 \min(\mathbf{E}\{\sigma(\Xi)\}, (1/1.34)\mathbf{E}\{\mathrm{iqr}(\Xi)\})$ the Sylverman rule based Gaussian kernel band-width estimation. Note that $\sigma(\cdot)$ computes the standard deviation and $\mathrm{iqr}(\cdot)$ the interquartile range of a provided set of features, respectively. Table 1 shows the best BCI system performance for each subject according to each training framework. Fig. 2 presents the system performance for the four tested frameworks, thus is, these figures show the accuracy as a function of the number of chosen features according to proposed relevance analysis. Finally Fig. 4 presents the distribution relevance information per channel extracted to each method.

Table 1. Classification results (average accuracy ± standard deviation, 10-fold cross validation)

Subject	Framework 1 Acc. (%)	Framework 2 Acc. (%)	Framework 3 Acc. (%)	Framework 4 Acc. (%)
S1	74.50±09.26	89.50±07.62	82.66±11.36	98.50±03.37
S2	67.00±15.49	86.50±04.74	67.74±09.41	95.97±03.95
S3	71.50±12.03	96.50±05.29	60.50±08.32	98.50±03.37
S4	59.00±13.70	93.00±06.75	63.37±08.48	91.84±04.84
S5	65.00±08.16	96.50±04.74	65.55±07.18	91.76±05.98
S6	73.50±12.03	93.50±04.74	75.29±05.38	93.42±06.28
S7	75.50±08.96	92.00±05.37	74.53±09.19	96.50±4.11
Mean	**69.43±11.38**	**92.50±5.61**	**69.95±08.47**	**95.21±04.42**

4 Discussion

According to Table 1, it is possible to notice that carry out a preprocessing stage improves both the performance and the BCI system stability. The best mean discrimination performances are obtained by FW4 and FW2, respectively. The above statement can be explained by the fact that FW4 and FW2 use signal decomposition methods (CSP, CSP-EMD) working as filters, which remove information that can decrease the classification performance. Although some subjects (S4 and S5) present lower classification performance in FW4 than FW2, it is explained by the fact that each subject presents different cognitive characteristics, not mentioning the non-stationary nature of the signals. Additionally, the quality of the EEG trials is perturbed by the artifacts, and by the brain

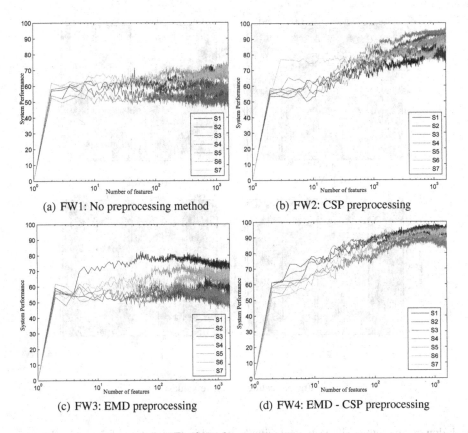

(a) FW1: No preprocessing method

(b) FW2: CSP preprocessing

(c) FW3: EMD preprocessing

(d) FW4: EMD - CSP preprocessing

Fig. 2. Performance curves

response capability of each subject. Even though FW1 and FW3 present a similar accuracy, the performance computed by FW3 is more stable than FW1, because FW3 is calculated only using the first three IMFs which are mainly related to the optimal informative frequency bands of interest (α and β rhythms) for MI classification [1].

Moreover, from the attained performance of each framework (Fig. 2), note that for FW1 (Fig.2(a)) and FW3 (Fig. 2(c)), overall, the first 10 relevant features achieved the maximum system performance without a notable gain by increasing the number of features. For FW2 (Fig. 2(b)) and FW4 (Fig. 2(d)) the performance notedly increases by adding features. Also, in some cases, the BCI performance curves present local minimums when adding new relevant features, and then the classification accuracy grows up again. This behavior is explained by the fact that some features may represent highly relevant attributes, but they involve redundant information, i.e. needling phenomenon.

Figure 3 shows the relevance of features to each one of the frameworks. In FW1 (Fig. 3(a)), FW2 (Fig. 3(b)), and FW4 (Fig. 3(d), both PSD and DWT methods provides a better relevance value than the other analyzed features. This is because the PSD features are estimated into a restricted frequency band-width (α and β bands), in order to take advantage of the prior knowledge about the studied phenomenon [4], besides, the signal

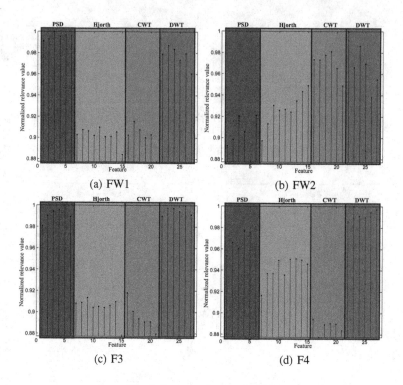

Fig. 3. Feature relevance values

windowing procedure allows to deal with the non-stationary nature of the EEG. Regarding to DWT, they also bring relevant information, since, this method allows to extract features of interest from μ and β rhythms. As expected, the DWT transformation provides a multi-resolution decomposition, which is able to deal with non-stationary signals. In this sense, the Hjorth features, generally, can not captured the non-stationarity behavior of the signals, because the just analyze second-order statistical moments.

From figures 4(a), 4(b), and 4(c) (FW1, FW2, and FW3, respectively) it is possible to see how most of the channels are considered as high relevance channels. In these frameworks the frontal cortex (i.e. the AF,F,FC electrodes) seems to present discriminant information. However, the Primary Motor Cortex – PMC (FC electrodes) is related to movement mode but not to imagery mode. On the other hand, the FW4 (Fig. 4(d)) exhibits low relevance on the frontal cortex (AF, F, and FC electrodes). A human lesion study suggests that PMC does not play a fundamental role in motor imagery process, although individual subjects may show PMC activity during motor imagery, depending on their thinking strategy. Besides the activity associated to the task performance for the imagery mode was localized in the precentral sulcus, indicating the significance of this region in motor imagery [9].

FW1, FW2, and FW3 show high relevance for the parietal cortex, however, activity in the anterior parts of the parietal cortex (i.e. CP electrodes), most likely reflects somatosensory-motor association and sensory feedback from movement mode, but not

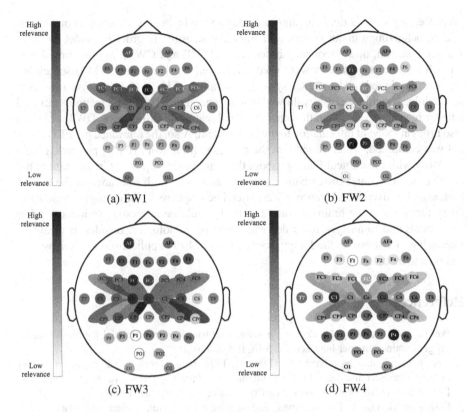

(a) FW1

(b) FW2

(c) FW3

(d) FW4

Fig. 4. Channels relevance

the imagery mode, which implies a decrease of the performance of the classification stage in these frameworks. Moreover, detailed neuropsychological examination supports the role of the parietal cortex in generating mental movement representations. The posterior part of the parietal cortex, including the precuneus (P and PO electrodes), has been reported to be active during tasks involving motor imagery [9], which corresponds with the relevance configuration found by FW4 (Fig. 4(d)). The imagery-predominant activity showing the overactivity of the posterior parietal cortex during motor imagery of finger movement. Thus, the channels relevance analysis, computed by FW4, resemble to the clinic findings of the state of the art.

5   Conclusions and Future Work

In this paper we develop a BCI based motor imagery classification framework using CSP and feature relevance analysis. The CSP is matched with EMD to reveal the dynamics of interest. We select a set of features representing the best as possible the studied process. For such purpose, a variability analysis is presented to identify relevant features. Four frameworks of training for MI classification were compared. Experimental results showed that the precision of the MI system significantly increases when a preprocessing stage is done, gaining accuracy and reducing the variance among experiments, achieving a major system reliability and stability. In order to model the studied phenomenon, three frequency-based (PSD, DWT, and CWT) and one time-based (Hjorth parameters) features were used. Moreover, a soft-margin SVM based classifier was employed, and the BCI-system was validated by a 10-fold cross validation methodology. Achieved results showed that in general the PSD based features provided a better relevance value than the other analyzed features. Furthermore, DWT attributes also brought relevant information to the BCI-system. Also, the relevance per EEG channel was computed, which found that the proposed framework (FW4) presents a great similarity with the clinical findings about the brain function pointed in the state of the art, that is, the frontal cortex shows low relevance, and high relevance in the central cortex, and an average relevance in the parietal hemisphere, highlighting the motor imagery functions of the brain. As future work, it would be interesting to identify other movements and to analyze other decomposition methodologies. Besides, it would be interesting to test our methodology in other kind of EEG applications as Epilepsy detection and monitoring.

References

1. Wei-He, Wei, P., Wang, L., Zou, Y.: A novel emd-based common spatial pattern for motor imagery brain-computer interface. In: IEEE EMBC (2012)
2. Ang, K.K., Chin, Z.Y., Zhang, H., Guan, C.: Filter bank common spatial pattern (fbcsp) in brain-computer interface. In: IEEE International Joint Conference on Neural Networks, IJCNN 2008 (IEEE World Congress on Computational Intelligence) (2008)
3. Rodríguez, G., García, P.J.: Automatic and adaptive classification of electroencephalographic signals for brain computer interfaces. Medical Systems 36(1), 51–63 (2012)
4. Corralejo, R., Hornero, R., Álvarez, D.: Feature selection using a genetic algorithm in a motor imagerybased brain computer interface. In: IEEE EMBC (2011)
5. Bankertz, B., Tomioka, R., Lemm, S., Kawanabe, M., Müller, K.-R.: Optimizing spatial filters for robust eeg single-trial analysis. IEEE Signal Processing Magazine 08, 41–56 (2008)
6. Teixeira, A.R., Tomé, A.M., Boehm, M., Puntonet, C., Lang, E.: How to apply nonlinear subspace techniques to univariate biomedical time series. IEEE Trans. on Instrument. and Measur. 58(8), 2433–2443 (2009)
7. Li, M.-A., Wang, R., Hao, D.-M., Yang, J.-F.: Feature extraction and classification of mental eeg for motor imagery. In: 5th Int. Conf. on Nat. Comp., ICNC 2009, vol. 2 (2009)
8. Daza-Santacoloma, G., Arias-Londoño, J.D., Godino-Llorente, J.I., Sáenz-Lechón, N., Osma-Ruíz, V., Castellanos-Domínguez, G.: Dynamic feature extraction: An application to voice pathology detection. Intel. Aut. and Soft Comp. (2009)
9. Hanakama, T., Immisch, I., Toma, K., Dimyan, M.A., Van Gelderen, P., Hallett, M.: Functional properties of brain areas associated with motor execution and imagery. J. Neurophysiol. 89, 989–1002 (2003)

High-Level Hardware Description of a CNN-Based Algorithm for Short Exposure Stellar Images Processing on a HPRC

Jose Javier Martínez-Álvarez, Fco. Javier Garrigós-Guerrero,
F. Javier Toledo-Moreo, Carlos Colodro-Conde,
Isidro Villó-Pérez, and José Manuel Ferrández-Vicente

Dpto. Electrónica, Tecnología de Computadoras y Proyectos,
Universidad Politécnica de Cartagena, 30202 Cartagena, Spain
`jjavier.martinez@upct.es`

Abstract. A CNN-based algorithm, adequate for short exposure image processing and an application-specific computing architecture developed to accelerate its execution are presented. Algorithm is based on a flexible and scalable CNN architecture specifically designed to optimize the projection of CNN kernels on a programmable circuit. The objective of the proposed algorithm is to minimize the adverse effect that atmospheric disturbance has on the images obtained by terrestrial telescopes. Algorithm main features are that it can be adapted to the detection of several astronomical objects and it supports multi-stellar images. The implementation platform made use of a High Performance Reconfigurable Computer (HPRC) combining general purpose standard microprocessors with custom hardware accelerators based on FPGAs, to speed up execution time. The hardware synthesis of the CNN model has been carried out using high level hardware description languages, instead of traditional Hardware Description Languages (HDL).

Keywords: CNN, HLS, HDL, HPRC, HW/SW Co-execution, FPGA.

1 Introduction

A CNN-based algorithm for short exposure image processing (Lucky-Imaging techniques) and an application specific computing architecture developed to accelerate the algorithm are presented in this paper. The computing platform is a High Performance Reconfigurable Computer (HPRC), also called hybrid supercomputer, combining general purpose standard microprocessors with custom hardware accelerators based on FPGAs. The proposed algorithm is adequate for image registration methods used in short exposure observational techniques and makes intensive use of CNN kernels at the pre-processing and post-processing stages. In this context, the use of CNNs provides the advantage of their versatility, which allows the modification of the algorithm and, through suitable

J.M. Ferrández Vicente et al. (Eds.): IWINAC 2013, Part II, LNCS 7931, pp. 375–384, 2013.

training, the adjustment of the instrument to each application, i.e., the detection of different astronomical objects or events. The proposed algorithm improves the spatial resolution of multi-stellar images reaching the diffraction limit of the telescope.

The hardware specification of the CNN has been carried out using High-Level Synthesis (HLS) languages, instead of traditional Hardware Description Languages (HDLs). This design methodology provides hardware designs from high level descriptions, using several flavors of well-known programming languages, such as C, C++ or Matlab. Both the HLS tool and the HPRC platform were successfully evaluated previously in [1, 2], where they were used to emulate Carthago CNN architecture [3], demonstrating their viability and effectiveness for quickly prototyping of video processing applications based on CNNs.

2 Algorithm to Improve the Spacial Resolution of Multi Stellar Images

The main objective of the proposed algorithm is to apply Lucky Imaging Techniques to minimize the adverse effect of the atmospheric disturbance has on the images obtained by telescopes.

The Lucky Exposures Technique or Lucky-Imaging (LI), originally proposed by Fried in 1966 [4–6], is based on the idea of registering just the moments of maximum atmospheric stability. According to this consideration, if astronomical images (specklegrams) are acquired with short exposure time, using a high sensitivity, fast and enough low-noise sensor (such as EMCCD devices), some ones out of thousands of images, typically 5-15%, offer a much lower distortion than the others. If only these lucky images are taken into account and they are processed and recombined to reduce the information and to reach the desired sensibility, the resulting image can nearly reach the maximum resolution of the telescope, i.e., the resulting image offers a quality in the same order that it would offer if the image had been acquired beyond the atmosphere or in absence of it. The difficulty here is that high performance processing is required, thus, keeping the adequate trade-off between quality of results and the computational cost of the algorithm is a major concern.

Several instruments have been developed based on LI principles, starting from the experiences of J.E. Baldwin's working group at Cambridge University [7, 8], or later the FASTCAM instrument from the Instituto de Astrofísica de Canarias, in collaboration with the Universidad Politécnica de Cartagena [9].

In a previous work [2], a platform for accelerating basic LI based algorithms was proposed. The processing flow included several CNN modules and implemented a modified *shift and add* registration algorithm using a single reference star. The system performed properly for images with a reduced number of punctual objects of interest. However, in case of wide field images (with tens of punctual objects), resolution enhancement decreased as the objects were farther from the reference star. Reaching the diffraction limit here requires more images, what means more observation and computation time. Another issue is related to the use of the maximum as the unique image quality factor. This might cause that highly-noisy sparse speckles surpass the quality criteria.

In order to overcome the mentioned limitations, in this paper a new algorithm is proposed, which is well suited for multi-stellar images and improves resolution in wider zones of interest. Moreover, instead of considering just its brightness level, we propose to use also a form factor (Point Spread Function, PSF) indicator, measure of the speckle distortion, to determine the image quality. This algorithm, depicted in Figure 1, requires however a greater computational effort, thus, it was necessary to redefine and optimize the architecture of the computing intensive stages.

The proposed algorithm works with stacks of images (typically 1000 specklegrams) that are stored in the system memory. Based on the first specklegram, the user defines several regions where there is confined an object of interest and threshold values for the quality criteria (minimum brightness and PSF deviation) on each region. Then, a computationally intensive step processes every speckle independently and is followed by a low effort step where the resulting images are combined and post-processed to obtain the final image. The first step begins with a pre-processing stage that makes use of two CNN modules in parallel. CNN1 performs a band-pass filter of every speckle, fine tuned to the size of the desired objects. CNN2 implements a low-pass filter that will be used to estimate every speckle PSF. From these parameters and the threshold values selected by the user, the algorithm determines automatically if every region in an image has the minimum quality to contribute to the averaged region. If a speckle surpasses the quality threshold, the whole specklegram is re-centered with respect to the brightest pixel in that region and accumulated with the other specklegrams selected to contribute to the same region. The algorithm is repeated until the last image in the stack is processed. As a result, for every region, a full image is obtained which accumulates those specklegrams that had better quality in that region. In a second step the obtained images are combined to form a unique image of better quality. First, all images are re-centered again, relatively to the brightest pixel of the first region, and then, combined using a weighted average, in which the contribution of an image pixel is inversely proportional to its distance to the maximum of every region. Finally, the obtained image is postprocessed by CNN3 module, which applies a smooth sharpening to highlight detected objects.

3   System Architecture

The most immediate solution is to implement the algorithm with a software approach using off-the-shelf PCs. It is the fastest and simplest approach, but it presents the shortcoming of performance, since it requires too much time to emulate the CNN modules. A full hardware implementation can overcome this major drawback and achieve very fast execution times, but it implies a considerably bigger effort in the design process and much more development time. We have adopted an intermediate approach, which combines the advantages of both

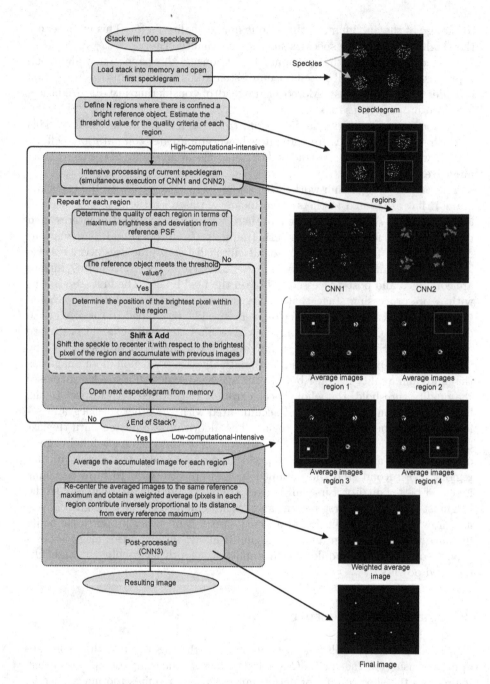

Fig. 1. Workflow and explanation of the proposed algorithm

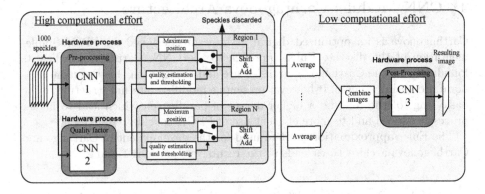

Fig. 2. Hardware/software partitioning for the DS1002 HPRC platform

software and hardware solutions, consisting in implementing the algorithm on a DS1002 platform from DRC Computer Corporation.

The main objective of this system is accelerating an external CPU by means of a FPGA board. The close coupling between both boards, using high performance dedicated buses, provide a communication channel characterized by its high bandwidth and low latency, that make it possible for the FPGA to work as an application-specific coprocessor. The hybrid mix of standard processors and FPGAs confers HPRC platforms a great versatility and the capacity of exploiting both coarse and fine grain parallelism. This may provide several order of magnitude improvements in performance, power consumption and cost when compared to standard computers. Some examples of different HPRC architectures are the SRC-7, the SGI Altix 350 and Cray's XD1.

Figure 2 depicts algorithm implementation on the DS1002 platform. The algorithm has been partitioned in several processes with the aim of executing the computing intensive tasks over dedicated hardware on the FPGA, while the less costly, decision oriented, or non-parallelizable tasks are executed as pure software on the Opteron microprocessor. Thus, image reading/writing, and control routines are executed in software, while the CNN blocks are implemented as dedicated hardware.

The difficult task of Software/Hardware partitioning has been resolved by using CoDeveloper, an HLS environment from Impulse Accelerated Technologies. This tool provides a "C-to-gates" workflow where the algorithms are described in a slightly modified C language and then used as software or translated to a hardware description language (VHDL or Verilog) for their hardware synthesis. Thus, once the algorithm has been coded, it can be compiled using any standard C compiler. The software processes defined are translated to software threads, while the hardware processes are used to configure the RPU (Reprogrammable Processing Unit) on the FPGA. Communication channels between software, hardware or mixed software/hardware processes are automatically inferred by CoDeveloper for a choice of supported platforms, including the DS1002 system.

4   CNN Module: Carhtagonova Architecture

Carthagonova is an optimized digital architecture designed for FPGA implementation of the discrete Euler approach to the CNN. Carthagonova has its foundation on the Carthago [3] architecture and shares with this the same temporal processing schema. It however presents a modified internal structure of the basic stage of the cell and a higher degree of parallelism, which lead to higher processing speed and to more efficient use of FPGA specific resources.

The Euler approximation method used for the development of Carthago and Carthagonova architectures leads to the recursive equations 1 and 2:

$$X_{ij}[n] = \sum_{k,l \in Nr(ij)} A_{kl}[n-1]Y_{kl}[n-1] + \sum_{k,l \in Nr(ij)} B_{kl}[n-1]U_{kl} + I_{ij}, \tag{1}$$

$$Y_{ij}[n] = \frac{1}{2}\left(|X_{ij}[n]+1| - |X_{ij}[n]-1|\right) \tag{2}$$

These equations imply an infinite feedback loop that in most applications must be approximated in a finite number N of iterations. In hardware implementation, instead of executing the N iterations on the same circuit it is possible to unfold the loop and execute its operation on N stages connected on cascade. The unfolding of a cell into stages allows to take advantage of streaming parallelism and so to achieve faster processing, since the execution of N iterations in a N-stage cell is N times faster than in a one-stage cell, at the expense of consuming N times more area, roughly.

At the same time, the unfolding approach implies that the data to be processed are shared out into the stages of the cell, making it possible to use small buffers for local storage instead of the external memories required with an iterative approach. With these local memories not only the design is simpler but also it is faster since the access time to data is lower than with external memories. Further, as it has been shown in [10], the Carthagonova architecture makes use of a simple I/O interface between stages, what makes it easy the expansion of the systems developed, and so to adapt the hardware to virtually any kind of applications. Figure 3 shows, exemplified with a radius of neighborhood equal to 1, the detailed internal architecture of the basic CNN stage.

5   High-Level Description of the CNN Modules

The Carthagonova stage has also been implemented using a HLS tool, CoDeveloper from Impulse Accelerated Tech. The architecture was coded using Impulse-C, a slightly modified C language that uses standard ANSI C syntax plus some custom pragmas, data types and specific functions for process intercommunication. The CNN module implemented performs 5 Euler iterations using 5 cascaded Carthagonova stages. each cell stage has been implemented as a single Impulse-C process.

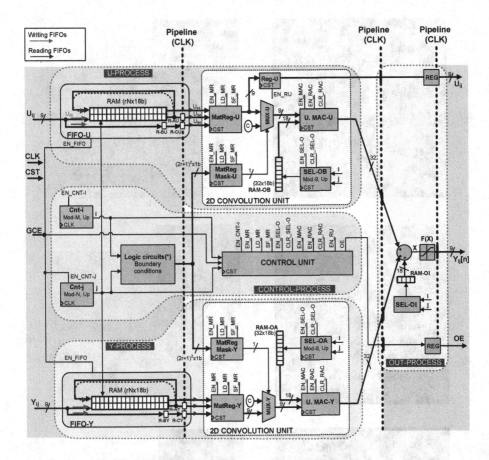

Fig. 3. Microarchitecture of the Carthagonova stage exemplified for a stage with a 3×3 neighborhood

In the code, Producer and Consumer software processes were merged in a single process to maximize efficiency, as it just has to read image data, send pixels to the hardware processes, receive processed pixels, and write images back to disk. Images were sized 512×512 pixels, coded 16 bits grey-scale. CoDeveloper hardware simulation tool, State Master Debugger, revealed that the CNN module takes 14 clock cycles to process a pixel, showing a perfect correspondence with the real execution of the algorithm. Maximum clock frequency for the CNN after P&R was 100 MHz, what met our specifications. Table 1 shows the processing time for a high-level description with 512×512 pixel images.

6   Implementation Results

Execution time in the DS1002 has been measured for two scenarios: a full software implementation in the Opteron processor and Linux operating system

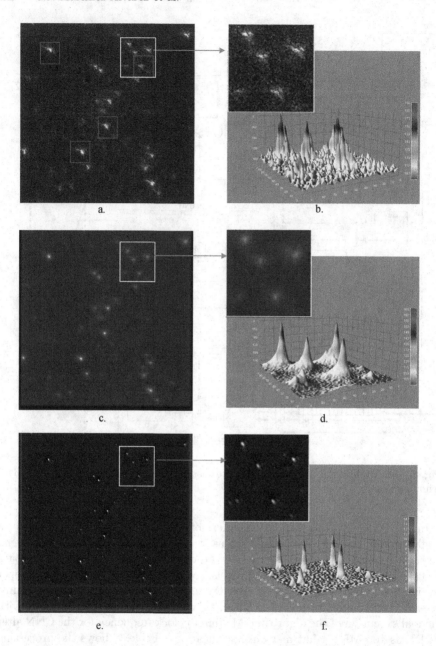

Fig. 4. Globular cluster M15, images acquired with the NOT telescope (2.5m diameter) at La Palma Island [11]. a) Example of one of the 1000 images (specklegram) obtained with the imaging sensor of the FASTCAM instrument. b) Detail of a region in the speckelgram and its 3D representation. c) Output of the algorithm after recentering and accumulation of best quality factor images d) Detail of a region in figure 4-c and its 3D representation. e) Output of the algorithm with CNN based pre-processing, quality factor and post-processing. f) Detail of a region in figure 4-e and its 3D representation.

Table 1. Processing performance for Impulse-C implementation of a CNN-stage on a XC4VLX200 device

XCV4LX200-11	
Size of the CNN stage	5×5
Precision	16 bits
Mult/Conv. Unit	5
N cycles/Conv. (Frec. MHz)	14(100)
Proc. Time $(5 \times 5) * 5Iter$ (ms)	36.7

(RPU disabled); and mixed hardware/software co-execution (RPU enabled). In this case, once the system design was completed, the RPU was reconfigured with the hardware implementation of the three Carthagonova CNNs modules. Next, the software part of the algorithm was loaded into the system processor. Then, the software/hardware co-execution of the algorithm was carried out. Timing information has been obtained for a 512×512-pixel image in a 16-bit fixed point greyscale format. The hardware/software co-execution was the fastest solution, consuming only 0.057 s. This was approximately 40 times faster than the software implementation, which required 2.24 s.

With the aim of verifying the processing results obtained, the system has been evaluated using real data with versions of growing algorithm complexity. These data are stacks of astronomical images acquired by terrestrial telescopes at the Roque de Los Muchachos observatory in the La Palma Island. A stack of images is made of 1000 specklegrams.

The algorithm first executes a software process which sequentially reads the set of specklegrams stored in the hard drive of the DS1002 and sends it to the RPU through the high speed HyperTransport bus. As data are been received in the RPU they are processed in data streaming schema by the different CNN modules implemented in the FPGA (see Figure 2). Once the hardware processing is carried out, the generated information is read back by the Opteron processor, where the last software tasks of the algorithm are run. Finally, output data are stored in the hard drive for further analysis.

Figure 4 shows the output obtained when trying to resolve a globular cluster (M15) using algorithms of increasing complexity. Figures 4-a and 4-b show one specklegram, included in the set of 1000 specklegrams captured with the NOT telescope [11]. Figures 4-c and 4-d show the results obtained with the algorithm without the pre-processing and post-processing stages. Finally, Figures 4-e and 4-f show the output of the algorithm when the described CNN based pre-processing is included in the algorithm. It clearly offers the best results.

7   Conclusions

In this paper, we have proposed an CNN-based algorithm and its application to short exposure stellar images processing on a HPRC. The CNN modules used by the algorithm have been implemented using a high-level hardware description language, which makes it much easier and faster the implementation of a whole system, while ensuring that performance requirements are satisfied.

In our case, the paper demonstrates the development of a CNN based accelerator for an astronomical application, which has been successfully implemented on a HPRC platform (DRC-DS1002). The HPRC system has shown higher performance when compared to a standard personal computer. Our initial results have shown 40x acceleration for the hardware/software co-execution on the DS1002 when compared with the algorithm running as software-only on the same machine.

These tools and design methodology have been validated as a valuable approach to the implementation of hardware-accelerated algorithms, while maintaining the flexibility of software based implementations. These platforms might be especially advisable for researchers not closely related to hardware design, involved in complex algorithm development.

Acknowledgements. This work has been partially supported by the Fundación Séneca de la Región de Murcia through the research projects 15419/PI/10.

References

1. Martínez, J.J., Garrigós, F.J., Toledo, F.J., Ferrández, J.M.: Using reconfigurable supercomputers and C-to-hardware synthesis for CNN emulation. In: Int. Work. on the Interplay between Natural and Artificial Computation, pp. 244–253 (2009)
2. Martínez, J.J., Garrigós, F.J., Villó, I., Toledo, F.J., Ferrández, J.M.: Hardware Acceleration on HPRC of a CNN-based Algorithm for Astronomical Images Reduction. In: Int. Work. on Cellular Nanoscale Networks ans their Applications, pp. 1–5 (2010)
3. Martínez, J.J., Garrigós, F.J., Toledo, F.J., Ferrández, J.M.: High Performance Implementation of an FPGA-Based Sequential DT-CNN. In: Int. Work. on the Interplay between Natural and Artificial Computation, pp. 1–9 (2007)
4. Fried, D.: Limiting resolution looking down through the atmosphere. Journal of the Optical Society of America, 1380–1384 (1966)
5. Fried, D.: Optical resolution through a randomly inhomogeneous medium for very long and very short exposures. Journal of the Optical Society of America, 1372–1379 (1966)
6. Fried, D.: Probability of getting a lucky short-exposure image through turbulence. Journal of the Optical Society of America, 1651–1658 (1978)
7. Baldwin, E., Tubbs, N., Cox, C., Mackay, D., Wilson, W.: Diffraction-limited 800 nm imaging with the 2.56 m Nordic Optical Telescope. Astronomy & Astrophysics, 1–4 (2001)
8. Tubbs, R., Baldwin, J., Mackay, C., Cox, G.: Diffraction-limited CCD imaging with faint reference stars. Astronomy & Astrophysics, 21–24 (2002)
9. FastCam homepage, http://www.iac.es/proyecto/fastcam/
10. Martínez, J.J., Toledo, F.J., Garrigós, F.J., Ferrández, J.M.: An efficient and expandable hardware implementation of multilayer cellular neural networks. Neurocomputing (In press)
11. Observing Schedules for NOT telescope, Period 37: April 2008 - September 2008 homepage, http://www.not.iac.es/observing/schedules

Analysis of Connection Schemes between the ICU and the DPU of the NISP Instrument of the Euclid Mission

Carlos Colodro-Conde[1], Rafael Toledo-Moreo[1], José Javier Díaz-García[2],
Óscar Manuel Tubío-Araujo[2], Isidro Villó-Pérez[1], Fco. Javier Toledo-Moreo[1],
José Javier Martínez-Álvarez[1], Fco. Javier Garrigós-Guerrero[1],
José Manuel Ferrández-Vicente[1], and Rafael Rebolo[2]

[1] Universidad Politécnica de Cartagena
rafael.toledo@upct.es
[2] Instituto de Astrofísica de Canarias
jdg@iac.es

Abstract. Euclid is a middle class ESA mission dedicated to the observation of the space, more in particular the dark matter and the dark energy. To do so, Euclid features two instruments: VIS (visible imager) and NISP (Near-Infrared Spectrometer and Photometer). Within the NISP, the NI-ICU (NISP Instrument Control Unit) is in charge of the control and monitoring of the instrument, and the communication with the spacecraft and the Data Processing Unit (NI-DPU). This paper summarizes the analysis carried out at the Universidad Politécnica de Cartagena (UPCT) and the Instituto de Astrofísica de Canarias (IAC) with regard to the communication link between NI-ICU and NI-DPU at the end of phase B1, 9 months before IPDR (Instrument Preliminary Design Review).

Keywords: space instrumentation, control electronics, communication interfaces.

1 Introduction

Euclid is the Medium Class mission of the ESA Cosmic Vision 2015-2025 programme selected in 2011, with a foreseen launch slot in 2020. The main objective of Euclid [1,2] is to understand the origin of the accelerating expansion of the Universe by studying the dark matter and the dark energy.

Among all the European institutions participating in this mission, the Universidad Politécnica de Cartagena (UPCT), in collaboration with the Instituto de Astrofísica de Canarias (IAC), is responsible for the design, manufacturing and validation of the control electronics of the NISP instrument, which is one of the two instruments of the Euclid spacecraft.

The purpose of this paper is to summarize the efforts done at the Universidad Politécnica de Cartagena (UPCT) to find the most appropriate physical interfaces for communicating the NISP Instrument Control Unit (NI-ICU) with the

J.M. Ferrández Vicente et al. (Eds.): IWINAC 2013, Part II, LNCS 7931, pp. 385–394, 2013.

other elements of the NISP Warm Electronics (NI-WE), namely the NISP Data Processing Unit (NI-DPU).

This paper is structured as follows: Section 2 briefly introduces the Euclid spacecraft, describing the elements which will be part of the study. Section 3 presents the candidate physical interfaces which may be suitable for the considered scenario. Section 4 describes the implications of selecting each possible interface on the NI-WE. Finally, Section 5 selects the most appropriate interface based on these studies and draws the main conclusions.

2   Description of the Euclid Spacecraft

The Euclid spacecraft consists of two main modules: the Payload Module (PLM) and the Service Module (SVM). The payload module includes a 1.2 m Korsch telescope which directs the light to the two instruments: the visual instrument [3] (VIS) and the near infrared instrument [4,5] (NISP). The SVM comprises all the conventional spacecraft subsystems, the instruments warm electronics, the sun shield and the solar arrays.

2.1   The NISP Instrument

NISP [4,5] is the infrared instrument on-board the Euclid mission. It performs imaging photometry and slitless spectroscopy measurements in sequence by selecting respectively a filter wheel and a grism wheel.

The NISP instrument consists of three main assemblies: NI-OMA (Opto-Mechanical Assembly) [5], NI-DS (Detector System Assembly) [6,7] and NI-WE (Warm Electronics Assembly) [8].

2.2   The NISP Warm Electronics

The NISP Warm Electronics (NI-WE) is located in a warm environment ($\sim$ 300°K) inside the SVM. It is composed of two functional units: the Instrument Control Unit (NI-ICU) and the Data Processing Unit (NI-DPU).

The NI-DPU consists of two DCU/DPU boxes, each containing half of the NI-DCU+NI-DPU functionality. It includes the necessary electronic modules for the management and operation of the NI-DS as well as the collection, processing and compression of the scientific data. The two DPUs boxes are operated in hot redundancy, meaning that both DPUs will be working concurrently in nominal operations. Each DPU box contains two Maxwell SCS750 CPU boards, so there will be a total of four CPU boards.

On the other hand, the NI-ICU performs the management of the housekeeping and telecommands, the thermal control of NI-DS and of the NI-OMA, and the control of the NI-OMA mechanisms themselves. These functions will be managed by a LEON2 processor running in a RTAX-S/SL radiation-tolerant FPGA. The same FPGA shall include the necessary interface controllers.

The NI-ICU will be a fully cold-redundant unit, made up by two identical and independent sections. Because of the fact that the NI-ICU is cold-redundant and the NI-DPU is hot-redundant, both sections of the NI-ICU (nominal and redundant) shall be able to communicate with both DPU boxes at the same time, more specifically, with the four CPU boards.

Since both units are in the WE, the expected length of the cables between them is less than 1 m.

3   Candidate Communication Interfaces

Not all of the existing communication interfaces are usable for intra-spacecraft communications. Common wired interfaces like USB or FireWire do not have radiation hardened components available in the market. On the other hand, wireless interfaces like Bluetooth or Wi-Fi are not desirable for space applications because they can produce high levels of electromagnetic interference (EMI) which can affect other equipment of the spacecraft.

In the last few years, there has been a lot of interest in finding alternative interfaces which provide low EMI, low mass, low volume, high flexibility and high data rates. The current research is focused mainly on wireless optical interfaces [9,10]. These new communication interfaces are still under development and cannot be used in space missions yet.

The objective of this work is to find the most suitable interface for communicating each NI-ICU (nominal or redundant) with the four CPU boards of the NI-DPU. This section describes the candidate communication interfaces which were initially selected for this purpose. Only interfaces which have been used or standardized by ESA have been considered, all of them having radiation hardened components available.

3.1   MIL-STD-1553B

The MIL-STD-1553B standard is used for many spacecraft onboard data links including commercial and scientific satellites, space exploration vehicles, launchers and in-orbit manned flight applications [11].

A MIL-STD-1553B system consists of a Bus Controller (BC) controlling up to 31 Remote Terminals (RT), all connected together by a data bus. The bus medium is composed of a nominal (Channel A) and a redundant (Channel B) physical interconnection. Messages can be sent either on channel A or on channel B by the BC, and the RT replies always on the same channel.

Figure 1 shows the most common way of connecting a remote terminal to a bus channel (A or B) [12]. Although there is only one RT shown in the figure, for a complete bus system more units are needed.

Systems based on MIL-STD-1553B are considered to be extremely reliable. However, the cost of these systems is relatively high due to the need of expensive components such as transformers. Another issue is the power consumption, which is very high compared to other communication interfaces.

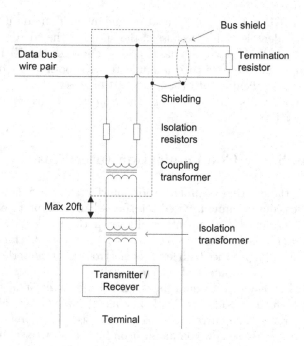

Fig. 1. MIL-STD-1553B data bus interface

3.2   LVDS

Low-voltage differential signaling, or LVDS, is a general-purpose communication interface standard used primarily for point-to-point and multi-drop cable driving applications [13].

LVDS uses balanced signals to provide very high-speed interconnection (>100 Mbits/s) using a low voltage swing (350 mV typical). The balanced or differential signalling provides adequate noise margin to enable the use of low voltages in practical systems. Low voltage swing means low power consumption at high speed. LVDS is appropriate for connections between boards in a unit, and unit to unit interconnections over distances of 10 m or more.

A typical LVDS driver and receiver are shown in Figure 2. The signaling levels used by LVDS are illustrated in Figure 3.

Since the LVDS standard covers the physical layer specification only, many data communication standards and applications use it adding a data link layer on top of it.

3.3   RS-422

RS-422 is a serial communications standard that specifies the electrical characteristics of a balanced voltage digital interface. This interface is capable of transmitting data over long distances (up to 1500 m) and in noisy environments. At short distances (<12 m), the data rate can reach 10 Mbits/s [14].

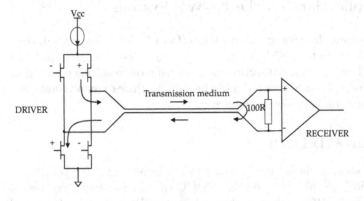

Fig. 2. LVDS operation

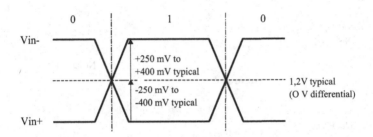

Fig. 3. LVDS signaling levels

The transmitting and receiving circuits for RS-422 are equivalent to the ones shown in Figure 2. Figure 3 is also valid for describing RS-422 signaling, but with a typical voltage swing of 3V and a common mode voltage of 1.8V. These higher voltage levels lead to higher power dissipation than LVDS.

3.4 SpaceWire

SpaceWire is a data-handling network for use on-board spacecraft. It provides serial, high-speed (2 Mbits/s to 200 Mbits/s), bi-directional, full-duplex data-links. Data is encoded using two differential signal pairs in each direction. That is a total of eight signal wires, four in each direction. Networks can be built to suit particular applications using point-to-point data links and routing switches [15,16].

The SpaceWire standard is divided into several protocol levels, ranging from the physical level to the packet level. Describing those levels is out of the scope of this paper.

The main advantages of the SpaceWire interface are its simple circuitry, low power consumption and low-error rate. These features make SpaceWire a very suitable interface for many space applications.

4   Implications on the NI-WE System

Once the candidate communication interfaces have been analyzed, the next step
is to apply them to the NI-WE in order to study the possible implications on the
system. The following subsections will describe the resulting connection schemes
for each one of the considered interfaces, highlighting any relevant consequences
which each configuration may have.

4.1   MIL-STD-1553B

Figure 4 shows a diagram of a connection scheme for the NI-WE based on MIL-
STD-1553B. In this diagram, the NI-ICU (nominal or redundant) would be the
bus controller (BC) of the 1553 bus, while the CPU boards in the NI-DPU would
act as remote terminals (RT).

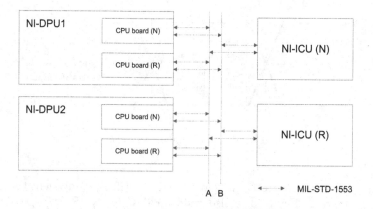

Fig. 4. NI-WE connection scheme based on MIL-STD-1553B

In the first stages of the Euclid mission, 1553 was considered a very convenient
interface for the communications between the NI-ICU and the NI-DPU, as the
bus topology seemed to naturally harmonize with the architecture of the NI-WE.
Besides, the Maxwell SCS750 board is available with an optional 1553 interface,
so there would not be the need to develop any extra hardware in the DPU side.

However, further analysis led to think that this solution was far from optimal.
Other alternative interfaces, when used properly, can provide the same functionality as 1553 with lower cost, lower power consumption, smaller PCB area and,
in some cases, higher transmission rates.

Another disadvantage of 1553 is that it requires lots of hardware resources
when implemented in an FPGA. These resources will be very constrained in the
NI-ICU side, because the FPGA will be hosting a full LEON2 processor and
some other interfaces apart from 1553. If there are not enough resources for
the 1553 IP core, an additional FPGA or an ASIC containing this functionality

should be added to the NI-ICU design, thus increasing development complexity, power consumption and economic cost dramatically. Selecting a more appropriate communication interface can avoid the need of this additional piece of hardware.

4.2 LVDS

A communication scheme based on a point-to-point serial interface like LVDS would need several links in order to provide features similar to those of the system presented in Figure 4. Such scheme is depicted in Figure 5.

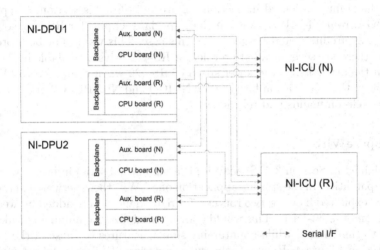

Fig. 5. NI-WE connection scheme based on a serial interface (LVDS, RS-422 or SpaceWire)

Figure 5 shows a new type of board called "Aux. board". These boards will contain any additional interface electronics needed by the DPUs but not provided by the Maxwell boards, which are not modifiable. In all the configurations presented in this paper, the auxiliary boards contain SpaceWire routers which will be used by the CPU boards to send science data to the spacecraft mass memory. The existence of SpaceWire routers allows multiple paths for sending data, as a preventive measure for link errors.

By inspecting Figure 5, one can conclude that the new scheme allows any ICU (nominal or redundant) to communicate with any of the CPU boards. Communication between the different CPU boards is not possible with this scheme, but this is not a problem because such functionality is not required.

Adding the two necessary LVDS interfaces to each Maxwell board is possible because these boards include two generic UARTs in the backplane. The electrical characteristics of these serial ports can be either LVDS or RS-422. The LVDS signals should then be routed through the backplane to the auxiliary board, where the necessary connectors shall be located. The LVDS drivers are already provided by the Maxwell board, so there is no need to add extra circuits in the auxiliary board.

On the NI-ICU side, there should not be any problems with the limited FPGA resources, as the UARTs are small circuits with very little hardware needs. Any number of UARTs can be attached to the LEON2 processor through the internal AMBA bus. The FPGA which will be used in the NI-ICU already provides some specific I/O pins for LVDS, so no additional driving circuitry would be needed.

4.3   RS-422

All of the points presented in Section 4.2 are valid for this section, except one: the FPGA from NI-ICU does not include special I/O pins for RS-422, so external driving circuits would be needed in this case. This would not be a problem because there are several radiation tolerant RS-422 drivers available in the market. On the other hand, the lower speed of RS-422 when compared to LVDS is tolerable in this application because the NI-ICU and NI-DPU will only exchange low-rate telecommands and telemetry.

4.4   SpaceWire

As explained in Section 3.4, SpaceWire is a high-speed serial interface with network capabilities. In the specific application of NI-WE, these network capabilities could be exploited if one or two routers (for redundancy) were added between the NI-ICU and the NI-DPU. This would allow to minimize the number of links connected to the NI-ICU while maintaining similar features to those of the system presented in Figure 4. However, the current design of NI-WE does not permit placing any additional electronics between NI-ICU and NI-DPU, apart from the cabling itself. The router(s) cannot be located inside the NI-ICU either, because some kind of cross-strapping would be needed inside the NI-ICU, thus breaking the concept of a fully cold-redundant unit.

Because of the reasons outlined in the paragraph above, the most preferable connection scheme would be one presented in Figure 5. The auxiliary boards shown in this figure already include SpaceWire routers, so there must be two free ports in each router in order to connect the two necessary links for each board. If there are not enough free ports, a bigger, more expensive router would have to be used. Communication between the auxiliary boards and their corresponding CPU boards would be done through a PCI bus located in every backplane.

Regarding the NI-ICU, four SpaceWire nodes would have to be implemented inside the FPGA. It is not necessary to implement a full router, because every packet that arrives at the NI-ICU is aimed at the NI-ICU itself. Nonetheless, it must be noted that SpaceWire IP cores are more resource demanding than simple UARTs, so there is a possibility that there are not enough resources in the FPGA to implement the four SpaceWire nodes. If that is the case, an additional FPGA or an ASIC containing this functionality should be added to the NI-ICU design, with a dramatic impact on the development complexity, power consumption and economic cost.

5   Conclusions

The discussion presented in Section 4 leads us to select LVDS as the most suitable interface for the communication links between the NI-ICU and the NI-DPU. This is not only because of the intrinsic advantages of this interface (high speed, low power, simple interface circuits), but also because of the benefits obtained in the specific case of the NI-WE.

LVDS has proven to be the easiest interface to implement in both the NI-ICU and the NI-DPU. This easiness ultimately results in faster development times and smaller development costs. In the case of NI-WE, this is specially true because the Maxwell boards already provide enough UART controllers for this application, and also because any number of UARTs can be added to the FPGA of the NI-ICU as long as there are enough hardware resources. What is more, both the Maxwell boards and the RTAX-S/SL FGPAs provide several LVDS channels which can be used in conjunction with the UARTs, eliminating the need of additional driving hardware.

RS-422 could be considered a backup alternative to LVDS, with lower transmission rates and higher power consumption. As explained in Section 4.3, the lower transmission rates would not be an issue in the case of NI-WE. Still, LVDS is the preferred solution because of the reasons explained in the paragraph above.

It is certainly possible to use SpaceWire as the communication interface between NI-ICU and NI-DPU. However, in practice, the NI-WE would not be benefiting from any of the advantages of this more complex interface (i.e., networking capabilities), and it would suffer anyway from the problems derived from the higher hardware needs.

MIL-STD-1553B seems to be least suitable option for the NI-WE communications. Among all the analyzed interfaces, it is the most demanding one in terms of power, economic cost and hardware utilization in a FPGA. No relevant benefits have been identified when using this interface.

Finally, it must be noted that both MIL-STD-1553B and SpaceWire have been standardized by ESA with very detailed specifications, which forces any unit using any of these interfaces to demonstrate that the standard is met. This results in extra time and effort on specific verification test plans, requiring dedicated ground equipment and software.

Acknowledgements. The authors want to acknowledge the contributions provided by the NISP system team of the Euclid Consortium to this work. This work has been supported by the Spanish Ministry of Economy under the projects AYA2011-14245-E and AYA2011-15855-E.

References

1. Laureijs, R.J., Duvet, L., Escudero Sanz, I., Gondoin, P., et al.: The Euclid Mission. In: Proceedings of the SPIE Conference on Astronomical Telescopes and Instrumentation, Proc. SPIE, vol. 7731 (2010)
2. Cimatti, A., Robberto, M., Baugh, C., et al.: SPACE: the spectroscopic all-sky cosmic explorer. Experimental Astronomy 23, 39–66 (2009)
3. Cropper, M., Cole, R.E., James, A., et al.: VIS: the visible imager for Euclid. In: Proceedings of the SPIE Conference on Astronomical Telescopes and Instrumentation, Proc. SPIE, vol. 8453 (2012)
4. Valenziano, L., Zerbi, F.M., Cimatti, A., et al.: The E-NIS instrument on-board the ESA Euclid Dark Energy Mission: a general view after positive conclusion of the assesment phase. In: Proceedings of the SPIE Conference on Astronomical Telescopes and Instrumentation, Proc. SPIE, vol. 7731 (2010)
5. Prieto, E., Amiaux, J., Auguères, J.L., et al.: Euclid near-infrared spectrophotometer instrument concept at the end of the phase A study. In: Proceedings of the SPIE Conference on Astronomical Telescopes and Instrumentation, Proc. SPIE, vol. 8453 (2012)
6. Bortoletto, F., Bonoli, C., D'Alessandro, M., et al.: Euclid ENIS Spectrograph Focal Plane Design. In: Proceedings of the SPIE Conference on Astronomical Telescopes and Instrumentation, Proc. SPIE, vol. 7731-55 (2010)
7. Cerna, C., Clemens, J.C., Ealet, A., et al.: The EUCLID NISP detectors system. In: Proceedings of the SPIE Conference on Astronomical Telescopes and Instrumentation, Proc. SPIE, vol. 8453 (2012)
8. Corcione, L., Ligori, S., Bortoletto, F., et al.: The on-board electronics for the near infrared spectrograph and photometer (NISP) of the EUCLID Mission. In: Proceedings of the SPIE Conference on Astronomical Telescopes and Instrumentation, Proc. SPIE, vol. 8453 (2012)
9. Martin-Gonzalez, J.A., Poves, E., Lopez-Hernandez, F.J.: Algorithm Optical Codes: An Alternative to Random Optical Codes in an Intra-Satellite Optical Wireless Network. Presented at the 10th Anniversary International Conference on Transparent Optical Networks, ICTON (2008)
10. Perez-Mato, J., Perez-Jimenez, R., Tristancho, J., Zechmeister, C.S.: Experimental approach to an optical wireless interface for an avionics data bus. In: 2012 IEEE/AIAA 31st Digital Avionics Systems Conference, DASC, pp. 7E4-1–7E4-10 (2012)
11. ECSS Space engineering: Interface and communication protocol for MIL-STD-1553B data bus onboard spacecraft. Standard ECSS-E-50-13C, ESA Requirements and Standards Division (November 2008)
12. Ljunggren, B.: Spacecraft Interface Standards Analysis and Simple Breadboarding. Institutionen för teknik och naturvetenskap (2005)
13. Parkes, S.M.: High-Speed, Low-Power, Excellent EMC: LVDS for On-Board Data-handling. In: 6th International Workshop on Digital Signal Processing Techniques for Space Applications, DSP 1998. ESTEC, Noordwijk (1998)
14. American National Standards Institute, Telecommunications Industry Association, and Electronic Industries Alliance: Electrical Characteristics of Balanced Voltage Digital Interface Circuits. TIA/EIA standard. Telecommunications Industry Association (2000)
15. ECSS: Space engineering: SpaceWire - Links, nodes, routers and networks. Standard ECSS-E-ST-50-12C, ESA Requirements and Standards Division (July 2008)
16. Parkes, S.: SpaceWire User's Guide. STAR-Dundee Limited (2012)

Discriminant Splitting of Regions
in Traffic Sign Recognition

Sergio Lafuente-Arroyo[1], Roberto J. López-Sastre[1],
Saturnino Maldonado-Bascón[1], and Rafael Martínez-Tomás[2]

[1] GRAM, Department of Signal Theory and Communications, UAH,
Alcalá de Henares, Spain
[2] Department of Artificial Intelligence, UNED, Madrid, Spain

Abstract. Mining discriminative spatial patterns in image data is a
subject of interest in traffic sign recognition. In this paper, we use an
approach for detecting spatial regions that are highly discriminative.
The main idea is to search the normalized size blobs for discriminative
regions by adaptively partitioning the space into progressively smaller
sub-regions. Thus, each cluster of signs is characterized by an unique
region pattern which consists of homogeneous and discriminant 2-D regions. The mean intensities of these regions are used as features. To evaluate the discriminative power of the attributes corresponding to detected
regions, we performed classification experiments using a classifier based
on Support Vector Machines. The proposed method has been tested in
a real traffic sign database. Results demonstrate that the method can
achieve a considerable reduction of features with respect to extraction
from raw images while maintaining accurate.

Keywords: feature extraction, adaptive partitioning, traffic sign classification.

1 Introduction

Within the context of smart cities, there are several applications which need a
robust system for the automated detection of traffics signs. A recognition system
is a module integrated in a complete traffic sign detection system (TSDS) oriented to applications related to intelligent vehicles or maintenance of highways.
When a TSDS is considered in full generality, the sign detection and classification
stages are usually distinguished. In a traffic scene image, a road-sign detector
identifies a set of candidate regions that have been segmented previously Each
region of interest (ROI) is then passed on to a classification module and either
assigned to one of the known road-sign classes or rejected as a nonsign. In this
paper, we focus on the design of the classification module.

In order to design a road sign classifier, candidate regions must be appropri-
ately represented for the given classification technique. After a basic preprocessing such as scaling of the regions to equal size, or masking out the general sign

J.M. Ferrández Vicente et al. (Eds.): IWINAC 2013, Part II, LNCS 7931, pp. 395–403, 2013.
© Springer-Verlag Berlin Heidelberg 2013

background a more specific data representation has to be constructed. So far, different data representations have been used for road sign classification. Each candidate region is represented by a vector of numerical characteristics (features). Examples of descriptors used in road sign classification are color histograms [2], wavelets [8], appearance-based features [10], or directly the subsampled pixel intensities [1][5][6].

Based on a database of labeled examples, a road-sign-recognition system is trained, minimizing the error expected on examples unseen in training. Apart from high accuracy in classification of different sign types, a recognition system should also avoid erroneous identification of nonsigns, i.e., limit the number of false alarms. Furthermore, it should be suited for real-time deployment.

In order to reduce the computational load in the recognition stage, we propose a novel approach based on the search of a descriptor which splits the sign in blocks and codifies them in different ways. The most relevant areas are encoded with a high number of features whereas other parts are described with less information. Motivated by the use of partitioning splitting in three dimensional (3D) image data [7], we propose a novel descriptor for traffic sign recognition that splits the image into significative regions. The main idea is the creation of a set of regions that are significative in the sense that these discriminant and homogeneous regions will provide adequate information in order to distinguish a certain set of images from another one. The region splitting is based on a variant of the classical image splitting technique. The features that this method uses are the mean intensities of the regions.

This paper is organized as follows. In the next section we present the database used to measure the quality of the proposed method. In section 3 we introduce the discriminant splitting approach to extract the descriptor features while in section 4 the method is evaluated. Finally, section 5 draws conclusions and prospectives.

2   Traffic Sign Dataset

Unlike other fields of pattern recognition, where researchers work with well-known datasets, there are not standard sets for training and testing for traffic sign detection and recognition system (TSDS). A few publicly available traffic sign data sets exist:

1. German TSR Benchmark (GTSRB) [11]
2. KUL Belgium Traffic Signs Data set (KUL Data set) [12]
3. Swedish Traffic Signs Data set (STS Data set) [3]

Most of these databases have emerged within the last two years and are not yet widely used. Some important limitations is that GTSRB and the KUL Data sets are oriented toward classification, rather than detection, since each image contains exactly one sign without background. However, complete scenes are necessary for detection.

As the objective of a complete TSDS is to analyze the accuracy detection of traffic signs as well as the identification of their pictogram, Recognition and

Multi-sensorial analysis group (GRAM) at the Universidad of Alcalá has collected a complete database of Spanish traffic signs. All the samples have been extracted from sequences acquired by different video-cameras under variable lighting conditions. In general, the sizes of individual local regions used within the trainable similarity measure may vary. However, we have fixed the region size as the external meta-parameter. Input color images of variable size were first converted to a gray-level and re-scaled to 32×32 pixel raster by a nearest neighbor interpolation.

Some examples of normalized patterns of our database are shown in Figure 1 for the case of triangular signs. Note that samples include noisy blobs. For each color-shape combination the data set comprises a different number of classes with unbalanced class frequencies due to some signs are more common in traffic scenes. In Table 1 we summarized the number of classes and the number of patterns of all classes with the same color and shape. It is important to note that many signs can be segmented by two colors. For example, prohibitory signs can be extracted by two criteria corresponding to the red outer rim and the inner white area.

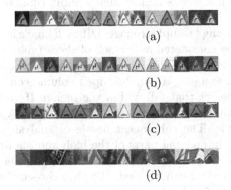

Fig. 1. Sample images of the GRAM dataset. (a), (b) and (c) Samples of red triangular signs. (d) Noisy blobs.

3   Extraction of Descriptors

We assume that there exist N subsets corresponding to the different combinations of color and shape regarding to Table 1. Thus, the dataset D is divided into subsets, $D = \cup_{i=1}^{N} U_i$, and in addition, each color-shape combination includes N_c classes.

Natural image information is not distributed uniformly over each pattern image that includes a sign: there are parts that are most relevant regarding to their contents, while other parts are far less relevant. In fact, the message of traffic signs is given by the pictogram, which covers different parts of the blob

Table 1. Number of classes for different color-shape combinations in the GRAM road-sign dataset

Combination	# Classes	# Samples
Red-Circular	62	18432
Red-Triangular	45	12498
Blue-Circular	54	2770
Blue-Rectangular	99	3600
White-Circular	114	8274
White-Triangular	44	9019
White-Rectangular	114	266
Yellow-Circular	47	1536
Yellow-Triangular	26	570

according to the class. Otherwise, background pixels of the sign does not provide relevant information for recognition task. Background of signs include big areas with pixels of the same color and texture that can be encoded by a representative value. We search homogeneous regions that are discriminant between two sets of images in order to build a map of features.

Let a set U_a containing l samples (images). If each image is of dimensions $h \times w$, these l images can be considered as a stack of slices (volume) with dimensions $l \times h \times w$, as illustrated in Figure 2. Thus for our purpose, a certain region B can be considered as being a parallelepiped volume comprising of the parts of every image in the set that fall within the region. If an image I is divided into R regions we can constitute a map of regions with a strategy similar to coarse-to-fine methods. The subsequent fine-level analysis can thus focus the attention just on the interesting parts of the blob and obtain the descriptor by considering the map. The criterium exploits the homogeneity and any volume homogeneity check method can be used. We have chosen the one based on the intensity range $|I_{max} - I_{min}|$, where I_{max}, I_{min} are the maximum and minimum intensity values of a region. If the range is smaller than a certain threshold, i.e.: $|I_{max} - I_{min}| \leq T_s$, then the region is regarded to be homogeneous, where the threshold T_s denotes the Otsu threshold [9] calculated for the current region. The homogeneity of a region is judged based on the pixels intensity values of the parts of all the training images that fall within the regions boundaries, i.e. on all pixels of the corresponding volume.

In order to determine the discriminant and homogeneous regions for each set, the classical splitting approach is applied to the l images of this set. For each set the stack of images is recursively split into four quadrants or regions (see Figure 2), until 2D homogeneous regions are encountered. The splitting is performed by bisecting the rectangular regions (in the entire image stack) in the vertical and horizontal directions. To deal with the problem of determining the right splitting resolution of the partitioning approach, it dynamically and adaptively partitions the space. For this reason we call the proposed approach Dynamic Recursive Partitioning (DRP).

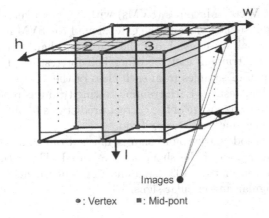

Fig. 2. Representation of a stack of images

An example of application of the method is shown in Figure 3. Note the splitting maps are obtained when the threshold T_s takes values from to 0.1 to 0.7 in steps of 0.1.

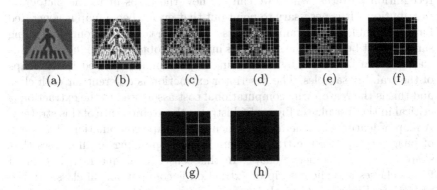

Fig. 3. Example of a traffic sign and its maps of splitting for different thresholds (T_s). (a) Original sign. (b) $T_s = 0.1$. (c) $T_s = 0.2$. (d) $T_s = 0.3$. (e) $T_s = 0.4$. (f) $T_s = 0.5$. (g) $T_s = 0.6$. (h) $T_s = 0.7$.

4   Results

4.1   Experimental Setup

We compared the performance of DRP with the descriptor used in [5], where we scanned directly the raw pixel values in gray-level images. We will refer to this last method as Raw Pixel Descriptor(RPD). Note that in the limit situation in which each block of the DRP method corresponds with a pixel of the image, DRP and RPD are the same. Both descriptors were used as inputs to an architecture based on Support Vector Machines (SVMs) with gaussian kernel. The problem under study is a multi-class one and the extension of the SVM method was done using a one-against-all strategy. From the database, half of the samples was chosen randomly for training and the other half for testing. The training was performed using half of samples from each class of the dataset. The remaining samples were used for testing. Each training-testing trial was repeated five times and we averaged the percentage of the correct predictions to obtain the reported accuracy. Processing time was averaged too.

In the RPD method, background sign is masked out and only those pixels that cover the inner area of the shape are evaluated. The number of features of 32×32 normalized blobs is 1024, 545 and 711 components, respectively, for rectangular, triangular and circular signs.

4.2   Architectures

The DRP descriptor needs the creation of a map of features in the training process. According to how we selected the sets from which we obtained the maps, two strategies were used:

- A map of features is created for each class. Figure 4 shows examples of recognition patterns, where we can see how the maps fit to the pictogram distributions. It is necessary to point out that even when a map is generated for each sample in the training process, we fusion the maps from all training samples that belong to the same class in order to obtain an unique map. This class pattern is created by considering regions more repeated in the maps of the training samples. The descriptor extraction is different for each class and this is the reason why computational cost associated to the extraction is critical in the test phase. Figure 5 illustrates the architecture of this strategy.
- A map of features is generated for each color-shape combination. The stack of images comprises the training samples corresponding to all classes that share color and geometric shape. In this sense an unique map is created for all classes and the descriptor extraction is common for all classes in the test stage. Figure 6 shows the maps associated to three specific sets: blue rectangular, red triangular and red circular. Note that since outer pixels are noisy in many samples, regions of the periphery are strongly split sometimes as it is illustrated in Figures 6(a) and 6(c) even when these areas are quite uniform. Figure 7 illustrates the architecture, which is chosen in this work due to the common process of descriptor extraction for all the classes.

4.3   Results

We compared the performance of the proposed method (DRP) using a map common to all the classes with that based on scanning raw pixels (RPD). The mean intensity values of regions were used as inputs to the classifier (SVMs). Table 2 summarizes the results obtained in the test stage. The number of support vectors (SV) obtained in the training phase is denoted in the third column.

Fig. 4. DRP maps for blue rectangular signs with $T_s=0.5$

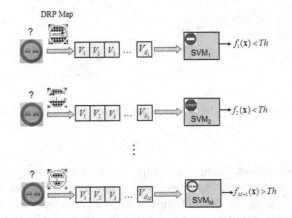

Fig. 5. Architecture DRP for the testing stage when a map of features is generated for each class

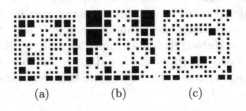

Fig. 6. Global maps for color-shape combinations. (a) Blue rectangular map. (b) Red triangular map. (c) Red circular map.

Table 2. Results with with blue rectangular road signs

Combination	Descriptor	# Features	# SV	$P_d(\%)$	T_{ext}	T_{svm}	T_{total}
Blue Rectangular	RPD	1024	2241	96.01	28.88	54.16	83.04
	DRP	433	1841	95.8	12.76	26.46	39.22
Red Triangular	RPD	545	7133	96.3	108.5	384.3	492.8
	DRP	352	6788	95.52	76.3	328.43	404.73
Red Circular	RPD	711	9440	94.9691	525.87	849.5	1375.37
	DRP	511	9419	94.839	433.29	739.56	1172.85

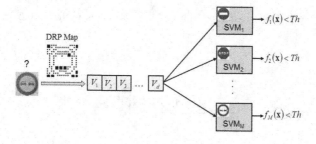

Fig. 7. Architecture DRP for the testing stage when a map of features is generated for each color-shape combination

The computational cost, including computation and storage, increases as the number of support vector does. The classification accuracy across the whole test dataset is included as the percentage of samples classified correctly by SVMs. Finally, the overall execution speed is quantified through the time (in seconds) that algorithm needs to extract the descriptors of training sets (T_{ext}) and classify them with SVMs (T_{svm}), respectively. The total time of classification is given in the last column of the table.

By inspecting the results we can conclude that the goal of this descriptor is the reduction of the number of features used to describe the signs, specially in the case of rectangular ones due to the geometry used in the decomposition. The reduction of processing time is approximately 52.7%, 17.8% and 14.7%, respectively, for blue rectangular, red triangular and red circular traffic signs even when the proposed algorithm was found to be able to recognize traffic signs with similar accuracy compared to RPD. Note that values are the average of 5 times of tests.

5   Conclusions

A scheme based on texture analysis is proposed in this paper in order to get a compact descriptor for the traffic sign recognition task. To show advantages of our approach we have conducted experiments on a real traffic sign dataset.

The mean intensity values of the different pattern blocks are integrated into a vector to characterize the images. Results show that the proposed technique is able to recognize the objects with very good accuracy and low computational burden. Further investigations are ongoing about the applicability of blocks with different geometry, such as radial descriptors.

Acknowledgment. This work was partially supported by projects TIN2010-20845-C03-03, IPT-2011-1366-390000 and IPT-2012-0808-370000.

References

1. de la Escalera, A., Armigol, J.M., Pastor, J.M., Rodriguez, F.J.: Visual sign information extraction and identification by deformable models for intelligent vehicles. IEEE Transactions on Intelligent Transportation Systems 5(2), 57–68 (2004)
2. Hsien, J.C., Liou, Y.S., Chen, S.Y.: Road sign detection and recognition using hidden Markov model. Asian Journal of Health and Information Sciences 1, 85–100 (2006)
3. Larsson, F., Felsberg, M.: Using fourier descriptors and spatial models for traffic sign recognition. In: Heyden, A., Kahl, F. (eds.) SCIA 2011. LNCS, vol. 6688, pp. 238–249. Springer, Heidelberg (2011)
4. Liu, Y., Ikenaga, T., Goto, S.: Geometrical, physical and text/symbol analysis based approach of traffic sign detection system. In: Proceedings of the IEEE Intelligent Vehicle Symposium, pp. 238–243 (2006)
5. Maldonado-Bascon, S., Lafuente-Arroyo, S., Gil-Jiménez, P., Gómez-Moreno, H., López-Ferreras, F.: Road-Sign Detection and Recognition Based on Support Vector Machines. IEEE Trans. on Intelligent Transportation Systems 8(2), 264–278 (2007)
6. Maldonado-Bascon, S., Acevedo-Rodríguez, J., Lafuente-Arroyo, S., Fernández-Caballero, A., López-Ferreras, F.: An optimization on pictogram identification for the road-sign recognition task using SVMs. Computer Vision and Image Understanding 114(3), 373–383 (2010)
7. Megalooikonomou, V., Kontos, D., Pokrajac, D., Lazarevic, A., Obradovic, Z.: An adaptive partitioning approach for mining discriminant regions in 3D image data. Journal of Intelligent Information Systems Archive 31(3), 217–242 (2008)
8. Nguwi, Y.Y., Cho, S.Y.: Two-tier self-organizing visual model for road sign recognition. In: Proceedings of the IEEE International Joint Conference on Neural Networks (IJCNN), pp. 794–799 (2008)
9. Otsu, N.: A threshold selection method from gray-level histograms. IEEE Transactions on Systems, Man, and Cybernetics 9, 62–66 (1979)
10. Shaposhnikov, W.D., Shaposhnikov, D.G., Lubov, N., Golovan, E.V., Shevtsova, A.: Road sign recognition by single positioning of space-variant sensor window. In: Proceedings of the 15th International Conference on Vision Interface (2002)
11. Stallkamp, J., Schlipsing, M., Salmen, J., Igel, C.: The german traffic sign recognition benchmark: a multi-class classification competition. In: Proceedings of the International Joint Conference on Neural Networks (IJCNN), pp. 1453–1460 (2011)
12. Timofte, R., Zimmermann, K., Van Gool, L.: Multi-view traffic sign detection, recognition and 3D localisation. Journal of Machine Vision and Applications, 1–15 (2011)

Detection of Fishes in Turbulent Waters
Based on Image Analysis

Alvaro Rodriguez[1], Juan R. Rabuñal[1,2],
Maria Bermudez[3], and Jeronimo Puertas[3]

[1] Dept. of Information and Communications Technologies,
[2] Centre of Technological Innovation in Construction and Civil Engineering
(CITEEC),
[3] Dept. of Hydraulic Engineering (ETSECCP),
University of A Coruña, Campus Elviña s/n 15071, A Coruña
{arodriguezta,juanra,mbermudez,jpuertas}@udc.es

Abstract. This paper analyses the automatic fish segmentation problem in turbulent waters. To this end, a *SOM* neural network is used to detect fishes in images from an underwater camera system built in a vertical slot fishway, an hydraulic structure built in obstructions in rivers to allow the upstream migration of fishes.

This technique allows the study of real fish behavior and may help to understand biological variables and swimming limitations of the fish species in high speed environments.

This knowledge, may be used to replace traditional techniques such as direct observation or placement of sensors on the specimens, which are impractical or affect the fish behavior.

To test the proposed technique, a ground true dataset was designed with experts and a series of assays have been performed where the results obtained with the proposed technique were compared with different segmentation techniques.

Keywords: Fish-Detection, Segmentation, SOM, Turbulent Water.

1   Introduction

While water and fish behavior has been widely studied, and flow properties of fluids are well known. The fish behavior and swimming performance of fishes in turbulent or high speed waters is an unsolved question.

This question is, however, critical when studying upstream migration of fishes and especially when a fish must pass through artificial structures such as slot fishways.

A vertical slot fishway (Fig. 1) is an hydraulic structure built in rivers to preserve fish biodiversity in places affected by the construction of engineering works, such as dams or weirs. These elements alter the ecosystem of rivers, causing changes in the fauna and flora and obstructing fish migration.

J.M. Ferrández Vicente et al. (Eds.): IWINAC 2013, Part II, LNCS 7931, pp. 404–412, 2013.
© Springer-Verlag Berlin Heidelberg 2013

This type of fishway is basically a channel divided into several pools separated by slots.

An effective vertical slot fishway must allow fish to enter, pass through, and exit safely with minimum cost to the fish in time and energy. Thus, biological requirements should drive design and construction criteria for this type of structures.

However, while flow in vertical slot fishways has been characterized in several works [1–4] and fish swimming performance has been studied by some authors [5, 6], the analysis of the real fish behavior in fishway models, or structures with a high velocity of water are scarce in the literature, so the actual behavior of the fish in these situations is practically unknown.

Fig. 1. Fishway model used in experiments, built at Center for Studies and Experimentation of Public Works CEDEX, in Madrid

The analysis and detection of fishes in real or controlled scenarios, is a problem widely studied.

Some early examples of these applications are the use of acoustic transmitters and a video camera for observing the behavior of various species [7], or the utilization of acoustic scanners for monitoring fish stocks [8].

More recently, different computer vision techniques have been used for the study of the fish behavior such as the study of swimming performance by analyzing the water with a particle image velocimetry [9].

In more related works, different techniques have been used for segmentation and fish tracking.

Some examples are techniques based on color segmentation such as [10] where fluorescent marks are used for the identification of fishes in a tank or in [11] where color properties and background subtraction are used to recognize live fish in clear water.

Other used techniques are stereo vision [12], background models [13], shape priors [14], local thresholding [15], moving average algorithms [16] or pattern classifiers applied to the changes measured in the background [17].

The works previously described, use color features, background or fish models, and stereo vision systems. These techniques are carried out in calm and low turbulent water where information about textures and color may be enough to discriminate the fish. However, these conditions are not fulfilled in the turbulent conditions of a vertical slot fishway.

In this work, the use of SOM Neural Networks will be studied to detect living fishes in vertical slot fishways from images obtained through a network of video cenital underwater video cameras (Fig. 2).

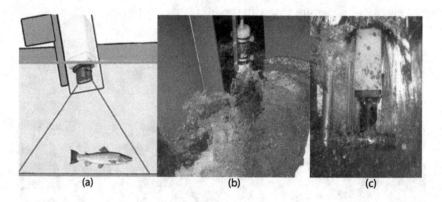

Fig. 2. Camera system. (a) Camera position in the fishway. (b,c) Recording conditions.

This work is an extension of the carried out in [18] where an image system is designed to record assays with real fishes and a methodology to study fish behavior is carried out, consisting in analysing fish trajectories with computer vision techniques and computing fish statistics mixing observed trajectories with hydraulic properties of the fishway.

As a part of the system proposed in [18] a *SOM* based segmentation technique was defined. However, the segmentation question was not in depth analysed, and this technique was not compared with other techniques or tested with ground true data.

This paper is focused in studying the fish segmentation problem in a vertical slot fishway as a part of a tracking system with the objective of generate the trajectory of the fish over time by locating its position in every frame of the video.

The behavior of different techniques is studied, analysing different approaches with a dataset recorded in experimental conditions and a true ground data created by experts.

The efficiency and accuracy of each technique has been measured with the defined data test.

Additionally a representation and filtering shape-based technique has been defined and tested.

2   Fish Detection

In order to study the behavior of the fish from the sequence of images acquired, the area occupied by the fish should be separated from the rest of the image or back-ground. This process is known as image segmentation.

The aim of image segmentation algorithms is to partition the image into perceptually similar regions. Every segmentation algorithm addresses two problems, the criteria for a good partition and the method for achieving efficient partitioning [22].

Therefore it is necessary to find a variable which allow a robust separation of the fish from the background. Then it is necessary to choose a technique to classify the image in different groups according to a criteria.

Most common criteria in segmentation processes are the color features, the analysis of edges and a priori knowledge of the background.

Additionally, underwater images are characterized by low levels of contrast [17] and images recorded in the high turbulent waters will be characterized by extreme luminosity changes, huge noise levels and poor contrast, being texture and color information useless (Fig. 3).

Fig. 3. Example of images recorded in the fishway during the assays

A technique based on self-organizing maps (SOM), has been used. A **SOM** is a type of Artificial Neural Network (ANN), [20] which has been widely used image segmentation [23–26].

This network establishes a correlation between the number patterns supplied as an input and a two-dimensional output space (Topological map); thus, the input data with common features activate areas close to the map.

The SOM networks have the advantage of not needing to define a set of patterns to supervise the learning process. They can be straightforward created by defining a representative image or set of images. additionally, they are highly operable in parallel and have error tolerance [21].

After preliminary assays, a three-layer topology with 3 processing elements in each layer was fixed for each network.

the input parameters of the network were the local average of the RGB values in a window centered in the neighborhood of the pixel (i, j) (1).

$$E_{i,j} = \{\mu_{x,y}\}_{(x,y)=[(i-N/2,j-N/2),(i+N/2,j+N/2)]} \tag{1}$$

Additionally, to obtain a technique robust to different light conditions, and additional background knowledge was introduced to avoid the detection of background objects present in the pool.

To this end, a selected frame of the background was used to normalize the current image values

3   Preprocessing and Postprocessing

The input images were normalized and preprocessed to enhance the contrast of the image by using a contrast-limited adaptive histogram equalization $(CLAHE)$.

Once the image segmentation is obtained with the SOM network, the objective of the segmentation system is to determine the position of the detected fish in the image.

Due to the characteristics of the image, where the fish is often partially hidden and where is expected the presence of shadows, bubbles and reflections, the algorithm should respond well to partial or abnormal detections.

A simple algorithm has been built, in order not to increase the computational bur-den of the process. The algorithm is based on obtaining the connected-body vector from the segmented image. Each body will be characterized by the vector of pixels that make it up and by a set of descriptive parameters: its area, its centroid and the minimum ellipse containing the body.

Then the detected object is classified as a potential fish or noise, according to the size of the object and the minimum and major axis of the ellipse. An iterative fusion operation has been defined to replace two close unconnected bodies for a new one, which will be formed by the points from the two previous ones while shape criteria are satisfied.

Thus, the algorithm determines the mass centers of the fishes and puts together those connected bodies whose characteristics can be matched to a fish or a part of it, and discards those bodies that, due to their size or shape, are regarded as noise. Subsequently, the mass center of each detected fish is obtained.

4   Experimental Results

To measure the performance of the SOM networks, a set of experiments were per-formed with living fishes of the salmo trutta specie in a 1:1 vertical slot fishway mod-el located at Center for Studies and Experimentation of Public Works *CEDEX*, in Madrid.

A data set of 1000 images from 10 different cameras selected from different pool and fishway regions was defined and the corresponding ground true data was manually created by experts.

To evaluate the dependence of the results with the selected training patterns, the networks were trained using a single image, and results were analysed using 3 different trainings (Fig. 4).

Obtained results are shown in Fig. 4.

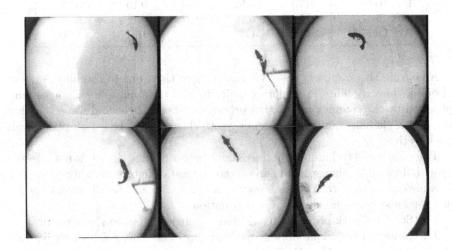

Fig. 4. Segmentation results

To establish a comparative background, two classic techniques have been designed based on the conventional approaches in segmentation, edge-based classification by means of analyzing the discontinuities in the image and the region-based classification by analyzing the similarity of pixels.

– A region technique based in the otsu method, which performs a region classification by automatically thresholding the histogram.
– An edge analysis technique based in the Canny edge detector.

Additionally, two different *SOM* based techniques were considered:

- A standard *SOM* using the intensity values of the image *I* from the neighborhood of the pixel *(i, j)*
- The SOM network proposed in [18] using the global and local average features from the current image I and a background reference image *I*

Processing and postprocessing stages have been applied with every technique. The average obtained results are shown in Table 1.

Table 1. The average results obtained with the different techniques

Avg. Results	Region	RGB SOM	Edge	Feature [18]	Proposed
Detections	2126	690	819	894	876
True Pos.	631	328	716	808	809
False Pos.	1495	362	103	85	67
True Neg.	12	40	47	49	50
False Neg.	705	748	270	189	187
Precission	0.30	0.48	0.87	0.90	0.92
Recall	0.47	0.30	0.73	0.81	0.81
F.P. Rate	0.70	0.52	0.13	0.10	0.08
T.P. Rate	0.53	0.70	0.27	0.19	0.19
Av. Time	404	1172	296	2881	3566

Analyzing the, obtained results it can be seen that, due to the huge amount of noise, the variability in camera conditions and the luminosity changes (emphasized by the complex interaction processes between light and water) neither the region classification technique nor *SOM* classification based on pixel levels are effective.

The edge based technique, obtained much better results, since it is not affected by global intensity changes and can overcome most of light variability processes. It is also the fastest technique, however, its precision and recall estimators are not good enough to be used in a real situation.

The SOM network proposed in [18] has obtained the second best result.

The proposed technique obtained the best results, with a false positive rate significant inferior to any other technique.

In general, it can be seen that a very low false positive rate is achieved, so obtained results are very reliable, and they will represent true positions of the fish with a high probability. Furthermore, precision is the most important factor in a detection system because a high rate of false positives would make the results useless.

However, the false negative rate is much higher. Showing that, in almost a 20% of the measurements, the system will not provide information about fish position. This is due to the fact that the fish will be frequently occluded by the turbulence of water. Nevertheless, it must be taken in account that the system can execute up to 25 measurements per second, so it will provide a correct position more than 15 times per second in average, which is far beyond any detection system used at present in this field.

5   Conclussions

In this work, a solution to detect fishes in vertical slot fishways based on SOM networks and an algorithm to filter anomalous detections is analysed.

The accuracy and performance of the proposed technique has been tested, analysing and comparing the obtained results with different segmentation techniques in dataset created with the help of experts.

The results obtained with this system have been very promising, as they allowed us to obtain the fish position in the image with a low error rate.

In future stages of this work, a tracking algorithm will be defined to manage the detected positions of the fish or fishes obtaining the trajectories of the specimens.

Although further research is needed, the results obtained will be used to elaborate a new methodology to study fish behavior inside vertical slot fishways so it can con-tribute to develop robust guidelines for future fishway designs and to establish more realistic criteria for the evaluation of biological performance of the current designs.

Acknowledgment. This work was supported by the Direccion Xeral de Investigacion, Desenvolvemento e Innovacion (General Directorate of Research, Development and Innovation) de la Xunta de Galicia (Galicia regional government) (Ref. 10MDS014CT) and from the Ministerio de Economa y Competitividad (Spanish Ministry of Economy and Competitiveness) Ref. CGL2012-34688. The authors would also like to thank the Center for Studies and Experimentation on Public Works (CEDEX).

References

1. Puertas, J., Pena, L., Teijeiro, T.: An Experimental Approach to the Hydraulics of Vertical Slot Fishways. Journal of Hydraulics Engineering 130 (2004)
2. Rajaratnam, N., Vinnie, G.V.D., Katopodis, C.: Hydraulics of Vertical Slot Fishways. Journal of Hydraulic Engineering 112, 909–927 (1986)
3. Tarrade, L., Texier, A., David, L.: Topologies and measurements of turbulent flow in vertical slot fishways. Hydrobiologia 609, 177–188 (2008)
4. Wu, S., Rajaratma, N., Katopodis, C.: Structure of flow in vertical slot fishways. Journal of Hydraulic Engineering 125, 351–360 (1999)
5. Dewar, H., Graham, J.: Studies of tropical tuna swimming performance in a large water tunnel – Energetics. Journal of Experimental Biology 192, 13–31 (1994)
6. Blake, R.W.: Fish functional design and swimming performance. Journal of Fish Biology 65, 1193–1222 (2004)
7. Photographic and acoustic tracking observations of the behavior of the grenadier Coryphaenoides (Nematonorus) armatus, the eel Synaphobranchus bathybius, and other abyssal demersal fish in the North Atlantic Ocean. Marine Biology 112, 1432–1793 (1992)
8. Steig, T.W., Iverson, T.K.: Acoustic monitoring of salmonid density, target strength, and trajectories at two dams on the Columbia River, using a split-beam scaning system. Fisheries Research 35, 43–53 (1998)

9. Deng, Z., Richmond, M.C., Guest, G.R., Mueller, R.P.: Study of Fish Response Using Particle Image Velocimetry and High-Speed, High-Resolution Imaging. US Department of Energy, Technical Report (2004)
10. Duarte, S., Reig, L., Oca, J., Flos, R.: Computerized imaging techniques for fish tracking in behavioral studies. European Aquaculture Society (2004)
11. Chambah, M., Semani, D., Renouf, A., Courtellemont, P., Rizzi, A.: Underwater color constancy enhancement of automatic live fish recognition. IS&T Electronic Imaging, SPIE (2004)
12. Petrell, R.J., Shi, X., Ward, R.K., Naiberg, A., Savage, C.R.: Determining fish size and swimming speed in cages and tanks using simple video techniques. Aquacultural Engineering 16, 63–84 (1997)
13. Morais, E.F., Campos, M.F.M., Padua, F.L.C., Carceroni, R.L.: Particle filter-based predictive tracking for robust fish count. In: Brazilian Symposium on Computer Graphics and Image Processing, SIBGRAPI (2005)
14. Clausen, S., Greiner, K., Andersen, O., Lie, K.-A., Schulerud, H., Kavli, T.: Automatic segmentation of overlapping fish using shape priors. In: Scandinavian Conference on Image Analysis (2007)
15. Chuang, M.-C., Hwang, J.-N., Williams, K., Towler, R.: Automatic fish segmentation via double local thresholding for trawl-based underwater camera systems. In: IEEE International Conference on Image Processing, ICIP (2011)
16. Spampinato, C., Chen-Burger, Y.-H., Nadarajan, G., Fisher, R.: Detecting, Tracking and Counting Fish in Low Quality Unconstrained Underwater Videos. In: Int. Conf. on Computer Vision Theory and Applications, VISAPP (2008)
17. Lines, J.A., Tillett, R.D., Ross, L.G., Chan, D., Hockaday, S., McFarlane, N.J.B.: An automatic image-based system for estimating the mass of free-swimming fish. Computers and Electronics in Agriculture 31, 151–168 (2001)
18. Rodriguez, A., Bermudez, M., Rabuñal, J.R., Puertas, J., Dorado, J., Balairon, L.: Optical Fish Trajectory Measurement in Fishways through Computer Vision and Artificial Neural Networks. Journal of Computing in Civil Engineering 25, 291–301 (2011)
19. Cheng, H.D., Jiang, X.H., Sun, Y., Wang, J.: Color image segmentation: advances and prospects. Pattern Recognition 34, 2259–2281 (2001)
20. Kohonen, T.: Self-organized formation of topologically correct feature maps. Biol. Cybernet. 43, 59–69 (1982)
21. Moya, F., Herrero, V., Guerrero, G.: La aplicación de redes neuronales artificiales (RNA) a la recuperación de la información. SOCADI Yearbook of Information and Documentation 1998(2), 147–164 (1998)
22. Yilmaz, A., Javed, O., Shah, M.: Object Tracking: A Survey. ACM Computing Surveys 38 (2006)
23. Verikas, A., Malmqvist, K., Bergman, L.: Color image segmentation by modular neural networks. Pattern Recognition Letters 18, 175–185 (1997)
24. Dong, G., Xie, M.: Color clustering and learning for image segmentation based on neural networks. IEEE Transactions on Neural Networks 16, 925–936 (2005)
25. Egmont-Petersen, M., Ridder, D., Handels, H.: Image processing with neural networks-a review. Pattern Recognition 35, 2279–2301 (2002)
26. Cristea, P.: Application of Neural Networks In Image Processing and Visualization. In: GeoSpatial Visual Analytics, pp. 59–71. Springer, Netherlands (2009)

Experimental Platform for Accelerate the Training of ANNs with Genetic Algorithm and Embedded System on FPGA

Jorge Fe, R.J. Aliaga, and R. Gadea

Universidad Politécnica de Valencia, Spain
jorfe@posgrado.upv.es
http:www.upv.es

Abstract. When implementing an artificial neural networks (ANNs) will need to know the topology and initial weights of each synaptic connection. The calculation of both variables is much more expensive computationally. This paper presents a scalable experimental platform to accelerate the training of ANN, using genetic algorithms and embedded systems with hardware accelerators implemented in FPGA (Field Programmable Gate Array). Getting a 3x-4x acceleration compared with Intel Xeon Quad-Core 2.83 Ghz and 6x-7x compared to AMD Optetron Quad-Core 2354 2.2Ghz.

1 Introduction

The structure most known and used in artificial neural networks ANN[1] is the multilayer perceptron (MLP) and backPropagation algorithm (BP) is used for training, then, one of the problems when implementing an ANN training with MP structure and BP as training algorithm is to determine the optimal topology. For the determination of the optimal topology requires a long time of experimentation. In [2] different methods are detailed topology optencion as, trial and error, empirical or statistical methods, hybrid methods, constructive and or pruning algorithms y evolutionary strategie. Another problem that has a high computational cost is the optimal determination of the initial weights of ANN training, in [3] shown that a proper selection of the initial weights reduces training times.

Available literature known developments in ANN applied to specific problems in FPGA devices as [4] where it develops a coprocessor for convolutional neural networks, made a comparison of CPU acceleration. Of [5]provides a training platform for reconfigurable topology. The drawback to this application is that it has to synthesize each time you change the topology and then implement it in FPGA. Another application implements a fixed-topology ANN and Backpropagation algorithm for training [6]. In [7] where proposed BP algorithm implementation in FPGA, this is reconfigurable by software. In [8] have implemented in FPGA accelerator for online training.

J.M. Ferrández Vicente et al. (Eds.): IWINAC 2013, Part II, LNCS 7931, pp. 413–420, 2013.

According to the previous review proposes a systematic search of hidden neurons number and value of initial weights in ANN. The search for both variables, initial weights and number of neurons, is done with genetic algorithms (GA) [9]. The platform consists of eight embedded systems (IS) development boards implemented. The board has a Cyclone IV EP4CE115F29CN FPGA Altera [10]. Each of the ES using the coprocessor [11] called Neural Network Processor (NNP) having a training system in FPGA based algorithm Resilient Backpropagation (RBP). ES each communicates with a host PC via Ethernet, this PC with Matlab [12], Global Optimization Toolbox and Parallel Computing Toolbox, manages training with GA and parallel tasks executed in each of the ES. Getting a 6x-7x acceleration compared with Quad-Core AMD Optetron 2354 and 3x-4x Porcesador Intel Xeon Quad-Core E5540. The paper is organized as follows: Section 2, describes the platform. Section 3, the implemented software. Section 4, details the embedded system. Section 5, presents experimental results. Section 6, conclusions

2   Platform Training

The experimental platform is presented in Figure 1, composed of a host PC multi-core, and eight ES, connected to the network via Ethernet. Use this mode of communication as it allows the scalability, and transfer data effectively, and allows to quickly add ES. As the number of variables for the determination of the

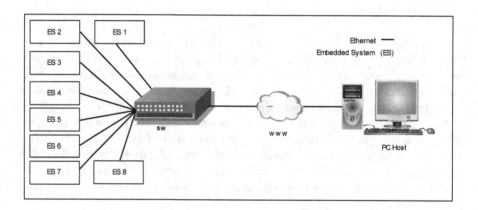

Fig. 1. ANN Platform training

topology or the initial weights vary, when used GA to the systematic search for both variables, generate an initial population of individuals n, the calculation of each individual is performed in each ES, allowing the calculation of eight individuals simultaneously.

3   Software

This section presents the execution flow of the software to achieve optimal topology and determination of the initial weights of an ANN. The software responsible for managing the training tasks are Matlab with Parallel Computing Toolbox (PCT), this allows you to manage the tasks running on each processor core and the execution of the tasks of higher computational cost run on SE. Besides using Global Optimization (GO), this provides different methods for finding solutions to different problems, one of them is with genetic algorithms are algorithms that are based on natural selection and the laws of genetics.

3.1   GA Tasks

In GA the Most Important parameters to configure are the number of individuals, generation number and the fitness function. These parameters determine the execution time of the GA in this aplication. The computation time depends of the individuals quantity to be computed in parallel, this parameter is configurable in the algorithm implemented by combining the PCT with GO, can be run from 1 to 8 individuals in parallel. Figure 2 shows a flowchart of the algorithm. Here is a brief description of each:

- Initialize the number of cores used in the host PC for ANN training.
- Transmission set of training vectors to SDRAM memory in ES.
- Configure the genetic algorithm (number of generations, numbers of individuals, use PCT, etc).
- Called the fitness function.
- The fitness function determines if there is a free core, then, generates an identifier and this identifier is assigned an IP address.
- The fitness function communicates with each of the SE through the IP address.
- So on until the end of the number of individuals and generations set in the GA is completed.
- The stopping method is by the number of generations.

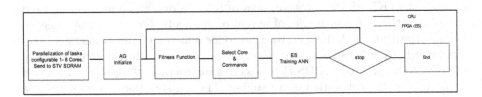

Fig. 2. Software flow diagram

4   Embedded System

Using Terasic development board DE2-115 to implement the embedded system, the ES is mainly composed of a real-time operating system RTOS MicroC/OS-II, processor NIOS II/f and NNP coprocessor, The NNP is a soft-core Single Instruction, Multiple Data Path (SIMD) in Figure 3 shows the hardware implemented.

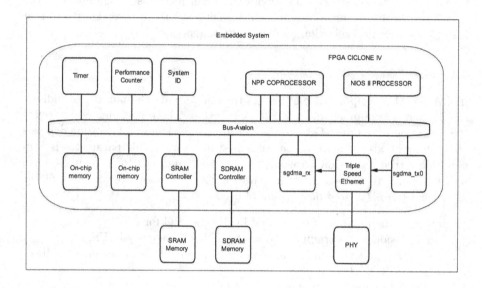

Fig. 3. Embedded System

4.1   Description

The Nios embedded processor is the master and controls the other system components, running from the program memory from External 2MB SSRAM to FPGA. The training vectors are stored in the external SDRAM. The remaining components are internal to the FPGA, besides the coprocessor are two DMAs that control the training vector transfer and instruction to the coprocessor, a memory of 16 KB which stores coprocessor instructions, and a memory of 4 KB (double the size of the memories of weight or gradients) for the variables needed for the algorithm RBP, these correspond to local increases and gradients of previous epoch. Figure 4 shows the connection diagram and internal architecture of the NNP. The processor NIOS II / f receives instructions through the Ethernet connection to the generation of a training. These instructions to configure the topology, if the generation of the initial weights is undertaken by the NIOS II, or should start with weights sent to memory via the Ethernet connection, the number of epoch, next, with all data received NIOS is responsible for generating the instructions for each new topology received, then, known [11] that the coprocessor is capable of processing a certain number n of parallel training vectors

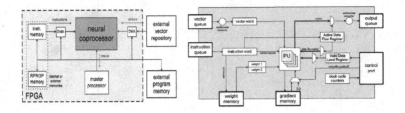

Fig. 4. Neural Network Coprocessor

and the gradients generated automatically accumulate, ultimately appearing in the memory of gradients at the end of the epoch. The reading and processing of each group of n vectors are getting by means of a set of instructions NNP.

5   Experimental Results

To show the validity of the platform as an accelerator expermiental to determine the initial weights of an ANN, has set a topology 6/5/3/1 with a sample set of 243,000 training vectors. Executed tests from one (ES) to eight (ES), also for each test also varied the number of epoch from 5 to 160 epoch. The result are show from figure 5 to figure 8.

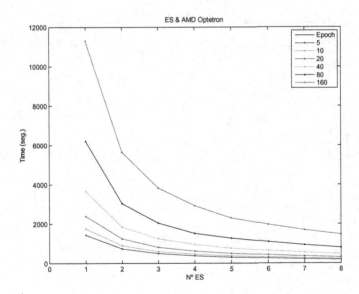

Fig. 5. Training, used from 1 ES to 8 ES, from 5 epoch to 160 epoch

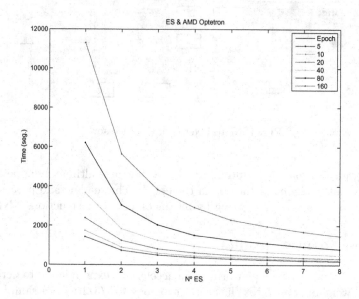

Fig. 6. Training, used from 1 ES to 8 ES, from 5 epoch to 160 epoch

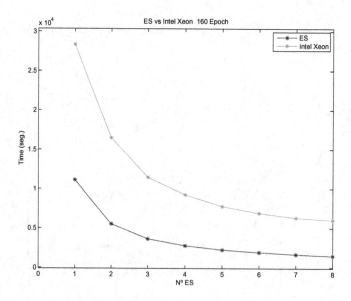

Fig. 7. ES vs CPU to 160 Epoch

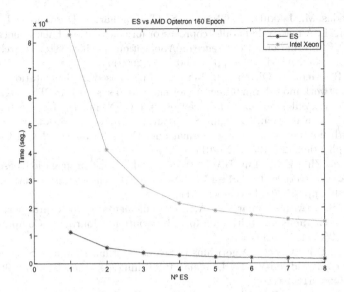

Fig. 8. ES vs CPU to 160 Epoch

6 Conclusions and Future Implementations

A scalable and reconfigurable platform for neural network training was proposed in this paper. It is scalable because it gives the possibility to connect different amounts of hardware accelerators (ES) and software reconfigurable topology. The method has been demostrated using the coprocessor for the computing task that have a high computational cost on ANN. achieving an acceleration of 3x-4x compared to Intel Xeon Quad-Core 2.83 Ghz and 6x-7x compared to AMD Optetron Quad-Core 2354 2.2Ghz.

In future applications will be implemented partial and dynamic reconfiguration. That is, several systems have loaded in Flash memory and to perform hardware reconfiguration. This will allow you to add multiple training algorithms and have more versatility on the platform. Also using the Distributed Computing Server Toolbox will not have the limitation of a PC, this Toolbox lets be scalable to several PC Host.

References

1. Haykin, S.: Neural Networks and Learning Machines, 3rd edn. Prentice Hall (November 2008)
2. Curteanu, S., Cartwright, H.: Neural networks applied in chemistry. i. determination of the optimal topology of multilayer perceptron neural networks. Journal of Chemometrics 25(10), 527–549 (2011)
3. Nguyen, D., Widrow, B.: Improving the learning speed of 2-layer neural networks by choosing initial values of the adaptive weights

4. Sankaradas, M., Jakkula, V., Cadambi, S., Chakradhar, S., Durdanovic, I., Cosatto, E., Graf, H.: A massively parallel coprocessor for convolutional neural networks. In: 20th IEEE International Conference on Application-specific Systems, Architectures and Processors, ASAP 2009, pp. 53–60 (July 2009)

5. Prado, R., Melo, J., Oliveira, J., Neto, A.: Fpga based implementation of a fuzzy neural network modular architecture for embedded systems. In: The 2012 International Joint Conference on Neural Networks, IJCNN, pp. 1–7 (June 2012)

6. Çavuşlu, M., Karakuzu, C., Şahin, S., Yakut, M.: Neural network training based on fpga with floating point number format and it's performance. Neural Computing and Applications 20, 195–202 (2011)

7. Wu, G.D., Zhu, Z.W., Lin, B.W.: Reconfigurable back propagation based neural network architecture. In: 2011 13th International Symposium on Integrated Circuits, ISIC, pp. 67–70 (December 2011)

8. Pinjare, S.L., Arun Kumar, M.: Article: Implementation of neural network back propagation training algorithm on fpga. International Journal of Computer Applications 52(6), 1–7 (2012)

9. Goldberg, D.E.: Genetic Algorithms in Search, Optimization and Machine Learning, 1st edn. Addison-Wesley Longman Publishing Co., Inc., Boston (1989)

10. http://www.altera.com

11. Aliaga, R., Gadea, R., Colom, R., Cerda, J., Ferrando, N., Herrero, V.: A mixed hardware-software approach to flexible artificial neural network training on fpga. In: International Symposium on Systems, Architectures, Modeling, and Simulation, SAMOS 2009, pp. 1–8 (July 2009)

12. http://www.matlab.com

Segmentation of Weld Regions in Radiographic Images: A Knowledge-Based Efficient Solution

Rafael Vilar and Juan Zapata

Depto. Electrónica, Tecnología de Computadoras y Proyectos
ETSIT- Escuela Técnica Superior de Ingeniería de Telecomunicación
Universidad Politécnica de Cartagena
Antiguo Cuartel de Antigones. Plaza del Hospital 1, 30202 Cartagena, Spain
rafael.vilar@upct.es,
juan.zapata@upct.es
http://www.detcp.upct.es

Abstract. The objective of the work consists in the design and implementation of technique of automatic recognition of weld region. It is aimed to complete a system of automatic inspection of radiographic images of welded joints. It deals with the problem of the delimitation of the weld regions according to the general scheme of image interpretation systems based on knowledge.

Keywords: Automated radiographic inspection, image processing, segmentation techniques.

1 Introduction

Radiographic inspection of welded joints is a quality control technique that today has become indispensable in industrial sectors such as nuclear, naval, chemical or aerospace. It is particularly important in critical applications where a weld failure can be catastrophic. Overall, the radiographic inspection problem has a high level of complexity, which has determined that this task is carried out exclusively by experts who base their opinion efficiency on experience of similar cases to test the over the years.

Both the difficulty of having experts to reliably detect and classify the defects in welded joints by means of radiographic inspection, and their costly training, justify the efforts for automation in this field. However, the desirable automation of this process encounters the obstacle of the necessary translation of expert's 'impressions' to computable information. Most of times these approaches are uncertain because that knowledge is a biological learning of not well-known mechanisms. The application of techniques typical of Artificial Intelligence allows to modify the expectations of the automation of these tasks. It is in this area of automating the radiographic inspection of welded joints where the work performed in this papaer is located.

The system of automatic inspection of digitised images consists of the following stages: digitalisation of the films, image pre-processing seeking mainly the

J.M. Ferrández Vicente et al. (Eds.): IWINAC 2013, Part II, LNCS 7931, pp. 421–430, 2013.

attenuation/elimination of noise; contrast improvement and discriminate feature enhancement, a multi-level segmentation of the scene to isolate the areas of interest (the interest region must be isolated from the rest the elements that compose the scene), detection of heterogeneities, feature extractions and, finally, classification in terms of individual and global features through neuro-computing tools of pattern recognition.

One of the critical stages to achieve good performance in identifying defects is segmentation or delineation of the weld regions. To perform this task a procedure have been developed and implemented which is presented n this paper. The procedure is located in the field of image interpretation based on knowledge. The analysis process is performed at three levels of abstraction, which will gradually injecting domain knowledge. This is used implicitly in the intermediate and explicitly in the highest level. This proposal has been called *progressive segmentation technique*.

The background of the paper is presented in Section 2. It includes work related to the localisation of weld regions in automated radiographic inspection applications.The proposed progressive segmentation technique is presented in Section 3. Results obtained are shown in Section 4. Conclusions are referenced in Section 5.

2   Related Work

It is interesting to note the relative lack of published work on the localisation of weld regions in automated radiographic inspection applications. The network training is achieved with a single image showing a typical weld in the run which is to be inspected, coupled to a very simple schematic weld template. Liao and Ni [1] proposed a methodology for the extraction of welds from digitised radiographic images. The method was based on the observation that the intensities of pixels in the weld area distribute more Gaussian distribution than other areas in the image. However this method has proved effective, only in segmenting linear welds. Subsequently, Liao and Tang [2] applied a multilayer perceptron (MLP) neural network procedure for the same application, which was successfully applied to segment both linear and curved welds. Another study by Liao, Li, and Li [3] employed fuzzy classifiers, specifically fuzzy k-nearest neighbour (K-NN) and fuzzy c-means, instead of neural networks as the pattern classifier. This method can also be applied to segment curved welds, and can handle a varying number of welds in one radiographic image. In a more recent work, Liao [4] has developed an expert system based on fuzzy reasoning for recognition of weld regions in radiographic images. Each object in the image is identified and described with a feature vector, and the fuzzy rules are learned from these characteristics based on a modification of *fuzzy C-Means* algorithm. Finally, another interesting work is published by Felisberto et al. [5]. The proposed methodology uses a genetic algorithm to guide the search for characteristic values (position, width, length, and angle) that best defines a window in the radiographic image, to fit the model image of a weld bead simple.

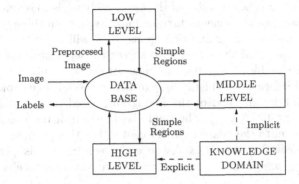

Fig. 1. Procedure for segmentation progressive

3 Methodology

In order to solve the problems appearing in the segmentation techniques previously mentioned, it is necessary to resort to the own previous knowledge about radiographs of welded joints. We have carried out, implemented and proposed a new segmentation technique in this work. This proposal resorts to the information provided by the radiographic image to carry out its analysis. This proposal is an adaptation of the procedure used by Carreira [6] for the delimitation of lungs contour on thorax images. Basically, this technique makes successive segmentations on the image to isolate the weld region, after a previous phase of contour elimination. This proposal has been called progressive segmentation technique.

This proposal to get the delimitation of the weld region assumes the formal framework of analysis and interpretation of images based on knowledge. The analysis process is carried out in three levels. In the low level, operations independent of the domain knowledge are performed on the image. This knowledge is implicitly used in the middle level and explicitly in the last phase or high level. The idea responds to a progressive segmentation scheme in which an initial segmentation is progressively refined using domain knowledge.

The control structure can be understood, on one hand, as a bottom-up control: we begin with a x-ray image on which several operations are developed. This action allows obtaining intermediate images until getting a final image on which the weld region appears isolated. But this is not the only type of control, a hybrid control is used in the last part of the system, where the algorithmic flow in search is determined by the region type. The control is ruled by the image characteristics regarding the contour identification of the weld region.

The previous three levels are connected through a database. This database store the original image, several intermediate results and the final image with the isolated welded region. Intermediate results are images which have been obtained from the original image by different operations, as binarization, thresholding, labelling, contouring and elimination of unnecessary information. The general block diagram of the system is illustrated in Figure 1

Low Level Block: The objective of the low level block is to segment the image in a group of homogeneous elementary regions. Even when these regions do not necessarily coincide with significant regions, they will constitute the elements for the interpretation of the scene. Two successive segmentations of the image are realized in this block. For both segmentations, we apply a technique that uses statistical information (histogram) to define the local centroids. Thus, we perform an iterative division of the histogram, based on an analysis of local centers clustering [6] [7] y [8]. The final classification of the pixels is done by associating a pixel to the nearest cluster center. For this purpose, calculate the distance *city-block*. The *city-block* distance between two pixels, $p(x, y)$ y $q(s, t)$ is defined as: $D_4(p, q) = |x - s| + |y - t|$ and where coordenatis refer to the space properties.

In the first segmentation, we pretend to divide the image into elementary regions. Initially, we do not aim to achieve a high degree of detail. We simply want to separate areas with relevant information in areas without this information. We need to evaluate a parameter called minimum percentage histogram. This parameter expresses the minimum size that has to be considered to be a cluster region. The choice of this parameter allows determining the degree of detail of the segmentation. As we said before, the objective of this first segmentation is to define the areas of interest, what can be achieved with parameter value near to 10. The first segmentation result, is a bit segmented image, with two regions that match the radiographic support and the area occupied by the base metal, which includes the weld bead. Once the first segmentation then carries out an operation of extraction of contours. The result will be the entrance to the stage of labelling of the regions detected.

The second segmentation is performed on the image contours obtained from the first cleavage. With it, we get an over-segmentation of regions classified as clear, because among them will be the weld bead. We use the same algorithm, but in this case the parameter which regulates the degree of segmentation is the minimum interval between centres of adjacent clusters. This parameter measures the minimum distance that must exist between adjacent clusters centres to be considered as a region. Since the weld seam will be between the bright regions of the histogram, the parameter value should be low, at around 8%. The process scheme of the first block is summarized in Figure 2. The result of this first block is an image segmented in a great number of regions and where a multiplicity of contours will have been generated in the weld region due to soft transitions of grey levels.

Middle Level Block: The objective of the second phase or middle level block is the refinement of the segmentation obtained in the previous block, confirming or not the presence of contours. We uses spatial information obtained from the image plane and we use to define the parameters that control the process. In the figure 3 the process is illustrated.

The image of contours obtained in the low level block is subjected to a process of vertexes codification to facilitate subsequent operations. This codification is

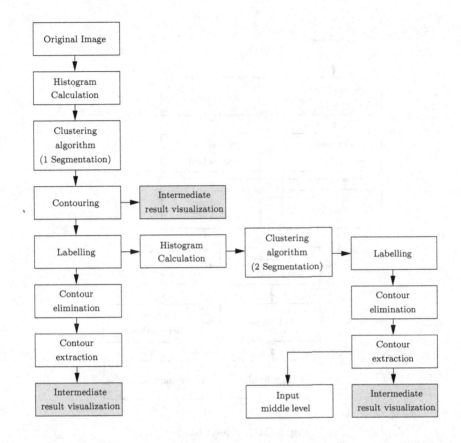

Fig. 2. Low level block

carried out on the base of an ordinal number between 1 and 4, taking into account the four main directions. Therefore, beginning or ending contour vertexes will be considered as vertexes of order 1, contour continuation vertexes as vertexes of order 2, and as vertexes of order 3 and 4, those that constitute crossing points of three or four contours. The information generated from vertexes codification is managed in a data structure (array of properties), where all the contours and their longitudes are stored.

After the process of vertexes codification, the contour filing proceeds with the objective of purifying the detected contours, so that they can be traced tracked, starting from a initial vertex, directly, until their end, except for the crossing points among contours.

Once finished the filing, the obtained image is prepared for the contour tracking that will give rise to the elaboration of the array of properties of coded contours. This array of dimensions numnber contours x maximum length, where every labelled contour and their length are stored, will let know what pixel belongs to the contour and its location within the contour.

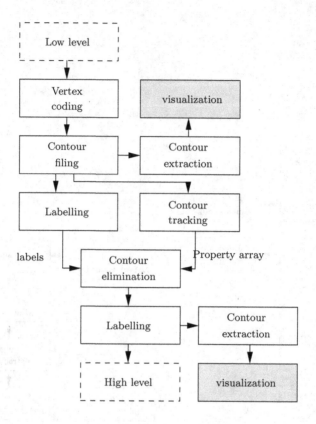

Fig. 3. Middle level block

Once obtained the properties array, the polished process of the initial segmentation is developed, determining if the obtained contours are to be eliminated, modified or maintained. The process will penalize parallel contours, establishing a unique contour in the area of greatest contrast. The refining process is to calculate a merit function for each contour. The merit function includes criteria contrast, length, and adjacent information. Thus, for each contour b_k between two regions R_i y R_j, define a merit function,$fm(b_k)$:

$$fm(b_k) = |vmg(R_i) - vmg(R_j)| \qquad (1)$$

being $vmg()$ the function represents the average gray value of a region.Thus, the contours with little merit function are deleted.

One problem that can occur when making the elimination of contours, is that topological restrictions may apply. This problem is solved by defining an global merit function to be maximized. Proceed as follows: if a contour is to be eliminated, contributing to the sum with $\theta - fm(b_i)$, being θ is a contrast threshold value; If however, the contour is preserved contributes with $fm(b_i) - \theta$.

If I_c and I_e are the sets of indexes for the two sets of contours: (c: conserved, e: deleted), then the global merit function F is defined as:

$$F(b_1, b_2, \cdots, b_n) = \sum_{i \in I_e} (\theta - fm(b_i) + \sum_{i \in I_c} (fm(b_i - \theta) \qquad (2)$$

Thus, the contours removal process is controlled by three parameters: the contrast threshold between two regionsθ, the percentage of similarity between two merit functions and the minimum relationship among longitudes of two contours.

High Level Block. The inputs to the block are the original image and the image of the earlier stage regions. The output will be an image with the weld completely isolated. The goal of the last stage or high-level block is to give meaning to each of the regions obtained. For this, the regions are characterized by a set of morphological properties, positional, and relational. This properties are: area ($area$), mean gray level (vmg), coordinates of the center of mass (cmx, cmy), and maximum ($xmax$, $ymax$) and minimum values ($xmin$, $ymin$) in two directions.

Next, define a series of thresholds of mean gray level of spatial position and size. The criterion for the determination of the threshold is based on the analysis of the results obtained after the application of the progressive segmentation algorithm to a collection about 50 radiographic images. For the gray level adopted a minimun threshold (u_{min}) of 110 and a maximum (u_{max}) of 190. For size threshold adopted 12000 pixels, representing 4% of the total image area. Regarding position thresholds, only the horizontally threshold was established; a minimum threshold to 50 and the maximum threshold to 580.

Production rules for identifying the weld bead are very simple, we look vertically centered light areas occupying much of the horizontal dimension of the image and they are among the largest. The possible weld bead must satisfy the following criteria.

1. Gray level criterion: $u_{min} < vmg_i < u_{max}$
2. Position criterion: $x_{inic} < x_1$ y $x_2 < x_{fin}$
3. Size criterion: $u_{tam} < area_i$

The basic scheme of the steps that constitute this block is shown in Figure 4.

4   Results

In order to test the reliability of the proposed methodology, a set of 86 radiograph images of the II/IIS were qualified by four skilled inspectors as a function of their difficulty. These were selected because they contain single o multiple typical defects (i.e., longitudinal and transverse cracks, blowhole and slag inclusions) and defect-free samples. Two different qualification schemes were used. In the first, the expert inspector qualified each of the radiographs from its radiographic appearance into three categories, radiographs of low, medium and high

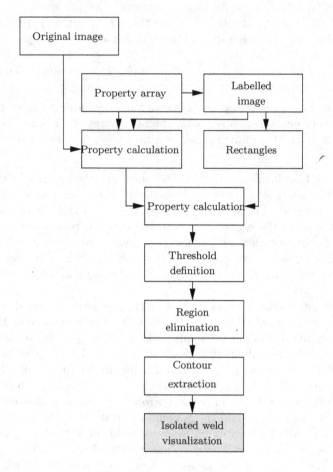

Fig. 4. High level block

complexity. In a second scheme, two quantitative measures were used to charac-
terise the image database. The first measure was contrast, whereby the database
was ordered in to three groups; radiographs of low, medium and high contrast.
And lastly, from the appearance of the histograms representing grey levels, the
images were grouped in to another two categories; images with overlapping or
without overlapping histograms.

In order to validate the proposed technique for automatic segmentation, the
weld region was used as comparative element. The area identified manually by
an experienced inspector of welded joints was termed real weld region. Similarity
is defined as the percentage of agreement between the weld region detected by
the system and the real weld region.

$$\text{Similarity}(\%) = \frac{\text{concordance area}}{\text{real area}} \times 100 \qquad (3)$$

Table 1. Test weld region results

Category	Similarity (%)
Radiographic Appearance	
Low Complexity	91
Medium Complexity	88
High Complexity	85
Contrast	
Low Contrast	83
Medium Contrast	88
High Contrast	95
Histogram	
Overlapping Histogram	91
Not Overlapping Histogram	94
Average Similarity	89%

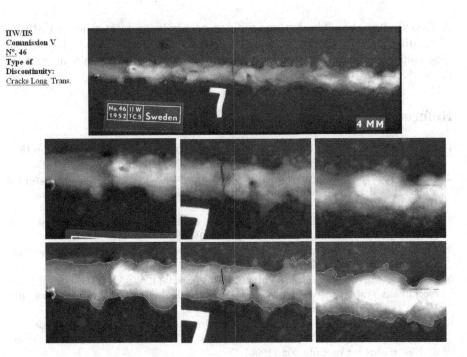

Fig. 5. Examples of segmentation results of the weld bead

The test results indicate that the proposed methodology achieves a similarity of about 87%. The results are shown in Table 1, for each one of the categories. Example results of the proposed method to test two images are shown in Figure 5.

5   Conclussions

The developed work is devoted to solving one of stages, maybe the most delicate, of a system of automatic weld defect recognition: the segmentation of weld regions. The conclusions and main contributions that in our opinion this work presents are the following: this paper presents and analysed the aspects related to the design and implementation of one specific techniques of segmentation to get the automatic recognition of weld regions. This technique, progressive segmentation, assumes the formal framework of images interpretation based on knowledge.

The validation process in 86 radiograph images of the standards of the International Institute of Welding, we conclude that the proposed technique offers good results, so was able to isolate the weld zone in all cases. Finally, we conclude that the progressive segmentation technique is close to the detection by one skilled in 87% on average.

Acknowledgment. The authors wish to acknowledge International Institute of Welding for the permission given to publish the present work using the radiographic standards. This work was supported by the Fundación Séneca under grant 15303/PI/10.

References

1. Liao, T.W., Ni, J.: An automated radiographic NDT system for weld inspection. Part I. Weld extraction. NDT & E Intern. 29(3), 157–162 (1996)
2. Liao, T.W., Tang, K.: Automated extraction of welds from digitized radiographic images based on mlp neural networks. Applications of Artificial Intelligence 11, 197–218 (1997)
3. Liao, T.W., Li, D.M., Li, Y.M.: Extraction of welds from radiographic images using fuzzy classifiers. Inform. Sci. 126, 21–42 (2000)
4. Liao, T.W.: Fuzzy reasoning based automatic inspection of radiographic welds: Weld recognition. Journal of Intel. Manuf. 15(1), 69–85 (2004)
5. Felisberto, M.K., Lopes, H.S., Centeno, T.M., Arruda, L.V.: An objet detection and recognition system for weld bead extraction from digital radiographs. Computer Vision and Image Understanding 102(3), 238–249 (2006)
6. Carreira, M.: Diagnóstico Asistido por Ordenador: Detección Automática de Nódulos Pulmonares. PhD thesis, Universidad de Santiago de Compostela, Depto. de Electrónica y Computación (1996)
7. Mosquera, A., Cabello, D., Carreira, M., Penedo, M.: Unsupervised textured image segmentation using markov random field and clustering algorithms. In: Klette, R. (ed.) Computer Analysis of Images and Patterns, vol. 5, pp. 139–147. Akademie Verlag (1991)
8. Wilson, R., Spann, M.: Image Segmentation and Uncertainty. Letchworth, Wiley, Herts, New York (1988)

Automatic Detection of Facial Landmarks
in Images with Different Sources of Variations

Ángel Sánchez, A. Belén Moreno, and José F. Vélez

Departamento de Ciencias de la Computación
Universidad Rey Juan Carlos
c/Tulipán s/n, 28933 Móstoles (Madrid), Spain

Abstract. Accurate and robust extraction of feature points in 2D facial images
has multiple applications in biometric face recognition, facial expression classification, facial animation or human-computer interaction, among others. This paper describes a methodology for the fully automatic identification of 20 relevant
facial points on static gray level images containing different types of variations.
To solve the problem considered, we mainly use the shape and texture information provided by the images. The main advantage of this approach is its precision
at point location, even for images with pronounced expressions. The presented
method is tolerant to moderate scale changes and pose variations, and also to
different illumination conditions. Our approach was tested on two of the most
common databases used for facial expression analysis: Cohn-Kanade and JAFFE
datasets, achieving respective average correct point detection rates of 94.2% and
96.15% on them. Our results were also compared to other related results presented in the literature on the same databases.

Keywords: Facial landmark detection, face region location, facial expressions,
anthropometric measurements, pattern recognition.

1 Introduction

The automatic and precise localization of interest regions and anthropometric points
from facial images is still an open and complex problem [1]. Facial landmark detection
is a subtask of a more general detection issue and aims to find the correct positions of
the facial regions, given that a face appears in the image [5]. Difficulties arise from the
acquisition conditions of face images, as well as from a high intra-personal and inter-
personal variability of the patterns. Among the acquisition conditions there can be some
differences in the illumination in the captured images, in the spatial and/or radiomet-
ric resolutions of the patterns, the camera view of the subject when the photo or video
is captured, etc. Inter-personal differences appear due to variability in gender, race or
age among the individuals of a database. Finally, there are intra-personal differences
corresponding to facial changes due to expressions and pose variations. A challenge is
to find algorithms for accurate facial regions and feature points detection which, at the
same time, remain robust with respect to variations in the facial appearance and to the
conditions mentioned previously. Most methods to extract the considered set of facial
landmarks follow a similar top-down approach. First, facial regions are extracted and

J.M. Ferrández Vicente et al. (Eds.): IWINAC 2013, Part II, LNCS 7931, pp. 431–440, 2013.
© Springer-Verlag Berlin Heidelberg 2013

next the fiducial points are detected in their corresponding regions. The eyes are usually located first, due to their importance when detecting other facial regions. However, the precise detection of eyes becomes difficult since they can have different shapes or degrees of opening in the image. Once the fiducial points in this region have been located, these points are used to normalize the pose and also to find other landmarks [13]. Due to their proximity to eyes, the eyebrow regions and their corresponding points are extracted next. Finally, the mouth and their relevant points (in a few of the works, the nose or chin region too) are localized. Regarding the problem of facial point detection under different illumination conditions, Lai et al [7] proposed a model that used skin color information and achieved a detection accuracy of around 86%. Gizatdinova and Surakka [5] proposed a feature-based method where edge maps were created at different resolution levels. However, their method was not fully automatic and required a manual classification of the edge regions located. In general, the proposed face point detection methods can be classified as color-based [2] and shape-based ones [3]. The first group tries to represent the facial features using only color information in different spaces and the second group searches for specific shapes in the image, adopting different approaches like templates, graph matching, active contour models, etc. Although color-based methods can be more efficient, they are not robust at all under realistic scene conditions. On the other hand, shape-based methods work well under more complex illumination, pose and/or scaling conditions but they are computationally more expensive [3].

This paper proposes a hybrid method which considers the shape of the face regions and, in same cases, its semantics (i.e. the intensity values of pixels) in order to locate the facial regions and the corresponding feature points automatically. We also seek a trade-off between the accuracy in the point location in real conditions and the time efficiency for this task. In this work, some inherent difficulties regarding this complex recognition problem have been handled; in particular, race and age differences between individuals, different illumination conditions and a slight variability in the pose. The current limitations of our approach are that only controlled face rotations (i.e. not more than 20 degrees) are allowed, and individuals can not partially occlude their facial regions where the searched-for fiducial points are located (i.e. when wearing glasses). We have tested our methods on two of the most used facial expression data sets: the Cohn-Kanade [6] and the Japanese Female Facial Expression (JAFFE) [8] databases.

2 Proposed Approach for Facial Region and Feature Point Extraction

We have developed an automatic, robust and accurate method to extract relevant facial landmarks in images with expressions, as well as to detect a set of 20 relevant anthropometric facial points from these regions. Facial images can have different types and intensities of expressions under varying illumination conditions. The proposed method consists of the following stages: (a) face region extraction, (b) eye regions extraction and detection of corresponding feature points, (c) eyebrow regions extraction and detection of corresponding feature points, and (d) mouth region extraction together with detection of corresponding feature points. All point detection stages include a

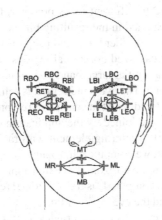

Fig. 1. Facial model with the landmark points considered

post-processing refinement that improves the accuracy of the point detections. The selected points correspond to a subset of those defined by the standard MPEG-4. The searched facial points are shown in Fig. 1.

2.1 Face Region Extraction

This task automatically crops a more adjusted and complete face region from the whole image, thus separating the region of interest from the background. The purpose of this step is to detect and extract a rough face region in order to delimit the further search of the considered facial points only to a smaller image region containing the face. This stage is performed using the cascade-based face detection method proposed by Viola and Jones [11]. Due to the simpler and reduced background region in the case of images in the JAFFE database, this pre-processing only needs to be computed for the case of the Cohn-Kanade database images.

2.2 Detection of the Eyes and Corresponding Feature Points

First, we extract a narrow rectangle region containing the eyes and next the corresponding feature points are detected. This task can be broken down into the following steps that are set out in detail below (the last three ones are applied separately to the segmented region corresponding to each independent eye): (a) segment eye band region (containing both eyes), (b) segment each eye region, (c) detect and refine pupil-center positions, (d) detect upper and lower eye points, and (e) detect inner and outer eye corners.

To automatically segment the eye band region, we applied a progressive thresholding algorithm which is based on an iterative application of a common thresholding algorithm (i.e. the Otsu method [10]) on the original image, until the percentage of black pixels required in the final binary image achieves a desired value. Consequently, the algorithm can be adapted to different races properly, by taking into account this percentage of black pixels searched in the resulting binary image. Once a binary facial image has been computed, we consider the facial proportions [4] to divide the face region horizontally into five equal-sized horizontal regions and to select the second and third regions (thus retaining an initial large band containing the eyes and also discarding the hair and mouth regions). The eye band region is refined as follows. Some noise pixels around the eyes are filtered. Next, as each eye region could be split into several connected components, these are joined by the application of a morphological dilation using a circular (disk-shaped) structuring element with radius $r = 1.5\%$ of the face's width. On the resulting image, we compute a horizontal projection histogram. Its maximum value determines an accurate horizontal line that approximates both eye positions. Finally, the eye band region is computed, as the horizontal region corresponding to 25% of the image, is cut out around the eye position line.

To segment each separate eye region, the graylevel image corresponding to the eye band computed previously is used. Each eye region is extracted as follows. Histogram equalization is applied on it, to reduce the effect of shadows or overexposed areas caused by illumination conditions or due to the skin colors when locating the remaining eye points. Next, we apply a Harris detector [10] on the equalized graylevel eye band region. By computing a vertical histogram around the most significant two hundred interest points on this image, a reduced rectangular eye region (containing only the pupil and eyelids) is accurately estimated.

The detection and refining of pupil-centers (LP and RP points of Fig. 1) is computed as follows. Each eye region is described by a set of five landmark points: the pupil-center, both eye corners, and the top and bottom eye points. Due to the relevance of pupil-centers [13] in the precise localization of other eye points and the remaining facial points considered, its position is first located approximately and is refined later. Firstly, a progressive thresholding algorithm is applied to binarize the eye region, thus enhancing the eye shape. Next, the little gaps on the pupils are filled using mathematical morphology, to obtain one connected-component pupil region. Later, as a first approximation, the pupil-center position (x_I, y_I) is located on this binary image, by seeking the high peaks on the respective horizontal and vertical-projection histograms. The corresponding column and row values with the highest number of 8-connected pixels in the corresponding histograms determine the initial pupil-center coordinates. To refine the pupil position, the following proposed fitness function f is minimized, aiming to reduce the difference between the number of black pixels that are above and below a given pupil-center estimation (x, y), and those placed to the left and right of this initial pupil point position:

$$\arg \min_{x,y \in N} f(x, y) = |hp_x - x| + |lp_x - x| + |hp_y - y| + |lp_y - y| \qquad (1)$$

where hp_x and lp_x, respectively, represent the highest and lowest x-coordinate pupil region points, and hp_y and lp_y represent the corresponding highest and lowest y-coordinate pupil region points. This minimization procedure is implemented through a local search starting from the initial pupil-center estimation (x_I, y_I) using a first-improvement searching strategy applied to a 3×3 neighborhood region.

To detect upper and lower eye points (LET, LEB, RET and REB points of Fig. 1), we start from the binary eye region image obtained by progressive thresholding at the previous stage. Given the refined pupil-center position (x_R, y_R), a set of vertical parallel lines in the interval $x_R \pm d_1$ (where d_1 is set to 12% of the inter-ocular distance) are thrown upwards from the pupil-center to the top of eye region. The highest and the most centered point cutting the upper eyelid is chosen as the top of the eye point. Analogously, the lower eye point is found using a similar approach to the applied for the upper eye point detection.

Finally, in order to detect the inner and outer eye corners (LEI, LEO, REI and REO points of Fig. 1), again the binary eye region image obtained by progressive thresholding and the two refined pupil-center positions (x_R, y_R) are used. A set of horizontal parallel lines in the interval $y_R \pm d_2$ (where d_2 is set to 12% of the inter-ocular distance) are thrown from each pupil-center to both sides of the eye regions. The longest and lowest intersection lines at both sides of the pupil-center determine the respective inner and outer right eye corner points for each of the eyes.

2.3   Detection of the Eyebrows and Corresponding Feature Points

This task is organized into the following steps: (a) segment each eyebrow region, (b) detect eyebrow middle point, and (c) detect inner and outer eyebrow corner points. As eyebrows are placed at a short distance above the corresponding eye, a vertical narrow region is cropped above each pupil (using the previously computed highest and lowest y-coordinates hp_y and lp_y of the pupil points). Next, this region is binarized using the average pixel intensity as threshold (since skin pixels prevail in the region), and it is later morphologically cleaned. Let I_X and I_Y be the horizontal and vertical sizes of the facial image, a rectangular region of size:

$$(cm_x \pm 0.06 \cdot I_X) \times (cm_y \pm 0.18 \cdot I_Y) \tag{2}$$

is cropped (taking into account facial proportions) around the center-of-mass point (cm_x, cm_y) of the eyebrow region previously extracted.

To detect the eyebrow middle point (LBC and RBC points of Fig. 1), the cropped binary eyebrow region image is dilated, using a disk-shaped structuring element with radius $r = 3\%$ of the inter-ocular distance to join possible split components in the eyebrow feature. After that, we estimate the eyebrow middle point as the highest position of this connected eyebrow region, again using a local search strategy.

Finally, for the detection of inner and outer eyebrow extremes (LBO, LBI, RBO and RBI points of Fig. 1), the contour of the binary right eyebrow feature is first tracked towards the left, starting from the previously detected eyebrow middle point. The first contour eyebrow pixel found whose local slope is higher than $\pm 45°$ is considered to be the inner eyebrow corner point. The same tracking approach is now applied towards the right, to find the outer eyebrow corner point. To detect the same feature points on the binary left eyebrow, the tracking direction is changed from what took place as regards the right eyebrow.

2.4 Detection of the Mouth and Corresponding Feature Points

This task can be broken down into the following steps: (a) segment the mouth region, (b) detect both mouth-corner points, and (c) detect top upper-lip and bottom lower-lip points. First, a rectangle region centered in the estimated mouth center position point (mc_x, mc_y) is defined as: $(mc_x \pm 0.1 \cdot I_X) \times (mc_y \pm 0.2 \cdot I_Y)$, where I_X and I_Y are the horizontal and vertical sizes of the facial image. Once this initial mouth region is obtained, its vertical size is refined as follows. Again, we apply the Harris interest point detector [10] on the previous region in order to locate around the most significant two hundred feature points on the graylevel cropped mouth region. As most of these points fit on the lips, we compute the horizontal-projection histogram of interest points. This histogram is dilated in the same way as for the eyes, thus obtaining a more precise mouth detection. Next, we search for the two most extreme non-isolated points in the y-coordinate which do not exceed a threshold distance from the set of interest points detected. These two points are initially labeled as both mouth-corners (i.e. ML and MR points of Fig. 1). Next, the position of these corner points is refined by template matching using two respective binary symmetric triangle-shaped masks with side 25% of the inter-ocular distance. The middle point of the upper lip (MT) is usually more difficult to locate due to the illumination conditions. Several edge detector algorithms [10] (in particular *Sobel*, *LoG* and *Canny*, respectively) were initially applied on the refined mouth region to locate these lip points, but all of the detectors produced unsatisfactory results in many test images. Consequently, a new contrast enhancement filter was defined (its pseudocode appears in Algorithm 1) to detect the upper (and lower) lip(s) for the images. Differently to the other edge detectors tested, our new filter produces a graylevel image where a lip is now a continuous line and its pixels have higher intensity values than those corresponding to other surrounding areas. Once the mouth region (upper lip) has been filtered, the pixel which is equidistant from both corner points in the brightest line of the upper lip is the sought point (MT). A similar approach has been folloved to find the correponding MB point of the lower lip.

Fig. 2 illustrates some examples of automatic eye, eyebrow and mouth feature point detections on different complex test images.

 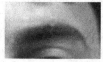 

Fig. 2. Some examples of feature point detections in difficult conditions

3 Experiments

We have tested our methods on two of the most used facial expression data sets: the Cohn-Kanade [6] and on the Japanese Female Facial Expression (JAFFE) [8] databases.

Algorithm 1. Proposed contrast enhacement filter

---

parameters *image_region*, *side*

    filt_image = *image_region*;
    for i = 1 **to** $ROWS$ **do**
      for j = 1 **to** $COLUMNS$ **do**
        // extract square of side 5 centered at filt_image(i,j)
        subregion =Mask(*filt_image*, *side*);
        subregion =sort_descending(*subregion*); // sort descending the values of subregion
        for k = 1 **to** $side * side/2 - 1$ **do**
          // subtract values of symmetric positions in subregion
          filt_image(i,j)+ = *subregion*(k)
            −*subregion*($side * side - k + 1$);
        end for
      end for
    end for
    return *normalize_to_grayscale*(*filt_image*); // return grayscale image

---

The proposed approach has been evaluated on 785 images: 572 images from the Cohn-Kanade database and 213 images from the JAFFE database, respectively. In the Cohn-Kanade database, the image data consist of approximately 500 frame sequences from 100 different subjects. The subjects included range in age from 18 to 30 years, 65% of whom were female, 15% were African-American and 3% were Asian or Latino. Sequences of frontal images representing a prototype facial expression always start with the neutral face and finish with the expression at its highest intensity (the corresponding frames are incrementally numbered in this way). These sequences were captured with a video camera, digitized into 640×480 pixel arrays with 8-bit precision for grayscale values, and stored in *jpeg* format. The JAFFE database contains a total of 213 static images with a spatial resolution of 256×256 pixels which were stored in *tiff* format without compression. These images describe 7 types of expressions (the 6 basic facial expressions and the neutral face) posed by 10 Japanese females. For each of the automatically-detected fiducial points, the test images were compared to the corresponding manually-marked points. The manual labeling of points in all the images was performed independently by the authors. After that, the corresponding markings for each point were compared with each other, in order to have an estimation of the precision of manual marking. As the variability of manual point selection was on average less than 4% of the inter-ocular distance, we considered the average of the authors' point selections to be the correct ones. As the different points correspond to variable-sized regions of pixels, a point found automatically is successfully detected if the Euclidean distance to the corresponding manual location of the same point is above its specific threshold (determined experimentally as in [12]). For example, a bias below of 30 pixels with respect to the manual detection for the eye corner points would is considered as a successful detection for the resolution of frame images (640×480) of the Cohn-Kanade database.

The second and third columns of Table 1 respectively show the average correct detection results and deviation average (measured as a % of the inter-ocular distance) achieved by our method using the Cohn-Kanade database for each different feature point (first column). The whole average correct detection result for all the points was 94.2%. The results are given as a percentage of the average inter-ocular distance (101.5 pixels) for all the Cohn-Kanade images and they represent a scale-invariant error measurement. We noticed that the points detected best were those placed in both eye regions, since these points are less sensitive as regards the morphological facial changes produced by expressions. Among these points, both of the pupil-centers were the most robustly located ones. Eyebrow central points were more precisely located than the corresponding end-of-eyebrow points. All points considered in the mouth region, except the central point of the lower lip, presented an acceptable and homogeneous detection results. This was also reported by different authors [9,12]), since this particular point (MR point of Fig. 1) is highly sensitive to positional changes in the presence of some facial expressions (i.e. surprise). Experimental results produced even a better performance for the JAFFE database (see fourth column of Table 1). The whole average correct detection result for all the points was 96.15%. The fifth column of this table also shows the corresponding deviation average between each type of automatically-detected facial point in relation to the manually-detected ones. These scale-invariant results are given as a percentage of the average inter-ocular distance (102.2 pixels) for all the JAFFE images. The best-detected points also correspond to those placed in both eye regions (improving the results achieved using Cohn-Kanade dataset by around 2%). The worst-detected points in eyebrow and mouth regions are the same ones as in the Cohn-Kanade database, mainly due to the effect of facial expression in morphological shape changes. In general, as JAFFE images have better illumination and focus conditions than the Cohn-Kanade ones, the average correct detection of the points increases with respect to this other database. The whole code was implemented in MATLAB using an Intel Pentium (R) Dual CPU T3200 2GHz with 3GB of memory. The method required on average less than 5 seconds to extract automatically from a face image the whole set of 20 facial points considered.

Our results were also compared to those reported by Vukadinovic and Pantic [12] on the Cohn-Kanade database, and by Sayeed et al[9] on the JAFFE database. This comparison was done with respect to the accuracy reported by the respective methods and was carried out for the same common subset of facial points. Columns 2 and 3 of Table 2 presents the average success for each of the 16 common detected facial points (our method detected 20 different facial points and 19 points were detected in Vukadinovic and Pantic's paper), by taking into account only neutral face images (since these authors only consirede neutral faces for detections of points). We show in bold typeface those points (6 of them) where our approach outperforms this work. We achieved an average success rate for the whole set of common points of the 90.77% compared to the 93.06% reported by Vukadinovic and Pantic's paper. Columns 4 and 5 of the same Table 2 presents the average success for each of the 16 common detected facial landmark points by Sayeed et al [9] and our method on the JAFFE database. We show in bold typeface those points (10 of them) where our approach outperforms Sayeed et al's work. We achieved an average success rate for the whole set of common points of 95.89% compared to the 91.91% reported by Sayeed et al.'s paper.

Table 1. Respective correct point detection rate and average deviation (both in %) of our method using both Cohn-Kanade and JAFFE databases

Feature Point	Detection success (Cohn-Kanade)	Deviation average (Cohn-Kanade)	Detection success (JAFFE)	Deviation average (JAFFE)
LBO	93.71	5.74	93.90	15.41
LBC	95.28	4.51	97.18	3.62
LBI	95.45	6.34	96.24	4.19
RBI	91.61	5.56	94.37	4.09
RBC	95.45	5.09	92.96	3.34
RBO	91.61	5.04	98.59	7.54
LP	97.03	4.79	99.06	5.12
LEO	94.93	3.52	99.53	6.48
LET	96.33	4.12	99.06	3.38
LEI	96.68	5.73	98.12	8.79
LEB	96.50	4.43	99.53	6.32
RP	97.03	5.26	99.53	4.97
REI	96.50	4.15	98.59	5.00
RET	96.85	4.47	98.59	3.17
REO	95.28	4.68	97.65	5.38
REB	97.03	4.47	99.53	5.90
ML	92.66	3.61	91.55	4.26
MT	94.23	6.10	91.08	4.90
MR	94.05	3.58	92.49	3.63
MB	75.70	7.75	85.45	6.45
Average	94.20	4.91	96.15	5.59

Table 2. Comparative accuracy results (in %) for the same automatic points detected (column 1) using the Cohn-Kanade (columns 2-3) and JAFFE (columns 4-5) databases

Point	Our method	Vukadinovic et al	Our method	Sayeed et al
LBO	88.10	96.00	93.90	98.57
LBI	86.51	96.00	**96.24**	92.17
RBI	87.30	95.00	94.37	96.38
RBO	**90.48**	90.00	**98.59**	90.35
LEO	**96.03**	92.00	**99.53**	89.62
LET	**96.83**	91.00	**99.06**	88.40
LEI	90.48	96.00	**98.12**	94.70
LEB	94.44	95.00	**99.53**	86.73
REI	96.03	99.00	**98.59**	92.83
RET	**94.44**	83.00	**98.59**	89.61
REO	91.27	96.00	**97.65**	91.46
REB	90.48	99.00	**99.53**	88.98
ML	**97.62**	97.00	91.55	95.32
MT	86.51	93.00	91.08	91.20
MR	**97.62**	91.00	92.49	97.89
MB	68.25	80.00	85.45	86.28

4   Conclusions

This paper presented new algorithms for the automatic extraction of the facial regions and a set of 20 representative feature points on gray level images containing variations in expressions, races and illumination. The proposed approach requires as input a gray level facial image and it uses mainly the shape information contained in it. For the more complex cases like when detecting the upper-lip and lower-lip middle points (the MT and MR points of Fig. 1), the method also uses some semantic information of the corresponding region. Our approach sought a trade-off between the accuracy in the point location in real conditions and the time efficiency. We achieved accurate point detection rates of 94.2% and 96.15% on the Cohn-Kanade and the JAFFE datasets, respectively. The proposed method is also robust to moderate scale and pose variations.

As future work, we will investigate the influence of other realistic factors on the feature point extraction results. In particular, we are interested in the point detection results depending on the type of facial expression of the subject, and how the degree of eye opening influences these results. Another interesting problem is to experiment with people wearing glasses or partially occluding their faces.

Acknowledgements. This work was funded by the Spanish MICINN project TIN2011-29827-C02-01.

References

1. Arca, S., Campadelli, P., Lanzarotti, R.: A face recognition system bases on automatically determined facial fiducial points. Pattern Recognition 39, 432–443 (2006)
2. Bagherian, E., Rahmat, R.W., Udzir, N.I.: Extraction of Facial Feature Points. International Journal of Computer Science and Network Security 9, 49–53 (2009)
3. Campadelli, P., Lanzarotti, R.: Fiducial point localization in color images of face fore-grounds. Image and Vision Computing 22, 863–872 (2004)
4. Farkas, L.G.: Anthropometry of the Head and Face. L. Williams and Wilkins (1994)
5. Gizatdinova, Y., Surakka, V.: Feature-based detection of facial landmarks from neutral and expressive facial images. IEEE Trans. PAMI 28, 135–139 (2006)
6. Kanade, T., Cohn, J.F., Tian, Y.: Comprehensive Database for Facial Expression Analysis. In: Proc. IEEE Intl. Conf. on Automatic Face and Gesture Recognition (FG 2000), pp. 46–53 (2000)
7. Lai, J.H., et al.: Robust Facial Feature Point Detection Under Nonlinear Illuminations. In: Proc. IEEE ICCV, pp. 168–174 (2001)
8. Lyons, J., et al.: Coding Facial Expressions with Gabor Wavelets. In: Proc. IEEE Intl. Conf. on Automatic Face and Gesture Recognition, pp. 200–205 (1998)
9. Sayeed, A., et al.: Detection of Facial Feature Points using Anthropometric Face Model. In: Signal Proc. for Image Enhancement and Multimedia Proc., pp. 189–200. Springer (2008)
10. Szeliski, R.: Computer Vision: Algorithms and Applications. Springer (2011)
11. Viola, P., Jones, M.: Rapid Object Detection using a Boosted Cascade of Simple Features. In: Proc. IEEE Conf. Computer Vision and Pattern Recognition, vol. 1, pp. 511–518 (2001)
12. Vukadinovic, D., Pantic, M.: Fully Automatic Facial Feature Point Detection using Gabor Feature Boosted Classifiers. In: Proc. IEEE SMC Conf., pp. 1692–1698 (2005)
13. Zhou, Z.-H., Geng, X.: Projection functions for eye detection. Pattern Recognition 37, 1049–1056 (2004)

An Empirical Study of Actor-Critic Methods for Feedback Controllers of Ball-Screw Drivers

Borja Fernandez-Gauna, Igor Ansoategui,
Ismael Etxeberria-Agiriano, and Manuel Graña

Grupo de Inteligencia Computacional (GIC), Universidad del País Vasco
(UPV/EHU), San Sebastian, Spain

Abstract. In this paper we study the use of Reinforcement Learning
Actor-Critic methods to learn the control of a ball-screw feed drive. We
have tested three different actors: Q-value based, Policy Gradient and
CACLA actors. We have paid special attention to the sensibility to sub-
optimal learning gain tuning. As a benchmark, we have used randomly-
initialized PID controllers. CACLA provides an stable control comparable to the best heuristically tuned PID controller, despite its lack of
knowledge of the actual error value.

1 Introduction

Machine tool manufacturers have embraced Proportional Integrative Derivative
(PID) controllers as a *de-facto* standard, because of their simple structure and
well-developed supporting theory [1,2]. Nevertheless, PID controllers have some
some weak points: the learning parameters must be empirically tuned (which
can be expensive and time consuming), the output is linear and symmetric to
the error to be minimized, and they are very sensitive to noise in the derivative,
and consequently to time variations of the control variables.

In this paper, we report our results approaching the control of a ball-screw
drive using Reinforcement Learning Actor-critic methods [3]. The aim of this
work is to obtain time-efficient realizations of the learning process. A ball-screew
driver is a mechanism transforming rotational movement into translational linear
movement, with minimal friction and energy loss. They are used in some specific tasks requiring very precise motion at wide diverse scales of load. We have
conducted computer simulations using an inertial model of a commercial Ideko
ball-screw feed drive (Figure 1). This workbench has the following components:
a *Fagor Fkm 42.30.A* motor, a *Rotex Gs 24/28* coupling, a *NSK-Rhp Bsb 025
062* screw-nut, a *Kondia D32x10* nut, and *Ina Kuse 45 850* guide-ways.

The paper is structured as follows: Section 2 states the problem at hand, presenting the model and the typical feedback control problem. Section 3 gives the
necessary background on PID controllers and Reinforcement Learning. Section
4 reports the results obtained in a simulation environment, and we offer our
conclusions in Section 5.

J.M. Ferrández Vicente et al. (Eds.): IWINAC 2013, Part II, LNCS 7931, pp. 441–450, 2013.

Fig. 1. Ball-screw drive setup

2   Problem Statement

2.1   Inertial Model

Complex and accurate mechatronic models have been proposed for ball-screw drives [4]. Nevertheless, we have used a simpler inertial model, which has been validated by error measurement experiments on the actual mechanic device. This model can be simulated faster than more sophisticated models, which is critical for Reinforcement Learning trials. The components of the specific ball-screw feed drive we are dealing with are depicted in Figure 2. The corresponding inertial model is expressed as:

$$u - (J_M + J_C + J_S) \cdot \ddot{\theta} = \frac{p}{2 \cdot \Pi} \cdot M \cdot \ddot{x}, \tag{1}$$

where u represents the torque of the motor, J_C, J_S and J_M are the coupling, ball-screw and motor inertia. θ represents the motor's angular position, p is the screw pitch, M is the mass of the table and the nut, and x is the lineal position

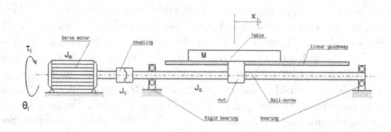

Fig. 2. Mechanical components of a ball-screw feed drive

of the table. The rotational movement is transformed to lineal movement due to the nut. This relation is expressed as

$$\ddot{\theta} = \frac{2 \cdot \Pi}{p} \cdot \ddot{x}. \tag{2}$$

Substituting Equation 2 into Equation 1, we obtain, after some manipulation, the acceleration of the table as a function of the servomotor torque τ:

$$\ddot{x} = \frac{u}{M \cdot \frac{p}{2 \cdot \Pi} + (J_C + J_S + J_M)\left(\frac{2\pi}{P}\right)}. \tag{3}$$

2.2 Feedback Control

Typical feedback controllers receive a changing set point $w(t)$ and a measured variable $x(t)$, and output some control variable $u(t)$. In the general case, all of these variables could be vectors, but we will assume them as scalars for simplicity without loss of generality. In the model defined by Equation 3, $w(t)$ corresponds to the desired position of the table, $x(t)$ is the measured position, and $u(t)$ is the torque of the motor, which is the output of the control algorithm. The scheme is depicted in Figure 2.2. The goal of the controller is to minimize the measured error $e(t) = |x(t) - w(t)|$. Usually, some tolerance range μ is defined around the set point $w(t)$; in such cases the goal of the controller is to keep the error within the tolerance range $(e(t) < \mu)$.

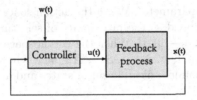

Fig. 3. Typical feedback control process scheme

3   Methods

3.1   Proportional Integrative Derivative Controller

The best known and widely used controller type is the *Proportional Integrative Derivative* (PID) controller, which is defined by the following Equation:

$$u(t) = K_p e(t) + K_i \int_0^t e(s)\,ds + K_d \frac{de(t)}{dt},$$

where K_p, K_i and K_d are, respectively, the proportional, integral and derivative gains. The output of the controller is calculated adding three terms: the *proportional* term, which is proportional to the current measured error, the integral term, which is proportional to the accumulated errors, and the derivative term, which is proportional to the error derivative. In this work, we have used a double-loop PID controller as the reference benchmark for comparison. Tuning the gain parameters is a difficult problem, that has been dealt with heuristic techniques providing suboptimal approaximate solutions. In this paper, we perform a random search for the PID parameter tuning algorithm.

3.2   Reinforcement Learning

Markov Decision Processes. RL methods [3] model environments as a Markov Decision Process (MDP), which is defined by a tuple $\langle S, A, P, R \rangle$, where S is the state space, A is the action space, P is the transition function $P : S \times A \times S \rightarrow [0, 1]$, and R is the reward function $R : S \times A \times S \rightarrow \mathbb{R}$. The action space S can be either a finite state set, or a possibly multi-dimensional Euclidean space $S \subseteq \mathbb{R}^{d_s}$, where d_s is the dimensionality of the state space. A can also be a finite action set or a possibly multi-dimensional Euclidean space $A \subseteq \mathbb{R}^{d_a}$ with d_a dimensions [5]. We will restrict our work to MDPs with continuous state spaces and discrete or continuous action space. Usually, RL agents observe state s_t at time-step t, select an action a_t, and then observe a new state s_{t+1} and the reward obtained r_{t+1}. Then agents use the tuple $\langle s_t, a_t, s_{t+1}, r_{t+1} \rangle$ to update its internal structures in order to learn.

The goal of the learning agent is to learn a policy $\pi (s, a)$ that maximizes the expected accumulated discounted rewards $R^\pi (s) = E^\pi \left\{ \sum_{k=1}^{\infty} r_{t+k} \gamma^{k-1} | s_t = s \right\}$, where γ is the discount parameter. When the action space is discrete and finite, a stochastic policy is defined as the probability of selecting each action in each observable state $\pi : S \times A \rightarrow [0, 1]$. When the action space is continuous, we will assume a deterministic policy $\pi (s) : S \rightarrow A \in \mathbb{R}^{d_a}$ whose output is an action value. For clarity, we assume a discrete set of states and actions in the reminder of this section.

The state value function of a given state s is given by the following *Bellman equation*:

$$V^\pi (s) = \sum_{a'} \pi (s, a) \sum_{s'} P (s, a, s') \cdot \{ R (s, a, s') + \gamma V^\pi (s') \} .$$

The state-action value function $Q^\pi (s, a)$ is defined as the expected accumulated rewards obtained after executing action a in state s and following policy π thereafter:

$$Q^\pi (s, a) = \sum_{s'} P (s, a, s') \cdot \{ R (s, a, s') + \gamma V^\pi (s') \} .$$

The optimal action-state value $Q^* (s, a)$ is defined as

$$Q^* (s, a) = \sum_{s'} P (s, a, s') \cdot \left\{ R (s, a, s') + \gamma \max_{a'} Q^* (s', a') \right\} ,$$

and the optimal state value function is defined as:

$$V^* (s) = \max_a Q^* (s, a) .$$

Actor-critic Algorithms. Actor-critic methods use different memory structures to represent the actor (implementing the action selection policy) and a
critic performing the role of the teacher. The critic learns V^* using some value
iteration method, such as $TD(0)$, which updates the state value function estimates using a similar update-rule:

$$V_t (s) \leftarrow V (s) + \alpha [r_t + \gamma V (s') - V (s)] .$$

After the actor selects and executes and action following policy $\pi (s, a)$, it receives
a scalar value from the critic assessing the value of this action. Then, the actor
can update its policy according to the critic. Usually, the critic corresponds to
the Time-Difference error $TD_t = r_t + \gamma V (s') - V (s)$. In this paper, we have
used the TD critic to teach three different kinds of actors:

- Q-AC: An actor which derives a policy from a Q-function [3]. Using a discrete
  action space, the actor executes an action a in state s, receives TD error from
  the critic, and updates the estimate $Q_t (s, a)$ using the following update-rule:

  $$Q_t (s, a) \leftarrow Q_{t-1} (s, a) + \alpha \cdot TD_t \{min + (1 - \pi (s, a))\} ,$$

  where min is a constant minimum amount that ensures that actions with
  high probabilities will also be updated.
- Policy gradient Actor-Critic (PG-AC). The actor explicitly represents a continuous policy $\pi (s)$ and updates it:

  $$\pi_t (s) \leftarrow \pi_t (s) + \alpha \cdot noise_t \cdot TD_t,$$

  where $noise_t$ is the noise signal added to the output of the policy.
- Continuous Action-Critic Learning Automaton (CACLA) [5,6]. The actor
  only updates its policy if the the critic is positive, that is, if the policy has
  been improved:

  $$\pi_t (s) \leftarrow \pi_t (s) + \alpha \cdot noise_t.$$

The main difference with respect to PG-AC is that the policy update is
done in action space, proportional to the noise added to the policy's output,
instead of being proportional to the TD error.

Need for Exploration. Agents must make a compromise between *exploitation*
(i.e. greedily selecting the action with the highest Q-value) and *exploration* (i.e.
selecting some suboptimal action). Without sufficient exploration of the action-
state space, agents may not be able to find an optimal policy, and pure random
exploration can be expected to perform very poorly.

Policies based on Q-values such as Q-AC often use Soft-Max action selection:

$$\pi\left(s,a\right) = \frac{\exp^{\beta \cdot Q(s,a)}}{\sum\limits_{a'} \exp^{\beta \cdot Q(s,a')}},$$

where β is a scalar control parameter. A null value of β gives all actions the same probability of being selected. As its value increases, the selection probability of actions with higher values also increases. When β tends towards infinity, the policy becomes greedy.

Continuous Actor-Critic methods such as PG-AC or CACLA, use a noise signal, which is added to the output of the policy, so that the actor can explore yet unknown policies. In this paper, we have used Gaussian noise following a normal distribution $N\left(\pi_t\left(s\right), \sigma^2\right)$ centered around the algorithm's output $\pi_t\left(s\right)$.

Function Approximation. The estimates for the functions can be stored as a look up-table, allowing to derive formal convergence proofs, but they are not feasible in problems with a great number of states or actions. To overcome this so-called *cursed of dimensionality*, researchers have studied the use of linear and non-linear function approximators [7], at the cost of losing some of the convergence properties offered by look-up tables.

Linear approximators use a feature-extraction function $\phi : S \to \Phi$ that maps the state space to some d_Φ-dimensional feature space $\Phi \subseteq \mathbb{R}^{d_\Phi}$, and approximate function $V_t\left(s\right)$ using the expression $V_t\left(s\right) = \theta_t^T \phi\left(s\right)$, where $\theta_t \in \Theta$ is the parameter vector updated by the learning algorithm. For practical reasons, only $N \ll d_\Phi$ features are activated (they are non-zero), and it is usually a good idea to use activation factors such that $\sum\limits_{i=1}^{d_\Phi} \phi_i\left(s\right) \leq 1; \; \forall s \in S$.

The most common linear approximators are Tile-coding and Radial Basis Functions (RBF). Tile-coding uses one or more overlapping tilings, each of which maps states to a single active feature per tiling. The joint-product of all the tiling's features is the feature space. The main limitation of this approach is that, because each state belongs to a single feature, the approximated function is non-smooth. RBF use fuzzy representation to overcome this problem. Using fuzzy activation factors, each state activates more than one feature. We have used Gaussian RBF.

4   Experiments

4.1   Experimental Design

In our experiments, we have compared the aforementioned Actor-Critic algorithms: Q-AC, PG-AC and CACLA. We have also performed a heuristic tuning of the PID controllers as follows: randomly initialized the gain parameters of 1000 double-loop PID controllers, tested them all, and selected the best configuration for comparison with the RL approaches. After the learning phase, an evaluation episode has been performed using a greedy policy and using the average rewards as a performance measurement.

Model parameters The parameter of the inertial model defined in Equation 3 were set as follows:

- $J_C = 170.5 \, kg \cdot m^2$
- $J_S = 6.25 \, kg \cdot m^2$
- $J_M = 85 \, kg \cdot m^2$
- $p = 0.01 \, m/rev$
- $M = 68.25 \, kg$

The maximum torque of the motor was set $\tau_{max} = 37 \, N \cdot m$. The length of the guide-ways was $1 \, m$, the initial position of the table was $x(0) = 0.0$ and the command signal was constant: $w(t) = 0.5$.

Learning parameters. During the learning process, $n_{episodes} = 1000$ episodes of 2 seconds were conducted with each algorithm ($\Delta t = 0.02$). Experiments have been conducted for different fixed learning rates $\alpha_0 = \{0.005, 0.025, 0.05, 0.075, 0.1\}$. The exploration parameter used with Q-AC was initially $\beta = 0.0$ and it was linearly increased each episode with $\Delta\beta = 1.0/n_{episodes}$. On the other hand, PG-AC and CACLA used $\sigma_0^2 = 5.0$ and decreased this parameter using $\Delta\sigma^2 = \sigma_0^2/n_{episodes}$.

The state was composed by the two variables of the system: the position $x(t)$ and the speed of the table $v(t)$. Each state variable was approximated by 50 Gaussian RBF, which were set so that only 3 features were active each time *per* variable. For Q-AC, actions were discretized using 100 uniformly distributed points in range $[-\tau_{max}, \tau_{max}]$.

The reward signal was defined as

$$r(t) = C \cdot \left[1 - \tanh^2\left(\frac{|e(t)|}{\mu}\right)\right],$$

where μ represents the size of the tolerance area around $w(t)$ and $C = 1000$. In our tests, $\mu = 0.005$. This reward function is similar to the one proposed in [8].

4.2   Results

We report here results obtained during the evaluation episodes. In Table 1 we show the best average rewards obtained by the RL controllers for different learning rates. As expected, the best PID obtained a higher score (701.72) because the error and the output need not be approximated. Taking only the best result into account, the two continuous action algorithms (PG-AC and CACLA) come next, and the worst results are obtained by Q-AC, presumably because the action space needs to be discretized. It is also very noticeable that CACLA was more robust than the others to sub-optimal tuning of the learning rate parameter α. Q-AC and PG-AC required iterative manual parameter tuning before the system was able to converge to an acceptable policy, while CACLA offered consistent results for every α_0 tested. Our educated guess is that this is because the formers update the policy in error space (bounded by the minimum and maximum TD error:$[-1000, 1000]$) while the later makes updates in action space.

Table 1. Average rewards per time-step during the evaluation episodes

α	Q-AC	PG-AC	CACLA
0.005	495.903	8.922	594.871
0.025	86.965	17.759	610.057
0.050	45.391	661.865	630.884
0.075	9.711	17.759	631.863
0.100	21.545	106.450	632.475

In Figures 4, 5 and 6 we have respectively plotted the position of the table $x\,(t)$, the output of the controller, and the rewards obtained each time-step during the evaluation episode for each of the controllers. Although it is hard to say from Figure 4, Figure 5 shows that the system stays within the tolerance range with PID and CACLA, but the system is not that stable with Q-AC and PG-AC. Also, Figures 5 and 6 suggest that Q-AC and PG-AC are behaving similar to a bang-bang controller (maximum output towards the error gradient), which is generally considered a poor control approach.

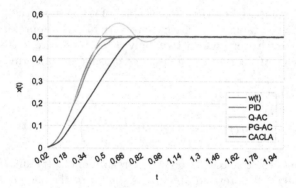

Fig. 4. Table position during the evaluation episode. $w\,(t)$ is the goal position signal.

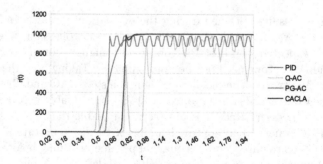

Fig. 5. Rewards obtained by the controllers during the evaluation episode

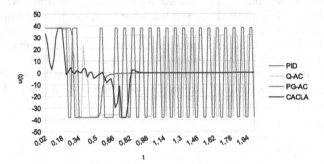

Fig. 6. Output of the controllers during the evaluation episode

5 Conclusions

In this paper, we have compared PID controllers and three different Reinforcement Learning Actor-Critic methods in a feed drive control problem. Results show that PID controllers offer a more accurate and stable response to impulse signals, and more work is required before they can be outperformed by RL methods in ideal environments. Before RL become a real alternative to PID controllers in feedback control tasks, the tuning of the learning parameters must be more thoroughly studied, so general adaptive or auto-tuning techniques are developed. We know of no such standard tuning procedure.

We consider RL-based feedback controllers as a promising area of research, because they offer adaptability to changing environments and are expected to perform better with more complex command signals $w\,(t)$ and noise perturbations [8]. During our experiments, CACLA seems to be the most robust and promising solution. Future work shall compare performance in noisy environments with changing set points.

References

1. Koren, Y., Lo, C.C.: Advanced controllers for feed drives. In: Annals of the CIRP, vol. 41 (1992)
2. Srinivasan, K., Tsao, T.C.: Machine feed drives and their control - a survey of the state of the art. Journal of Manufacturing Science and Engineering 119, 743–748 (1997)
3. Sutton, R., Barto, A.G.: Reinforcement Learning I: Introduction. MIT Press (1998)
4. Chen, J.-S., Huang, Y.-K., Cheng, C.C.: Mechanical model and contouring analysis of high-speed ball-screw drive systems with compliance effect. Int. J. Adv. Manuf. Technol. 24, 241–250 (2004)
5. Hasselt, H.: Reinforcement Learning in Continuous State and Action Spaces. In: Reinforcement Learning: State of the art. Adaptation, Learning, and Optimization, pp. 207–251. Springer (2012)

6. van Hasselt, H., Wiering, M.A.: Reinforcement learning in continuous action spaces. In: Proceedings of the 2007 IEEE Symposium on Approximate Dynamic Programming and Reinforcement Learning (2007)
7. Busoniu, L., Babuska, R., De Schutter, B., Ernst, D.: Reinforcement Learning and Dynamic Programming using Function Approximation. CRC Press (2010)
8. Hafner, R., Riedmiller, M.: Reinforcement learning in feedback control: Challenges and benchmarks from technical process control. Machine Learning 84(1-2), 137–169 (2011)

Robust Solutions for a Robotic Manipulator Optimization Problem

Ricardo Soto[1,2], Stéphane Caro[3], Broderick Crawford[1,4], and Eric Monfroy[5]

[1] Pontificia Universidad Católica de Valparaíso, Chile
[2] Universidad Autónoma de Chile, Chile
[3] IRCCyN, Ecole Centrale de Nantes, France
[4] Universidad Finis Terrae, Chile
[5] CNRS, LINA, Université de Nantes, France
{ricardo.soto,broderick.crawford}@ucv.cl,
stephane.caro@irccyn.ec-nantes.fr,
eric.monfroy@univ-nantes.fr

Abstract. In robotics, pose errors are known as positional and rotational errors of a given mechanical system. Those errors are commonly produced by the so-called joint clearances, which are the play between pairing elements. Predicting pose errors can be done via the formulation of two optimization models holding continuous domains, which belong to the NP-Hard class of problems. In this paper, we focus on the use of constraint programming in order to provide rigorous and reliable solution to this problem.

1 Introduction

Accuracy is one of the key features that favor robotic manipulators for many industrial applications. Superior levels of accuracy are achieved by controlling or measuring all possible sources of errors on the pose of the moving platform of a robotic manipulator. Joint clearance is one of most important sources of errors. It introduces extraneous degrees of freedom between two connected links. When present, they generally contribute importantly to the degradation of the performance of a mechanism. Various approaches have been proposed to compute and quantify the errors due to joint clearances [20,11,16,7,18,17,13], however none of them focuses on the reliability of solutions, which is mandatory to provide an accurate prediction of the pose error.

In this paper, we present a more complete version of our previous work using constraint programming in conjunction with interval analysis with the aim of guaranteeing the reliability of solutions [4,15]. To this end, we combine a branch and bound algorithm with interval analysis, which allow drawing firm bounds on the pose errors given possible ranges for the clearances. We illustrate experimental results where the proposed approach generally outperforms the well-known solvers GAMS/BARON [1] and Ecl^{i}ps^e [19], while providing reliable solutions.

J.M. Ferrández Vicente et al. (Eds.): IWINAC 2013, Part II, LNCS 7931, pp. 451–460, 2013.

The remaining of this paper is organized as follows. Section 2 presents an error-prediction model used to characterize the influence of joint clearances on the end-effector pose of any robotic mechanical system. An overview of constraint programming —including the implemented approach— is presented in Sect. 3. The experiments are presented in Sect. 4, followed by the conclusions and future works.

2   The Problem Formulation

In order to generalize the application of our approach, in this paper we consider robotic mechanical systems with the following configuration: n revolute joints, n links and an end-effector. Hence, the manipulator is assumed to be composed of $n+1$ links and n joints; where 0 is the fixed base, while link n is the end-effector. Next, a coordinate frame $\mathcal{F}_j$ is defined with origin O_j and axes X_j, Y_j, Z_j. This frame is attached to the $(j-1)$st link for $j = 1, \ldots, n+1$.

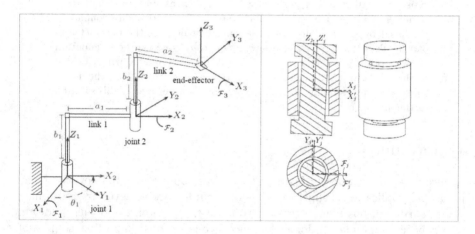

Fig. 1. Left: A serial manipulator composed of two revolute joints. Right: Clearance-affected revolute joint.

Figure 1 depicts an instance of such system considering $n = 2$ involving joints with join clearance as illustrated on the right side of Figure 1. Our goal is to find the maximal positional (translational) and rotational error for a given robotic configuration. Let $\delta\mathbf{p}_r$ and $\delta\mathbf{p}_t$ representing the rotational and translational errors of the manipulator end-effector, respectively. Then, the maximal pose error of the end-effector for a given manipulator configuration can be obtained by solving the following two optimization problems:

$$\text{maximize} \quad \delta p_r, \tag{1}$$
$$\text{subject to} \quad \delta r_{j,X}^2 + \delta r_{j,Y}^2 - \Delta \beta_{j,XY}^2 \leq 0,$$
$$\delta r_{j,Z}^2 - \Delta \beta_{j,Z}^2 \leq 0,$$
$$j = 1, \ldots, n$$

$$\text{maximize} \quad \delta p_t, \tag{2}$$
$$\text{subject to} \quad \delta r_{j,X}^2 + \delta r_{j,Y}^2 - \Delta \beta_{j,XY}^2 \leq 0,$$
$$\delta r_{j,Z}^2 - \Delta \beta_{j,Z}^2 \leq 0,$$
$$\delta t_{j,X}^2 + \delta t_{j,Y}^2 - \Delta \gamma_{j,XY}^2 \leq 0,$$
$$\delta t_{j,Z}^2 - \Delta \gamma_{j,Z}^2 \leq 0,$$
$$j = 1, \ldots, n$$

where $\delta r_{a,b}$ corresponds to the small rotation in joint a with respect to axis b, and $\delta t_{a,b}$ corresponds to the translation in joint a with respect to axis b. β and γ are simply constants used to limit the pose errors depending on the given configuration (an extended explanation of this model can be found in [5]). Let us notice that the two problems are nonconvex quadratically constrained quadratic (QCQPs). Indeed, all constraints of both problems are convex, their feasible sets are convex, and their objectives functions are non-convex. Then, both problems belong to the category of NP-Hard problems [10].

3 Constraint Programming for Global Optimization

3.1 Definitions

A Constraint Satisfaction Problem (CSP) is a formal problem representation, which mainly consists in a sequence of variables holding a domain and a set of relations over those variables called constraints. The idea is to find values for those variables so as to satisfy the constraints. The software technology devoted to tackle this problem is named Constraint Programming (CP). In this work, due to the presence of continuous decision variables, we focus on Numerical CSP (NCSP), which is an extension of a CSP devoted to continuous domains. Formally, a NCSP $\mathcal{P}$ is defined by a triple $\mathcal{P} = \langle \mathbf{x}, [\mathbf{x}], \mathcal{C} \rangle$ where:

- $\mathbf{x}$ is a finite sequence of variables $(x_1, x_2, \ldots, x_n)$.
- $[\mathbf{x}]$ is a finite set of real intervals $([x_1], [x_2], \ldots, [x_n])$ such that $[x_i]$ is the domain of x_i.
- $\mathcal{C}$ is a set of constraints $\{c_1, c_2, \ldots, c_m\}$.

A solution to a NCSP is a set of real intervals that satisfy all the constraints. Optimization problems are handled in the same way. Hence, the 4-tuple $\mathcal{P} = \langle \mathbf{x}, [\mathbf{x}], \mathcal{C}, f(x) \rangle$ is employed in this case, where $f(x)$ is the cost function to be maximized or minimized.

3.2   NCSP Solving

In order to guarantee accurate solutions, NCSPs cannot be handled in the same way that CSPs mainly due to the presence of constraints over real numbers. Indeed, the representation of reals in numerical computations is not exact since it is commonly done by means of floating-point numbers, which are a finite set of rational numbers. This inaccuracy may lead to rounding errors and as a consequence to reach wrong solutions. One solution for rigorously dealing with real numbers relies on the integration of interval analysis on the solving process. The idea is to compute approximations over domains represented by intervals bounded by floating-point numbers [3]. A detailed presentation of interval analysis can be seen in [14], and some examples devoted to robotics in [8].

Then, the core idea for solving NCSPs relies in combining a branch and prune algorithm with interval analysis for handling continuous domains. A tree-data structure that holds intervals as the potential solutions is built on the fly by interleaving branching and pruning phases. The branching phase is responsible for creating the branches of the tree by splitting real intervals, while the pruning tries to filter from domain intervals that do not conduce to any feasible solution. The idea is to speed-up the solving process. This is possible by applying consistency techniques for continuous domains such as the hull and the box consistency [10,2], which are similar to the arc-consistency [12] for finite domain CSPs.

Figure 2 depicts an algorithm for rigorously handling NCSPs. The procedure begins by receiving as input the set of constraint and domains of the problem. Then, four actions are embedded in a while loop. The **Contract** operator is responsible for pruning the tree, and **Split** applies a dichotomic division of intervals in order to carry out the branching process. Every computation of elementary operations $\{+, -, \times, /\}$ is done by using interval arithmetic. The process stops when the real values of the solution have reached the precision required of the problem.

A slight modification to the previous algorithm is required to handle optimization problems. Indeed, here a CP-based branch and bound algorithm is combined with interval analysis (see Figure 2). The corresponding cost function f has been added to the input set. A variable m is initialized to $+\infty$ in order to maintain an upper bound on the global minimum. In this way, potential solutions exceeding this bound are discarded by adding to the set of constraints. Five instructions are embedded in the same while loop. Now, **Contract** takes into account the cost function in order to prune the tree. The **Update** function has been added for updating the upper bound once better solutions are found.

Algorithm 1	**Algorithm 2**
Input: $\mathcal{C} = \{c_1, \ldots, c_m\}$, $[\mathbf{x}]$	**Input:** f, $\mathcal{C} = \{c_1, \ldots, c_m\}$, $[\mathbf{x}]$
1 $\mathcal{L} \leftarrow \{[\mathbf{x}]\}$	1 $\mathcal{L} \leftarrow \{[\mathbf{x}]\}$
2 **While** $\mathcal{L} \neq \emptyset$ **and** $\neg$stop_criteria **do**	2 $m \leftarrow +\infty$
3 $([\mathbf{x}], \mathcal{L}) \leftarrow$**Extract**$(\mathcal{L})$	3 **While** $\mathcal{L} \neq \emptyset$ **and** $\neg$stop_criteria **do**
4 $[\mathbf{x}] \leftarrow$**Contract**$_{\mathcal{C}}([\mathbf{x}])$	4 $([\mathbf{x}], \mathcal{L}) \leftarrow$**Extract**$(\mathcal{L})$
5 $\{[\mathbf{x}'], [\mathbf{x}'']\} \leftarrow$**Split**$([\mathbf{x}])$	5 $[\mathbf{x}] \leftarrow$**Contract**$_{\mathcal{C} \cup \{f(\mathbf{x}) \leq m\}}([\mathbf{x}])$
6 $\mathcal{L} = \mathcal{L} \cup \{[\mathbf{x}'], [\mathbf{x}'']\}$	6 $m \leftarrow$**Update**$([\mathbf{x}], f)$
7 **End While**	7 $\{[\mathbf{x}'], [\mathbf{x}'']\} \leftarrow$**Split**$([\mathbf{x}])$
8 **Return**$(\mathcal{L})$	8 $\mathcal{L} = \mathcal{L} \cup \{[\mathbf{x}'], [\mathbf{x}'']\}$
	9 **End While**
	10 **Return**$(\mathcal{L}, m)$

Fig. 2. Left: The branch and prune algorithm. Right: The branch and bound algorithm

The branch and bound algorithm described above has been implemented on top of the RealPaver solver [6]. This implementation is used for the experiments presented in the next section.

4 Experiments

As previously mentioned the RealPaver solver has been used as base for our branch and bound algorithm. In this section, we verify the reliability of solutions and we compare the performance of such implementation with the state-of-the-art solvers GAMS/BARON [1] and $Ecl^i ps^e$ [19]. $Ecl^i ps^e$ is a widely used solver in CP and one of the few having support for continuous domains. GAMS/BARON is a popular solver from the mathematical programming field devoted to particularly hard optimization problems.

The following experiments take into account 4 rotation and 4 translation models (see Table 1). We consider from two to five joints for each model. The information related to the problem size (number of variables, number of constraints, and number of operators) is also given. Then, we provide the solving times in milliseconds for BARON; $Ecl^i ps^e$, and for the proposed algorithm. The experiments were run on a 3 Ghz Intel Pentium D Processor 925 with 2 GB of RAM running Ubuntu 9.04. All solving times are the best of ten runs.

The results show that our implementation is in general faster than its competitors. For smaller problems involving two and three joints, our algorithm exhibits excellent performance, being 100 times faster that BARON. This performance is clearly reached by the efficient filtering work done by the HC4. This is also influenced by the fact that there is no need for highly accurate computations for this particular problem: we actually do not have to reach the usual precision, which makes the search process less costly than they usually are.

In the presence of rotational problems considering four and five joints, our approach remains faster. However, once the number of joints increases, in particular when the number of variables is greater than 20, the searching begins to be slow. This is obviously explained by the exponential complexity in the problem size that leads to complete search methods like our branch and bound to a slower convergence. Despite of this common phenomenon, we estimate that the presented solving times are reasonable regarding the problem complexity as well as the solving techniques involved.

Table 1. Type, problem size and solving times (in seconds)

| #joints | Type | Problem Size | | | BARON | RealPaver | Ecl^{i}ps^e |
		#var	#ctr	#op			
2	T	12	8	28	0.08	0.004	>60
2	R	6	4	18	0.124	0.004	>60
3	T	18	12	135	0.124	0.008	t.o.
3	R	9	6	90	0.952	0.004	t.o.
4	T	24	16	374	0.144	0.152	t.o.
4	R	12	8	205	2.584	0.02	t.o.
5	T	30	20	1073	0.708	3.71	t.o.
5	R	15	10	480	9.241	0.26	t.o.

Let us remark that although RealPaver as well as Ecl^{i}ps^e use CP-based solving techniques. The performance of our approach is significantly better, this is not surprising due to RealPaver was originally designed for tackling problems with continuous domains.

Besides, solving times may actually not be the only point to emphasize here: It may also be important to discuss the reliability of the solutions given by BARON. We could indeed verify using RealPaver that the optimum enclosure computed by BARON is unfeasible on some of the above tested problem instances. It was already noted in [9] that BARON may not always be reliable, where the following example illustrates how BARON can fail to give the correct value and return a an inconsistent point instead. Let x, y be two real variables, $x, y \in [-10, 10]$, and two constraints $y - x^2 \geq 0$ and $y - x^2(x - 2) + 10^{-5} \leq 0$. When BARON tries to find the lowest x for which both constraints are satisfied, it returns the point $(0, 0)$ although it is inconsistent w.r.t. the constraints of the problem (cf Figure 3, reproduced from [9]).

Finally, let us focus on the results obtained for the manipulator composed of two revolute joints illustrated in Fig. 1. Let us assume its geometric parameters are defined as follows:

$$a_1 = 1 \text{ m}$$
$$b_1 = 0 \text{ m}$$
$$\alpha_1 = 0 \text{ rad}$$
$$a_2 = 0.7 \text{ m}$$
$$b_2 = 0 \text{ m}$$
$$\alpha_2 = 0 \text{ rad}$$

and the joint clearances are equal to:

$$\Delta\beta_{j,XY} = 0.01 \text{ rad}$$
$$\Delta\beta_{j,Z} = 0.01 \text{ rad}$$
$$\Delta b_{j,XY} = 2 \text{ mm}$$
$$\Delta b_{j,Z} = 2 \text{ mm}$$

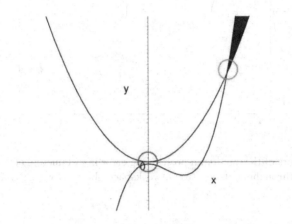

Fig. 3. BARON fails to find the global minimum and returns an inconsistent point instead

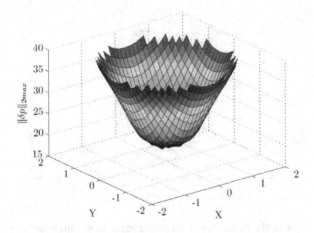

Fig. 4. 3D plot of the maximum positioning error of the manipulator with two joints throughout its Cartesian workspace

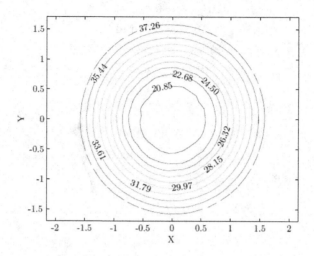

Fig. 5. Isocontours of the maximum positioning error in millimeters of the manipulator with two joints throughout its Cartesian workspace obtained with RealPaver

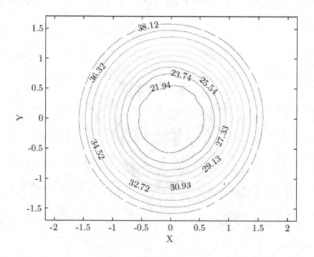

Fig. 6. Isocontours of the maximum positioning error in millimeters of the manipulator with two joints throughout its Cartesian workspace obtained with BARON

RealPaver and BARON were used to solve the optimization problem for all manipulator poses. Figures 5 and 6 show the value of the maximum positioning error of the end-effector obtained with RealPaver and BARON, respectively. This value is expressed in millimeters and plotted throughout the manipulator workspace. We can notice that the maximum obtained with BARON is always higher than the one obtained with RealPaver. The difference in any point between the results given by the two solvers is around one millimeter and decreases when positioning error rises: As we already mentioned it, we could notice that the solutions obtained with BARON may not be feasible.

5   Conclusions and Future Works

In this paper we have modeled and solved the complex problem of predicting the maximum pose error in robotic mechanical systems. To this end we have employed constraint programming techniques and interval analysis. Experimental results demonstrate the effectiveness of our approach where it is able to in general outperform well-known solvers such as BARON and Ecl^{i}ps^e, providing reliable solutions. Let us also mention that the presented approach is not only devoted to robotics, it in fact applies to whatever problem requiring rigorous computation and reliability in the solutions.

In this context, there are several directions for future work. In the future we plan to design new pruning criteria for accelerating the convergence of solutions. In this way, it would be easier to handle the exponential growth of the space of solutions. Finally, it would be interesting to pay an extended attention to the unreliability of solutions obtained by BARON. Such a problem could be also present in additional solvers.

References

1. GAMS, http://www.gams.com/ (Visited October 2009)
2. Benhamou, F., Mc Allester, D., Van Hentenryck, P.: CLP(Intervals) Revisited. In: Proceedings of ILPS, pp. 124–138. MIT Press, Cambridge (1994)
3. Benhamou, F., Older, W.J.: Applying Interval Arithmetic to Real, Integer and Boolean Constraints. Journal of Logic Programming 32(1), 1–24 (1997)
4. Berger, N., Soto, R., Goldsztejn, A., Caro, S., Cardou, P.: Finding the maximal pose error in robotic mechanical systems using constraint programming. In: García-Pedrajas, N., Herrera, F., Fyfe, C., Benítez, J.M., Ali, M. (eds.) IEA/AIE 2010, Part I. LNCS, vol. 6096, pp. 82–91. Springer, Heidelberg (2010)
5. Binaud, N., Cardou, P., Caro, S., Wenger, P.: The kinematic sensitivity of robotic manipulators to joint clearances. In: Proceedings of ASME Design Engineering Technical Conferences, Montreal, Canada, August 15-18 (2010)
6. Granvilliers, L., Benhamou, F.: Algorithm 852: RealPaver: an Interval Solver Using Constraint Satisfaction Techniques. ACM Trans. Math. Softw. 32(1), 138–156 (2006)
7. Innocenti, C.: Kinematic clearance sensitivity analysis of spatial structures with revolute joints. ASME Journal of Mechanical Design 124(1), 52–57 (2002)

8. Jaulin, L., Kieffer, M., Didrit, O., Walter, E.: Applied Interval Analysis with Examples in Parameter and State Estimation, Robust Control and Robotics. Springer (2001)
9. Lebbah, Y., Michel, C., Rueher, M.: An Efficient and Safe Framework for Solving Optimization Problems. Journal of Computational and Applied Mathematics 199(2), 372–377 (2007); Special Issue on Scientific Computing, Computer Arithmetic, and Validated Numerics (SCAN 2004)
10. Lhomme, O.: Consistency Techniques for Numeric CSPs. In: Proceedings of IJCAI, pp. 232–238 (1993)
11. Lin, P.D., Chen, J.F.: Accuracy analysis of planar linkages by the matrix method. Mechanism and Machine Theory 27(5), 507–516 (1992)
12. Mackworth, A.: Consistency in Networks of Relations. Artificial Intelligence 8(1), 99–118 (1977)
13. Meng, J., Zhang, D., Li, Z.: Accuracy analysis of parallel manipulators with joint clearance. ASME Journal of Mechanical Design 131(1), 011013-1–011013-9 (2009)
14. Moore, R.: Interval Analysis. Prentice-Hall, Englewood Cliffs (1966)
15. Soto, R., Caro, S., Crawford, B.: On the pursuit of reliable solutions for a robotic optimization problem. AASRI Procedia (to appear, 2013)
16. Zhu, J., Ting, K.-L.: Uncertainty analysis of planar and spatial robots with joint clearances. Mechanism and Machine Theory 35(9), 1239–1256 (2000)
17. Venanzi, S., Parenti-Castelli, V.: A new technique for clearance influence analysis in spatial mechanisms. ASME Journal of Mechanical Design 127(3), 446–455 (2005)
18. Voglewede, P., Ebert-Uphoff, I.: Application of workspace generation techniques to determine the unconstrained motion of parallel manipulators. ASME Journal of Mechanical Design 126(2), 283–290 (2004)
19. Wallace, M., Novello, S., Schimpf, J.: ECLiPSe: A Platform for Constraint Logic Programming. Technical report, IC-Parc, Imperial College, London (1997)
20. Wang, H.H.S., Roth, B.: Position errors due to clearances in journal bearings. ASME Journal of Mechanisms, Transmissions, and Automation in Design 111(3), 315–320 (1989)

On the Identification and Establishment
of Topological Spatial Relations

Sergio Miguel-Tomé*

Grupo de Investigación en Minería de Datos (MiDa),
Universidad de Salamanca, Salamanca, Spain
sergiom@usal.es

Abstract. Human beings use spatial relations to describe many daily
tasks in their language. However, to date in robotics the navigation problem has been thoroughly investigated as the task of guiding a robot from
one spatial coordinate to another. Therefore, there is a difference of abstraction between the language of human beings and the algorithms used
in robot navigation. This article introduces the research performed on
the use of topological relations for the formalization of spatial relations
and navigation. The main result is a new heuristic, called Heuristic of
Topological Qualitative Semantic (HTQS), which allows the identification and establishment of spatial relations.

Keywords: qualitative navigation, spatial relations, Heuristic of Topological Qualitative Semantic (HTQS).

1 Introduction

Robots have proven to be useful tools for police officers, surgeons and cleaning staff. However, in each of these cases robots are designed for specific tasks.
Currently, the construction of multifunctional robots is studied: robots able to
navigate in environments where the behavior must change such as offices or
homes, and to directly work with humans in various activities. Until now, many
successes in robotics are related to the reactive paradigm [2]. But it seems difficult for the reactive paradigm to enable the development of multifunctional
robots, as the manner to interact in the environment will constantly change depending on the task at hand. It should be kept in mind that the natural way of
communicating spatial tasks by humans is by making use of spatial relations. A
multifunctional robot would continually receive tasks such as:
"Take the package that is on the table and leave it in the closet."
This way, it seems that it is necessary for a multifunctional robot to function in
human's life to possess a high degree of spatial reasoning, and concretely about
spatial relations.

---

* I wish to thank Dr. Antonio Fernández-Caballero for his huge help to write this
  article, and Obra Social de Caja de Burgos and Fundación Gutiérrez Manrique for
  the economical support to present this work.

J.M. Ferrández Vicente et al. (Eds.): IWINAC 2013, Part II, LNCS 7931, pp. 461–470, 2013.
© Springer-Verlag Berlin Heidelberg 2013

I believe that within the symbolic paradigm it is possible to develop decision-making algorithms founded on the topological relations and apply these to navigation. So, I have begun a research program to study the usefulness of topological notions for navigation and get multifunctional robots. The starting hypothesis of this work is that topological relations are a useful and effective tool in achieving navigation based on identifying and establishing spatial relations. In this article, I present the first theoretical results about identifying and establishing spatial relations through topological notions of my program research.

This paper is structured as follows. In section 2 spatial representation methods used in AI and their application are examined. Then, section 3 introduces the Heuristic of Topological Qualitative Semantics. Afterwards, section 4 describes a method for identifying topological relations. Finally, in section 5 we will discuss the results and future work.

2   Representation of Space for Navigation

The methods of robot decision making can be divided into quantitative and qualitative, depending on the type of spatial representation used. In qualitative methods, space is not represented by a metric space. But specifically the best way to represent the space is not a resolved issue yet in qualitative methods. Even when the first years of research on general methods of qualitative spatial reasoning were unsuccessful, "the poverty conjecture" was set out [13]. This conjecture states that: "There is no purely qualitative, general-purpose representation of spatial properties". But research conducted in recent times seems to refute or at least weaken "the poverty conjecture" [3].

In AI the problems of satisfiability, decidability, complexity of different variants, composition of spatial relations, and its mixture with fuzzy techniques for handling uncertainty given a qualitative description of a static space have been largely treated [20,18,22]. In the last times the same kind of problems also for dynamic spaces and reasoning about the movement [23,4]. However, the issue is the creation of an algorithm to make decisions.

Depending on the complexity of the environment and the objective assigned to a robot, the architecture for decision making may come to consist of three levels: implementation, navigation (or local navigation) and planning (or global navigation). Among the objects of a space there are three types of spatial information: topological, orientational and positional. Proposals for building qualitative spatial representations are based on any of the above types of spatial information in a qualitative basis. There is the opinion in the robotics field that the qualitative topological information is appropriate for planning, but too abstract to allow the realization of navigation [21]. So, until now the problem of qualitative planning has been also largely treated [12][19][16][17]but there are few works about navigation with qualitative methods and they are focus on soccer[11][5]. So, there is not general methods to translate the information captured by the robot sensors to a high-level representation with a full group of spatial relations. Therefore, in the way for getting multifunctional robots it is necessary to investigate the use of spatial relations for navigation. Based on before we define a new type of navigation problem, which will be called "navigation problem based on spatial relations", which can be expressed as follows:

Given a starting point A establish the spatial relation(s) $G(G_1, G_2, ...)$ among objects $[O, O']$ ($[O_1, O_1'], [O_2, O_2'], ...$) by using [the robot's] knowledge and sensory information received.

3 Heuristic of Topological Qualitative Semantic

Initially, Freksa proposed the idea of representing the relation between relations by a graph, it is called conceptual neighborhood[15], and uses it to reason with incomplete or coarse knowledge about temporal relations and spatial relations [14]. Independently, Egenhofer propose a similar idea to reasoning about changes in topological relations[6]. Our research has come from the work of Egenhofer and Al-Taha[6]. They introduced a way to relate the topological relations by defining a distance between them and their representation in a graph. In the graph, each one of the topological relations is represented by a node and each node has an arc with that node with a minimal distance to their respective topological relations. This type of graph is called the Closest Topological Relationship Graph (CTRG). The same authors also created a revised version that includes arcs between nodes when there is a transformation that enables passing two objects from a topological relation to another. In their work, Egenhofer and Al-Taha studied the possible use of CTRG for inference and prediction. The conclusions were that the CTRG could be used to infer and predict with certainty only in a few cases. The problem discovered by Egenhofer and Al-Taha is the CTRG threw more than one possible deformation diagram in some cases.

The first target has been develop an algorithm of the kind of analysis of means and ends for topological relations. Thanks to Ernst's investigations [10], we know that an analysis of means and ends will find a solution if you can define a complete order relation on the differences. A graph is a binary relation on a set. Therefore, as the CTRG is a graph with cycles, it is impossible to create an analysis of means and ends from it. So, it is needed to obtain a linear binary relation on the topological relations. The reason for the CTRG to be non-linear is the wide range of changes that the objects may suffer. Clearly, Geography Information System takes into account changes in the size of an object, as rivers, lakes, sea or forest stands may increase or decrease surface drastically. However, in a common navigation environment, such as a building, this does not happen to objects (that is, to robots). Therefore, one can take into account the following two conditions: temporal isosize and temporal isoshape to generate different lineal graphs. But the linealization is not the final solution. There are cases for which the 8 relations of the "9-intersection model" [8] are not enough to make decisions because it keeps multiple trajectories from one an initial node and another final node. Hence, more relation are necessary to distinguish that trajectories. I propose the following 13 relations: Disjoint-0, Meet-0, Overlap-0, CoveredBy-0, Covers-0, Inside, Equal, Contains, Disjoint-1,

Meet-1, Overlapping-1, CoveredBy-1, Covers-1. The relative topological relations are jointly exhaustive and pairwise disjoint. The formal definition of topological relations will be seen later on. In order to distinguish between the set of relations to be defined and R_8, the new set of relations is called *relative topological relations* and denoted by S_{13}. This relations will be used to create a kind of lineal graphs. Each of those graphs is called *topological reasoning linear graph*(TRLG).

The decision-making method must indicate what action must be done to establish a spatial relation. The method found here is a heuristic. The heuristic finds a solution to establishing the topological relations problem. Let us call it Heuristic of Topological Qualitative Semantic (HTQS). To explain in what consists the HTQS a generic example to illustrate its operation will be given. Imagine you have two objects: an object that has the capacity to act, o_A, and a reference object, o_R, in respect of which one wants that o_A be able to fix a concrete relative topological relation. Thus, suppose that o_A can work with three actions:

$$\Lambda(o_A) = \{f_1|_x(x) = x + 1, \quad f_2|_x(x) = x, \quad f_3|_x(x) = x - 1\}$$

and o_R can not move and remains static.

$$\Lambda(o_R) = \{f_4|_x(x) = x\}$$

The first data structure used is a table that is constructed by applying a means-ends analysis on the actions. Thus, the functions are labeled with the way in which the quantitative positions in space are changed, assigning an order relation to be satisfied when the action applies. If isosize and isoshape conditions are met, there are only three possible cases. Thus, the functions of o_A are labeled creating the results contained in the next table.

function	label
f_1	<
f_2	=
f_3	>

The second data structure used are the TRLG. The nodes of each TRLG are labeled with an incremental enumeration. In this example we take that o_A and o_R are bigger than one unit and equal in size. Since we are in $\mathbb{Z}$ the form of an object is necessarily the same if the size is the same. Therefore, the TRLG used is shown in Fig. 1.

Once you have the above data structures, the algorithm to select the function has the next steps:

1. The relative topological relation between the objects o_A and o_R is calculated from their current positions, s_c, and the number associated to the node of s_c, n_{s_c}, is stored.
2. The number associated to the node of the relative topological relation which is the target between the objects o_A and o_R, n_{s_t}, is gotten.

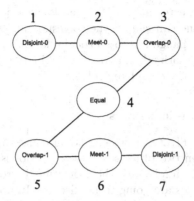

Fig. 1. Incremental enumeration on an TRLG

3. It is checked what order relation holds between n_{s_c} and n_{s_t} from the following:
   - $n_{s_c} < n_{s_t}$
   - $n_{s_c} = n_{s_t}$
   - $n_{s_c} > n_{s_t}$
4. It is looked at the means-ends table using the order relation that holds between n_{s_c} and n_{s_t} to pick the action that is labeled with the same order relation. This is the action selected by heuristics.

The heuristic is for a one dimensional space but it can be easily generalized to a topological n-dimensional space. To do this, several results from topology can be used to build the topological space $\mathbb{Z}^n$. It is known that, given a topology on X and another on Y, there is a canonical way to create a topology on the cartesian product $X \times Y$, the so-called product topology. Thus, from the base $\mathcal{B} = \{(x_1, x_2) : x_1 < x_2 \quad x_1, x_2 \in \mathbb{Z}\}$, which generates the order topology on $\mathbb{Z}$, we can construct the product topology on $\mathbb{Z} \times \mathbb{Z}$. The definition of the product topology on $X \times Y$ can be applied equally to $X_1 \times X_2 \times \cdots \times X_n$. Therefore, there is no problem in building a generalization to dimension n of the topological space.

The next subject is the topological relations in a space $X_1 \times X_2 \times \cdots \times X_n$. Because space is constructed by the Cartesian product, the n-dimensional topological relation between two objects is defined by a tuple of n components, being the component k the topological relation that occurs in dimension k between the two objects. Thus:

$$\forall A, B \in X_1 \times \cdots \times X_n \quad Ar_x B \Leftrightarrow A\langle r_{x'}, r_{x''}, ..., r_{x^n} \rangle B$$

where r_{x^k} is the relation between A and B in dimension k.

Therefore, the generalization to n dimensions of the HTQS consists in applying it successively on each dimension. Nevertheless, this generalization has several restrictions on the real world because for the generalization it is necessary that the sides of each object have right angles to each other. This way,

2-dimensional objects are rectangles, 3-dimensional objects are cubes, and so on. This ensures that one can change the corresponding topological relation in one dimension without change the topological relation in another one. That is, it meets the principle of optimality, and the combination of a solution for every dimension is solution to the complete problem.

3.1   HTQS Over Non-dense Sets

The criteria to assign a TRLG to two objects in dense topological spaces is the size relationship between two objects . However we focus on non-dense spaces, because the intention is implement the algorithm in robots which as sensor systems as data structures are going to be discrete. In no-dense spaces the size ratio between objects is not sufficient to assign a TRLG, since in each case subcases arise by the difference between the dense and non-dense sets. Table 1 presents all subcases for non-dense topological spaces, with their corresponding TRLG, assuming isosize and isoshape in the objects.

Table 1. This table shows the conditions between two objects to assign a TRLG in a topological space generated with a non-dense set. The notation can be consulted int the 3.

Case	Subcase	Lineal topological reasoning graph
$|A| > |B|$	$|A| = (|B| + 1)$	$\langle s_1, s_2, s_3, s_5, s_{10}, s_{11}, s_{12}, s_{13} \rangle$
	$|A| > (|B| + 1)$	$\langle s_1, s_2, s_3, s_5, s_8, s_{10}, s_{11}, s_{12}, s_{13} \rangle$
$|A| < |B|$	$(|A| + 1) = |B|$	$\langle s_1, s_2, s_3, s_4, s_9, s_{11}, s_{12}, s_{13} \rangle$
	$|A| > (|B| + 1)$	$\langle s_1, s_2, s_3, s_4, s_6, s_9, s_{11}, s_{12}, s_{13} \rangle$
$|A| = |B|$	$|A| = |B| = 1$	$\langle s_1, s_7, s_{13} \rangle$
	$|A| = |B| > 1$	$\langle s_1, s_2, s_3, s_7, s_{11}, s_{12}, s_{13} \rangle$

4   Calculation of Relative Topological Relations

The previous section described the heuristic to solve a establishing topological relations problem. The first step of the heuristic says that the relative topological relation is calculated using the positions of the two objects. But so far how to perform the calculation has not been explained. This step is important because it connects the mechanisms of perception of the environment of a mobile robot with the navigation algorithm. Thus, this section introduces a method to calculate the relative topological relation of two regions of space.

At first, one might consider the use of the "9-intersection model" for the calculation[7]. But there are two impediments that prevent implementation. The first obstacle is that the "9-intersection model" was proposed for dense topological spaces, such as $\mathbb{R}^n$, and fails for non-dense spaces. To solve this problem it has been proposed to use interior and border definitions different from those contained in the topology [9]. But these new definitions from the field of digital topology lose the simplicity that makes the "9-intersection model" so interesting. The second and most important impediment is that the number of relative topological relations is 13, compared to the 8 relations that the "9-intersection model" can define for regions in $\mathbb{R}^2$. Therefore, the "9-intersection model" reconstructed with the definitions of digital topology has no ability to define the 13 relative topological relations that are necessarily to be defined. Due to the two obstacles just mentioned, we have created a variant of an old formalism. The old formalism is the method for finding relations between time intervals used in the Allen algebra [1]. Freksa noted that the 13 relationships defined by Allen's formalism could be interpreted as spatial in a spatial context [14]. Certainly, Allen applies his method to convex intervals in $\mathbb{R}$, and these do not correspond with opens sets of the topological connected space over $\mathbb{R}$. Nevertheless, the convex intervals over $\mathbb{Z}$ coincide with the open sets of the order topology on $\mathbb{Z}$. Thus, instead of using the Allen method to calculate the relations between two time intervals, it is used to calculate a topological relation. Indeed, the interpretation given to $\mathbb{Z}$ is that of a spatial dimension. Allen's method also hasn't the second impediment cited for the "9-intersection model" because it can characterize the relative topological relations. Although it is needed to do some variations due that Allen was using $\mathbb{R}$, a dense set, as model of time, but $\mathbb{Z}$ is a not dense set. The Allen's formalism is the next set of conditions

$$\begin{pmatrix} \min(X) < \min(Y) & \min(X) < \max(Y) \\ \max(X) < \min(Y) & \max(X) < \max(Y) \\ \min(X) = \min(Y) & \min(X) = \max(Y) \\ \max(X) = \min(Y) & \max(X) = \max(Y) \end{pmatrix}$$

Now, we want to use the formalism on a no-dense set as model of space. For this reason we introduce two changes to modify the definition of the relation "Meets". This changes are in my opinion a better definition of the relation Meets for space because two objects do not share points of space, but they have adjacent points. The new formalism, called *order propositions matrix* and denoted by $P^{\leqq}(X,Y)$ consists of the following matrix:

$$P^{\leqq}(X,Y) = \begin{pmatrix} \min(X) < \min(Y) & \min(X) < \max(Y) \\ \max(X) < \min(Y) & \max(X) < \max(Y) \\ \min(X) = \min(Y) & \min(X) - 1 = \max(Y) \\ \max(X) + 1 = \min(Y) & \max(X) = \max(Y) \end{pmatrix}$$

Each of the elements of the matrix is a proposition that takes a value of $\mathbb{B} = \{0,1\}$, depending on whether the proposition is false(0) or true(1). Table 2 shows the characterization of each of the relative topological relations through the order propositions matrix.

Thus, the values taken by matrix $P^{\leqq}(X,Y)$ directly show the relative topological relations between two sets, and therefore, the node of the TRLG they deserve. The reader should realize that Allen's method enables characterizing 13 binary relations (see Table 3). These thirteen relations are composed of 6 binary

Table 2. Table containing the relative topological relations defined by means of the order propositions matrix

Disjoint-0	Meet-0	Overlap-0	CoveredBy-0	Covers-0
$P^< = \begin{pmatrix} 1\ 1 \\ 1\ 1 \\ 0\ 0 \\ 0\ 0 \end{pmatrix}$	$P^< = \begin{pmatrix} 1\ 1 \\ 1\ 1 \\ 0\ 0 \\ 1\ 0 \end{pmatrix}$	$P^< = \begin{pmatrix} 1\ 1 \\ 0\ 1 \\ 0\ 0 \\ 0\ 0 \end{pmatrix}$	$P^< = \begin{pmatrix} 0\ 1 \\ 0\ 1 \\ 1\ 0 \\ 0\ 0 \end{pmatrix}$	$P^< = \begin{pmatrix} 1\ 1 \\ 0\ 0 \\ 0\ 0 \\ 0\ 1 \end{pmatrix}$
Inside	Equal	Contains		
$P^< = \begin{pmatrix} 0\ 1 \\ 0\ 1 \\ 0\ 0 \\ 0\ 0 \end{pmatrix}$	$P^< = \begin{pmatrix} 0\ 1 \\ 0\ 0 \\ 1\ 0 \\ 0\ 1 \end{pmatrix}$	$P^< = \begin{pmatrix} 1\ 1 \\ 0\ 0 \\ 0\ 0 \\ 0\ 0 \end{pmatrix}$		
Covers-1	CoveredBy-1	Overlap-1	Meet-1	Disjoint-1
$P^< = \begin{pmatrix} 0\ 1 \\ 1\ 0 \\ 0\ 0 \\ 0\ 1 \end{pmatrix}$	$P^< = \begin{pmatrix} 0\ 1 \\ 0\ 0 \\ 0\ 0 \\ 0\ 1 \end{pmatrix}$	$P^< = \begin{pmatrix} 0\ 1 \\ 0\ 0 \\ 0\ 0 \\ 0\ 0 \end{pmatrix}$	$P^< = \begin{pmatrix} 0\ 0 \\ 0\ 0 \\ 0\ 1 \\ 0\ 0 \end{pmatrix}$	$P^< = \begin{pmatrix} 0\ 0 \\ 0\ 0 \\ 0\ 0 \\ 0\ 0 \end{pmatrix}$

Table 3. The 13 binary relations enabled by Allen's method

Allen's Relations	Symbol	Relative Topological Relations (S_{13})	Symbol
Before	r_1	Disjoint-0	s_1
Meets	r_2	Meet-0	s_2
Overlaps	r_3	Overlap-0	s_3
Starts	r_4	CoveredBy-0	s_4
Finished-by	r_6^-	Covers-0	s_5
During	r_5	Inside	s_6
Equal	r_7	Equal	s_7
Includes	r_5^-	Contains	s_8
Finishes	r_6	CoveredBy-1	s_9
Started-by	r_4^-	Covers-1	s_{10}
Overlapped-by	r_3^-	Overlapping-1	s_{11}
Meet-by	r_2^-	Meet-1	s_{12}
After	r_1^-	Disjoint-1	s_{13}

relations, their corresponding 6 inverse binary relations, and the binary relation "equal". The inverse relation of "equal" is itself. These relations defined for the temporal dimension can also be used in spatial dimension.

If we recall the definition of inverse

$$R^- = \{(y, x) : (x, y) \in R\}$$

one realizes that any spatial location between two objects meets a relation and its inverse. When describing the temporal relationship between two events, it is indifferent to use a relation or its inverse, since the observer is describing from the outside, and both relations contain the same information. But when a particular agent must make a decision to move, the observer is the agent; so it matters if you use a relation or its inverse, since actions are labeled for a specific order within pairs of objects in the relations. Thus, for HTQS only one of two relations is useful, the other leads to a misunderstanding of the algorithm.

So, what is the relation that must be used in the HTQS? The answer is that it is chosen according to what object is applying HTQS to make decisions. In $P^{\leqq}(X,Y)$, X must always be the agent making the decision, and Y is the object respect of which the agent must make a decision.

5   Discussion and Future Work

Language and human thought make use of spatial relations for the description of many tasks performed daily. But so far, the approaches most commonly used for representing space in navigation algorithms have been through numeric spatial coordinates. The creation of algorithms to perform navigation tasks, identifying and establishing spatial relations, is a logical step in the objective of getting multifunctional robots that can be integrated into human society. The application of topological relations for the representation of spatial relations has been used in cartography and geography for the construction of commercial GIS. However, topological relations to date have not been investigated in robot navigation, despite the importance of spatial relations for human navigation and communication. Therefore, I have decided to start a research to study the usefulness of topological notions to achieve multifunctional robots. This article has presented the first results of my research in decision making to establish spatial relations.

The next step in my research programa will be the creation of an architecture will use HTQS to make decisions. The architecture must implement a knowledge base to store information to choose what will be its targets and it let's to avoid the physical restrictions of the environment. Also another issue will be the mathematical method to identify the topological relation between two objects. The actual representation imposes important restrictions in the generalization to a n-dimensional space, an deep analysis can let improve the situation finding another method to representation. One important issue will be the technique of computer vision which will be linked with the spatial representation.

To sum up, my research has got a method to making decisions based on identifying and establishing spatial relations. I have linked concepts of different fields. Namely the work of Egenhofer and colleagues that belongs to the realm of GIS, the research of Ernst on heuristic analysis of means and ends, and Allen's algebras for temporal reasoning. Three results, which apparently were not related, are the foundation that has allowed the development of the HTQS heuristics. The new theoretical results presented here are the basis on which I am developing new navigation algorithms. Thus, the work presented here offers the first results in the agenda of my research program to develop a navigation architecture based on topological notions for the creation of robots that use spatial relations in their interactions with humans and methods to make decisions.

References

1. Allen, J.F.: Maintaining knowledge about temporal intervals. Communications of the ACM 26(11), 832–843 (1983)
2. Brooks, R.: A robust layered control system for a mobile robot. IEEE Journal of Robotics and Automation 2(1), 14–23 (1986)
3. Cohn, A.G., Renz, J.: Qualitative spatial reasoning. In: Handbook of Knowledge Representation, pp. 581–584 (2008)
4. Delafontaine, M., et al.: Qualitative relations between moving objects in a network changing its topological relations. Information Sciences 178(8), 1997–2006 (2008)
5. Dylla, F., et al.: Approaching a formal soccer theory from behaviour specifications in robotic soccer. Computer Science and Sports (2008)
6. Egenhofer, M.J., Al-Taha, K.K.: Reasoning about gradual changes of topological relationships. In: Frank, A.U., Formentini, U., Campari, I. (eds.) GIS 1992. LNCS, vol. 639, pp. 196–219. Springer, Heidelberg (1992)
7. Egenhofer, M.J., Herring, J.: A mathemathical framework for the definition of topological relathionships. In: Fourth ISSDH, pp. 803–813 (1990)
8. Egenhofer, M.J., et al.: A critical comparison of the 4-intersection and 9-intersection models for spatial relations. Auto-Carto 11, 1–11 (1993)
9. Egenhofer, M.J., Sharma, J.: Topological relations between regions in $\mathbb{R}^2$ and $\mathbb{Z}^2$. In: Abel, D.J., Ooi, B.-C. (eds.) SSD 1993. LNCS, vol. 692, pp. 316–336. Springer, Heidelberg (1993)
10. Ernst, G.W.: Sufficient conditions for the success of GPS. Journal of the ACM 16(4), 517–533 (1969)
11. Ferrein, A., Fritz, C., Lakemeyer, G.: Using Golog for deliberation and team coordination in robotic soccer. Künstliche Intelligenz 19, 24–30 (2005)
12. Fikes, R.E., Nilsson, N.J.: Strips: a new approach to the application of theorem proving to problem solving. Artificial Intelligence 2(3), 189–208 (1971)
13. Forbus, K.D., Nielsen, P., Faltings, B.: Qualitative spatial reasoning: the clock project. Artificial Intelligence 51(1-3), 417–471 (1991)
14. Freksa, C.: Conceptual neighborhood and its role in temporal and spatial reasoning (1991)
15. Freksa, C.: Temporal reasoning based on semi-intervals. Artificial Intelligence 54(1-2), 199–227 (1992)
16. Kowalski, R., Sergot, M.: A logic-based calculus of events. New Generation Computing 4(1), 67–95 (1986)
17. Levesque, H.J., et al.: Golog: a logic programming language for dynamic domains. The Journal of Logic Programming 31(1-3), 59–83 (1997)
18. Li, Y., Li, S.: A fuzzy sets theoretic approach to approximate spatial reasoning. IEEE Transactions on Fuzzy Systems 12(6), 745–754 (2004)
19. McCarthy, J., Hayes, P.J.: Some philosophical problems from the standpoint of artificial intelligence. In: Readings in Nonmonotonic Reasoning, pp. 26–45 (1987)
20. Renz, J.: Qualitative Spatial Reasoning with Topological Information. LNCS (LNAI), vol. 2293. Springer, Heidelberg (2002)
21. Schlieder, C.: Representing visible locations for qualitative navigation. In: Qualitative Reasoning and Decision Technologies, pp. 523–532 (1993)
22. Schockaert, S., De Cock, M., Kerre, E.E.: Spatial reasoning in a fuzzy region connection calculus. Artificial Intelligence 173(2), 258–298 (2009)
23. Van de Weghe, N., et al.: A qualitative trajectory calculus as a basis for representing moving objects in geographical information systems. Control and Cybernetics 35(1), 97–119 (2006)

Author Index